T0137920

Lecture Notes in Computer Science 13186

More information about this series at https://link.springer.com/bookseries/558

Matthias Hagen · Suzan Verberne ·
Craig Macdonald · Christin Seifert ·
Krisztian Balog · Kjetil Nørvåg ·
Vinay Setty (Eds.)

Advances in Information Retrieval

44th European Conference on IR Research, ECIR 2022
Stavanger, Norway, April 10–14, 2022
Proceedings, Part II

 Springer

Editors
Matthias Hagen 🆔
Martin Luther University Halle-Wittenberg
Halle, Germany

Suzan Verberne 🆔
Leiden University
Leiden, The Netherlands

Craig Macdonald 🆔
University of Glasgow
Glasgow, UK

Christin Seifert 🆔
University of Duisburg-Essen
Essen, Germany

Krisztian Balog 🆔
University of Stavanger
Stavanger, Norway

Kjetil Nørvåg 🆔
Norwegian University of Science
and Technology
Trondheim, Norway

Vinay Setty 🆔
University of Stavanger
Stavanger, Norway

ISSN 0302-9743   ISSN 1611-3349   (electronic)
Lecture Notes in Computer Science
ISBN 978-3-030-99738-0   ISBN 978-3-030-99739-7   (eBook)
https://doi.org/10.1007/978-3-030-99739-7

This Springer imprint is published by the registered company Springer Nature Switzerland AG
The registered company address is: Gewerbestrasse 11, 6330 Cham, Switzerland

Preface

The 44th European Conference on Information Retrieval (ECIR 2022) was held in Stavanger, Norway, during April 10–14, 2022, and brought together hundreds of researchers from Europe and abroad. The conference was organized by the University of Stavanger, in cooperation with the British Computer Society's Information Retrieval Specialist Group (BCS IRSG).

These proceedings contain the papers related to the presentations, workshops, and tutorials given during the conference. This year's ECIR program boasted a variety of novel work from contributors from all around the world. In total, 395 papers from authors in 53 countries were submitted to the different tracks.

The final program included 35 full papers (20% acceptance rate), 29 short papers (22% acceptance rate), 12 demonstration papers (55% acceptance rate), 11 reproducibility papers (61% acceptance rate), 12 doctoral consortium papers (71% acceptance rate), and 13 invited CLEF papers. All submissions were peer-reviewed by at least three international Program Committee members to ensure that only submissions of the highest relevance and quality were included in the final program. The acceptance decisions were further informed by discussions among the reviewers for each submitted paper, led by a senior Program Committee member.

The accepted papers cover the state of the art in information retrieval: advances in ranking models, applications of entities and knowledge graphs, evaluation, multimodal retrieval, recommender systems, query understanding, user simulation studies, etc. As in previous years, the ECIR 2022 program contained a high proportion of papers with students as first authors, as well as papers from a variety of universities, research institutes, and commercial organizations.

In addition to the papers, the program also included three keynotes, four tutorials, five workshops, a doctoral consortium, the presentation of selected papers from the 2021 issues of the Information Retrieval Journal, and an industry day. Keynote talks were given by Isabelle Augenstein (University of Copenhagen), Peter Flach (University of Bristol), and this year's BCS IRSG Karen Spärck Jones Award winner, Ivan Vulić (University of Cambridge & PolyAI). The tutorials covered a range of topics including high recall retrieval, incrementally testing for online advertising, information extraction from social media, and keyphrase identification, while the workshops brought together participants to discuss algorithmic bias in search and recommendation (BIAS), bibliometrics (BIR), online misinformation (ROMCIR), narrative extraction (Text2Story), and technology-assisted review systems (ALTARS).

The success of ECIR 2022 would not have been possible without all the help from the team of volunteers and reviewers. We wish to thank all the reviewers and meta-reviewers who helped to ensure the high quality of the program. We also wish to thank the reproducibility chairs: Faegheh Hasibi and Carsten Eickhoff; the demo chairs: Theodora Tsikrika and Udo Kruschwitz; the workshop chairs: Lucie Flek and Javier Parapar; the tutorial chairs: Nazli Goharian and Shuo Zhang; the industry chairs: Jiyin

He and Marcel Worring; the doctoral consortium chairs: Asia Biega and Alistair Moffat; and the awards chair: Maarten de Rijke. We would like to thank our local administration chair, Russel Wolff, along with all the student volunteers who helped to create an excellent online and offline experience for participants and attendees.

ECIR 2022 was sponsored by Amazon, Bloomberg, Cobrainer, Elsevier, Google, the L3S Research Center, MediaFutures, the Norwegian University of Science and Technology, NorwAI, Schibsted, SIGIR, Signal AI, Spotify, Springer, Textkernel, Thomson Reuters, the University of Stavanger, Vespa AI, and Wayfair. We thank them all for their support.

Finally, we wish to thank all the authors and contributors to the conference.

April 2022

Matthias Hagen
Suzan Verberne
Craig Macdonald
Christin Seifert
Krisztian Balog
Kjetil Nørvåg

Organization

General Chairs

Krisztian Balog   University of Stavanger, Norway
Kjetil Nørvåg   NTNU, Norway

Program Chairs – Full Papers

Suzan Verberne   Leiden University, The Netherlands
Matthias Hagen   Martin-Luther-Universität Halle-Wittenberg, Germany

Program Chairs – Short Papers

Christin Seifert   University of Duisburg-Essen, Germany
Craig Macdonald   University of Glasgow, UK

Reproducibility Track Chairs

Faegheh Hasibi   Radboud University, The Netherlands
Carsten Eickhoff   Brown University, USA

Demo Chairs

Theodora Tsikrika   Information Technologies Institute, CERTH, Greece
Udo Kruschwitz   University of Regensburg, Germany

Workshop Chairs

Lucie Flek   University of Marburg, Germany
Javier Parapar   Universidade da Coruña, Spain

Tutorials Chairs

Nazli Goharian   Georgetown University, USA
Shuo Zhang   Bloomberg, UK

Industry Day Chairs

Jiyin He   Signal AI, UK
Marcel Worring   University of Amsterdam, The Netherlands

Doctoral Consortium Chairs

Asia Biega   Max Planck Institute for Security and Privacy,
                                  Germany
Alistair Moffat   University of Melbourne, Australia

Awards Chair

Maarten de Rijke   University of Amsterdam, The Netherlands

Local Organization Chairs

Vinay Setty   University of Stavanger, Norway
Russel Wolff   University of Stavanger, Norway

Program Committee

Qingyao Ai   University of Utah, USA
Dyaa Albakour   Signal AI, UK
Mohammad Aliannejadi   University of Amsterdam, The Netherlands
Satya Almasian   Heidelberg University, Germany
Omar Alonso   Instacart, USA
Sophia Althammer   TU Vienna, Austria
Ismail Sengor Altingovde   Bilkent University, Turkey
Giuseppe Amato   ISTI-CNR, Italy
Enrique Amigó   UNED, Spain
Avishek Anand   L3S Research Center, Germany
Linda Andersson   Artificial Researcher IT GmbH and TU Wien, Austria
Negar Arabzadeh   University of Waterloo, Canada
Ioannis Arapakis   Telefonica Research, Spain
Jaime Arguello   University of North Carolina at Chapel Hill, USA
Arian Askari   Shahid Beheshti University, Iran
Maurizio Atzori   University of Cagliari, Italy
Sandeep Avula   Amazon, USA
Leif Azzopardi   University of Strathclyde, UK
Mossaab Bagdouri   Walmart Labs, USA
Ebrahim Bagheri   Ryerson University, Canada
Seyed Ali Bahreinian   IDSIA, Swiss AI Lab, Switzerland
Georgios Balikas   Salesforce Inc, France
Valeriia Baranova   RMIT University, Australia
Alvaro Barreiro   University of A Coruña, Spain
Alberto Barrón-Cedeño   University of Bologna, Italy
Alejandro Bellogin   Universidad Autonoma de Madrid, Spain
Patrice Bellot   CNRS, LSIS, Aix-Marseille Université, France
Michael Bendersky   Google, USA
Alessandro Benedetti   Sease, UK

Klaus Berberich	Saarbruecken University of Applied Sciences, Germany
Sumit Bhatia	Adobe Inc., India
Paheli Bhattacharya	Indian Institute of Technology Kharagpur, India
Roi Blanco	Amazon, Spain
Alexander Bondarenko	Martin-Luther-Universität Halle-Wittenberg, Germany
Ludovico Boratto	University of Cagliari, Italy
Gloria Bordogna	CNR, Italy
Mohand Boughanem	IRIT, Université Toulouse III - Paul Sabatier, France
Leonid Boytsov	BCAI, USA
Alex Brandsen	Leiden University, The Netherlands
Pavel Braslavski	Ural Federal University, Russia
Timo Breuer	TH Köln, Germany
Fidel Cacheda	Universidade da Coruña, Spain
Jamie Callan	Carnegie Mellon University, USA
Rodrigo Calumby	State University of Feira de Santana, Brazil
Ricardo Campos	Polytechnic Institute of Tomar and INESC TEC, Portugal
Zeljko Carevic	GESIS Leibniz Institute for the Social Sciences, Germany
Ben Carterette	Spotify, USA
Shubham Chatterjee	University of New Hampshire, USA
Tao Chen	Google Research, USA
Xuanang Chen	University of Chinese Academy of Sciences, China
Adrian-Gabriel Chifu	CNRS, LIS, Aix-Marseille Université and Université de Toulon, France
Charles Clarke	University of Waterloo, Canada
Maarten Clements	TomTom, The Netherlands
Stephane Clinchant	Xerox Research Centre Europe, France
Paul Clough	University of Sheffield, UK
Juan Soler Company	Pompeu Fabra University, Spain
Alessio Conte	University of Pisa, Italy
Gordon Cormack	University of Waterloo, Canada
Anita Crescenzi	University of North Carolina at Chapel Hill, USA
Fabio Crestani	University of Lugano, Switzerland
Bruce Croft	University of Massachusetts Amherst, USA
Arthur Câmara	Delft University of Technology, The Netherlands
Arjen de Vries	Radboud University, The Netherlands
Yashar Deldjoo	Polytechnic University of Bari, Italy
Elena Demidova	University of Bonn, Germany
Emanuele Di Buccio	University of Padua, Italy
Giorgio Maria Di Nunzio	University of Padua, Italy
Gaël Dias	Normandy University, France
Laura Dietz	University of New Hampshire, USA
Anne Dirkson	Leiden University, The Netherlands
Vlastislav Dohnal	Masaryk University, Czech Republic

Gilles Hubert	IRIT, France
Bogdan Ionescu	Politehnica University of Bucharest, Romania
Radu Tudor Ionescu	University of Bucharest, Romania
Adam Jatowt	University of Innsbruck, Austria
Faizan Javed	Kaiser Permanente, USA
Shiyu Ji	University of California, Santa Barbara, USA
Jiepu Jiang	University of Wisconsin-Madison, USA
Hideo Joho	University of Tsukuba, Japan
Gareth Jones	Dublin City University, Ireland
Joemon Jose	University of Glasgow, UK
Chris Kamphuis	Radboud University, The Netherlands
Jaap Kamps	University of Amsterdam, The Netherlands
Nattiya Kanhabua	SCG CBM, Thailand
Sumanta Kashyapi	NIT Hamirpur, India
Liadh Kelly	Maynooth University, Ireland
Roman Kern	Graz University of Technology, Austria
Oren Kurland	Technion - Israel Institute of Technology, Israel
Mucahid Kutlu	TOBB University of Economics and Technology, Turkey
Saar Kuzi	Amazon, USA
Jochen L. Leidner	Refinitiv Labs and University of Sheffield, UK
Mark Levene	Birkbeck, University of London, UK
Elisabeth Lex	Graz University of Technology, Austria
Xiangsheng Li	Tsinghua University, China
Shangsong Liang	Sun Yat-sen University, China
Jimmy Lin	University of Waterloo, Canada
Matteo Lissandrini	Aalborg University, Denmark
Haiming Liu	University of Bedfordshire, UK
Yiqun Liu	Tsinghua University, China
Sean MacAvaney	University of Glasgow, UK
Andrew Macfarlane	City, University of London, UK
Joel Mackenzie	University of Melbourne, Australia
Eddy Maddalena	King's College London, UK
Joao Magalhaes	Universidade NOVA de Lisboa, Portugal
Maria Maistro	University of Copenhagen, Denmark
Antonio Mallia	New York University, USA
Behrooz Mansouri	University of Tehran, Iran
Jiaxin Mao	Renmin University of China, China
Stefano Marchesin	University of Padua, Italy
Mirko Marras	University of Cagliari, Italy
Monica Marrero	Europeana Foundation, The Netherlands
Bruno Martins	University of Lisbon, Portugal
Yosi Mass	IBM Haifa Research Lab, Israel
Jeanna Matthews	Clarkson University, USA
David Maxwell	TU Delft, The Netherlands

Philipp Mayr	GESIS - Leibniz-Institute for the Social Sciences, Germany
Richard McCreadie	University of Glasgow, UK
Graham McDonald	University of Glasgow, UK
Edgar Meij	Bloomberg L.P., UK
Ida Mele	IASI-CNR, Italy
Massimo Melucci	University of Padua, Italy
Zaiqiao Meng	University of Glasgow, UK
Donald Metzler	Google, USA
Stefano Mizzaro	University of Udine, Italy
Ali Montazeralghaem	University of Massachusetts Amherst, USA
Jose Moreno	IRIT, Université Toulouse III - Paul Sabatier, France
Yashar Moshfeghi	University of Strathclyde, UK
Josiane Mothe	IRIT, France
Philippe Mulhem	LIG-CNRS, France
Cristina Ioana Muntean	ISTI-CNR, Italy
Vanessa Murdock	Amazon, USA
Henning Müller	HES-SO, Switzerland
Franco Maria Nardini	ISTI-CNR, Italy
Wolfgang Nejdl	L3S Research Center, Germany
Jian-Yun Nie	University de Montreal, Canada
Michael Oakes	University of Wolverhampton, UK
Doug Oard	University of Maryland, USA
Harrie Oosterhuis	Radboud University, The Netherlands
Salvatore Orlando	Università Ca' Foscari Venezia, Italy
Iadh Ounis	University of Glasgow, UK
Pooja Oza	University of New Hampshire, USA
Deepak Padmanabhan	Queen's University Belfast, Ireland
Panagiotis Papadakos	FORTH-ICS, Greece
Javier Parapar	Universidade da Coruña, Spain
Pavel Pecina	Charles University in Prague, Czech Republic
Gustavo Penha	Delft University of Technology, The Netherlands
Giulio Ermanno Pibiri	ISTI-CNR, Italy
Karen Pinel-Sauvagnat	IRIT, France
Florina Piroi	TU Wien, Austria
Benjamin Piwowarski	CNRS, Sorbonne Université, France
Martin Potthast	Leipzig University, Germany
Chen Qu	Google and University of Massachusetts Amherst, USA
Pernilla Qvarfordt	FX Palo Alto Laboratory, USA
Filip Radlinski	Google, UK
Gábor Recski	TU Wien, Austria
David Reiley	Google, USA
Zhaochun Ren	Shandong University, China
Jean-Michel Renders	Naver Labs Europe, France

Yannis Tzitzikas	University of Crete and FORTH-ICS, Greece
Md Zia Ullah	CNRS, France
Manisha Verma	Amazon, UK
Vishwa Vinay	Adobe Research, India
Marco Viviani	Università degli Studi di Milano-Bicocca, Italy
Michael Völske	Bauhaus-Universität Weimar, Germany
Xi Wang	University of Glasgow, UK
Zhihong Wang	Tsinghua University, China
Zhijing Wu	Tsinghua University, China
Xiaohui Xie	Tsinghua University, China
Eugene Yang	Johns Hopkins University, USA
Andrew Yates	University of Amsterdam, The Netherlands
Ran Yu	GESIS - Leibniz Institute for the Social Sciences, Germany
Eva Zangerle	University of Innsbruck, Austria
Richard Zanibbi	Rochester Institute of Technology, USA
Fattane Zarrinkalam	University of Guelph, Canada
Sergej Zerr	L3S Research Center, Germany
Junqi Zhang	Tsinghua University, China
Min Zhang	Tsinghua University, China
Rongting Zhang	Amazon, USA
Ruqing Zhang	Chinese Academy of Sciences, China
Yongfeng Zhang	Rutgers, The State University of New Jersey, USA
Liting Zhou	Dublin City University, Ireland
Steven Zimmerman	University of Essex, UK
Justin Zobel	University of Melbourne, Australia
Guido Zuccon	University of Queensland, Australia

Additional Reviewers

Aumiller, Dennis	Gerritse, Emma
Bartscherer, Frederic	Ghauri, Junaid
Belém, Fabiano	Gottschalk, Simon
Bigdeli, Amin	Gémes, Kinga
Biswas, Debanjali	Gérald, Thomas
Cheema, Gullal	Haak, Fabian
Chen, Fumian	Hamidi Rad, Radin
Cunha, Washington	Hoppe, Anett
Dashti, Arman	Huang, Jin
Ebrahimzadeh, Ehsan	Kamateri, Eleni
Engelmann, Björn	Kanase, Sameer
Feher, Gloria	Khawar, Farhan
Ferreira, Thiago	Knyazev, Norman
Fortes, Reinaldo	Leonhardt, Jurek
França, Celso	Liu, Siwei

Lodhia, Zeeshan Ahmed
Mahdavimoghaddam, Jalehsadat
Mamedov, Murad
Mangaravite, Vítor
Marcia, Diego
Mayerl, Maximilian
Mountantonakis, Michalis
Müller-Budack, Eric
Navarrete, Evelyn
Nguyen, Hoang
Paiva, Bruno
Pérez Vila, Miguel Anxo
Ramos, Rita
Renders, Jean-Michel

Saier, Tarek
Sanguinetti, Manuela
Santana, Brenda
Seyedsalehi, Shirin
Shliselberg, Michael
Soprano, Michael
Springstein, Matthias
Stamatis, Vasileios
Su, Ting
Tempelmeier, Nicolas
Viegas, Felipe
Vo, Duc-Thuan
Zhang, Yue
Ziaeinejad, Soroush

Sponsors

Platinum Sponsors

amazon | science

Bloomberg
Engineering

Google

SIGIR
Special Interest Group
on Information Retrieval

Gold Sponsor

wayfair

Silver Sponsors

ELSEVIER

Media Futures ●

Spotify

University of Stavanger

vespa

Bronze Sponsors

L3S Forschungszentrum · Research Center

◉ **NTNU**
Department of Computer Science

NorwAI

Schibsted

Industry Impact Award Sponsor

With Generous Support From

Contents – Part II

Demonstration Papers

CLEF 2022 Lab Descriptions

Doctoral Consortium

Workshops

Tutorials

Contents – Part I

Full Papers

Short Papers

Improving BERT-based Query-by-Document Retrieval with Multi-task Optimization

Amin Abolghasemi[1(✉)], Suzan Verberne[1], and Leif Azzopardi[2]

[1] Leiden University, Leiden, The Netherlands
{m.a.abolghasemi,s.verberne}@liacs.leidenuniv.nl
[2] University of Strathclyde, Glasgow, UK
leif.azzopardi@strath.ac.uk

Abstract. Query-by-document (QBD) retrieval is an Information Retrieval task in which a seed document acts as the query and the goal is to retrieve related documents – it is particular common in professional search tasks. In this work we improve the retrieval effectiveness of the BERT re-ranker, proposing an extension to its fine-tuning step to better exploit the context of queries. To this end, we use an additional document-level representation learning objective besides the ranking objective when fine-tuning the BERT re-ranker. Our experiments on two QBD retrieval benchmarks show that the proposed multi-task optimization significantly improves the ranking effectiveness without changing the BERT re-ranker or using additional training samples. In future work, the generalizability of our approach to other retrieval tasks should be further investigated.

Keywords: Query-by-document retrieval · BERT-based ranking · Multi-task optimization

1 Introduction

Query by document (QBD) [37,38], is a widely-used practice across professional, domain-specific retrieval tasks [33,35] such as scientific literature retrieval [9,25], legal case law retrieval [2,3,30,34], and patent prior art retrieval [13,28]. In these tasks, the user's information need is based on a seed document of the same type as the documents in the collection. Taking a document as query results in long queries, which can potentially express more complex information needs and provide more context for ranking models [16]. Transformer-based ranking models have proven to be highly effective at taking advantage of context [10,11,24], but the long query documents pose challenges because of the maximum input length for BERT-based ranking models. Recent work showed that transformer-based models which handle longer input sequences are not necessarily more effective when being used in retrieval tasks on long texts [3]. We, therefore,

© The Author(s), under exclusive license to Springer Nature Switzerland AG 2022
M. Hagen et al. (Eds.): ECIR 2022, LNCS 13186, pp. 3–12, 2022.
https://doi.org/10.1007/978-3-030-99739-7_1

direct our research towards improving retrieval effectiveness while acting within the input length limitation of ranking models based on large scale pre-trained BERT models [12]. We posit that the representations learned during pre-training have been tailored toward smaller sequences of text – and additional tuning the language models to better represent the documents in this specific domain and query setting, could lead to improvements in ranking.

To investigate this, we first focus on the task of Case Law Retrieval (i.e. given a legal case find the related cases), and employ multi-task optimisation to improve the standard BERT-based cross-encoder ranking model [15] for QBD retrieval. We then explore the generalizability of our approach by evaluating our approach on four QBD retrieval tasks in the academic domain. Our approach draws upon multi-task learning to rank – where a shared structure across auxiliary, related tasks is used [1,8,20,29]. Specifically, in our method, we employ document-level representation learning as an auxiliary objective for multi-task fine-tuning (MTFT) of the BERT re-ranker. To our knowledge, there is no prior work on using representation learning directly as an auxiliary task for fine-tuning a BERT re-ranker. We show that optimizing the re-ranker jointly with document-level representation learning leads to consistently higher ranking effectiveness over the state-of-the-art with greater efficiency i.e., with the same training instances on the same architecture.

2   Preliminaries

BERT-Based Ranking. Pre-trained transformer-based language models [12] have shown significant improvement in ranking tasks [10,11,18,24]. In this work, we use the BERT re-ranker proposed by Nogueira and Cho [26], which is a pre-trained BERT model followed by a projection layer W_p on top of its $[CLS]$ token final hidden states. The BERT re-ranker, which is a cross-encoder neural ranking model, uses the concatenation of a query and candidate document as the input to a fine-tuned pre-trained BERT model. The output of the model is used to indicate the relevance score s of the document d for the input query q, such that:

$$s(q,\ d) = BERT\ ([CLS]\ q\ [SEP]\ d\ [SEP])_{[CLS]} * W_p \qquad (1)$$

BERT-Based Representation Learning. BERT was originally pre-trained on two tasks, namely Masked Language Modeling and Next Sentence Prediction [12]. These tasks, however, are not meant to optimize the network for document-level information representation [9] which may make the model less effective in representation-focused [14] downstream tasks [10]. Previous works have shown that leveraging a Siamese or triplet network structure for fine-tuning BERT could optimize the model for document-level representation [9]. Following Devlin et al. [12], we use the final hidden state corresponding to the $[CLS]$ token to encode the query q and the document d into their representations r_q, and r_d:

$$r_q = BERT(\ [CLS]\ q\ [SEP]\ )_{[CLS]} \quad r_d = BERT(\ [CLS]\ d\ [SEP]\ )_{[CLS]} \quad (2)$$

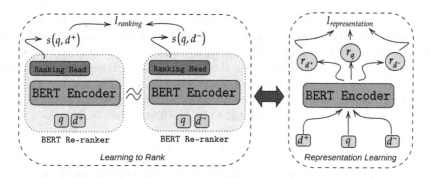

Fig. 1. The fine-tuning process. The same training triples (q, d^+, d^-) are used in each step. The BERT re-rankers are the same, and the BERT encoder is shared between the ranking and representation learning tasks.

Pairwise Ranking Loss. In Learning-To-Rank tasks, pairwise loss minimizes the average number of pairwise errors in a ranked list [5,6]. Here, we aim to optimize the BERT re-ranker with a pairwise cross-entropy softmax loss function [5]:

$$l_{rank} = -log \frac{e^{score(q,d^+)}}{e^{score(q,d^+)} + e^{score(q,d^-)}} \tag{3}$$

where the *score* function represents the degree of relevance between a query and a document computed as described in Eq. 1. In fact, this pairwise loss frames the ranking task as a binary classification problem in which, given a query (q) and a pair of relevant (d^+) and non-relevant (d^-) documents, the fine-tuned ranking model predicts the relevant one. However, at inference time the model is used as a point-wise score function.

Triplet Representation Learning Loss. In the context of representation learning with pre-trained transformers, a triplet loss function fine-tunes the weights of the model such that given an anchor query q, the representations of the query r_q and the document r_d (obtained as described in Eq. 2) are closer for a relevant document d^+ than for a non-relevant document d^-:

$$l_{representation} = max\{(f(r_q, r_{d+}) - f(r_q, r_{d-}) + margin), 0\} \tag{4}$$

Here, f indicates a distance metric and *margin* ensures that d^+ is at least *margin* closer to q than d^- [31].

3   Multi-task Fine-Tuning of the BERT Re-ranker

Our proposed re-ranker aims to jointly optimise both the l_{rank} and $l_{representation}$ – we shall refer to our BERT re-ranker model as MTFT-BERT. As shown in Fig. 1, the Multi Task Fine Tuning (MTFT) is achieved by providing training instances consisting of triples (q, d^+, d^-). To do so, we first feed the concatenation of q and d^+, and the concatenation of q and d^- separately to the MTFT-BERT re-ranker, as described in Sect. 2, to compute the pairwise loss l_{rank} following Eq. 3. In the next step, we feed each of q, d^+, and d^- separately to the shared encoder of the re-ranker to compute the $l_{representation}$ following Eq. 4. As distance metric f we use the L2-norm and we set $margin = 1$ in our experiments. The shared encoder is then fine-tuned with the aggregated loss as shown in Eq. 5 while the ranking head is only fine-tuned by the first term:

$$l_{aggregated} \; = \; l_{rank} \; + \; \lambda \, l_{representation} \tag{5}$$

The λ parameter balances the weight between the two loss functions. Later, we investigate the stability of our model under different values of λ. Since ranking is the target task, and the ranking head is only optimized by the ranking loss, we assign the regularization weight $(0 < \lambda < 1)$ only to the representation loss. It is noteworthy that at inference time, we only use the ranking head of the MTFT-BERT re-ranker.

4   Experimental Setup

Datasets. We first evaluate our proposed method on legal case retrieval. The goal of case law retrieval is to retrieve the relevant prior law cases which could act as supporting cases for a given query law case. This professional search task is a query-by-document (QBD) retrieval task, where both the query and the documents are case law documents. We use the test collection for the case law retrieval task of COLIEE 2021 [30]. This collection contains a corpus with 4415 legal cases with a training and a test set consisting of 650 and 250 query cases respectively. In addition, to evaluate the generalizability of our approach, we use another domain-specific QBD retrieval benchmark, called SciDocs [9]. SciDocs was originally introduced as a representation learning benchmark in the scientific domain while framing the tasks as ranking; we use the four SciDocs tasks: {citation, co-citation, co-view, and co-read}-prediction to evaluate our method. It is worth mentioning that while the original paper trains the model on a citation graph of academic papers, we take the validation set provided for each task and use 85% of it as training set and the rest as the validation set for tuning purposes.

Implementation. We use Elasticsearch[1] to index and retrieve the initial ranking list using a BM25 ranker. It was shown in prior work that BM25 is a strong baseline [32], and it even holds the state-of-the-art in case law retrieval on COL-IEE 2021 [3]. Therefore, to make our work comparable, we use the configuration provided by [3] to optimize the BM25 with Elasticsearch for COLIEE 2021 case law retrieval. For query generation, following the effectiveness of term selection using Kullback-Leibler divergence for Informativeness (KLI) in prior work in case law retrieval [3,21], we use the top-10% of a query document terms scored

---

[1] https://github.com/elastic/elasticsearch.

Table 1. The reranking results with BM25 and BM25$_{optimized}$ as initial rankers for the COLIEE 2021 test data. † indicates the statistically significant improvements over BM25$_{optimized}$ according to a paired t-test (p < 0.05). TLIR achieved the highest score in the COLIEE 2021 competition.

Model	Initial ranker	Precision %	Recall %	F1 %
BM25	–	8.8	16.51	11.48
TLIR [23]	–	15.33	25.56	19.17
BM25$_{optimized}$ [3]	–	17.00	25.36	20.35
BERT	BM25	10.48	18.80	13.46
MTFT-BERT	BM25	12.08	21.59	15.49
BERT	BM25$_{optimized}$	14.40	24.63	18.17
MTFT-BERT	BM25$_{optimized}$	17.44†	29.99†	22.05†

Table 2. Ranking results on the SciDocs benchmark. HF is Huggingface. † indicates the statistically significant improvements according to a paired t-test (p < 0.05).

Model	Co-view		Co-read		Cite		Co-cite	
	MAP	nDCG	MAP	nDCG	MAP	nDCG	MAP	nDCG
SPECTER [9]	83.6%	0.915	84.5%	0.924	88.3%	0.949	88.1%	0.948
SPECTER w/ HF[36]	83.4%	0.914	85.1%	0.927	92.0%	0.966	88.0%	0.947
BM25	75.4%	0.874	75.6%	0.881	73.5%	0.876	76.3%	0.890
BM25$_{optimized}$	76.26%	0.877	76.09%	0.881	75.3%	0.884	77.41%	0.896
BERT	85.2%	0.925	87.5%	0.940	94.0%	0.975	89.7%	0.955
MTFT-BERT	86.2%†	0.930†	87.7%	0.940	94.2%	0.976	91.0%†	0.961†

with KLI^2 as the query for BM25 in our experiments. As the BERT encoders, we use LegalBERT [7], and SciBERT[4], which are domain-specific BERT models pre-trained on the legal and scientific domains respectively. We train our neural ranking models for 15 epochs with a batch size of 32, and AdamW optimizer [22] with a learning rate of 3×10^{-5}. All of our models are implemented and fine-tuned using PyTorch [27] and the HuggingFace library [36].

5 Results and Analysis

Ranking Quality. Table 1 displays the ranking quality of the MTFT-BERT reranker in comparison to BM25, TLIR [23], BM25$_{optimized}$, and the original BERT re-ranker on COLIEE 2021. The cut-off k for all rankers is set to 5 during both validation and test since the train queries in COLIEE 2021 have 5 relevant documents on average. We report precision and recall besides F1, which is the official

---

^{2} Implementation from https://github.com/suzanv/termprofiling/.

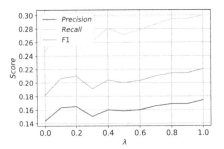

Fig. 2. The evaluation results of MTFT-BERT+BM25$_{optimized}$ with various λ in the COLIEE 2021 case law retrieval task. $\lambda = 0$ indicates the BERT re-ranker.

metric used in the COLIEE competition. It can be seen that the BERT re-ranker and the MTFT-BERT re-ranker can both achieve better quality over BM25 with default parameters as initial ranker. In contrast, when we use BM25$_{optimized}$ as the initial ranker, the BERT re-ranker fails to yield improvement, while MTFT-BERT outperforms the state-of-the-art BM25$_{optimized}$ [3] by a statistically significant margin of 8.3% relative improvement.

For comparability reasons on the SciDocs benchmark, we have included both the original paper results and the results reported in their official code repository[3], which is achieved using Huggingface models like our implementations. As Table 2 shows, while both the BERT re-ranker, and the MTFT-BERT re-ranker yield improvement over the SPECTER method, the MTFT-BERT re-ranker outperforms the BERT re-ranker which confirms the effectiveness of our method in an additional domain-specific QBD retrieval setting.

Robustness to Varying λ. Task weighting is a widely used method in multi-task learning algorithms [17,19] where a static or dynamic weight is assigned to the loss of different tasks. Figure 2 displays the ranking quality of the MTFT-BERT re-ranker over different values of λ on the COLIEE test set, using BM25$_{optimized}$ as the initial ranker. We bound λ at 1 since our target task is ranking, and we do not want the representation loss rate to have higher impact in the training. We can see that our model quality is relatively consistent across different values above 0.5 which indicates the robustness of our model in tuning this parameter.

Effect of Re-ranking Depth. We experimented with the ranking depth, i.e., number of documents re-ranked from the initial ranker result, by increasing it from 15 to 100 in steps of 5. We then analyzed the MTFT-BERT re-ranking quality relative to depth. We found that the ranking quality decreases rapidly after the lower ranking depths, to $F1 = 17.3$ at 100, which is lower than the original BM25$_{optimized}$ ranking. While MTFT-BERT can improve over BM25 with a shallow re-ranking set, we confirm the findings by previous studies that BM25 is a strong baseline for case law retrieval [3,32].

---

[3] https://github.com/allenai/specter.

6 Conclusion

This paper shows that it is possible to improve the BERT cross-encoder re-ranker quality using multi-task optimization with an auxiliary representation learning task. We showed that the resulting model named MTFT-BERT re-ranker obtains consistently better retrieval quality than the original BERT re-ranker using the same training instances and structure. While our focus was on query-by-document retrieval in professional search domains (legal and academic), as a future work, it would be interesting to study the effectiveness of MTFT-BERT re-ranker in other retrieval tasks where we have shorter queries.

References

1. Ahmad, W.U., Chang, K.W., Wang, H.: Multi-task learning for document ranking and query suggestion. In: International Conference on Learning Representations (2018)
2. Althammer, S., Hofstätter, S., Sertkan, M., Verberne, S., Hanbury, A.: Paragraph aggregation retrieval model (parm) for dense document-to-document retrieval. In: Advances in Information Retrieval, 44rd European Conference on IR Research, ECIR 2022 (2022)
3. Askari, A., Verberne, S.: Combining lexical and neural retrieval with longformer-based summarization for effective case law retrieval. In: DESIRES (2021)
4. Beltagy, I., Lo, K., Cohan, A.: SciBERT: a pretrained language model for scientific text. In: Proceedings of the 2019 Conference on Empirical Methods in Natural Language Processing and the 9th International Joint Conference on Natural Language Processing (EMNLP-IJCNLP), pp. 3615–3620. Association for Computational Linguistics, Hong Kong (2019). https://doi.org/10.18653/v1/D19-1371, https://aclanthology.org/D19-1371
5. Burges, C., Shaked, T., Renshaw, E., Lazier, A., Deeds, M., Hamilton, N., Hullender, G.: Learning to rank using gradient descent. In: Proceedings of the 22nd international conference on Machine learning - ICML 2005, pp. 89–96. ACM Press, Bonn (2005). https://doi.org/10.1145/1102351.1102363, http://portal.acm.org/citation.cfm?doid=1102351.1102363
6. Cao, Z., Qin, T., Liu, T.Y., Tsai, M.F., Li, H.: Learning to rank: from pairwise approach to listwise approach. In: Proceedings of the 24th International Conference on Machine Learning, pp. 129–136 (2007)
7. Chalkidis, I., Fergadiotis, M., Malakasiotis, P., Aletras, N., Androutsopoulos, I.: LEGAL-BERT: the muppets straight out of law school. In: Findings of the Association for Computational Linguistics: EMNLP 2020, pp. 2898–2904. Association for Computational Linguistics (2020). https://doi.org/10.18653/v1/2020.findings-emnlp.261, https://aclanthology.org/2020.findings-emnlp.261
8. Cheng, Q., Ren, Z., Lin, Y., Ren, P., Chen, Z., Liu, X., de Rijke, M.D.: Long short-term session search: joint personalized reranking and next query prediction. In: Proceedings of the Web Conference 2021, pp. 239–248 (2021)

9. Cohan, A., Feldman, S., Beltagy, I., Downey, D., Weld, D.: SPECTER: document-level representation learning using citation-informed transformers. In: Proceedings of the 58th Annual Meeting of the Association for Computational Linguistics, pp. 2270–2282. Association for Computational Linguistics (2020). https://doi.org/10.18653/v1/2020.acl-main.207, https://www.aclweb.org/anthology/2020.acl-main.207

10. Dai, Z., Callan, J.: Deeper text understanding for IR with contextual neural language modeling. In: Proceedings of the 42nd International ACM SIGIR Conference on Research and Development in Information Retrieval, pp. 985–988 (2019)

11. Dai, Z., Callan, J.: Context-aware term weighting for first stage passage retrieval. In: Proceedings of the 43rd International ACM SIGIR Conference on Research and Development in Information Retrieval, pp. 1533–1536 (2020)

12. Devlin, J., Chang, M.W., Lee, K., Toutanova, K.: BERT: pre-training of deep bidirectional transformers for language understanding. In: Proceedings of the 2019 Conference of the North American Chapter of the Association for Computational Linguistics: Human Language Technologies, vol. 1 (Long and Short Papers), pp. 4171–4186. Association for Computational Linguistics, Minneapolis (2019). https://doi.org/10.18653/v1/N19-1423, https://aclanthology.org/N19-1423

13. Fujii, A., Iwayama, M., Kando, N.: Overview of the patent retrieval task at the ntcir-6 workshop. In: NTCIR (2007)

14. Guo, J., Fan, Y., Ai, Q., Croft, W.B.: A deep relevance matching model for ad-hoc retrieval. In: Proceedings of the 25th ACM International on Conference on Information and Knowledge Management, pp. 55–64 (2016)

15. Humeau, S., Shuster, K., Lachaux, M.A., Weston, J.: Poly-encoders: architectures and pre-training strategies for fast and accurate multi-sentence scoring. In: International Conference on Learning Representations (2020). https://openreview.net/forum?id=SkxgnnNFvH

16. Huston, S., Croft, W.B.: Evaluating verbose query processing techniques. In: Proceedings of the 33rd International ACM SIGIR Conference on Research and Development in Information Retrieval, pp. 291–298 (2010)

17. Kongyoung, S., Macdonald, C., Ounis, I.: Multi-task learning using dynamic task weighting for conversational question answering. In: Proceedings of the 5th International Workshop on Search-Oriented Conversational AI (SCAI), pp. 17–26 (2020)

18. Lin, J., Nogueira, R., Yates, A.: Pretrained transformers for text ranking: bert and beyond (2021)

19. Liu, S., Liang, Y., Gitter, A.: Loss-balanced task weighting to reduce negative transfer in multi-task learning. In: Proceedings of the AAAI Conference on Artificial Intelligence, vol. 33, pp. 9977–9978 (2019)

20. Liu, X., He, P., Chen, W., Gao, J.: Multi-task deep neural networks for natural language understanding. In: Proceedings of the 57th Annual Meeting of the Association for Computational Linguistics, pp. 4487–4496. Association for Computational Linguistics, Florence (2019). https://doi.org/10.18653/v1/P19-1441, https://aclanthology.org/P19-1441

21. Locke, D., Zuccon, G., Scells, H.: Automatic query generation from legal texts for case law retrieval. In: Asia Information Retrieval Symposium, pp. 181–193. Springer (2017). https://doi.org/10.1007/978-3-319-70145-5_14

22. Loshchilov, I., Hutter, F.: Decoupled weight decay regularization. In: International Conference on Learning Representations (2019). https://openreview.net/forum?id=Bkg6RiCqY7

23. Ma, Y., Shao, Y., Liu, B., Liu, Y., Zhang, M., Ma, S.: Retrieving legal cases from a large-scale candidate corpus. In: Proceedings of the Eighth International Competition on Legal Information Extraction/Entailment, COLIEE2021 (2021)
24. MacAvaney, S., Yates, A., Cohan, A., Goharian, N.: Cedr: contextualized embeddings for document ranking. In: Proceedings of the 42nd International ACM SIGIR Conference on Research and Development in Information Retrieval, pp. 1101–1104 (2019)
25. Mysore, S., O'Gorman, T., McCallum, A., Zamani, H.: Csfcube-a test collection of computer science research articles for faceted query by example. arXiv preprint arXiv:2103.12906 (2021)
26. Nogueira, R., Cho, K.: Passage re-ranking with bert. arXiv preprint arXiv:1901.04085 (2019)
27. Paszke, A., et al.: Pytorch: an imperative style, high-performance deep learning library. In: Wallach, H., Larochelle, H., Beygelzimer, A., d' Alché-Buc, F., Fox, E., Garnett, R. (eds.) Advances in Neural Information Processing Systems, vol. 32, pp. 8024–8035. Curran Associates, Inc. (2019). http://papers.neurips.cc/paper/9015-pytorch-an-imperative-style-high-performance-deep-learning-library.pdf
28. Piroi, F., Hanbury, A.: Multilingual patent text retrieval evaluation: CLEF–IP. In: Information Retrieval Evaluation in a Changing World. TIRS, vol. 41, pp. 365–387. Springer, Cham (2019). https://doi.org/10.1007/978-3-030-22948-1_15
29. Qu, C., Yang, L., Chen, C., Qiu, M., Croft, W.B., Iyyer, M.: Open-retrieval conversational question answering. In: Proceedings of the 43rd International ACM SIGIR Conference on Research and Development in Information Retrieval, pp. 539–548 (2020)
30. Rabelo, J., Kim, M.-Y., Goebel, R., Yoshioka, M., Kano, Y., Satoh, K.: COLIEE 2020: methods for legal document retrieval and entailment. In: Okazaki, N., Yada, K., Satoh, K., Mineshima, K. (eds.) JSAI-isAI 2020. LNCS (LNAI), vol. 12758, pp. 196–210. Springer, Cham (2021). https://doi.org/10.1007/978-3-030-79942-7_13
31. Reimers, N., Gurevych, I.: Sentence-BERT: Sentence embeddings using Siamese BERT-networks. In: Proceedings of the 2019 Conference on Empirical Methods in Natural Language Processing and the 9th International Joint Conference on Natural Language Processing (EMNLP-IJCNLP), pp. 3982–3992. Association for Computational Linguistics, Hong Kong (2019). https://doi.org/10.18653/v1/D19-1410, https://aclanthology.org/D19-1410
32. Rosa, G.M., Rodrigues, R.C., Lotufo, R., Nogueira, R.: Yes, bm25 is a strong baseline for legal case retrieval. arXiv preprint arXiv:2105.05686 (2021)
33. Russell-Rose, T., Chamberlain, J., Azzopardi, L.: Information retrieval in the workplace: a comparison of professional search practices. Inf. Process. Manag. **54**(6), 1042–1057 (2018)
34. Shao, Y., et al.: Bert-pli: modeling paragraph-level interactions for legal case retrieval. In: Bessiere, C. (ed.) Proceedings of the Twenty-Ninth International Joint Conference on Artificial Intelligence, IJCAI-20, pp. 3501–3507. International Joint Conferences on Artificial Intelligence Organization (2020). https://doi.org/10.24963/ijcai.2020/484
35. Verberne, S., et al.: First international workshop on professional search. In: ACM SIGIR Forum, vol. 52, pp. 153–162. ACM, New York (2019)
36. Wolf, T., et al.: Transformers: state-of-the-art natural language processing. In: Proceedings of the 2020 Conference on Empirical Methods in Natural Language Processing: System Demonstrations, pp. 38–45. Association for Computational Linguistics (2020). https://www.aclweb.org/anthology/2020.emnlp-demos.6

37. Yang, E., Lewis, D.D., Frieder, O., Grossman, D.A., Yurchak, R.: Retrieval and richness when querying by document. In: DESIRES, pp. 68–75 (2018)
38. Yang, Y., Bansal, N., Dakka, W., Ipeirotis, P., Koudas, N., Papadias, D.: Query by document. In: Proceedings of the Second ACM International Conference on Web Search and Data Mining, pp. 34–43 (2009)

Passage Retrieval on Structured Documents Using Graph Attention Networks

Lucas Albarede[1,2(✉)], Philippe Mulhem[1], Lorraine Goeuriot[1], Claude Le Pape-Gardeux[2], Sylvain Marie[2], and Trinidad Chardin-Segui[2]

[1] Univ. Grenoble Alpes, CNRS, Grenoble INP, LIG, 38000 Grenoble, France
{lucas.albarede,philippe.mulhem,lorraine.goeuriot}@imag.fr
[2] Schneider Electric Industries SAS, Rueil-Malmaison, France
{lucas.albarede,claude.lepape,sylvain.marie,
trinidad.chardin-segui}@se.com

Abstract. Passage Retrieval systems aim at retrieving and ranking small text units according to their estimated relevance to a query. A usual practice is to consider the context a passage appears in (its containing document, neighbour passages, etc.) to improve its relevance estimation. In this work, we study the use of Graph Attention Networks (GATs), a graph node embedding method, to perform passage contextualization. More precisely, we first propose a document graph representation based on several *inter-* and *intra*-document relations. Then, we investigate two ways of leveraging the use of GATs on this representation in order to incorporate contextual information for passage retrieval. We evaluate our approach on a Passage Retrieval task for structured documents: CLEF-IP2013. Our results show that our document graph representation coupled with the expressive power of GATs allows for a better context representation leading to improved performances.

Keywords: Passage Retrieval · Graph Attention Networks · Experiments · Document representation

1 Introduction

Passage Retrieval is a long lasting topic for Information Retrieval (IR) that is concerned with the retrieval of passages, i.e. small textual elements. This task is faced with one key problem: as passages are small excerpts of longer documents, their content is not always sufficient to adequately estimate their relevance. To cope with this phenomenon, current approaches resort to contextualization [1,5–7,21,28]; that is, the consideration of a passage's context in its relevance estimation. We study here how to perform passage contextualization with methods akin to neural IR, as their expressive power have created a gap in performances compared with classical methods [10]. Multiple approaches represent a passage's

© The Author(s), under exclusive license to Springer Nature Switzerland AG 2022
M. Hagen et al. (Eds.): ECIR 2022, LNCS 13186, pp. 13–21, 2022.
https://doi.org/10.1007/978-3-030-99739-7_2

context based on the various types of relations it has with other parts of the document [1,4,7,15,23,24,28]. We investigate how such a representation, encoded as a graph, may be leveraged by graph neural networks. Graph neural networks, successfully applied on different Information Retrieval tasks [11,16,34,35], aim at computing the embedding of nodes in a graph by considering their relations with other nodes More precisely, we investigate the use of attention-based graph neural networks, known as *Graph Attention Networks* (GATs) [30]. GATs rely on dense representation (embeddings) of the nodes' content. Many embeddings approaches have been proposed for text retrieval [12,13,32] and multiple embeddings [13,18] have been proved to be the most effective. To our knowledge, this is the first time GATs have been exploited in a Passage Retrieval task. We present our proposal in Sect. 2, describing the graph document representation and our two models leveraging GATs to perform passage contextualization. In Sect. 3, we conduct an evaluation on a patent passage retrieval task and conclude in Sect. 4.

2   Proposal: Merged and Late Interaction Models

2.1   Document Graph Representation

We represent a document corpus as a graph where nodes are document parts and edges are the *intra* and *inter* relations between these documents parts. The document parts come from the logical structure of documents. To cope with the variation of such structure, we define two types of document parts: *sections*, i.e., non-textual units with a title, and *passages*, textual units without titles. The *intra*-document relations considered are: **(1)** the order of passages [4,7,15,28], **(2)** the hierarchical structure [1,23,24] and **(3)** internal citations. The *inter*-document relation considered is: the **(4)** citation of one document by another one. We also include the inverse relations of these four relations [24,25].

Our graph document representation is therefore composed of two types of node: *passage* nodes that represent textual units and *section* nodes that represent titled structural units – and eight types of edge (one for each relation and its symmetrical): *order* characterizing the relation order between *passage* nodes ($order_i$ its symmetrical), *structural* characterizing the composition between a *passage* node and a *section* node or between two *section* nodes ($structural_i$ its symmetrical), *internal* characterizing the *intra*-document citations between nodes ($internal_i$ its symmetrical) and *external* characterizing the *inter*-document citations between nodes ($external_i$ its symmetrical). Formally, the document corpus is a directed graph $G = (V, E, A, R)$ where each node $v \in V$ and each edge $e \in E$ are associated with their type mapping functions $\tau(v) : V \rightarrow A$ and $\phi(e) : E \rightarrow R$, respectively. We have $A = \{passage, section\}$ and $R = \{order, order_i, structural, structural_i, internal, internal_i, external, external_i\}$.

2.2   Models Architecture

We explore ways to compute a passage's score by taking into account information about its content and information about its context using GATs. We derive two models leveraging the power of GATs: Merged Interaction Model (MiM) and Late Interaction Model (LiM). They both follow a ColBERT-inspired [13] efficient design and they differ in the way they consider the content of a passage. We first describe the elements that are shared by our two models before focusing on the motivations behind their differences.

Encoder. We use the multiple representation text-encoder taken from the ColBERT model [13,18] to embed text into a dense semantic space. A passage is embedded using its text, a section using its title and a query is using its text.

Graph Attention Network. GATs are multi-layer graph neural networks which compute an embedding for each node in a graph by taking into account information from its neighbours [30]. Each layer aggregates, for each node, its embedding with the embedding of its neighbours using attention functions [3]. Stacking n layers allows a node to gather information about nodes that are at a distance of n hops in the graph. One element worth mentioning is that GATs implicitly add *self-edges* connecting each node to itself to build the embedding of a node.

Our model uses *Attention is all you need* [29] definition of attention and uses one attention function $MultiHead_r$ per type of edge r, so the model treats the interaction between nodes differently according to the type of their relation [31]. The model defines a learnable weight vector W_r for each type of edge in the graph, representing the global importance of the relation. For a node i, we define its neighbour nodes N_i, and the edge between nodes i and j e_{ij} (with j in N_i) . For h_i and h_j, their respective intermediate representation, the output of a layer is computed as:

$$h'_i = \sum_{j \in N_i} softmax(W_{\phi(e_{ij})}) * MultiHead_{\phi(e_{ij})}(h_i, h_j, h_j) \qquad (1)$$

Query Similarity Measure. As described in [13], the similarity between E_q and E_p, the multiple representation embeddings from a query q and a passage p is:

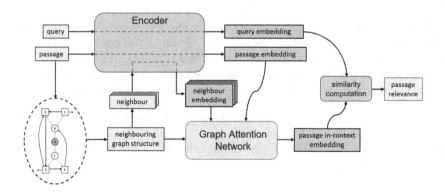

Fig. 1. Overview of the merged interaction model.

$$sim(E_q, E_p) = \sum_{i \in [1,|E_q|]} \max_{j \in [1,|E_p|]} E_{q_i} \cdot E_{p_j}^T \qquad (2)$$

Merged Interaction Model. The MiM, described in Fig. 1, proposes a *in-context* representation by simultaneously considering the content and the context for each passage. We compute such a representation of each passage by feeding its embedding, its neighbours' embedding and the neighbouring graph structure to the GAT. To obtain a passage's relevance to a query, the model computes using the above-mentioned query similarity measure (2).

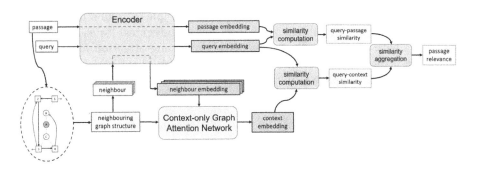

Fig. 2. Overview of the late interaction model.

Late Interaction Model. The LiM, shown in Fig. 2, computes two embedding-based representations for a passage: one based on its content and one solely based on its context. Its relevance to a query is estimated using two similarity measures, computed using each representation. We are questioning here if mixing content and context information inside a single representation cannot impair its expressive power. For a passage p, the LiM computes a *content-based* embedding by feeding its text to the encoder, and a *context-based* embedding by making use of a Context-only Graph Attention Network (CGAT), a **modified Graph Attention Network** that does not consider p's content to compute its embedding.

To do that, a CGAT removes every edge going out of p's node in the graph. Note that this does not affect Eq. (1). To obtain a passage's relevance to a query, the model first computes the *content-based* and the *context-based* query similarities with the above-mentioned similarity measure (2). Then, it aggregates the two similarities using a linear combination with a parameter $\lambda \in [0, 1]$:

$$relevance(q, p) = (1 - \lambda) * sim(E_q, E_p) + \lambda * sim(E_q, E_{context_p}) \qquad (3)$$

where E_p is the *content-based* embedding and $E_{context_p}$ is the *context-based* embedding of passage p.

3   Experiments

3.1   Experimental Setup

We use a classical neural approach [13] to re-rank passages extracted from the top 1000 documents retrieved by BM25 [27] (with Terrier [17] default parameters).

Dataset. We use the CLEF-IP2013 Passage Retrieval task [26]. This dataset, split between a training and testing set, contains structured patent documents with both internal and external citations. To derive the structure of documents and the internal citations, we use handcrafted features either based on XML tags, case or number of characters. Each query is a set of "claims" along with a full patent document, that must be transformed into more refined queries [2,19, 20,33]. We use a state-of-the-art method [20]: let d be a query patent document with a set of claims, we build a first form of the query with the top-10 words with highest *tf-idf* in d's abstract to perform document retrieval, and a second form composed of d's natural language claims to perform passage reranking. We report the five official evaluation measures of the task: *PRES@100,Recall@100*, and *MAP@100* at the document-level; *MAP(D)* and *PREC(D)* at the passage-level.

Model Characteristics. We use a text encoder from the ColBERT model trained on MSMARCO [22] with a maximum passage length of 180 tokens and a maximum query length of 120 tokens. Our GAT and CGAT are composed of 3 layers, each layer having several attention function $MultiHead_r$ with 8 heads with dropout $= 0.7$. The retrieval filters documents based on the International Patent Classification codes they share with the query patent document [9].

Learning Process. The Adam optimizer [14] is used to jointly learn the GATs or CGATs parameters and to fine-tune the encoder. For the encoder, we use the advised learning rate of $3 * 10^{-6}$ [13] and freeze the first six layers. For the graph-based model, the weight vectors W_r and the parameter λ, we use a learning rate of $1 * 10^{-3}$. Our learning process is as follows: given a triple $\langle q, p^+, p^- \rangle$ with query second form q, positive passage p^+ and negative passage p^-, the model is optimised via pairwise softmax cross-entropy loss over the computed scores of p^+ and p^-. The negative passage sampling for a pair $\langle q, p^+ \rangle$ is done as follows: we randomly sample a passage from the corpus with a probability of 0.6 and randomly sample a non-relevant passage from the set of relevant documents with a probability of 0.4. The model is therefore confronted with positive and negative passages having similar contexts. This sampling process is repeated 1000 times for each $\langle q, p^+ \rangle$, yielding a training on $3.5M$ triplets.

3.2   Results

We evaluate the performances of our two models (MiM, LiM), a SoTA non-neural passage contextualization model ($QSF_{sectionPropagateAVG}$ [1]), and two baseline models that do not consider context (fine-tuned ColBERT [13], BM25 [27]). For the sake of comparison, we also evaluate the performances of a model

($Mix\text{-}iM$) that mixes our two proposed approaches: it computes a MiM in-context passage embedding along with a LiM content-based passage embedding and combines them the same way as the LiM model. As explained earlier, each of these approaches has been used to rank the passages of the documents retrieved initially by the BM25 model. We also report the results of a state-of-the-art approach focusing on the query generation [2], namely $Query\text{-}gen$. Additionally, we report the learned value of parameter λ for the LiM and $Mix\text{-}iM$ models. Table 1 shows that $QSF_{sectionPropagateAVG}$ and our three approaches (MiM, LiM, $Mix\text{-}iM$) improve upon BM25 and ColBERT, which focus solely on the content of passages, confirming the importance of taking context into account during passage relevance estimation [1]. Comparing our models, we see that MiM falls behind both LiM and $Mix\text{-}iM$, especially on the passage-level metrics. We hypothesize that MiM fails to correctly condense content and context information into a single representation leading to an over-prioritization of the context, which causes the model to rank high every (relevant or not-relevant) passages appearing in a relevant document. LiM outperforms $Mix\text{-}iM$, indicating that the model benefits from fully separating content and context information. Finally, we see that two of our approaches (LiM, $Mix\text{-}iM$) outperform $QSF_{sectionPropagateAVG}$ (significantly) and $Query\text{-}gen$ on every evaluation measures, showing the strength of our methods for the patent passage retrieval task.

Table 1. Performance over CLEF-IP2013 (in boldface: best result in a column). o, i, j, k and l represent statistical significance (two tailed Student paired t-test, p$\leq$5%) over BM25, ColBERT, $QSF_{sectionPropagateAVG}$, MiM and $Mix\text{-}iM$ respectively. Statistical significance over $Query\text{-}gen$ could not be computed (data not available from [2]).

	Model	PRES@100	Recall@100	MAP@100	MAP(D)	PREC(D)
Content only	$BM25$ [27]	0.385	0.482	0.125	0.142^{k}	0.21^{k}
	$ColBERT$ [13]	0.402	0.518^{o}	0.161^{o}	0.145^{k}	0.214^{k}
	$Query\text{-}gen$ [2]	0.444	0.560	0.187	0.146	0.282
Content & Context	$QSF_{sectionPropagateAVG}$ [1]	0.460^{oi}	0.609^{oi}	0.169^{o}	0.201^{oik}	0.237^{ok}
	MiM	0.470^{oi}	0.568^{oi}	0.181^{o}	0.104	0.141
	$Mix\text{-}iM$ ($\lambda = 0.518$)	0.541^{oijk}	0.631^{oijk}	0.257^{oijk}	0.246^{oijk}	0.299^{oijk}
	LiM ($\lambda = 0.307$)	$\mathbf{0.564}^{oijk}$	$\mathbf{0.651}^{oijk}$	$\mathbf{0.296}^{oijkl}$	$\mathbf{0.270}^{oijkl}$	$\mathbf{0.322}^{oijkl}$

4   Conclusion

In this work, we investigated the use of GATs to perform passage contextualization. First, we proposed a document graph representation based on document parts and their relations. Then, we presented two models leveraging GATs on this representation to estimate the relevance of a passage (to a query) according to both its content and its context. We evaluated our proposals on the CLEF-IP2013 patent passage retrieval task. Our results show that, while the use of GATs for passage contextualization improves the results, separately considering the content and context information of a passage leads to a significant gap in performances. In the future, we would like to conduct a parameter study on both our models in order to analyse more precisely how the context is taken into consideration. It would also be interesting to investigate other negative sampling techniques [32]. Finally, we plan to extend our experiments on more datasets such as the *INEX* Wikipedia dataset [8].

Acknowledgement. This work has been partially supported by MIAI@Grenoble Alpes (ANR-19-P3IA-0003), as well as the Association Nationale de la Recherche et de la Technologie (ANRT).

References

1. Albarede, L., Mulhem, P., Goeuriot, L., Le Pape-Gardeux, C., Marie, S., Chardin-Segui, T.: Passage retrieval in context: experiments on patents. In: Proceedings of CORIA 2021, Grenoble, France (2021). https://hal.archives-ouvertes.fr/hal-03230421
2. Andersson, L., Lupu, M., Palotti, J.A., Hanbury, A., Rauber, A.: When is the time ripe for natural language processing for patent passage retrieval? In: Proceedings of the 25th ACM International on Conference on Information and Knowledge Management, CIKM 2016, pp. 1453–1462. Association for Computing Machinery, New York (2016). https://doi.org/10.1145/2983323.2983858
3. Bahdanau, D., Cho, K., Bengio, Y.: Neural machine translation by jointly learning to align and translate (2016)
4. Beigbeder, M.: Focused retrieval with proximity scoring. In: Proceedings of the 2010 ACM Symposium on Applied Computing, SAC 2010, pp. 1755–1759. Association for Computing Machinery, New York (2010). https://doi.org/10.1145/1774088.1774462
5. Bendersky, M., Kurland, O.: Utilizing passage-based language models for document retrieval. In: Macdonald, C., Ounis, I., Plachouras, V., Ruthven, I., White, R.W. (eds.) ECIR 2008. LNCS, vol. 4956, pp. 162–174. Springer, Heidelberg (2008). https://doi.org/10.1007/978-3-540-78646-7_17
6. Callan, J.P.: Passage-level evidence in document retrieval. In: Proceedings of the 17th Annual International ACM SIGIR Conference on Research and Development in Information Retrieval, SIGIR 1994, pp. 302–310. Springer-Verlag, Heidelberg (1994). https://doi.org/10.1007/978-1-4471-2099-5_31
7. Fernández, R., Losada, D., Azzopardi, L.: Extending the language modeling framework for sentence retrieval to include local context. Inf. Retr. **14**, 355–389 (2011). https://doi.org/10.1007/s10791-010-9146-4
8. Geva, S., Kamps, J., Lethonen, M., Schenkel, R., Thom, J.A., Trotman, A.: Overview of the INEX 2009 ad hoc track. In: Geva, S., Kamps, J., Trotman, A. (eds.) INEX 2009. LNCS, vol. 6203, pp. 4–25. Springer, Heidelberg (2010). https://doi.org/10.1007/978-3-642-14556-8_4
9. Gobeill, J., Ruch, P.: Bitem site report for the claims to passage task in CLEF-IP 2012. In: Forner, P., Karlgren, J., Womser-Hacker, C. (eds.) CLEF 2012 Evaluation Labs and Workshop, Online Working Notes, Rome, Italy, 17–20 September 2012, CEUR Workshop Proceedings, vol. 1178. CEUR-WS.org (2012). http://ceur-ws.org/Vol-1178/CLEF2012wn-CLEFIP-GobeillEt2012.pdf

10. Guo, J., et al.: A deep look into neural ranking models for information retrieval. Inf. Process. Manag. **57**(6), 102067 (2020)
11. Han, F., Niu, D., Lai, K., Guo, W., He, Y., Xu, Y.: Inferring search queries from web documents via a graph-augmented sequence to attention network. In: The World Wide Web Conference, WWW 2019, pp. 2792–2798. Association for Computing Machinery, New York (2019). https://doi.org/10.1145/3308558.3313746
12. Karpukhin, V., et al.: Dense passage retrieval for open-domain question answering (2020)
13. Khattab, O., Zaharia, M.: Colbert: efficient and effective passage search via contextualized late interaction over BERT. CoRR abs/2004.12832 (2020). https://arxiv.org/abs/2004.12832
14. Kingma, D.P., Ba, J.: Adam: a method for stochastic optimization (2017)
15. Krikon, E., Kurland, O., Bendersky, M.: Utilizing inter-passage and inter-document similarities for reranking search results. ACM Trans. Inf. Syst. **29**(1) (2011). https://doi.org/10.1145/1877766.1877769
16. Li, X., et al.: Learning better representations for neural information retrieval with graph information. In: Proceedings of the 29th ACM International Conference on Information & Knowledge Management, CIKM 2020, pp. 795–804. Association for Computing Machinery, New York (2020). https://doi.org/10.1145/3340531.3411957
17. Macdonald, C., McCreadie, R., Santos, R.L., Ounis, I.: From puppy to maturity: experiences in developing terrier. In: Proceedings of OSIR at SIGIR, pp. 60–63 (2012)
18. Macdonald, C., Tonellotto, N., Ounis, I.: On single and multiple representations in dense passage retrieval. CoRR abs/2108.06279 (2021). https://arxiv.org/abs/2108.06279
19. Mahdabi, P., Gerani, S., Huang, J.X., Crestani, F.: Leveraging conceptual lexicon: query disambiguation using proximity information for patent retrieval. In: Proceedings of the 36th International ACM SIGIR Conference on Research and Development in Information Retrieval, SIGIR 2013, pp. 113–122. Association for Computing Machinery, New York (2013). https://doi.org/10.1145/2484028.2484056
20. Mahdabi, P., Keikha, M., Gerani, S., Landoni, M., Crestani, F.: Building queries for prior-art search. In: Hanbury, A., Rauber, A., de Vries, A.P. (eds.) IRFC 2011. LNCS, vol. 6653, pp. 3–15. Springer, Heidelberg (2011). https://doi.org/10.1007/978-3-642-21353-3_2
21. Murdock, V., Croft, W.B.: A translation model for sentence retrieval. In: Proceedings of Human Language Technology Conference and Conference on Empirical Methods in Natural Language Processing, pp. 684–691. Association for Computational Linguistics, Vancouver (2005). https://www.aclweb.org/anthology/H05-1086
22. Nguyen, T., et al.: MS MARCO: a human generated machine reading comprehension dataset. CoRR abs/1611.09268 (2016). http://arxiv.org/abs/1611.09268
23. Norozi, M.A., Arvola, P.: Kinship contextualization: Utilizing the preceding and following structural elements. In: Proceedings of the 36th International ACM SIGIR Conference on Research and Development in Information Retrieval, SIGIR 2013, pp. 837–840. Association for Computing Machinery, New York (2013). https://doi.org/10.1145/2484028.2484111

24. Norozi, M.A., Arvola, P., de Vries, A.P.: Contextualization using hyperlinks and internal hierarchical structure of wikipedia documents. In: Proceedings of the 21st ACM International Conference on Information and Knowledge Management, CIKM 2012, pp. 734–743. Association for Computing Machinery, New York (2012). https://doi.org/10.1145/2396761.2396855
25. Norozi, M.A., de Vries, A.P., Arvola, P.: Contextualization from the bibliographic structure (2012)
26. Piroi, F., Lupu, M., Hanbury, A.: Overview of CLEF-IP 2013 lab. In: Forner, P., Müller, H., Paredes, R., Rosso, P., Stein, B. (eds.) CLEF 2013. LNCS, vol. 8138, pp. 232–249. Springer, Heidelberg (2013). https://doi.org/10.1007/978-3-642-40802-1_25
27. Robertson, S., Walker, S., Jones, S., Hancock-Beaulieu, M., Gatford, M.: Okapi at trec-3, pp. 109–126 (1996)
28. Sheetrit, E., Shtok, A., Kurland, O.: A passage-based approach to learning to rank documents (2019)
29. Vaswani, A., et al.: Attention is all you need (2017)
30. Veličković, P., Cucurull, G., Casanova, A., Romero, A., Liò, P., Bengio, Y.: Graph attention networks (2018)
31. Wang, X., et al.: Heterogeneous graph attention network. CoRR abs/1903.07293 (2019). http://arxiv.org/abs/1903.07293
32. Xiong, L., et al.: Approximate nearest neighbor negative contrastive learning for dense text retrieval (2020)
33. Xue, X., Croft, W.B.: Automatic query generation for patent search. In: Proceedings of the 18th ACM Conference on Information and Knowledge Management, CIKM 2009, pp. 2037–2040. Association for Computing Machinery, New York (2009). https://doi.org/10.1145/1645953.1646295
34. Yu, J., et al.: Modeling text with graph convolutional network for cross-modal information retrieval (2018)
35. Zhang, T., Liu, B., Niu, D., Lai, K., Xu, Y.: Multiresolution graph attention networks for relevance matching. Proceedings of the 27th ACM International Conference on Information and Knowledge Management (2018). https://doi.org/10.1145/3269206.3271806

Expert Finding in Legal Community Question Answering

Arian Askari[1(✉)], Suzan Verberne[1], and Gabriella Pasi[2]

[1] Leiden Institute of Advanced Computer Science, Leiden University,
Leiden, Netherlands
{a.askari,s.verberne}@liacs.leidenuniv.nl
[2] Department of Informatics, Systems and Communication, University
of Milano-Bicocca, Milan, Italy
gabriella.pasi@unimib.it

Abstract. Expert finding has been well-studied in community question answering (QA) systems in various domains. However, none of these studies addresses expert finding in the legal domain, where the goal is for citizens to find lawyers based on their expertise. In the legal domain, there is a large knowledge gap between the experts and the searchers, and the content on the legal QA websites consist of a combination formal and informal communication. In this paper, we propose methods for generating query-dependent textual profiles for lawyers covering several aspects including sentiment, comments, and recency. We combine query-dependent profiles with existing expert finding methods. Our experiments are conducted on a novel dataset gathered from an online legal QA service. We discovered that taking into account different lawyer profile aspects improves the best baseline model. We make our dataset publicly available for future work.

Keywords: Legal expert finding · Legal IR · Data collection

1 Introduction

Expert finding is an established problem in information retrieval [4] that has been studied in a variety of fields, including programming [8,30], social networks [13,17], bibliographic networks [16,25], and organizations [26]. Community question answering (CQA) platforms are common sources for expert finding; a key example is Stackoverflow for expert finding in the programming domain [21].

Until now, no studies have addressed expert finding in the legal domain. On legal CQA platforms, citizens search for lawyers with specific expertise to assist them legally. A lawyer's impact is the greatest when they work in their expert field [22]. In terms of expertise and authority, there is a large gap between the asker and the answerer in the legal domain, compared to other areas. For instance, an asker in programming CQA is someone who is a programmer at least on the junior level, and the answerer could be any unknown user. In legal CQA,

© The Author(s), under exclusive license to Springer Nature Switzerland AG 2022
M. Hagen et al. (Eds.): ECIR 2022, LNCS 13186, pp. 22–30, 2022.
https://doi.org/10.1007/978-3-030-99739-7_3

the asker knows almost nothing about law, and the answerer is a lawyer who is a professional user. The content in legal CQA is a combination of formal and informal language and it may contain emotional language (e.g., in a topic about child custody). As a result, a lawyer must have sufficient emotional intelligence to explain the law clearly while also being supportive [19].

A lawyer's expertise(s) is crucial for a citizen to be able to trust the lawyer to defend them in court [23]. Although there are some platforms in place for legal expert finding (i.e., Avvo, Nolo, and E-Justice), there is currently no scientific work addressing the problem.

In this paper, we define and evaluate legal expert finding methods on legal CQA data. We deliver a data set that consists of legal questions written by anonymous users, and answers written by professional lawyers. Questions are categorized in different categories (i.e., bankruptcy, child custody, etc.), and each question is tagged by one or more expertises that are relevant to the question content. Following prior work on expert finding in other domains [7–9], we select question tags as queries. We represent the lawyers by their answers' content. For a given query (required expertise), the retrieval task is to return a ranked list of lawyers that are likely to be experts on the query topic. As ground truth, we use lawyers' answers that are marked as best answers as a sign of expertise.

Our contributions are three-fold: (1) We define the task of *lawyer finding* and release a test collection for the task;[1] (2) We evaluate the applicability of existing expert finding methods for lawyer finding, both probabilistic and BERT-based; (3) We create query-dependent profiles of lawyers representing different aspects and show that taking into account query-dependent expert profiles have a great impact on BERT-based retrieval on this task.

2   Related Work

The objective of expert finding is to find users who are skilled on a specific topic. The two most common ways to expert finding in CQA systems are topic-based and network-based. Because there is not a network structure between lawyers in legal CQA platforms, we focus on topic-based methods. The main idea behind topic-based models [3,10,11,14,20,27,29] is to rank candidate experts according to the probability $p(ca|q)$, which denotes the likelihood of a candidate ca being an expert on a given topic q. According to Balog et al. [3], expert finding can be approached by generative probabilistic modelling based on candidate models and document models. Recently, Nikzad et al. [18] introduces a multimodal method on academic expert finding that takes into account text similarity using transformers, the author network, and h-index of the author. We approach lawyer finding differently since a lawyer does not have an h-index, there is not a sufficiently dense network of lawyers in the comment sections of legal CQA platforms, and the content style in academia is different than in legal.

---

[1] The data and code is available on https://github.com/EF_in_Legal_CQA.

3   Data Collection and Preparation

Data Source and Sample. Our dataset has been scraped from the Avvo QA forum, which contains $5,628,689$ questions in total. In order to preserve the privacy of users, we stored pages anonymously without personal information and replaced lawyer names by a number. Avvo is a legal online platform where anyone could post their legal problem for free and receive responses from lawyers. It is also possible to read the answers to prior questions. Lawyers' profiles on Avvo have been identified with their real name, as opposed to regular users. The questions are organised in categories and each category (i.e. 'bankruptcy') includes questions with different category tags (i.e. 'bankruptcy homestead exemption'). For creating our test collection, we have selected questions and their associated answers categorised as 'bankruptcy' for California, which is the most populated state of the USA. We cover the period July 2016 until July 2021 which covers $9,897$ total posts and $3,741$ lawyers. The average input length of a candidate answer is 102 words.

Relevance Labels and Query Selection. We mark attorneys as experts on a *category tag* when two conditions are met. The first is engagement filtering: Similar to the definition proposed in [9], a lawyer should have ten or more of their answers marked as accepted by the asker on a *category*, and a more than average number of best answers among lawyers on that *category tag*. A best answer is either labelled as the most useful by the question poster or if more than three lawyers agree that the answer is useful. Second, following the idea proposed in [28], the acceptance ratio (count of best answers/count of answers) of their answers should be higher than the average acceptance ratio (i.e. 4.68%) in the test collection on a category. Based on the two conditions, we select 61 lawyers as experts, who combined have given $5,614$ answers and $1,917$ best answers. From the top 20 percent tags which co-occur with 'bankruptcy', we select tags (84) as queries that at least have two experts. There are on average 5 experts (lawyers who met expert conditions on a *category tag*) per query in the test collection. Our data size is comparable with four TREC Expert Finding test collections between 2005–2008, that have 49–77 queries and $1,092$–$3,000$ candidates [2,5,6,24].

Evaluation Setup. We split our data into train, validation, and test sets based on the relevant expert lawyers – instead of queries – to avoid our models being overfitted on previously seen experts. By splitting on experts, the retrieval models are expected to be more generalized and be able to detect new experts in

Table 1. Statistics on the counts of queries, answers, and relevant experts in our data.

	Train	Validation	Test	Train ∩ Validation	Train ∩ Test
Number of relevant experts	20	20	21	0	0
Number of queries	76	69	71	61	65
Number of answers	39,588	34,128	35,057	7,290	7,918

the system. The distribution of relevant experts and queries in each set is shown in Table 1. For each train/valid/test set, in retrieval, we have all non-relevant lawyers (3680 in total) plus relevant lawyers (experts) (20/20/21) to be ranked.

4   Methods

Lawyer finding is defined as finding the right legal professional lawyer(s) with the appropriate skills and knowledge within a state/city. Cities are provided by Avvo as metadata; we only keep the city of the asker in our ranking and filter out lawyers' answers from other cities. For relevance ranking, lawyers are represented by their answers, like in prior work on expert finding in other domains [3].

4.1   Baseline 1: Probabilistic Language Modelling

Following [7–9], we replicate two types of probabilistic language models to rank lawyers: document-level (model 1), and candidate-level (model 2) that were originally proposed by Balog et al. [3] In these models, the set of answers written by a lawyer is considered the proof of expertise.

In the **Candidate-based model,** we create a textual representation of a lawyer's knowledge based on the answers written by them. Following Balog et al. [3], we estimate $p(ca|q)$ by computing $p(q|ca)$ based on Bayes' Theorem. We call this model hereinafter *model 1*. In *model 1*, $P(q|ca)$ is estimated by:

$$p(q|ca) = \prod_{t \in q} \left\{ (1 - \lambda_{ca}) \times \left(\sum_{d \in D_{ca}} p(t|d) \times p(d|ca) \right) + \lambda_{ca} \times p(t) \right\} \quad (1)$$

Here, D_{ca} consists of documents (answers) that have been written by lawyer ca; $p(t|d)$ is the probability of the term t in document d; $p(t)$ is the probability of a term in the collection of documents; and $p(d|ca)$ is the probability of document d is written by candidate ca. In the legal CQA platform answers are written by one lawyer. Therefore, $p(d|ca)$ is constant.

In the **Document-based model,** the document-centric model builds a bridge between a query and lawyers by considering documents in the collection as link. Given a query q, and collection of answers ranked according to q, lawyers are ranked by aggregating the sum over the relevance scores of their retrieved answers:

$$p(q|ca) = \sum_{d \in D_{ca}} \left(\prod_{t \in q} \left\{ (1 - \lambda_d) \times p(t|d) + \lambda_d \times p(t) \right\} \times p(d|ca) \right) \quad (2)$$

λ_d and λ_{ca} are smoothing parameters that are dynamically computed per query and candidate lawyer document (lawyer's answer)/representation following [3]. Besides of the original *model 1*, and *model 2* based on probabilistic language modelling, we experiment with BM25 to rank expert candidates' profiles and documents and refer to those by *model 1 BM25*, and *model 2 BM25*.

Table 2. Baselines and proposed model results on the test set. Significant improvements over the probabilistic baselines (Model 1 LM/Bm25, Model 2 LM/BM25), and over the Vanilla BERT Document-based models are marked with ∗, and • respectively.

Model	MAP	MRR	P@1	P@2	P@5
Model 1 (Candidate-based) LM	22.8%	40.9%	23.9%	19.7%	13.2%
Model 1 (Candidate-based) BM25	3.7%	7.0%	2.9%	1.4%	2.6%
Model 2 (Document-based) LM	19.4%	21.9%	13.5%	12.7%	7.8%
Model 2 (Document-based) BM25	21.0%	36.6%	22.5%	18.3%	11.5%
Vanilla BERT Document-based (VBD)	37.3%∗	70.7%∗	60.5%∗	55.6%∗	25.9%∗
VBD + Profiles (weighted)	**39.3%∗•**	**73.2%**	**64.9%∗•**	**57.1%**	**27.7%∗•**

4.2 Baseline 2: Vanilla BERT

By Vanilla BERT, we mean a pre-trained BERT model (BERT-Base, Uncased) with a linear combination layer stacked atop the classifier [CLS] token that is fine-tuned on our dataset in a pairwise cross-entropy loss setting using the Adam optimizer. We used the implementation of MacAvaney et al. [15] (CEDR).

After initial ranking with *model 2*, we fine-tune Vanilla BERT to estimate the relevance between query and answer terms. We select retrieved answers of the top-k(50) lawyers to re-calculate their relevance score by Vanilla BERT according to the query q. Finally, we re-rank the top-k by these relevance scores. Given a query and an answer, we train Vanilla BERT to estimate the relevance that the answer was written by an expert: "[CLS]query[SEP]candidate answer[SEP]".

4.3 Proposed Method

Given a query q and a collection of answers D that are written by different lawyers, we retrieve a ranked list of answers (D_q) using *model 1*. We create four query-dependent profiles for the lawyers L_q who have at least one answer in D_q. Each profile consists of text, and that text is sampled to represent different aspects of a lawyer's answers. The aspects are comments, sentiment-positive, sentiment-negative, and recency.

On the CQA platform it is possible to post comments in response to lawyer's answer. Therefore, there is a collection of comments C_{D_q} with regard to the query. We consider the comments as possible signals for the asker's satisfaction (i.e., a "thank you" comment would indicate that the asker received a good answer). Thus, for **comment-based profiles** (CP), we shuffle the comments to l_i's answers and concatenate the first sentence of each comment. For **sentiment-positive** (PP) **and negative** (NP) **profiles**, we shuffle positive (negative) sentences from l_i's answers and concatenate them. Since our data in legal CQA is similar in genre to social media text, we identify answer sentiment using Vader [12], a rule-based sentiment model for social media text. For the **recency-based profile** (RP), we concatenate the most recent answers of l_i. For each profile we sample the text until it exceeds 512 tokens.

We fine-tune Vanilla BERT on each profile. We represent the query as sentence A and the lawyer profile as sentence B in the BERT input: " [CLS] query [SEP] lawyer profile [SEP]" Finally, we aggregate the scores of the four profile-trained BERT models and *BERT Document-based* using a linear combination of the five models' scores inspired by [1]: $aggr_S(d, q) = w_1 S_{BD} + w_2 S_{CP} + w_3 S_{PP} + w_4 S_{NP} + w_5 RP$, where $aggr_S$ is the final aggregated score; the weights w_i are optimized using grid search in the range $[1..100]$ on the validation set. BD refers to the *BERT Document-based* score, and CP, PP, NP, RP to the four profile-trained BERT models.

5   Experiments and Results

Experimental Setup We replicate [3] using Elasticsearch for term statistics, indexing, and BM25 ranking. Following the prior work on expert finding, we report MAP, MRR, and Precision@k ($k = 1, 2, 5$) as evaluation metrics.

Retrieval Results The ranking results for models are shown in Table 2. The best candidate-based and document-based lexical models are the original *model 1 LM* [3], and *model 2 BM25* respectively. We used *model 2 BM25* as our initial ranker for Vanilla BERT. *Vanilla BERT Document-based* outperforms all lexical models by a large margin. The best ranker in terms of all evaluation metrics is the weighted combination of BERT and the lawyer profiles. This indicates that considering different aspects of a lawyer's profile (comments, sentiment, recency) is useful for legal expert ranking. We employed a one-tailed t-test ($\alpha = 0.05$) to measure statistical significance.

Analysis of Models' Weights. We found $20, 13, 2, 4, 1$ as optimal weights for BERT, Comment, Recency, Sentiment positive and negative based models respectively. As expected, the BERT score plays the largest role in the aggregation as it considers all retrieved answers of a lawyer. The second weight is for the Comment profile which confirms our assumption that the content of askers' comments are possible signals for the relevance of the lawyer's answer. The Sentiment profile's weight shows positive sentiment is more informative than negative on this task.

Analysis of Differences on Seen and Unseen Queries. In Sect. 3, we argued that in our task, being robust to new lawyers is more important than being robust to new expertises (queries). We therefore split our data on the expert level and as a result there are overlapping queries between train and test set. We analyzed the differences in model effectiveness between seen and unseen queries. We found small differences: $p@5$ is 27% on seen queries, and 25% on unseen queries. This indicates the model generalizes quite well to unseen queries.

6   Conclusions

In this paper, we defined the task of legal expert finding. We experimented with baseline probabilistic, BERT-based, and proposed expert profiling methods on our novel data. BERT-based method outperformed probabilistic methods, and the proposed methods outperformed all models.

For future work, there is a need to study more in-depth the robustness of proposed methods on different legal categories. Moreover, by providing this dataset we facilitate other tasks such as legal question answering, duplicate question detection, and finding lawyers who will reply to a question.

References

1. Althammer, S., Askari, A., Verberne, S., Hanbury, A.: DoSSIER@ COLIEE 2021: leveraging dense retrieval and summarization-based re-ranking for case law retrieval. arXiv preprint arXiv:2108.03937 (2021)
2. Bailey, P., De Vries, A.P., Craswell, N., Soboroff, I.: Overview of the TREC 2007 enterprise track. In: TREC. Citeseer (2007)
3. Balog, K., Azzopardi, L., de Rijke, M.: A language modeling framework for expert finding. Inf. Process. Manage. **45**(1), 1–19 (2009)
4. Balog, K., Fang, Y., De Rijke, M., Serdyukov, P., Si, L.: Expertise retrieval. Found. Trends Inf. Retrieval **6**(2–3), 127–256 (2012)
5. Balog, K., Thomas, P., Craswell, N., Soboroff, I., Bailey, P., De Vries, A.P.: Overview of the TREC 2008 enterprise track. Amsterdam Univ (Netherlands), Tech. rep. (2008)
6. Craswell, N., De Vries, A.P., Soboroff, I.: Overview of the TREC 2005 enterprise track. In: TREC, vol. 5, pp. 1–7 (2005)
7. Dargahi Nobari, A., Sotudeh Gharebagh, S., Neshati, M.: Skill translation models in expert finding. In: Proceedings of the 40th International ACM SIGIR Conference on Research and Development in Information Retrieval, pp. 1057–1060 (2017)
8. Dehghan, M., Abin, A.A., Neshati, M.: An improvement in the quality of expert finding in community question answering networks. Decis. Supp. Syst. **139**, 113425 (2020)
9. van Dijk, D., Tsagkias, M., de Rijke, M.: Early detection of topical expertise in community question answering. In: Proceedings of the 38th International ACM SIGIR Conference on Research and Development in Information Retrieval, pp. 995–998 (2015)
10. Fu, J., Li, Y., Zhang, Q., Wu, Q., Ma, R., Huang, X., Jiang, Y.G.: Recurrent memory reasoning network for expert finding in community question answering. In: Proceedings of the 13th International Conference on Web Search and Data Mining, pp. 187–195 (2020)
11. Guo, J., Xu, S., Bao, S., Yu, Y.: Tapping on the potential of Q& A community by recommending answer providers. In: Proceedings of the 17th ACM Conference on Information and Knowledge Management, pp. 921–930 (2008)
12. Hutto, C., Gilbert, E.: Vader: a parsimonious rule-based model for sentiment analysis of social media text. In: Proceedings of the International AAAI Conference on Web and Social Media, vol. 8 (2014)
13. Li, G., et al.: Misinformation-oriented expert finding in social networks. World Wide Web **23**(2), 693–714 (2019). https://doi.org/10.1007/s11280-019-00717-6

14. Liu, X., Ye, S., Li, X., Luo, Y., Rao, Y.: ZhihuRank: a topic-sensitive expert finding algorithm in community question answering websites. In: Li, F., Klamma, R., Laanpere, M., Zhang, J., Manjôn, B., Lau, R. (eds.) Advances in Web-Based Learning – ICWL 2015. ICWL 2015. Lecture Notes in Computer Science, vol. 9412. Springer, Cham (2015). https://doi.org/10.1007/978-3-319-25515-6_15

15. MacAvaney, S., Yates, A., Cohan, A., Goharian, N.: CEDR: contextualized embeddings for document ranking. In: Proceedings of the 42nd International ACM SIGIR Conference on Research and Development in Information Retrieval, pp. 1101–1104 (2019)

16. Neshati, M., Hashemi, S.H., Beigy, H.: Expertise finding in bibliographic network: topic dominance learning approach. IEEE Trans. Cybern. **44**(12), 2646–2657 (2014)

17. Neshati, M., Hiemstra, D., Asgari, E., Beigy, H.: Integration of scientific and social networks. World Wide Web **17**(5), 1051–1079 (2013). https://doi.org/10.1007/s11280-013-0229-1

18. Nikzad-Khasmakhi, N., Balafar, M., Feizi-Derakhshi, M.R., Motamed, C.: Berters: multimodal representation learning for expert recommendation system with transformers and graph embeddings. Chaos, Solitons Fractals **151**, 111260 (2021)

19. Pekaar, K.A., van der Linden, D., Bakker, A.B., Born, M.P.: Emotional intelligence and job performance: the role of enactment and focus on others' emotions. Human Perf. **30**(2–3), 135–153 (2017)

20. Riahi, F., Zolaktaf, Z., Shafiei, M., Milios, E.: Finding expert users in community question answering. In: Proceedings of the 21st International Conference on World Wide Web, pp. 791–798 (2012)

21. Rostami, P., Neshati, M.: Intern retrieval from community question answering websites: a new variation of expert finding problem. Expert Syst. Appl. **181**, 115044 (2021)

22. Sandefur, R.L.: Elements of professional expertise: understanding relational and substantive expertise through lawyers' impact. Am. Soc. Rev. **80**(5), 909–933 (2015)

23. Shanahan, C.F., Carpenter, A.E., Mark, A.: Lawyers, power, and strategic expertise. Denv. L. Rev. **93**, 469 (2015)

24. Soboroff, I., de Vries, A.P., Craswell, N., et al.: Overview of the TREC 2006 enterprise track. In: TREC, vol. 6, pp. 1–20 (2006)

25. Torkzadeh Mahani, N., Dehghani, M., Mirian, M.S., Shakery, A., Taheri, K.: Expert finding by the dempster-shafer theory for evidence combination. Expert Syst. **35**(1), e12231 (2018)

26. Wang, Q., Ma, J., Liao, X., Du, W.: A context-aware researcher recommendation system for university-industry collaboration on R&D projects. Decis. Support Syst. **103**, 46–57 (2017)

27. Xu, F., Ji, Z., Wang, B.: Dual role model for question recommendation in community question answering. In: Proceedings of the 35th International ACM SIGIR Conference on Research and Development in Information Retrieval, pp. 771–780 (2012)

28. Yang, J., Tao, K., Bozzon, A., Houben, G.J.: Sparrows and owls: characterisation of expert behaviour in stackoverflow. In: Dimitrova, V., Kuflik, T., Chin, D., Ricci, F., Dolog, P., Houben, G.J. (eds.) User Modeling, Adaptation, and Personalization. UMAP 2014. Lecture Notes in Computer Science, vol. 8538. Springer, Cham (2014). https://doi.org/10.1007/978-3-319-08786-3_23

29. Yang, L., Qiu, M., Gottipati, S., Zhu, F., Jiang, J., Sun, H., Chen, Z.: CQARank: jointly model topics and expertise in community question answering. In: Proceedings of the 22nd ACM International Conference on Information & Knowledge Management, pp. 99–108 (2013)
30. Zhou, G., Zhao, J., He, T., Wu, W.: An empirical study of topic-sensitive probabilistic model for expert finding in question answer communities. Knowl.-Based Syst. **66**, 136–145 (2014)

Towards Building Economic Models
of Conversational Search

Leif Azzopardi[1(⊠)], Mohammad Aliannejadi[2], and Evangelos Kanoulas[2]

[1] University of Strathclyde, Glasgow, UK
leif.azzopardi@strath.ac.uk
[2] University of Amsterdam, Amsterdam, NL, The Netherlands
{m.aliannejadi,e.kanoulas}@uva.nl

Abstract. Various conceptual and descriptive models of conversational
search have been proposed in the literature – while useful, they do not
provide insights into how interaction between the agent and user would
change in response to the costs and benefits of the different interactions.
In this paper, we develop two economic models of conversational search
based on patterns previously observed during conversational search sessions, which we refer to as: *Feedback First* where the agent asks clarifying questions then presents results, and *Feedback After* where the agent
presents results, and then asks follow up questions. Our models show
that the amount of feedback given/requested depends on its efficiency at
improving the initial or subsequent query and the relative cost of providing said feedback. This theoretical framework for conversational search
provides a number of insights that can be used to guide and inform the
development of conversational search agents. However, empirical work is
needed to estimate the parameters in order to make predictions specific
to a given conversational search setting.

1 Introduction

Conversational Search is an emerging area of research that aims to couch the
information seeking process within a conversational format [5,11] – whereby
the system and the user interact through a dialogue, rather than the traditional query-response paradigm [16,17]. Much like interactive search, conversational search presents the opportunity for the system to ask for feedback, either
through clarifying questions or follow up questions in order to refine or progress
the search [4,9,10]. While there have been numerous studies trying to develop
methods to improve the query clarifications or follow up questions, and to better
rank/select results to present/use during a conversational search (e.g. [2,3,12–
14,18]) – less attention has been paid to modelling conversational search and
understanding the trade-offs between querying, assessing, and requesting feedback [8]. In this paper, we take as a reference point the work of Vakulenko
et al. [15] and the work of Azzopardi [6,7]. The former is an empirically derived
model of conversational search, called **QRFA**, which involves querying (Q),
receiving/requesting feedback (RF) and assessing (A), while the latter works

© The Author(s), under exclusive license to Springer Nature Switzerland AG 2022
M. Hagen et al. (Eds.): ECIR 2022, LNCS 13186, pp. 31–38, 2022.
https://doi.org/10.1007/978-3-030-99739-7_4

are economic models of querying (Q) and assessing (A). In this paper, we consider two ways in which we can extend the economic model of search to include feedback. And then we use these models to better understand the relationships and trade-offs between querying, giving/requesting feedback and assessing given their relative efficiencies and their relative costs. We do so by analysing what the optimal course of interaction would be given the different models in order to minimise the cost of the conversation while maximising the gain. As a result, this work provides a number of theoretical insights that can be used to guide and inform the development of conversational search agents.

2   Background

In their analysis of conversational search sessions, Vakulenko et al. [15] found that two common conversational patterns emerged:

1. the user issues a (Q), the system would respond by requesting feedback (RF), the user would provide the said feedback, and the system would either continue to ask further rounds of feedback (RF), or present the results where user assess items (A), and then repeats the process by issuing a new query (or stops); or,
2. the user issues a (Q), the system would present results where the user assesses items (A), and then, the system requests feedback given the results (RF), the user would provide the said feedback, and the system would present more results (A), the user would assess more items (A), and the system would request further feedback (RF) until the user issues a new query (or stops).

Inspired by our previous work [1], we can think of these two conversational patterns as **Model 1**: Feedback First, and **Model 2**: Feedback After. There, of course, are many other possible patterns i.e. feedback before and after, or not at all, and combinations of. In this work, we shall focus on modelling these two "*pure*" approaches for conversational search. But, before doing so, we first present the original economic model of search.

An Economic Model of Search: In [6], Azzopardi proposed a model of search focused on modelling the traditional query response paradigm, where a user issues a query (Q), the system presents a list of results where the user assesses A items. The user continues by issuing a new query (or stops). We can call this **Model 0**. During the process, the user issues Q queries and is assumed to assess A items per query (on average). The gain that the user received was modelled using a Cobbs-Douglas production function (see Eq. 1). The exponents α and β denote the relative efficiency of querying and assessing. If $\alpha = \beta = 1$ then it suggests that for every item assessed per query, the user would receive one unit of gain (i.e. an ideal system). However, in practice, α and β are less than one, and so the more querying or more assessing naturally leads to diminishing returns. That is, assessing one more item is less likely to yield as much gain as the previous item, and similarly issuing another query will retrieve less new information (as the pool of relevant items is being reduced for a given topic).

$$g_0(Q_0, A_0) = Q_0^\alpha . A_0^\beta \qquad (1)$$

Given A_0 and Q_0, a simple cost model was proposed (Eq. 2), where the total cost to the user is proportional to the number of queries issued and the total number of items inspected (i.e. $Q_0 \times A_0$), where the cost per query is C_q and the cost per assessment is C_a.

$$c_0(Q_0, A_0) = Q_0 . C_q + Q_0 . A_0 . C_a \qquad (2)$$

By framing the problem as an optimisation problem [7], where the user wants a certain amount of gain, then the number of assessments per query that would minimise the overall cost to the user is given by:

$$A_0^\star = \frac{\beta . C_q}{(\alpha - \beta) . C_a} \qquad (3)$$

Here, we can see that as the cost of querying increases, then, on average, users should assess more items. While if the cost of assessing increases, on average, users should assess fewer items per query. If the relative efficiency of querying (α) increases, then users should assess fewer items per query, while if the relative efficiency of assessing (β) increases then the users should assess more. Moreover, there is a natural trade-off between querying and assessing, such that to obtain a given level of gain $g(Q_0, A_0) = G$ then as the A_0 increases, Q_0 decreases, and vice versa. In the following sections, we look at how we can extend this model to also include the two different forms of feedback (first and after).

3   Models

Below we present two possible models for conversational search given the two pure strategies of feedback first and feedback after. In discussing the models, we will employ a technique called *comparatives statics* – that is we make statements regarding the changes to the outcomes (i.e. how users would change their behaviour) given a change in a particular variable assuming all other variables remain the same.

3.1   An Economic Model of Conversational Search - Feedback First

Under Model 1 (Feedback First), each round of feedback, aims to improve the initial query – and thus drive up the efficiency of querying. To model this, we need to introduce a function, called $\Gamma(F_1)$ where the efficiency of queries is directly related to how many rounds of feedback are given. We assume for simplicity that the relationship is linear, each round of feedback increases the efficiency of querying by a fixed amount, which we denote as γ_1. We can then set: $\Gamma(F_1) = \gamma_1 . F_1 + \alpha$. If no rounds of feedback are given/requested, then $\Gamma(F_1) = \alpha$ – which results in the original gain function. The gain function under feedback first is:

$$g_1(Q_1, F_1, A_1) = Q_1^{\Gamma(F_1)} . A_1^\beta \qquad (4)$$

Since the rounds of feedback are only given per query, then the cost model becomes:

$$c_1(Q_1, F_1, A_1) = Q_1.C_q + Q_1.F_1.C_f + Q_1.A_1.C_a \tag{5}$$

Again, we can frame the problem as an optimisation problem, where we would like to minimise the cost given a desired level of gain G. Then by using a Lagrangian multiplier, differentiating and solving the equations we are at the following expressions:

$$A_1^\star = \frac{\beta.C_q + F_1.C_f}{(\gamma_1.F_1 + \alpha - \beta).C_a} \tag{6}$$

$$F_1^\star = \frac{\beta.C_q + (\alpha - \beta)A_1.C_a}{\gamma_1.A_1.C_a + \beta.C_f} \tag{7}$$

From these expressions, we can see that there is a dependency between the two main actions of assessing and providing feedback (unfortunately, we could not reduce down the expression any further). With respect to the optimal number of assessments, we can see that when $F = 0$ the model reverts back to the original model. And similarly, if C_q increases, then $A_1^\star$ increases. If C_a increases, $A_1^\star$ decreases. And if C_f increases and $F_1 > 0$, then $A_1^\star$ increases. If the number of rounds of feedback increases, then the model stipulates that a user should inspect fewer items (as the query after feedback is more efficient – and so more effort should be invested into assessing). For feedback, we see that if C_q or C_a increases, then the user should provide more feedback, or alternatively the system should request more feedback (i.e. $F_1^\star$ increases). While if the cost of feedback increases then users should provide less feedback (or the system should request less feedback before returning results). Interestingly, if γ_1 increases, then fewer rounds of feedback are required. This makes sense because if the clarifying question improves the quality of the original query sufficiently, then there is no need for further clarifications of the information need. More clarifications would only increase the cost of the conversation but not necessarily increase the amount of gain. On the other hand, if γ_1 is very small, then it suggests that the clarifying questions are not increasing the quality of the query such that it retrieves more relevant material. Thus, asking clarifying questions, in this case, is unlikely to be worthwhile as it will drive up the cost of conversation without increasing the reward by very much.

3.2   An Economic Model of Conversational Search - Feedback After

Given Model 2 (Feedback After), we can extended the original economic model of search by adding in feedback (F_2), such that a user issues a query, assess A_2 items, then provides feedback to refine the query, followed by assessing another A items. The process of giving feedback and assessing, then repeats this F_2 times. For this kind of conversational interaction, we can define the gain to be proportional to the number of queries and the number of rounds of feedback per query and the number of assessments per query or round of feedback:

$$g_2(Q_2, F_2, A_2) = Q_2^\alpha.(1 + F_2)^{\gamma_2}.A_2^\beta \tag{8}$$

As before, α and β denote the efficiency of querying and assessing, while γ_2 expresses the efficiency of providing feedback. We can see that if $F = 0$ (e.g. no feedback), then the model reverts back to the original model in Eq. 1.

$$c_2(Q_2, F_2, A_2) = Q_2.C_q + Q_2.F_2.C_f + Q_2.(1 + F_2).A_2.C_a \tag{9}$$

The cost function is also extended to include the rounds of feedback, and the additional assessments per feedback round, where if $F_2 = 0$ then the cost model reverts back to the original cost model. The amount of feedback per query will depend on the cost of the feedback, and its relative efficiency (γ_2). So if $\gamma_2 = 0$ and $F_2 > 0$ then there is no additional benefit for providing feedback – only added cost.

To determine the optimal number of assessments and feedback, for a given level of gain G where we want to minimise the total cost to the user given G we formulated the problem as an optimisation problem. Then by using a Lagrangian multiplier, differentiating and solving the equations we are at the following expressions for the optimal number of assessments ($A^\star$), and the optimal number of rounds of feedback ($F_2^\star$) :

$$A_2^\star = \frac{\gamma.(C_q + C_f) - \alpha.(1 + F_2).C_f}{(\alpha - \gamma).(1 + F_2).C_a} \tag{10}$$

$$F_2^\star = \frac{(\gamma - \beta).C_q + (\beta - \alpha).C_f}{(\alpha - \gamma).C_f} \tag{11}$$

First, we can see that under this model, it is possible to solve the equations fully. While this may seem quite different – during the intermediate steps the optimal $A_2^\star$ was:

$$A_2^\star = \frac{\beta.(C_q + F_2.C_f)}{(\alpha - \beta).(F_2 + 1)C_a} \tag{12}$$

where we can see the relationship between assessing and giving feedback. And, if we set $F_2 = 0$, then the model, again, falls back to the original model in Eq. 1. Specifically, assuming feedback is to be given/requested (i.e. $F_2 > 0$), then if the cost of performing the feedback (C_f) increases, on average, a user should examine more items per query and per round of feedback. If the cost of assessing (C_a) increases, then, on average, a user should examine fewer items per query/feedback. Intuitively, this makes sense, as it suggests a user should invest more in refining their need to bring back a richer set of results, than inspecting additional results for the current query or round of feedback. Now, if the relative efficiency of assessing (β_2) increases, then a user should examine more items per query/feedback.

In terms of feedback, we can see that as the cost of querying (C_q) increases, then it motivates giving more feedback. While if the cost of feedback (C_f) increases, it warrants providing less feedback. This is because querying is a natural alternative to providing another round of feedback. We can also see that the relative efficiencies of querying (α_2) to feedback (γ_2) also play a role in determining the optimal amount of feedback – such that as γ_2 increases they users should give more feedback, while if α_2 increase they should query more.

4   Discussion and Future Work

In this paper, we have proposed two economic models of conversational search that encode the observed conversational patterns of *feedback first* and *feedback after* given the QFRA work. While these models represent two possible conversational strategies, they do, however, provide a number of interesting observations and hypotheses regarding conversational search paradigms and the role of feedback during conversational sessions. We do, of course, acknowledge that in practice conversational sessions are likely to be more varied. Nonetheless, the insights are still applicable. Firstly, and intuitively, if the cost of giving feedback either before/first or after increases, then the number of rounds of feedback will be fewer. While if the cost of querying increases, then it motivates/warrants requesting/giving more feedback. The amount of which depends on the relative costs and efficiencies of each action. However, a key difference arises between whether feedback is given before (first) or after. Under feedback first, if the relative efficiency γ_1 (e.g. answering clarifying questions, etc.) increases, then perhaps ironically less feedback is required. This is because the initial query will be enhanced quicker than when γ_1 is low. Moreover, if the relative efficiency of querying is initially high, then it also suggests that little (perhaps even no) feedback would be required because the query is sufficiently good to begin with and clarifications or elaborations will only result in increased costs. These are important points to consider when designing and developing a conversational search system. As the decision to give/request feedback is decided by both the gains and the costs involved (i.e. is it economically viable?). And thus both need to be considered when evaluating conversational agents and strategies.

Of course, such discussions are purely theoretical. More analysis is required, both computationally through simulations and empirically through experiments to explore and test these models in practice. With grounded user data, it will be possible to estimate the different parameters of the proposed models – to see whether they provide a reasonable fit and valid predictions in real settings. It is also worth noting that another limitation of these models is that they model the average conversational search process over a population of user sessions – rather than an individual's conversational search process. For example, the quality of clarifying questions asked during the feedback first model is likely to vary depending on the question, this, in turn, suggests that each question will result in different improvements to the original query (i.e. γ_1 is not fixed, but is drawn from a distribution). However, the model is still informative, because we can consider what would happen for different values of γ_1 and determine when feedback would be viable, and at what point it would not be. Once estimates of the costs and relative efficiencies are obtained for a given setting, it will also be possible to further reason about how or what in the conversational process needs to be improved. Finally, more sophisticated models of conversational search could also be further developed to analyse different possible mixed strategies. However, we leave such the empirical investigations and further modelling for future work.

Acknowledgements. This research was supported by the NWO (No. 016.Vidi 189.039 and No. 314-99-301), and the Horizon 2020 (No. 814961).

References

1. Aliannejadi, M., Azzopardi, L., Zamani, H., Kanoulas, E., Thomas, P., Craswell, N.: Analysing mixed initiatives and search strategies during conversational search. In: CIKM, pp. 16–26. ACM (2021)
2. Aliannejadi, M., Kiseleva, J., Chuklin, A., Dalton, J., Burtsev, M.S.: Convai3: generating clarifying questions for open-domain dialogue systems (ClariQ). CoRR abs/2009.11352 (2020)
3. Aliannejadi, M., Zamani, H., Crestani, F., Croft, W.B.: Asking clarifying questions in open-domain information-seeking conversations. In: SIGIR, pp. 475–484. ACM (2019)
4. Allen, J., Guinn, C.I., Horvtz, E.: Mixed-initiative interaction. IEEE Intell. Syst. Their Appl. **14**(5), 14–23 (1999)
5. Anand, A., Cavedon, L., Joho, H., Sanderson, M., Stein, B.: Conversational search (dagstuhl seminar 19461). Dagstuhl Rep. **9**(11), 34–83 (2019)
6. Azzopardi, L.: The economics in interactive information retrieval. In: SIGIR, pp. 15–24. ACM (2011)
7. Azzopardi, L.: Modelling interaction with economic models of search. In: SIGIR, pp. 3–12. ACM (2014)
8. Azzopardi, L., Dubiel, M., Halvey, M., Dalton, J.: Conceptualizing agent-human interactions during the conversational search process. The Second International Workshop on Conversational Approaches to Information Retrieval, CAIR (2018)
9. Belkin, N.J., Cool, C., Stein, A., Thiel, U.: Cases, scripts, and information-seeking strategies: on the design of interactive information retrieval systems. Expert Syst. Appl. **9**(3), 379–395 (1995)
10. Croft, W.B., Thompson, R.H.: I^3r: a new approach to the design of document retrieval systems. JASIS **38**(6), 389–404 (1987)
11. Culpepper, J.S., Diaz, F., Smucker, M.D.: Research frontiers in information retrieval: report from the third strategic workshop on information retrieval in Lorne (SWIRL 2018). SIGIR Forum **52**(1), 34–90 (2018)
12. Hashemi, H., Zamani, H., Croft, W.B.: Guided transformer: leveraging multiple external sources for representation learning in conversational search. In: SIGIR, pp. 1131–1140. ACM (2020)
13. Kiesel, J., Bahrami, A., Stein, B., Anand, A., Hagen, M.: Toward voice query clarification. In: SIGIR, pp. 1257–1260 (2018)
14. Rao, S., Daumé, H.: Learning to ask good questions: ranking clarification questions using neural expected value of perfect information. In: ACL (1), pp. 2736–2745 (2018)
15. Vakulenko, S., Revoredo, K., Di Ciccio, C., de Rijke, M.: QRFA: a data-driven model of information-seeking dialogues. In: Azzopardi, L., Stein, B., Fuhr, N., Mayr, P., Hauff, C., Hiemstra, D. (eds.) ECIR 2019. LNCS, vol. 11437, pp. 541–557. Springer, Cham (2019). https://doi.org/10.1007/978-3-030-15712-8_35
16. Vtyurina, A., Savenkov, D., Agichtein, E., Clarke, C.L.A.: Exploring conversational search with humans, assistants, and wizards. In: CHI Extended Abstracts, pp. 2187–2193 (2017)

17. Yan, R., Song, Y., Wu, H.: Learning to respond with deep neural networks for retrieval-based human-computer conversation system. In: SIGIR, pp. 55–64 (2016)

18. Zamani, H., Dumais, S.T., Craswell, N., Bennett, P.N., Lueck, G.: Generating clarifying questions for information retrieval. In: WWW, pp. 418–428. ACM / IW3C2 (2020)

Evaluating the Use of Synthetic Queries for Pre-training a Semantic Query Tagger

Elias Bassani[1,2]($\boxtimes$) and Gabriella Pasi[1]

[1] Consorzio per Il Trasferimento Tecnologico - C2T, Milan, Italy
gabriella.pasi@unimib.it
[2] University of Milano-Bicocca, Milan, Italy
e.bassani3@campus.unimib.it

Abstract. Semantic Query Labeling is the task of locating the constituent parts of a query and assigning domain-specific semantic labels to each of them. It allows unfolding the relations between the query terms and the documents' structure while leaving unaltered the keyword-based query formulation. In this paper, we investigate the pre-training of a semantic query-tagger with synthetic data generated by leveraging the documents' structure. By simulating a dynamic environment, we also evaluate the consistency of performance improvements brought by pre-training as real-world training data becomes available. The results of our experiments suggest both the utility of pre-training with synthetic data and its improvements' consistency over time.

Keywords: Semantic query labeling · Query generation · Vertical search.

1 Introduction

Nowadays, many different kinds of vertical online platforms, such as media streaming services (e.g., Netflix, Spotify), e-commerce websites (e.g., Amazon), and several others, provide access to domain-specific information through a search engine. This information is usually organized in structured documents. In this context, like in Web Search, users typically convey their information needs by formulating short keyword-based queries. However, in Vertical Search, users' queries often contain references to specific structured information contained in the documents. Nevertheless, Vertical Search is often managed as a traditional retrieval task, treating documents as unstructured texts and taking no advantage of the latent structure carried by the queries.

Semantic Query Labeling [12], the task of locating the constituent parts of a query and assigning domain-specific predefined semantic labels to each of them, allows unfolding the relations between the query terms and the documents' structure, thus enabling the search engine to leverage the latter during retrieval while leaving unaltered the keyword-based query formulation. We invite the reader to refer to Balog [1] for analogies and differences with other tasks.

M. Hagen et al. (Eds.): ECIR 2022, LNCS 13186, pp. 39–46, 2022.
https://doi.org/10.1007/978-3-030-99739-7_5

Recently, Bassani et al. [3] proposed to alleviate the need for manually labeled training data for this task, by leveraging a rule-based query generator that only requires a structured document collection for producing synthetic queries.

This paper enriches the evaluation originally performed by Bassani et al. [3] by conducting further experiments in a pre-training/fine-tuning perspective, not previously considered by the authors. The work reported in this paper aims at gathering new practical insights on the use of synthetic queries in a real-world scenario, and we specifically aim to answer the following research questions:

1. Can we improve the performance of a semantic query tagger by pre-training it with *synthetic* data before fine-tuning it with *real-world* queries?
2. Can pre-training with many synthetic queries solve the inconsistency of a model in predicting semantic classes under-represented in the training set?
3. Is the performance boost given by pre-training, if any, consistent over time while new real-world training data become available?
4. When does fine-tuning with real-world data become effective for achieving performance improvements over a model trained only on synthetic queries?

For answering these questions and for conducting the experiments, we rely on both the synthetic query generator and the semantic query tagger proposed in [3]. The query generator leverages the structure of domain-specific documents and simple query variation techniques to produce annotated queries. The tagging model is based on BERT [5], gazetteers-based features, and Conditional Random Fields [8]. We invite the reader to refer to the original paper for further details.

2   Related Work

Although the semantic tagging of query terms could play a key role in Vertical Search, this task has not been sufficiently studied until recently, mainly due to the lack of publicly available datasets. Because of that, the majority of past research efforts in this context come from private companies [7,9–12,15]). Mashadi et al. [12] proposed a combination of a rule-based probabilistic grammar, lexicon features and Support Vector Machine [4] to produce and rank all the possible sequences of labels for the query terms. Li et al. [10] focused on a semi-supervised approach for training a Conditional Random Fields-based [8] query tagger by leveraging fuzzy-matching heuristics. Li [9] employed both lexical and semantic features of the query terms and proposed a method based on semi-Markov Conditional Random Fields [8] to label the queries. Sarkas et al. [15] proposed to compute the likelihood of each possible sequence of labels generated w.r.t. a collection of structured tables using a generative model. Liu et al. [11] focused on enriching the lexicons used to derive lexical features for Semantic Query Tagging. Kozareva et al. [7] employed several features and Word2Vec embeddings [13] as the input to a combination of LSTM network [6] and Conditional Random Fields [8]. Recently, Bassani et al. [3] trained a semantic query tagger based on BERT [5], gazetteers-based features, and Conditional Random Fields [8] with synthetic queries generated by leveraging structured documents, without the need for manually annotated data nor a query log. In this paper, we investigate the potential of pre-training a semantic query tagger with synthetic data. We also evaluate the model performance over time to assess whether this kind of pre-training is always beneficial.

3   Experimental Setup

In this section, we describe the experimental settings, the training setup, and the evaluation metrics used for comparing the effectiveness of three query semantic taggers: 1) one trained on synthetic data, 2) one trained on real-world queries, and 3) one pre-trained on synthetic data and fine-tuned on real-world queries.

3.1   Dataset and Structured Corpus

To compare the models' performances, we rely on the Semantic Query Labeling benchmark dataset[1] proposed in [2,3]. The dataset is composed of 6749 manually annotated real-world unique queries in the movie domain, which originally come from the 2006 AOL query logs [14]. The queries are labeled as either *Title*, *Country*, *Year*, *Genre*, *Director*, *Actor*, *Production company*, *Tag* (mainly topics and plot features), or *Sort* (e.g., *new*, *best*, *popular*, etc.). The dataset proposes three different evaluation scenarios of increasing difficulty that allow conducting a fine-grained evaluation (*Basic*, *Advanced*, and *Hard*). The queries in each scenario come already divided into *train*, *dev*, and *test sets*, following a temporal splitting approach. We invite the reader to refer to [3] for further details. For each of these scenarios, we generated 100 000 unique training queries following the query generation method proposed in [3]. We discarded generated queries present in the *dev* and *test sets*. For generating synthetic queries and computing gazetteers-based features, we employ a publicly available dataset hosted on Kaggle. This dataset contains metadata for many movies, such as title, country, and genre. The structured information originally comes from a collaborative online database for movies and TV shows called The Movie Database. For consistency with the query set, we filtered out every movie released after 2006.

3.2   Training Setup

In each of the conducted experiments, we trained the models for 50 epochs using Stochastic Gradient Descent, batch size of 64, and a starting learning rate of 0.1. We used a starting learning rate of 0.01 only for the *Pre-trained* model on the *Basic* and *Advanced* scenarios, as the pre-trained model already achieved good performances (see Sect. 4). We halved the learning rate when the training loss did not decrease for 5 consecutive epochs. We applied Dropout [16] with a probability of 0.5 on the Conditional Random Field's input to help with regularization. As Semantic Query Labeling is a multi-class classification problem, we optimize our models using softmax cross-entropy. We selected the final models basing on their performances on the *dev set* in the best epoch.

---

[1] https://github.com/AmenRa/ranx.

3.3 Evaluation Metrics

To comparatively evaluate the proposed models, we employed F1, Micro-F1 and Macro-F1 scores. F1 score is defined as the harmonic mean of Precision and Recall. Micro-F1 is the average of the F1 scores computed independently for each class and weighted by their number of samples. For imbalanced datasets, the Micro-F1 score can be skewed towards the most populated classes. Macro-F1 is the average of the F1 scores computed independently for each class. Each class contributes equally to this score. A noticeable discrepancy between the Micro-F1 score (high) and the Macro-F1 score (low) highlights inconsistency in the model predictions, suggesting that the model is skewed towards specific classes, usually the most popular ones in the training set.

4 Results and Discussion

4.1 Experiment I

The first experiment we conducted aimed to evaluate the performance gains we can achieve by pre-training a semantic query tagger with synthetic data generated by leveraging the information contained in the documents of a structured corpus and fine-tuning it with real-world queries. Besides the overall improvements, a secondary aim of the experiment is to assess whether pre-training with many synthetic queries can solve the inconsistency of the predictions of a model trained with limited real-world training data, *i.e.*, the inconsistency of the model in predicting different semantic classes. We conducted this evaluation by comparing the obtained Micro-F1 and Macro-F1 scores as described in Sect. 3.3.

Tables 1 and 2 show the results obtained by training the query tagger proposed in [3] in three different ways on each of the proposed scenarios: 1) *Real* refers to the models trained with queries from the real-world dataset, 2) *Synthetic* refers to the models trained with 100k synthetic queries obtained following the query generation procedure proposed in [3], and 3) *Pre-trained* refers to the models pre-trained on the training data of *Synthetic* and fine-tuned with the real-world training queries of *Real*.

Table 1. Overall effectiveness of the models. Best results are in boldface.

| Model | *Basic* | | *Advanced* | | *Hard* | |
	Micro-F1	Macro-F1	Micro-F1	Macro-F1	Micro-F1	Macro-F1
Synthetic	0.909	0.884	0.903	0.865	0.765	0.756
Real	0.927	0.903	0.896	0.776	0.816	0.756
Pre-trained	**0.934**	**0.910**	**0.925**	**0.893**	**0.840**	**0.828**

As shown in the Table 1, *Pre-trained* model consistently outperforms the considered baselines in all the evaluation scenarios, achieving considerable improvements over both the *Synthetic* and the *Real* models. Interestingly, we registered the most noticeable benefits of pre-training/fine-tuning on *Hard*, the most complex scenario among the three. Furthermore, the discrepancies in Micro-F1 and Macro-F1 scores that affected the *Real* model—highlighting inconsistency in the model predictions and suggesting it is skewed towards the most popular classes in the real-world training set—do not affect the *Pre-trained* model. The latter gets its consistency from the large synthetically generated query sets it was pre-trained on, where each semantic class is represented evenly.

Table 2. F1 scores for each model and semantic class. Best results are in boldface.

Scenario	Model	Actor F1	Country F1	Genre F1	Title F1	Year F1	Director F1	Sort F1	Tag F1	Company F1
Basic	Synthetic	0.898	0.811	0.867	0.917	0.928	N/A	N/A	N/A	N/A
Basic	Real	0.865	**0.857**	**0.897**	**0.949**	0.945	N/A	N/A	N/A	N/A
Basic	Pre-trained	**0.905**	**0.857**	0.862	0.945	**0.978**	N/A	N/A	N/A	N/A
Advanced	Synthetic	0.885	0.833	**0.923**	0.914	0.983	0.667	0.853	N/A	N/A
Advanced	Real	0.844	0.765	0.880	0.921	0.975	0.111	**0.937**	N/A	N/A
Advanced	Pre-trained	**0.890**	**0.849**	0.895	**0.937**	1.000	**0.750**	0.929	N/A	N/A
Hard	Synthetic	0.857	0.773	0.855	0.777	0.971	0.550	0.876	0.522	0.623
Hard	Real	0.831	**0.837**	0.873	0.854	0.956	0.222	0.883	0.576	0.771
Hard	Pre-trained	**0.884**	0.809	**0.897**	**0.857**	**0.985**	**0.667**	**0.931**	**0.600**	**0.817**

Table 2 shows the F1 scores computed for each semantic class. As shown in the table, the obtained results suggest that the synthetically generated queries can play a complementary role w.r.t. real-world queries in effectively training a semantic query tagger. In fact, by pre-training the semantic query tagger with many synthetic queries, we can expose the model to abundant in-domain and task-related information and achieve the best performances across the line.

4.2 Experiment II

The second experiment we conducted aimed at two goals. First, we evaluate the consistency over time of the improvements brought by pre-training with synthetic data to assess whether it is always beneficial. Then, we assess *when* fine-tuning the pre-trained model with real-world data becomes effective to achieve a performance boost. By relying on the time-stamps from the original AOL query logs, we simulated a dynamic environment where new labeled queries are collected over time. During the simulation, we evaluated the performances of the models from the previous experiment at regular intervals. At each evaluation step, we used the corresponding week worth of queries as our *test set*, the queries from the week before as *dev set*, and all the queries submitted in the antecedent weeks as *train set*.

Figure 1 depicts the graphs of the results of the *over time evaluation* we conducted for the *Hard* scenario (*we omitted the other graphs because of space constraints*). As shown in the figure, the *Pre-trained* model consistently achieves better performances than both the *Real* model and the *Synthetic* model, corroborating the results of the first experiment and giving us new insights about the model. The results highlight that the improvements brought by pre-training are consistent over time. Furthermore, the *Pre-trained* model overtakes *Real* as soon as it is fine-tuned.

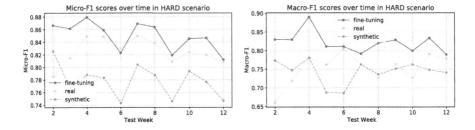

Fig. 1. Over time effectiveness of the models in the *HARD* scenario.

Table 3. Average performances of the models over time. Best results are in boldface.

Model	Basic		Advanced		Hard	
	Micro-F1	Macro-F1	Micro-F1	Macro-F1	Micro-F1	Macro-F1
Synthetic	0.916	0.892	0.896	0.851	0.778	0.743
Real	0.911	0.845	0.899	0.777	0.825	0.753
Pre-trained	**0.936**	**0.907**	**0.924**	**0.877**	**0.850**	**0.821**

In Table 3 are reported Micro-F1 scores and Macro-F1 scores averaged across all the evaluation steps, for all the evaluation scenarios. As shown in the table, the *Synthetic* model registered only a 6% decrease in Micro-F1 on average w.r.t. the *Real* model in the worst-case scenario, *Hard*. As we already highlighted, as soon as we fine-tune with real-world queries the model pre-trained on synthetic data, it achieves top performances. Because of that, the performance penalty between *Synthetic* and *Real* will never take effect as we can replace *Synthetic* with *Pre-trained* as soon as we gather real-world queries. These results suggest that, while we collect real-world training data for conducting fine-tuning, we can employ *Synthetic* with no actual performance loss w.r.t. *Real*.

5   Conclusion and Future Works

In this paper, we studied the effects of pre-training a semantic query tagger using synthetic data and subsequently fine-tuning the model with real-world queries, aiming at reducing the need for manually labeled data. Our experimental evaluation shows that 1) a semantic query tagger can be improved if first pre-trained on synthetic data, 2) the performance of a pre-trained/fine-tuned tagger is more consistent than a model trained only on real-world queries, 3) pre-training is beneficial regardless the amount of real-world training queries, and 4) fine-tuning a pre-trained model is beneficial even with minimal real-world data. To conclude, leveraging the already available information contained in a structured corpus is a valuable—*and cheap*—option for achieving performance gains without the need for additional real-world data, which, conversely, is very costly. As future work, we plan to build a dataset for the evaluation of retrieval models designed to take advantage of the output of a semantic query tagger.

References

1. Balog, K.: Entity-Oriented Search, The Information Retrieval Series, vol. 39. Springer, Cham (2018). https://doi.org/10.1007/978-3-319-93935-3
2. Bassani, E., Pasi, G.: On building benchmark datasets for understudied information retrieval tasks: the case of semantic query labeling. In: Anelli, V.W., Noia, T.D., Ferro, N., Narducci, F. (eds.) Proceedings of the 11th Italian Information Retrieval Workshop 2021, Bari, Italy, 13–15 September 2021. CEUR Workshop Proceedings, vol. 2947. CEUR-WS.org (2021). http://ceur-ws.org/Vol-2947/paper16.pdf
3. Bassani, E., Pasi, G.: Semantic query labeling through synthetic query generation. In: SIGIR 2021: The 44th International ACM SIGIR Conference on Research and Development in Information Retrieval, Virtual Event, Canada, 11–15 July 2021, pp. 2278–2282. ACM (2021). https://doi.org/10.1145/3404835.3463071, https://doi.org/10.1145/3404835.3463071
4. Cortes, C., Vapnik, V.: Support-vector networks. Mach. Learn. **20**(3), 273–297 (1995)
5. Devlin, J., Chang, M., Lee, K., Toutanova, K.: BERT: pre-training of deep bidirectional transformers for language understanding. In: Burstein, J., Doran, C., Solorio, T. (eds.) Proceedings of the 2019 Conference of the North American Chapter of the Association for Computational Linguistics: Human Language Technologies, NAACL-HLT 2019, Minneapolis, MN, USA, 2–7 June 2019, Volume 1 (Long and Short Papers), pp. 4171–4186. Association for Computational Linguistics (2019). https://doi.org/10.18653/v1/n19-1423, https://doi.org/10.18653/v1/n19-1423
6. Hochreiter, S., Schmidhuber, J.: Long short-term memory. Neural Comput. **9**(8), 1735–1780 (1997)
7. Kozareva, Z., Li, Q., Zhai, K., Guo, W.: Recognizing salient entities in shopping queries. In: Proceedings of the 54th Annual Meeting of the Association for Computational Linguistics, ACL 2016, 7–12 August 2016, Berlin, Germany, Volume 2: Short Papers. The Association for Computer Linguistics (2016). https://doi.org/10.18653/v1/p16-2018, https://doi.org/10.18653/v1/p16-2018
8. Lafferty, J.D., McCallum, A., Pereira, F.C.N.: Conditional random fields: probabilistic models for segmenting and labeling sequence data. In: Brodley, C.E., Danyluk, A.P. (eds.) Proceedings of the Eighteenth International Conference on Machine Learning (ICML 2001), Williams College, Williamstown, MA, USA, June 28–July 1 2001, pp. 282–289. Morgan Kaufmann (2001)
9. Li, X.: Understanding the semantic structure of noun phrase queries. In: Hajic, J., Carberry, S., Clark, S. (eds.) ACL 2010, Proceedings of the 48th Annual Meeting of the Association for Computational Linguistics, 11–16 July 2010, Uppsala, Sweden, pp. 1337–1345. The Association for Computer Linguistics (2010). https://www.aclweb.org/anthology/P10-1136/

10. Li, X., Wang, Y., Acero, A.: Extracting structured information from user queries with semi-supervised conditional random fields. In: Allan, J., Aslam, J.A., Sanderson, M., Zhai, C., Zobel, J. (eds.) Proceedings of the 32nd Annual International ACM SIGIR Conference on Research and Development in Information Retrieval, SIGIR 2009, Boston, MA, USA, 19–23 July 2009, pp. 572–579. ACM (2009). https://doi.org/10.1145/1571941.1572039, https://doi.org/10.1145/1571941.1572039

11. Liu, J., Li, X., Acero, A., Wang, Y.: Lexicon modeling for query understanding. In: Proceedings of the IEEE International Conference on Acoustics, Speech, and Signal Processing, ICASSP 2011, 22–27 May 2011, Prague Congress Center, Prague, Czech Republic, pp. 5604–5607. IEEE (2011). https://doi.org/10.1109/ICASSP.2011.5947630, https://doi.org/10.1109/ICASSP.2011.5947630

12. Manshadi, M., Li, X.: Semantic tagging of web search queries. In: Su, K., Su, J., Wiebe, J. (eds.) ACL 2009, Proceedings of the 47th Annual Meeting of the Association for Computational Linguistics and the 4th International Joint Conference on Natural Language Processing of the AFNLP, 2–7 August 2009, Singapore, pp. 861–869. The Association for Computer Linguistics (2009). https://www.aclweb.org/anthology/P09-1097/

13. Mikolov, T., Sutskever, I., Chen, K., Corrado, G.S., Dean, J.: Distributed representations of words and phrases and their compositionality. In: Burges, C.J.C., Bottou, L., Ghahramani, Z., Weinberger, K.Q. (eds.) Advances in Neural Information Processing Systems 26: 27th Annual Conference on Neural Information Processing Systems 2013. Proceedings of a meeting held December 5–8 2013, Lake Tahoe, Nevada, United States, pp. 3111–3119 (2013). https://proceedings.neurips.cc/paper/2013/hash/9aa42b31882ec039965f3c4923ce901b-Abstract.html

14. Pass, G., Chowdhury, A., Torgeson, C.: A picture of search. In: Jia, X. (ed.) Proceedings of the 1st International Conference on Scalable Information Systems, Infoscale 2006, Hong Kong, May 30-June 1 2006. ACM International Conference Proceeding Series, vol. 152, p. 1. ACM (2006). https://doi.org/10.1145/1146847.1146848, https://doi.org/10.1145/1146847.1146848

15. Sarkas, N., Paparizos, S., Tsaparas, P.: Structured annotations of web queries. In: Proceedings of the 2010 ACM SIGMOD International Conference on Management of Data, pp. 771–782 (2010)

16. Srivastava, N., Hinton, G.E., Krizhevsky, A., Sutskever, I., Salakhutdinov, R.: DropOut: a simple way to prevent neural networks from overfitting. J. Mach. Learn. Res. **15**(1), 1929–1958 (2014). http://dl.acm.org/citation.cfm?id=2670313

A Light-Weight Strategy for Restraining Gender Biases in Neural Rankers

Amin Bigdeli[1]([✉]), Negar Arabzadeh[2], Shirin Seyedsalehi[1], Morteza Zihayat[1], and Ebrahim Bagheri[1]

[1] Ryerson University, Toronto, Canada
{abigdeli,shirin.seyedsalehi,mzihayat,bagheri}@ryerson.ca
[2] University of Waterloo, Waterloo, Canada
narabzad@uwaterloo.ca

Abstract. In light of recent studies that show neural retrieval methods may intensify gender biases during retrieval, the objective of this paper is to propose a simple yet effective sampling strategy for training neural rankers that would allow the rankers to maintain their retrieval effectiveness while reducing gender biases. Our work proposes to consider the degrees of gender bias when sampling documents to be used for training neural rankers. We report our findings on the MS MARCO collection and based on different query datasets released for this purpose in the literature. Our results show that the proposed light-weight strategy can show competitive (or even better) performance compared to the state-of-the-art neural architectures specifically designed to reduce gender biases.

1 Introduction

With the growing body of literature on the prevalence of stereotypical gender biases in Information Retrieval (IR) and Natural Language Processing (NLP) techniques [1,2,5,9,12,19,25], researchers have specifically started to investigate how the existence of such biases within the algorithmic and representational aspects of a retrieval system can impact retrieval outcomes and as a result the end users [3–5,12,23]. For instance, Bigdeli et al. [4] investigated the existence of-stereotypical gender biases within relevance judgment datasets such as MS MARCO [17]. The authors showed that stereotypical gender biases are observable in relevance judgements. In another study, Rekabsaz et al. [23] showed that gender biases can be intensified by neural ranking models and the inclination of bias is towards the male gender. In line with the work of Rekabsaz et al., Fabris et al. [8] proposed a gender stereotype reinforcement metric to measure gender inclination within a ranked list of documents. The results of their study revealed that neural retrieval methods reinforce gender stereotypical biases.

To address such gender biases in neural rankers, Rekabsaz et al. [22] have been the first to focus on de-biasing neural rankers by removing gender-related information encoded in the vector representation of the query-document pairs specifically in the BERT reranker model. However, effectively de-biasing neural rankers is still a challenging problem due to the following major challenges: (1) The model by Rekabsaz et al., referred to as ADVBERT, can reduce bias but this comes at the cost of reduction in retrieval effectiveness. (2) ADVBERT introduces new adversarial components within the BERT reranker loss function, requiring structural changes in the architecture of the model. Such changes may not be generalizable to other neural rankers that have alternative loss functions.

In this paper, we address these challenges by proposing a novel training strategy, which 1) can decrease the level of gender biases in neural ranking models, while maintaining a comparable level of retrieval effectiveness, and 2) does not require any changes to the architecture of SOTA neural rankers. Our work is inspired by the findings of researchers such as Qu et al. [21] and Karpukhin et al. [15] who effectively argue that the performance of neural rankers are quite sensitive to the adopted negative sampling strategy where in some cases retrieval effectiveness of the same neural ranker can be increased by as much as 17% on MRR@10 by changing the negative sampling strategy. On this basis, we hypothesize that it would be possible to control gender biases in neural rankers by adopting an effective negative sampling strategy. We propose a systematic negative sampling strategy, which would expose the neural ranker to representations of gender bias that need to be avoided when retrieving documents. We then empirically show that SOTA neural rankers are able to identify and avoid stereotypical biases based on our proposed negative sampling strategy to a greater extent compared to models such as ADVBERT. At the same time, our work exhibits competitive retrieval effectiveness to strong SOTA neural rankers.

2   Problem Definition

Let $D_{q_i}^{\mathcal{M}} = [d_1^{q_i}, d_2^{q_i}, ..., d_m^{q_i}]$ be a list of initial retrieved documents for a query q_i by a first-stage retrieval method $\mathcal{M}$. Also, let us define $\mathcal{R}$ as a neural ranking model that accepts q_i and its initial retrieved list of documents $D_{q_i}^{\mathcal{M}}$ and generates $\Psi_{q_i}^{\mathcal{R}}$, which is the re-ranked version of $D_{q_i}^{\mathcal{M}}$ based on the neural ranker $\mathcal{R}$. The objective of this paper is to train a neural ranker $\mathcal{R}'$ through a bias-aware negative sampling strategy in a way that the following conditions are met:

$$\frac{1}{|Q|} \sum_{q \in Q} Bias(\Psi_q^{\mathcal{R}'}) < \frac{1}{|Q|} \sum_{q \in Q} Bias(\Psi_q^{\mathcal{R}}), \tag{1}$$

$$\frac{1}{|Q|} \sum_{q \in Q} Utility(\Psi_q^{\mathcal{R}'}) \simeq \frac{1}{|Q|} \sum_{q \in Q} Utility(\Psi_q^{\mathcal{R}}) \tag{2}$$

where $Bias(\Psi_q^{\mathcal{R}})$ is a level of bias of the top-k retrieved list of documents for query q_i by Ranker $\mathcal{R}$, as defined in [23], and $Utility(\Psi_q^{\mathcal{R}})$ is the retrieval effectiveness of ranker $\mathcal{R}$ based on metrics such as MRR. We are interested in training $\mathcal{R}'$ based on the same neural architecture used by $\mathcal{R}$, only differing in the negative sampling strategy such that the retrieval effectiveness of $\mathcal{R}$ and $\mathcal{R}'$ remain comparable while the level of bias in $\mathcal{R}'$ is significantly reduced.

3 Proposed Approach

The majority of SOTA neural rankers utilize the set of top-k documents retrieved by a fast and reliable ranker such as BM25 as a weakly-supervised strategy for negative sampling [10,11,13–16,20,21]. In this paper, instead of randomly sampling $N \leq m$ negative samples from top-k retrieved documents by BM25, we systematically select N negative samples such that the neural ranker is exposed to stereotypical gender biases that need to be avoided when ranking documents. Let β be a non-negative continuous function that measures the genderedness of any given document. An implementation of β function can be obtained from [23]. Given β and N, the retrieved document set $D_q^{\mathcal{M}}$ is sorted in descending order based on $\beta(d_i)$ $(d_i \in D_q^{\mathcal{M}})$ and the top-N documents form a non-increasing list S_q^β such that $\{\forall i \in [1, \ldots, N], \beta(s_i) \geq \beta(s_{i+1})\}$.

As the documents in S_q^β exhibit the highest degree of gender bias compared to the rest of the documents in $D_q^{\mathcal{M}}$, we suggest that S_q^β can be served as the negative sample set due to two reasons: (1) S_q^β is a subset of $D_q^{\mathcal{M}}$, as such, when using the random negative sampling strategy, S_q^β may have been chosen as the negative sample set; therefore, it is unlikely that the choice of S_q^β as the negative sample set results in decreased retrieval effectiveness; and (2) S_q^β consists of documents with the highest degree of gender bias and hence, the neural ranker would not only have a chance to learn that these documents are not relevant but also to learn to avoid biased gender affiliated content within these documents and hence avoid retrieving gender-biased documents at retrieval time.

Considering the highest gender-biased documents as the negative sample set may be a strict requirement and is not desirable as it might cause the neural ranker forgets the need of learning document relevance and only focuses on learning to avoid gender-biased documents. In order to avoid interpreting all gendered-biased documents as irrelevant during the training process, we relax the negative sampling strategy through a free-parameter λ. According to λ, a subset of negative documents is selected from S_q^β (NS_{Biased}) and the rest of the negative document set is randomly selected from the original pool $D_q^{\mathcal{M}}$ (NS_{Rnd}).

$$NS_{Biased} = \{d_i \in S_q^\beta | i \leq \lambda \times N\}$$

$$NS_{Rnd} = \{Rand(d \in D_q^{\mathcal{M}}) | d \notin NS_{Biased}\},$$

such that $|NS_{Rnd}| + |NS_{Biased}| = N$ and the final set of negative samples would be $NS = NS_{Rnd} \cup NS_{Biased}$.

4 Experiments

Document Collection. We adopt the MS MARCO collection consisting of over 8.8M passages and over 500k queries with at least one relevant document.

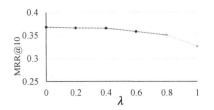

Fig. 1. Impact of λ on neural ranker performance on MS MARCO Dev Set. The red points indicate statistically significant drop in performance. (Color figure online)

Table 1. Comparison between the performance (MRR@10) of the base ranker and the ranker trained based on our proposed negative sampling strategy when $\lambda = 0.6$ on MS MARCO Dev Set.

Neural ranker	Training Schema		Change
	Original	Ours	
BERT (base)	0.3688	0.3583	−2.84%
DistilRoBERTa (base)	0.3598	0.3475	−3.42%
ELECTRA (base)	0.3332	0.3351	+0.57%

Query Sets. We adopt two query sets in our experiments that have been proposed in the literature for evaluating gender bias: (1) The first query set (QS1) includes 1,765 neutral queries [23]. (2) The second query set (QS2) includes 215 fairness-sensitive queries that are considered as socially problematic topics [22]. **Bias Metrics.** We adopt two bias measurement metrics from the literature to calculate the level of biases within the retrieved list of documents: (1) The first metric is introduced by Rekabsaz et al. [23] and measures the level of bias in a document based on the presence and frequency of gendered terms in a document, referred to as Boolean and TF ARaB metrics. (2) The Second metric is NFaiRR which calculates the level of fairness within the retrieved list of documents by calculating each document's neutrality score proposed in [22]. We note that less ARaB and higher NFaiRR metric values are desirable.

Neural Rankers. To train the neural rerankers, we adopted the cross-encoder architecture as suggested in SOTA ranking literature [18,21] in which two sequences (Query and candidate document) are passed to the transformer and the relevance score is predicted. As suggested by the Sentence Transformer Library[1], we fine tuned different pre-trained transformer models, namely, BERT-base-uncased [7], DistilRoBERTa-base [24], ELECTRA-base [6], BERT-Mini [26], and BERT-Tiny [26]. For every query in the MS MARCO training set, we considered 20 negative documents ($N = 20$) from the top-1000 unjudged documents retrieved by BM25 [17] (based on random sampling and our proposed sampling strategy).

Results and Findings. The objective of our work is to show that a selective negative sampling strategy can systematically reduce gender bias while maintaining retrieval effectiveness. As such, we first investigate how our proposed negative sampling strategy affects retrieval effectiveness. We note statistical significance is measured based on paired t-test with $\alpha = 0.05$. In Fig. 1, we demonstrate the performance of a SOTA BERT-base-uncased neural ranker trained with our proposed negative strategies when changing λ from [0,1] with 0.2 increments. Basically, when $\lambda = 0$ the model is trained with all randomly negative samples

---

[1] https://www.sbert.net/.

Table 2. Retrieval effectiveness and the level of fairness and bias across three neural ranking models trained on query sets QS1 and QS2 when $\lambda = 0.6$ at cut-off 10.

Query set	Neural ranker	Training schema	MRR@10	NFaiRR		ARaB			
				Value	Improvement	TF	Reduction	Boolean	Reduction
QS1	BERT (base)	Original	0.3494	0.7764	–	0.1281	–	0.0956	–
		Ours	0.3266	0.8673	11.71%	0.0967	24.51%	0.0864	9.62%
	DistilRoBERTa (base)	Original	0.3382	0.7805	–	0.1178	–	0.0914	–
		Ours	0.3152	0.8806	12.83%	0.0856	27.33%	0.0813	11.05%
	ELECTRA (base)	Original	0.3265	0.7808	–	0.1273	–	0.0961	–
		Ours	0.3018	0.8767	12.28%	0.0949	25.45%	0.0855	11.03%
QS2	BERT (base)	Original	0.2229	0.8779	–	0.0275	–	0.0157	–
		Ours	0.2265	0.9549	8.77%	0.0250	9.09%	0.0156	0.64%
	DistilRoBERTa (base)	Original	0.2198	0.8799	–	0.0338	–	0.0262	–
		Ours	0.2135	0.9581	8.89%	0.0221	34.62%	0.0190	27.48%
	ELECTRA (base)	Original	0.2296	0.8857	–	0.0492	–	0.0353	–
		Ours	0.2081	0.9572	8.07%	0.0279	43.29%	0.0254	28.05%

Table 3. Comparing ADVBERT training strategy and our approach at cut-off 10.

Neural ranker	Training schema	MRR@10	NFaiRR		ARaB			
			Value	Improvement	TF	Reduction	Boolean	Reduction
BERT-Tiny	Original	0.1750	0.8688	–	0.0356	–	0.0296	–
	ADVBERT	0.1361	0.9257	6.55%	0.0245	31.18%	0.0236	20.27%
	Ours	0.1497	**0.9752**	**12.25%**	0.0099	**72.19%**	0.0115	**61.15%**
BERT-Mini	Original	0.2053	0.8742	–	0.0300	–	0.0251	–
	ADVBERT	0.1515	0.9410	7.64%	0.0081	**73.00%**	0.0032	**87.26%**
	Ours	0.2000	**0.9683**	**10.76%**	0.0145	51.67%	0.0113	54.98%

from BM25 retrieved documents (baseline) and when $\lambda = 1$, the N negative samples are the most gendered documents in $D_q^{\mathcal{M}}$. Based on Fig. 1, we observe that gradual increase in λ will come at the cost of retrieval effectiveness. However, the decrease is only statistically significant when $\lambda > 0.6$. Thus we find that when up to 60% of negative samples are selected based on our proposed negative sampling strategy, the retrieval effectiveness remains comparable to the base ranker. As mentioned earlier, this drop in performance is due to the fact that the model would learn the concept of avoiding gender-biased documents and not the concept of relevance due to the large number of gender-biased negative samples. We further illustrate the performance of adopting our proposed negative sampling strategy with $\lambda = 0.6$ on other pre-trained language models including ELECTRA and DistilRoBERTa in Table 1. For these pre-trained language models, similar to BERT, we observe that retrieval effectiveness remains statistically comparable to the base ranker and no statistically significant changes occur in terms of performance. Thus we conclude that it is possible to adopt our proposed bias-aware negative sampling strategy (e.g. at $\lambda = 0.6$) and maintain comparable retrieval effectiveness. We note all our code and the run files are publicly available[2].

We now investigate the impact of our proposed negative sampling strategy on reducing gender biases. To this end, using each of the SOTA rankers, we re-rank the queries in each two query sets (QS1 and QS2) and report their retrieval

---

[2] https://github.com/aminbigdeli/bias_aware_neural_ranking.

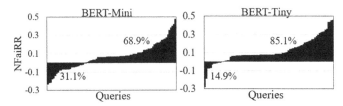

Fig. 2. Comparative analysis between ADVBERT and our proposed approach based on NFaiRR at cut-off 10 on a per query basis.

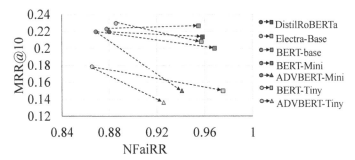

Fig. 3. Comparing the base rankers vs our proposed approach and ADVBERT in terms of performance and bias when using different pre-trained language models on QS2.

effectiveness as well as their bias metrics (ARaB and NFaiRR) measurements in Table 2. As shown, we observe that for both of the query sets the level of fairness (NFaiRR) increases, while the level of bias (ARaB) decreases across all of the three neural ranking models. In addition and more importantly, the decrease in gender biases does not come at the cost of significant reduction in retrieval effectiveness. In summary, our proposed negative sampling strategy is able to maintain retrieval effectiveness while reducing bias and increasing fairness.

It is important to also compare our work against the most recent neural ranking model designed to increase fairness, namely ADVBERT [22]. Unlike our proposed work which retains the same neural architecture of the original ranker and only changes the negative sampling strategy, ADVBERT proposes an adversarial neural architecture to handle gender biases. The authors of ADVBERT have publicly shared their trained models based on **BERT-Tiny** and **BERT-Mini** and only for QS2. For the sake of comparison, we compare our work with ADVBERT based on these two models and on QS2. Based on the results reported in Table 3, we make the following observations: (1) For the models based on **BERT-Tiny**, neither our model nor ADVBERT significantly drop retrieval effectiveness; however, the fairness (NFaiRR) and bias (ARaB) measures are notably more favorable for our proposed approach. (2) Similar observations can be made for **BERT-Mini** as well. In this case, the retrieval effectiveness of our proposed approach is substantially higher than ADVBERT and at the same time the reported level of fairness (NFaiRR) is also higher. However, in terms of bias metrics, ADVBERT has decreased both TF ARaB and Boolean ARaB more than our proposed approach.

We further compare the level of fairness between our proposed approach and ADVBERT on a per-query basis. To this end, for BERT-Tiny and BERT-Mini we calculate the level of fairness within the ranked list of documents returned by our method and ADVBERT. Followed by that, we subtract the level of fairness of each query and report the results in Fig. 2. As shown, the number of queries that have seen improvement in their fairness metric (NFaiRR) based on our approach compared to ADVBERT as well as the degree fairness has been impacted. Positive values show improved fairness by our approach compared to ADVBERT while negative values show otherwise. As shown, 69% and 85% of the queries have seen increased fairness based on our proposed approach on BERT Mini and Tiny, respectively. We contextualize this by mentioning that on both models, the retrieval effectiveness of our proposed approach is also higher than ADVBERT.

Now, let us illustrate the robustness of our proposed approach across all the neural rankers by showing the level of their effectiveness and fairness on QS2 (and not QS1 since ADVBERT is not available on QS1) in Fig. 3. As shown, when adopting our proposed approach, the level of fairness increases notably, while retrieval effectiveness remains at a comparable level. We further observe that while ADVBERT is able to increase fairness (not to the extent of our proposed approach), it does so at the cost of a notable decrease in retrieval effectiveness.

5 Concluding Remarks

We have shown that it is possible to adopt a simple yet effective sampling strategy for training neural rankers such that gender biases are reduced while retrieval effectiveness is maintained. Through our experiments, we show that a light-weight strategy is able to show competitive (or even better) tradeoff between bias reduction and retrieval effectiveness compared to adversarial neural rankers that are specifically designed for restraining gender biases.

References

1. Baeza-Yates, R.: Bias on the web. Commun. ACM **61**(6), 54–61 (2018)
2. Baeza-Yates, R.: Bias in search and recommender systems. In: Fourteenth ACM Conference on Recommender Systems, pp. 2–2 (2020)
3. Bigdeli, A., Arabzadeh, N., Seyersalehi, S., Zihayat, M., Bagheri, E.: On the orthogonality of bias and utility in ad hoc retrieval. In: Proceedings of the 44rd International ACM SIGIR Conference on Research and Development in Information Retrieval (2021)
4. Bigdeli, A., Arabzadeh, N., Zihayat, M., Bagheri, E.: Exploring gender biases in information retrieval relevance judgement datasets. In: Hiemstra, D., Moens, M.-F., Mothe, J., Perego, R., Potthast, M., Sebastiani, F. (eds.) ECIR 2021. LNCS, vol. 12657, pp. 216–224. Springer, Cham (2021). https://doi.org/10.1007/978-3-030-72240-1_18
5. Caliskan, A., Bryson, J.J., Narayanan, A.: Semantics derived automatically from language corpora contain human-like biases. Science **356**(6334), 183–186 (2017)

6. Clark, K., Luong, M.T., Le, Q.V., Manning, C.D.: Electra: pre-training text encoders as discriminators rather than generators. arXiv preprint arXiv:2003.10555 (2020)

7. Devlin, J., Chang, M.W., Lee, K., Toutanova, K.: BERT: pre-training of deep bidirectional transformers for language understanding. arXiv preprint arXiv:1810.04805 (2018)

8. Fabris, A., Purpura, A., Silvello, G., Susto, G.A.: Gender stereotype reinforcement: measuring the gender bias conveyed by ranking algorithms. Inf. Process. Manage. **57**(6), 102–377 (2020)

9. Font, J.E., Costa-Jussa, M.R.: Equalizing gender biases in neural machine translation with word embeddings techniques. arXiv preprint arXiv:1901.03116 (2019)

10. Gao, L., Callan, J.: Unsupervised corpus aware language model pre-training for dense passage retrieval. arXiv preprint arXiv:2108.05540 (2021)

11. Gao, L., Dai, Z., Callan, J.: Coil: revisit exact lexical match in information retrieval with contextualized inverted list. arXiv preprint arXiv:2104.07186 (2021)

12. Gerritse, E.J., Hasibi, F., de Vries, A.P.: Bias in conversational search: the double-edged sword of the personalized knowledge graph. In: Proceedings of the 2020 ACM SIGIR on International Conference on Theory of Information Retrieval, pp. 133–136 (2020)

13. Han, S., Wang, X., Bendersky, M., Najork, M.: Learning-to-rank with BERT in TF-ranking. arXiv preprint arXiv:2004.08476 (2020)

14. Izacard, G., Grave, E.: Leveraging passage retrieval with generative models for open domain question answering. arXiv preprint arXiv:2007.01282 (2020)

15. Karpukhin, V., Oğuz, B., Min, S., Lewis, P., Wu, L., Edunov, S., Chen, D., Yih, W.T.: Dense passage retrieval for open-domain question answering. arXiv preprint arXiv:2004.04906 (2020)

16. Macdonald, C., Tonellotto, N.: On approximate nearest neighbour selection for multi-stage dense retrieval. arXiv preprint arXiv:2108.11480 (2021)

17. Nguyen, T., et al.: Ms marco: a human generated machine reading comprehension dataset. In: CoCo@ NIPS (2016)

18. Nogueira, R., Cho, K.: Passage re-ranking with BERT. arXiv preprint arXiv:1901.04085 (2019)

19. Olteanu, A., et al.: Facts-IR: fairness, accountability, confidentiality, transparency, and safety in information retrieval. In: ACM SIGIR Forum, vol. 53, pp. 20–43. ACM New York, NY, USA (2021)

20. Pradeep, R., Nogueira, R., Lin, J.: The expando-mono-duo design pattern for text ranking with pretrained sequence-to-sequence models. arXiv preprint arXiv:2101.05667 (2021)

21. Qu, Y., et al.: RocketQA: an optimized training approach to dense passage retrieval for open-domain question answering. In: Proceedings of the 2021 Conference of the North American Chapter of the Association for Computational Linguistics: Human Language Technologies, pp. 5835–5847 (2021)

22. Rekabsaz, N., Kopeinik, S., Schedl, M.: Societal biases in retrieved contents: measurement framework and adversarial mitigation for BERT rankers. arXiv preprint arXiv:2104.13640 (2021)

23. Rekabsaz, N., Schedl, M.: Do neural ranking models intensify gender bias? In: Proceedings of the 43rd International ACM SIGIR Conference on Research and Development in Information Retrieval, pp. 2065–2068 (2020)

24. Sanh, V., Debut, L., Chaumond, J., Wolf, T.: Distilbert, a distilled version of BERT: smaller, faster, cheaper and lighter. arXiv preprint arXiv:1910.01108 (2019)

25. Sun, T., et al.: Mitigating gender bias in natural language processing: literature review. arXiv preprint arXiv:1906.08976 (2019)
26. Turc, I., Chang, M.W., Lee, K., Toutanova, K.: Well-read students learn better: on the importance of pre-training compact models. arXiv preprint arXiv:1908.08962 (2019)

Recommender Systems: When Memory Matters

Aleksandra Burashnikova[1,3]([⊠]), Marianne Clausel[2], Massih-Reza Amini[3],
Yury Maximov[4], and Nicolas Dante[2]

[1] Skolkovo Institute of Science and Technology, Moscow, Russia
aleksandra.burashnikova@skoltech.ru
[2] University of Lorraine, Nancy, France
[3] University Grenoble-Alpes, Grenoble, France
[4] Los Alamos National Laboratory, Los Alamos, USA

Abstract. In this paper, we study the effect of non-stationarities and memory in the learnability of a sequential recommender system that exploits user's implicit feedback. We propose an algorithm, where model parameters are updated user per user by minimizing a ranking loss over blocks of items constituted by a sequence of unclicked items followed by a clicked one. We illustrate through empirical evaluations on four large-scale benchmarks that removing non-stationarities, through an empirical estimation of the memory properties, in user's behaviour interactions allows to gain in performance with respect to MAP and NDCG.

1 Introduction

Recently there has been a surge of interest in the design of personalized recommender systems (RS) that adapt to user's taste based on their *implicit feedback*, mostly in the form of clicks. The first works on RS assume that users provide an *explicit feedback* (as scores) each time an item is shown to them. In many real scenarios, however, it generally takes time to provide a score, and when users do not have an interest in any products they are shown, they may not provide feedback, or may click on items of their own interest.

In the last few years, most works were interested in taking into account the sequential nature of user/item interactions in the learning process [5]. These approaches are mainly focused in the design of sequential neural networks for predicting, in the form of posterior probabilities, the user's preference given the items [14]. Models from the feedback history of a given user his next positive feedback [4]. All these strategies consider only the sequence of viewed items that are clicked or purchased; and rely on the underlying assumption that to be predictible user/item interactions have to be *homogeneous* in time, motivating the design of RS based on stationary neural networks. Non-stationarity is in many

A. Burashnikova is supported by the Analytical center under the RF Government (subsidy agreement 000000D730321P5Q0002, Grant No. 70-2021-00145 02.11.2021). Y. Maximov is supported by LANL LDRD projects.

M. Hagen et al. (Eds.): ECIR 2022, LNCS 13186, pp. 56–63, 2022.
https://doi.org/10.1007/978-3-030-99739-7_7

situations related to another property, *long-range dependence*, which basically model the impact of the whole history of the time series on its near future.

In this paper, we put in evidence (*a*) the effectiveness of taking into account negative feedback along with positive ones in the learning of models parameters, (*b*) the impact of homogeneous user/items interactions for prediction, after removal of non-stationarities and (c) the need of designing specific strategies to remove non-stationarities due to a specificity of RS, namely the presence of memory in user/items interactions. Thereafter, we turn this preliminary study into a novel and successful strategy combining sequential learning per blocks of interactions and removing user with non–homogeneous behavior from the training.

The remainder of the paper is organized as follows. In Sect. 2.1, we present the mathematical framework, used to model stationarity in RS data. Thereafter, we explain that in the case, where we have presence of long-memory in the data removing non-stationarites is specially tricky. We present our novel strategy combining the efficiency of sequential learning per block of interactions and the knowledge of the memory behavior of each user in Sect. 2.2 to remove non-stationarities. We then illustrate that memory is intrinsically present in RS user/items interactions in Sect. 3 and that we have to take it into account to remove non-stationarities and improve generalization. We then prove through experiments on different large-scale benchmarks the effectiveness of our approach.

2 A Memory Aware Sequential Learning Strategy for RS

2.1 Framework

Our claim is that all user/items interactions may not be equally relevant in the learning process. We prove in the sequel that we can improve the learning process, considering only the subset of users whose interactions with the system are *homogeneous in time*, meaning that the user feedback is statistically the same, whatever the time period is. Unfortunately, non-stationarities are not easy to detect, since we have to take into account another additional effect in RS, which is *long-range dependence*. Indeed, in RS the choice of a given user may be influenced not only by its near past but by the whole history of interactions.

We suggest modeling these two natural aspects of user feedbacks, using *stationarity* and *memory*, that are two popular and traditional mathematical tools developed for sequential data analysis. We recall that a time series $X = \{X_t, t \in \mathbb{Z}\}$, here the sequence of user's feedback, is said to be (wide-sense) stationary (see Sect. 2.2 in [2]) if its two first orders moments are homogeneous with time:

$$\forall t, k, l \in \mathbb{Z}, \ \mathbb{E}[X_t] = \mu, \text{ and } Cov(X_k, X_l) = Cov(X_{k+t}, X_{l+t}) \tag{1}$$

Under such assumptions the autocovariance of a stationary process only depends on the difference between the terms of the series $h = k - l$. We set $\gamma(h) = Cov(X_0, X_h)$.

Our other concept of interest, memory arouses in time series analysis to model memory that can be inherently present in sequential data. It provides a quantitative measure of the persistence of information related to the history of the time series in the long-run and it can be related to presence of non-stationarities in the data. Its definition is classically done in the Fourier domain and is based on the so-called spectral density. The spectral density is the discrete Fourier transform of the autocovariance function:

$$f(\lambda) = \frac{1}{2\pi} \sum_{h=-\infty}^{+\infty} \gamma(h)e^{-ih\lambda}, \qquad \lambda \in (-\pi, \pi]. \tag{2}$$

and reflects the energy contains at each frequency λ if the times series. A time series X admits memory parameter $d \in \mathbb{R}$ iff its spectral density satisfies:

$$f(\lambda) \sim \lambda^{-2d} \text{ as } \lambda \to 0 . \tag{3}$$

In the time domain, the memory parameter is related to the decay of the autocovariance function. The more it is large, the more the past of the time series has an impact on its next future. Interestingly, when the memory parameter is large, the time series tends to have a sample autocorrelation function with large spikes at several lags which is well known to be the signature of non-stationarity for many practitioners. It can then be used as a measure of non-stationarity.

In order to infer this memory parameter, we use one of the most classical estimators of the memory parameter, the GPH estimator introduced in [6]. It consists of a least square regression of the log-periodogram of X. One first defines a biased estimator of the spectral density function, the periodogram $I(\lambda)$ and evaluate it on the Fourier frequencies $\lambda_k = \frac{2\pi k}{N}$ where N is the sample length:

$$I_N(\lambda_k) = \frac{1}{N} \left| \sum_{t=1}^{N} X_t e^{it\lambda_k} \right|^2 \tag{4}$$

The estimator of the memory parameter is therefore as follows :

$$\hat{d}(m) = \frac{\sum_{k=1}^{m}(Y_k - \bar{Y})\log(I(\lambda_k))}{\sum_{k=1}^{m}(Y_k - \bar{Y})^2}, \tag{5}$$

where $Y_k = -2\log|1 - e^{i\lambda_k}|$, $\bar{Y} = (\sum_{k=1}^{m} Y_k)/m$ and m is the number of used frequencies.

While there are alternative long memory parameter estimators, such as Monte Carlo analysis, GPH is by far the most computationally efficient [1]. We then classify the time series as non-stationary if $d \geq 1/2$, and as stationary otherwise.

2.2   Learning Scheme

We now present our learning scheme. Here, the aim is to take into account the sequence of negative feedback along with positive ones for learning, and select users characteristics as stationarity and short-dependence. In the following, we first present our SequentiAl RecOmmender System for implicit feedback (called SAROS [3]) and then detail the explicit inclusion of memory in the algorithm (that we refer to as MOSAIC).

User preference over items depend mostly on the context where these items are shown to the user. A user may prefer (or not) two items independently one from another, but within a given set of shown items, he or she may completely have a different preference over these items. By randomly sampling triplets constituted by a user and corresponding clicked and unclicked items selected over the whole set of shown items to the user, this effect of local preference is not taken into account. Furthermore, triplets corresponding to different users are non uniformly distributed, as interactions vary from one user to another user, and for parameter updates; triplets corresponding to low interactions have a small chance to be chosen. In order to tackle these points; we propose to update the parameters for each user u; after each sequence of interactions t; constituted by blocks of non-preferred items, N_u^t, followed by preferred ones Π_u^t.

In a classical way [9], each user u and each item i are represented respectively by low dimensional vectors U_u and V_i living in the same latent space of dimension k. The goal of the sequential part of our algorithm is to learn a relevant representation of the couples users/items $\omega = (U, V)$ where $U = (U_u)$, $V = (V_i)$. Weights are updated by minimizing the ranking loss corresponding to this block:

$$\hat{L}_{\mathcal{B}_u^t}(\omega_u^t) = \frac{1}{|\Pi_u^t||N_u^t|} \sum_{i\in\Pi_u^t} \sum_{i'\in N_u^t} \ell_{u,i,i'}(\omega_u^t) , \tag{6}$$

where $\ell_{u,i,i'}$ is the logistic loss:

$$\ell_{u,i,i'} = \log\left(1 + e^{-y_{u,i,i'}U_u(V_i-V_{i'})}\right) + \lambda\left(\|U_u\|_2^2 + \|V_i\|_2^2 + \|V_{i'}\|_2^2\right)$$

with $y_{u,i,i'} = 1$ if the user u prefers item i over item i', $y_{u,i,i'} = -1$ otherwise.

We now describe the inclusion of the Memory-Aware step of our algorithm, allowing to include stationarity in the pipeline (called MOSAIC). In the first step we train SAROS on the full dataset. Thereafter we remove non-stationary embeddings, using a preliminary estimation of the memory parameter of each time series. Finally we train once more this filtered dataset and return the last updated weights.

3   Experiments and Results

In this section, we provide an empirical evaluation of our approach on some popular benchmarks proposed for evaluating RS. All subsequently discussed components were implemented in Python3 using the TensorFlow library.

3.1   Datasets

Description of the Datasets. We have considered four publicly available benchmarks, for the task of personalized Top–N recommendation: Kassandr [12], ML–1M [7], a subset out of the OUTBRAIN dataset from of the Kaggle challenge[1], and, Pandor [11]. Tables 1 presents some detailed statistics about the datasets and the blocks for each collection. Among these, we report the number of users, $|U|$, and items, $|I|$, the remaining number of users after filtering based on stationarity in embeddings, $|Stat\_U|$, and the average numbers of positive (clicks) and negative feedback (viewed but not clicked).

Table 1. Statistics on datasets used in our experiments.

Data	$|U|$	$|Stat\_U|$	$|I|$	Sparsity	Avg. # of +	Avg. # of −
Kassandr	2,158,859	26,308	291,485	.9999	2.42	51.93
Pandor	177,366	9,025	9,077	.9987	1.32	10.36
ML-1M	6,040	5,289	3,706	.9553	95.27	70.46
Outbrain	49,615	36,388	105,176	.9997	6.1587	26.0377

We keep the same set of users in both train and test sets. For training, we use the 70% oldest interactions of users and the aim is to predict the 30% most recent user interactions.

Identifying Stationary Users. We keep only users whose embeddings have four stationary components, using a preliminary estimation of the memory parameter. The output subset is much more small for Kassandr and Pandor than the full dataset whereas for ML-1M and OUTBRAIN we succeed in keeping a large part of the full dataset. Our filtering approach is then expected to be much more successful on the latter.

3.2   Evaluation

We consider the following classical metrics for the comparison of the models. The Mean Average Precision at rank K (MAP@K) over all users defined as $MAP@K = \frac{1}{N}\sum_{u=1}^{N} AP_K(u)$, where $AP_K(u)$ is the average precision of preferred items of user u in the top K ranked ones. The Normalized Discounted Cumulative Gain at rank K that computes the ratio of the obtained ranking to the ideal case and allow to consider not only binary relevance as in Mean Average Precision, $NDCG@K = \frac{1}{N}\sum_{u=1}^{N} \frac{DCG@K(u)}{IDCG@K(u)}$, where $DCG@K(u) = \sum_{i=1}^{K} \frac{2^{rel_i}-1}{\log_2(1+i)}$, and rel_i is the graded relevance of the item at position i; and $IDCG@K(u)$ is $DCG@K(u)$ with an ideal ordering equals to $\sum_{i=1}^{K} \frac{1}{\log_2(1+i)}$.

---

[1] https://www.kaggle.com/c/outbrain-click-prediction.

Table 2. Comparison of different models in terms of MAP@5 and MAP@10(top), and NDCG@5 and NDCG@10(down).

	ML-1M	KASANDR	Pandor	OUTBRAIN	ML-1M	KASANDR	Pandor	OUTBRAIN
	MAP@5				MAP@10			
BPR	.826	.522	.702	.573	.797	.538	.706	.537
Caser	.718	.130	.459	.393	.694	.131	.464	.397
GRU4Rec	.777	.689	.613	.477	.750	.688	.618	.463
SAROS	.832	.705	.710	.600	.808	.712	.714	.563
MOSAIC	**.842**	**.706**	**.711**	**.613**	**.812**	**.713**	**.715**	**.575**
	NDCG@5				NDCG@10			
BPR	.776	.597	.862	.560	.863	.648	.878	.663
Caser	.665	.163	.584	.455	.787	.198	.605	.570
GRU4Rec	.721	.732	.776	.502	.833	.753	.803	.613
SAROS	.788	**.764**	**.863**	.589	.874	**.794**	.879	.683
MOSAIC	**.794**	**.764**	**.863**	**.601**	**.879**	**.794**	**.880**	**.692**

Models. To validate our approach described in the previous sections, we compared SAROS and MOSAIC[2] with the following approaches.

BPR [10] corresponds to a stochastic gradient-descent algorithm, based on bootstrap sampling of training triplets, for finding the model parameters $\omega = (U, V)$ by minimizing the ranking loss over all the set of triplets simultaneously (without considering the sequence of interactions). GRU4Rec [8] is an extended version of GRU for session-based recommendation. The approach considers the session as the sequence of clicks of the user and learns model parameters by optimizing a regularized approximation of the relative rank of the relevant item. Caser [13] is a CNN based model that embeds a sequence of clicked items into a temporal image and latent spaces and find local characteristics of the temporal image using convolution filters. Hyper-parameters of different models and the dimension of the embedded space for the representation of users and items; as well as the regularisation parameter over the norms of the embeddings for all approaches were found by cross-validation.

Table 2 presents the comparison of BPR, Caser and Sequential Learning approaches over the logistic ranking loss. In boldface, we indicate best results for each dataset. These results suggest that compared to BPR which does not model the sequence of interactions, sequence models behave generally better. Furthermore, compared to Caser and GRU4Rec which only consider the positive feedback; our approach which takes into account positive interactions with respect to negative ones performs better.

Furthermore, as suspected results on OUTBRAIN and ML are better with MOSAIC than SAROS in these collections than the two other ones due to the fact that we have more LRD users. Keeping only in the dataset, *stationary* users, for

---

[2] The source code will be made available for research purpose.

which the behavior is consistent with time, is an effective strategy in learning recommender systems. The predictable nature of the behavior of stationary users makes the sequence of their interactions much exploitable than those of generic users, who may be erratic in their feedback and add noise in the dataset.

4   Conclusion

The contribution of this paper is a new way to take into account implicit feedback in recommender systems. In this case, system parameters are updated user per user by minimizing a ranking loss over sequences of interactions where each sequence is constituted by negative items followed by one or more positive ones. The main idea behind the approach is that negative and positive items within a local sequence of user interactions provide a better insight on user's preference than when considering the whole set of positive and negative items independently one from another; or, just the sequence of positive items. In addition, we introduce a strategy to filter the dataset with respect to homogeneity of the behavior in the users when interacting with the system, based on the concept of memory. From our results, it comes out that taking into account the memory in the case where the collection exhibits long range dependency allows to enhance the predictions of the proposed sequential model. As future work, we propose to encompass the analysis of LRD and the filtering phase in the training process.

References

1. Boutahar, M., Marimoutou, V., Nouira, L.: Estimation methods of the long memory parameter: Monte Carlo analysis and application. J. Appl. Stat. **34**(3), 261–301 (2007)
2. Brillinger, D.R.: Time series: data analysis and theory. In: SIAM (2001)
3. Burashnikova, A., Maximov, Y., Clausel, M., Laclau, C., Iutzeler, F., Amini, M.-R.: Learning over no-preferred and preferred sequence of items for robust recommendation. J. Artif. Intell. Res. **71**, 121–142 (2021)
4. Donkers, T., Loepp, B., Ziegler, J.: Sequential user-based recurrent neural network recommendations. In: Proceedings of the Eleventh ACM Conference on Recommender Systems, pp. 152–160 (2017)
5. Fang, H., Zhang, D., Shu, Y., Guo, G.: Deep learning for sequential recommendation: algorithms, influential factors, and evaluations. ACM Trans. Inf. Syst. **39**(1), 1–42 (2020)
6. Geweke, J., Porter-Hudak, S.: The estimation and application of long memory time series models. J. Time Ser. Anal. **4**(4), 221–238 (1983)
7. Harper, F.M., Konstan, J.A.: The movielens datasets: history and context. ACM Trans. Interact. Intell. Syst. **5**(4), 1–19 (2015)
8. Hidasi, B., Karatzoglou, A.: Recurrent neural networks with top-k gains for session-based recommendations. In: ACM International Conference on Information and Knowledge Management, pp. 843–852 (2018)
9. Hu, Y., Koren, Y., Volinsky, C.: Collaborative filtering for implicit feedback datasets. In: International Conference on Data Mining, pp. 263–272 (2008)

10. Rendle, S., Freudenthaler, C., Gantner, Z., Schmidt-Thieme, L.: BPR: Bayesian personalized ranking from implicit feedback. In: Proceedings of the Conference on Uncertainty in Artificial Intelligence, pp. 452–461 (2009)
11. Sidana, S., Laclau, C., Amini, M.: Learning to recommend diverse items over implicit feedback on PANDOR. In: Proceedings of the 12th ACM Conference on Recommender Systems, pp. 427–431 (2018)
12. Sidana, S., Laclau, C., Amini, M., Vandelle, G., Bois-Crettez, A.: KASANDR: a large-scale dataset with implicit feedback for recommendation. In: Proceedings of the International ACM SIGIR Conference on Research and Development in Information Retrieval, pp. 1245–1248 (2017)
13. Tang, J., Wang, K.: Personalized top-N sequential recommendation via convolutional sequence embedding. In: Proceedings of the ACM International Conference on Web Search and Data Mining, pp. 565–573 (2018)
14. Zhang, S., Yao, L., Sun, A., Tay, Y.: Deep learning based recommender system: a survey and new perspectives. ACM Comput. Surv. 52(1), 1–38 (2019)

Groupwise Query Performance Prediction with BERT

Xiaoyang Chen[1,2], Ben He[1,2(✉)], and Le Sun[2]

[1] University of Chinese Academy of Sciences, Beijing, China
chenxiaoyang19@mails.ucas.ac.cn , benhe@ucas.ac.cn
[2] Institute of Software, Chinese Academy of Sciences, Beijing, China
sunle@iscas.ac.cn

Abstract. While large-scale pre-trained language models like BERT have advanced the state-of-the-art in IR, its application in query performance prediction (QPP) is so far based on pointwise modeling of individual queries. Meanwhile, recent studies suggest that the cross-attention modeling of a group of documents can effectively boost performances for both learning-to-rank algorithms and BERT-based re-ranking. To this end, a BERT-based groupwise QPP model is proposed, in which the ranking contexts of a list of queries are jointly modeled to predict the relative performance of individual queries. Extensive experiments on three standard TREC collections showcase effectiveness of our approach. Our code is available at https://github.com/VerdureChen/Group-QPP.

Keywords: Groupwise ranking · BERT ranking · Information retrieval

1 Introduction

Query performance prediction (QPP) aims to automatically estimate the search results quality of a given query. While the pre-retrieval predictors enjoy the low computational overhead [15,23,24,29], the post-retrieval methods are in general more effective by considering sophisticated query and document features [3,7,15,17,20,34,41,44,47,48,51,54,55]. Recently, the large-scale pre-trained transformer based language models, e.g. BERT [19], has shown to advance the ranking performance, which provides a new direction for task of QPP.

Indeed, recent results demonstrate that BERT effectively improves the performance of post-retrieval QPP [4,22]. For instance, training with a large number of sparse-labeled queries and their highest-ranked documents, BERT-QPP [4] examines the effectiveness of BERT on the MS MARCO [30] and TREC DL [13,14] datasets, by pointwise modeling of query-document pairs. Beyond learning from single query-document pairs, the groupwise methods have achieved superior performance on both learning-to-rank [1,2,32,33] and BERT re-ranking [8] benchmarks. To this end, we propose an end-to-end BERT-based QPP model, which employs a groupwise predictor to jointly learn from multiple queries and documents, by incorporating both cross-query and cross-document information. Experiments conducted on three standard TREC collections show that our model improves significantly over state-of-the-art baselines.

© The Author(s), under exclusive license to Springer Nature Switzerland AG 2022
M. Hagen et al. (Eds.): ECIR 2022, LNCS 13186, pp. 64–74, 2022.
https://doi.org/10.1007/978-3-030-99739-7_8

2   Related Work

Query Performance Prediction (QPP). Early research in QPP utilizes linguistic information [29], statistical features [15,23,24] in pre-retrieval methods, or analyses clarity [15,16], robustness [7,20,48,54,55], retrieval scores [34,41, 44,47,55] for post-retrieval prediction, which further evolves into several effective frameworks [17,20,28,38,40,45,46]. The QPP techniques have also been explored and analyzed in [3,5,6,10,18,21,22,25,27,35,36,39,42,43,52,53]. With the recent development deep learning techniques, NeuralQPP [51] achieves promising results by training a three-components deep network under weak supervision of existing methods. Recently, while NQA-QPP [22] uses BERT to generate contextualized embedding for QPP in non-factoid question answering, BERT-QPP [4] directly applies BERT with pointwise learning in the prediction task, outperforming previous methods on the MS MARCO dev set [30] and TREC Deep Learning track query sets [13,14].

Groupwise Ranking. Beyond pointwise loss, pairwise and listwise losses are proposed to learn the relative relationships among documents [9]. Recently, Ai et al. [1] propose to represent documents into embedding with an RNN and refine the rank lists with local ranking context. Thereafter, a groupwise scoring function is proposed by Ai et al. [2] to model documents jointly. In the learning-to-rank context, both Pasumarthi et al. [33] and Pang et al. [32] use self-attention mechanism with groupwise design to improve retrieval effectiveness. Furthermore, Co-BERT [8] incorporates cross-document ranking context into BERT-based re-ranking models, demonstrating the effectiveness of using groupwise methods in boosting the ranking performance of BERT. In brief, while previous works are carried out on single query-document pairs with BERT, the groupwise methods have shown useful in multiple studies. To this end, this work proposes a groupwise post-retrieval QPP model based on pre-trained language models which simultaneously takes multiple queries and documents into account.

3   Method

Figure 1 shows our model architecture. Give an underlying retrieval method M and a corpus C, in response to a query q, a document set D is composed by the top_k documents retrieved from C with M. As aforementioned, existing BERT-based QPP methods only use the text from individual query-document pairs; however, considering information from different queries and documents is necessary for QPP tasks, which aim to obtain relative performance among queries. Inspired by Co-BERT [8], to boost the performance of BERT-based QPP methods, a groupwise predictor is integrated to learn from multiple queries and documents simultaneously on the basis of a BERT encoder.

Encoding Query-Document Pairs. Following Arabzadeh et al. [4], we first encode each query-document pair with BERT. As documents are frequently long enough to exceed BERT's 512 token limit, similar to Co-BERT [8], we split long

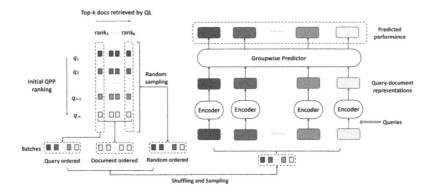

Fig. 1. Model architecture of the proposed groupwise framework.

texts into equal-sized passages. We use a BERT checkpoint fine-tuned on MS MARCO [30] to predict the relevance between each query and its corresponding passages. Each document used in the next steps is represented by its top-1 ranked passage. Consistent with common practices for text categorization using BERT, the token sequences $[CLS]Query[SEP]Document[SEP]$ are put into BERT to get encoded. We use the $[CLS]$ representation in the following groupwise step to further integrate the cross-query as well as cross-document information.

Groupwise Predictor. To incorporate cross-document and cross-query context, we regard each batch as a single group of query-document pairs. Suppose the batch size is n $(n \leq k)$, the $[CLS]$ vectors in a batch are reshaped into a sequence of length n, and we denote the sequence as $z_1, z_2, z_3, \cdots, z_n$. For $i \in [1, \cdots, n]$, each z_i is a d-dimensional vector, for example, $d = 768$ for BERT-Base. Similar to Chen et al. [8], we use a four-layers transformer as the groupwise predictor, which enables the cross attention among the $[CLS]$ vectors in each batch, and then produces n predicted performances of each query-document pair. During inference, suppose top$_t$ documents of q are used, we will get t predicted scores for q. We use three aggregation methods to get the final QPP score of q: max-pooling, average-pooling, and the direct use of the predicted performation of the first-ranked retrieved document for query q. In our experiments, the aggregation method with the best performance on the training set is chosen.

By assigning different positional ids to z_i, our model can be designed to incorporate with different types of ranking context. Thus, several **variants of our models** are investigated. (**Random order**) denotes that all query-document pairs are shuffled before being fed into the model in both training and inference. (**Query order**) denotes for BERT groupwise model considering only the *cross-query* context. For a batch of n samples, the ith ranked documents from n queries are grouped together in the batch, and position ids are assigned by the initial query order derived by $n(\sigma_{X\%})$. We leave other choices of the initial QPP for future study. (**Doc order**) denotes for BERT groupwise model considering only the *cross-document* context. A batch consists of n documents returned for a query, and the position ids are assigned by the initial document ranking. (**Query+Doc**) denotes for BERT groupwise model considering both cross-document and cross-query context. Batches containing one of the above two contexts appear randomly during training. (**R+Q+D**) denotes for BERT groupwise model with all three types of orders mentioned above. According to the maximum batch size allowed by the hardware, we use the batch size of 128/64/16 for Small, Base and Large BERT models, respectively. Note that the training data is still shuffled among batches to avoid overfitting.

4 Experiment Setup

Dataset and Metrics. We use three popular datasets, namely, Robust04 [49], GOV2 [11], and ClueWeb09-B [12], with 249, 150 and 200 keyword queries, respectively. Following [3], we use the Pearson's ρ and Kendall's τ correlations to measure the QPP performance, which is computed using the predicted ordering of the queries with the actual ordering of average precision for the top 1000 documents (AP@1000) per query retrieved by the Query Likelihood (QL) model implemented in Anserini [50]. Following [51], we use 2-fold cross-validation and randomly generate 30 splits for each dataset. Each split has two folds, the first fold is used for model training and hyper-parameter tuning . The ultimate performance is the average prediction quality on the second test folds over the 30 splits. Statistical significance for paired two-tailed t-test is reported.

Baselines. Akin to [3], we compare our model with several popular baselines including **Clarity** [15], **Query Feedback (QF)** [55], **Weighted Information Gain (WIG)** [55], **Normalized Query Commitment (NQC)** [44], **Score Magnitude and Variance (SMV)** [47], **Utility Estimation Framework (UEF)** [45], σ_k [34], $n(\sigma_{X\%})$ [17], **Robust Standard Deviation (RSD)** [41], **WAND**$[n(\sigma_{X\%})]$ [3], and **NeuralQPP** [51]. We also compare to **BERT-Small/Base/Large** [37] baselines, which are configured the same as our model except that they do not have a groupwise predictor. Note that the BERT baselines share the same structures with BERT-QPP except we use more documents for each query in training due to the small number of queries. Following [3], our proposed predictor is linearly combined with $n(\sigma_{X\%})$. The BERT baselines perform the same linear interpolation.

Data Preparation and Model Training. Akin to [8], for the BERT-based models, documents are sliced using sliding windows of 150 words with an overlap of 75 words. The max sequence length of the concatenated query-document pair is 256. We use MSE loss for individual documents and explore two kinds of training labels: P@k and AP@1000. According to our pilot study on the BERT-Base baseline, we use P@k as the supervision signals on Robust04 and GOV2, and use AP@1000 on ClueWeb09-B. All BERT models are trained for 5 epochs. Due to the memory limit, BERT-based models are trained with top-100 documents and tested on the last checkpoint with the top-25 documents for each query retrieved by QL. We use Adam optimizer [26] with the learning rate schedule from [31]. We select the initial learning rate from {1e–4, 1e–5, 1e–6}, and set the warming up steps to 10% of the total steps.

5   Results

Table 1. Evaluation results. Statistical significance at 0.05 relative to BERT baselines of the same model size (e.g. (R+Q+D)-Large vs. BERT-Large) is marked with *.

Method	Robust04		GOV2		ClueWeb09-B	
	P-ρ	K-τ	P-ρ	K-τ	P-ρ	K-τ
Clarity	0.528	0.385	0.428	0.291	0.300	0.213
QF	0.390	0.324	0.447	0.314	0.163	0.072
WIG	0.546	0.379	0.502	0.346	0.316	0.210
NQC	0.516	0.388	0.381	0.323	0.127	0.138
SMV	0.534	0.378	0.352	0.303	0.236	0.183
UEF	0.502	0.402	0.470	0.329	0.301	0.211
σ_k	0.522	0.389	0.381	0.323	0.234	0.177
$n(\sigma_{X\%})$	0.589	0.386	0.556	0.386	0.334	0.247
RSD	0.455	0.352	0.444	0.276	0.193	0.096
WAND[$n(\sigma_{X\%})$]	0.566	0.386	0.580	0.411	0.236	0.142
NeuralQPP	0.611	0.408	0.540	0.357	0.367	0.229
BERT-Small	0.591	0.391	0.615	0.436	0.394	0.278
BERT-Base	0.585	0.423	0.637	0.454	0.447	0.321
BERT-Large	0.579	0.422	0.645	0.461	0.342	0.251
(Random order)-base	0.608*	0.449*	0.665*	0.479*	0.481*	0.353*
(Query order)-base	**0.615***	0.456*	0.676*	0.486*	0.455	0.327
(Doc order)-base	0.563	0.383	0.660*	0.476*	0.365	0.262
(Query+Doc)-base	0.598	0.452*	0.682*	0.496*	0.438	0.317
(R+Q+D)-small	0.590	0.419*	0.680*	0.500*	0.437*	0.305*
(R+Q+D)-base	0.608*	0.460*	0.676*	0.489*	0.449	0.324
(R+Q+D)-large	0.612*	**0.470***	**0.688***	**0.508***	**0.545***	**0.399***

Overall Effectiveness. According to Table 1, the proposed model outperforms all the baselines on all three collections. Compared with the previous state-of-the-art results without using BERT, except for the ρ on Robust04, our groupwise model trained on BERT-Base with the random input order has an improvement on all metrics by at least 10%. In general, our (R+Q+D) outperforms the BERT baselines with all three different model sizes. Additionally, varying the type of ranking contexts incorporated with the groupwise models leads to different observations on the three datasets. The query-level ranking context marginally improves the effectiveness on Robust04 and GOV2, while it decreases the result on CluWeb09-B. Using document-level ranking context alone greatly harms the model performance on Robust04 and ClueWeb09-B. This may be due to the fact that the model has only learned the sequence information inside each query, but not the relative relations between the queries. Relative to the random case, using both contexts slightly elevates the performance on GOV2, while it has little effect on Robust04 and decreases the results on ClueWeb09-B. As the simultaneous use of all three types of context appears to be the best variant, we only report results of (R+Q+D)-Base in the following analysis.

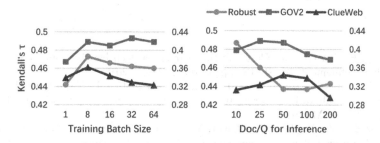

Fig. 2. Performance of (R+Q+D) with different training batch sizes and numbers of documents per query for inference. In each figure, the left axis represents the Kendall's τ of Robust04 and GOV2, and the right axis represents ClueWeb09-B.

Impact of Factors. We examine the impact of training batch size and the number of top-k documents per query for inference on the model performance. We first evaluate with different training batch sizes in $\{1, 8, 16, 32, 64\}$. The greater the batch size is, the more query-document pairs are jointly modeled. A special case is to set batch size to 1, which is equivalent to the pointwise learning without any context from other queries or documents. The results in Fig. 2 show that the cross-attention among queries is effective and improves upon the pointwise method by a large margin. The groupwise method works best with a group size of 8, which means the model may learn better with a relatively smaller group of queries. We also explore the impact of different numbers of documents per query used during inference, namely $\{10, 25, 50, 100, 200\}$. Results in Fig. 2 indicate that inference with less than 100 documents per query on all three collections yields the best results. The reason might be that there are more positive samples in the top-ranked documents which contribute more to the target metric, i.e. AP@1000, while considering more negative examples not only have little impact on the target metric, but also are more likely to introduce noise.

Limitations. We count the number of floating-point operations for all BERT-based models. It turns out our model can predict the retrieval performance with less than 1% additional computational cost compared to its BERT counterpart. However, for document retrieval with BERT MaxP, the passage selection brings an approx. 1 min extra computational overhead, which is more expensive than the non-BERT baselines.

6 Conclusion

In this paper, we propose a BERT-based groupwise query performance prediction method, which simultaneously incorporates the cross-query and cross-document information within an end-to-end learning framework. Evaluation on three standard TREC test collections indicates the groupwise model significantly outperforms the BERT baselines nearly in all cases. In further research, we plan to work on the efficiency, as well as adoption of our approach to more advanced experimentation framework [21].

References

1. Ai, Q., Bi, K., Guo, J., Croft, W.B.: Learning a deep listwise context model for ranking refinement. In: SIGIR, pp. 135–144. ACM (2018)
2. Ai, Q., Wang, X., Bruch, S., Golbandi, N., Bendersky, M., Najork, M.: Learning groupwise multivariate scoring functions using deep neural networks. In: ICTIR, pp. 85–92. ACM (2019)
3. Arabzadeh, N., Bigdeli, A., Zihayat, M., Bagheri, E.: Query Performance Prediction Through Retrieval Coherency. In: Hiemstra, D., Moens, M.-F., Mothe, J., Perego, R., Potthast, M., Sebastiani, F. (eds.) ECIR 2021. LNCS, vol. 12657, pp. 193–200. Springer, Cham (2021). https://doi.org/10.1007/978-3-030-72240-1_15
4. Arabzadeh, N., Khodabakhsh, M., Bagheri, E.: BERT-QPP: contextualized pre-trained transformers for query performance prediction. In: Demartini, G., Zuccon, G., Culpepper, J.S., Huang, Z., Tong, H. (eds.) CIKM 2021: The 30th ACM International Conference on Information and Knowledge Management, Virtual Event, Queensland, Australia, 1–5 November 2021, pp. 2857–2861. ACM (2021). https://doi.org/10.1145/3459637.3482063
5. Arabzadeh, N., Zarrinkalam, F., Jovanovic, J., Al-Obeidat, F.N., Bagheri, E.: Neural embedding-based specificity metrics for pre-retrieval query performance prediction. Inf. Process. Manag. 57(4), 102248 (2020)
6. Arabzadeh, N., Zarrinkalam, F., Jovanovic, J., Bagheri, E.: Neural embedding-based metrics for pre-retrieval query performance prediction. In: Jose, J.M., et al. (eds.) ECIR 2020. LNCS, vol. 12036, pp. 78–85. Springer, Cham (2020). https://doi.org/10.1007/978-3-030-45442-5_10
7. Aslam, J.A., Pavlu, V.: Query hardness estimation using Jensen-Shannon divergence among multiple scoring functions. In: Amati, G., Carpineto, C., Romano, G. (eds.) ECIR 2007. LNCS, vol. 4425, pp. 198–209. Springer, Heidelberg (2007). https://doi.org/10.1007/978-3-540-71496-5_20
8. Chen, X., Hui, K., He, B., Han, X., Sun, L., Ye, Z.: Co-bert: a context-aware BERT retrieval model incorporating local and query-specific context. CoRR abs/2104.08523 (2021). https://arxiv.org/abs/2104.08523
9. Chen, Z., Eickhoff, C.: Poolrank: Max/min pooling-based ranking loss for listwise learning & ranking balance. CoRR abs/2108.03586 (2021). https://arxiv.org/abs/2108.03586
10. Chifu, A., Laporte, L., Mothe, J., Ullah, M.Z.: Query performance prediction focused on summarized letor features. In: Collins-Thompson, K., Mei, Q., Davison, B.D., Liu, Y., Yilmaz, E. (eds.) The 41st International ACM SIGIR Conference on Research & Development in Information Retrieval, SIGIR 2018, Ann Arbor, MI, USA, 08–12 July 2018, pp. 1177–1180. ACM (2018). https://doi.org/10.1145/3209978.3210121

11. Clarke, C.L.A., Craswell, N., Soboroff, I.: Overview of the TREC 2004 terabyte track. In: Proceedings of the Thirteenth Text REtrieval Conference. NIST Special Publication, vol. 500–261, pp. 1–9. National Institute of Standards and Technology (2004)

12. Clarke, C.L.A., Craswell, N., Soboroff, I.: Overview of the TREC 2009 web track. In: Voorhees, E.M., Buckland, L.P. (eds.) Proceedings of The Eighteenth Text REtrieval Conference, TREC 2009, Gaithersburg, Maryland, USA, 17–20 November 2009. NIST Special Publication, vol. 500–278. National Institute of Standards and Technology (NIST) (2009). http://trec.nist.gov/pubs/trec18/papers/WEB09.OVERVIEW.pdf

13. Craswell, N., Mitra, B., Yilmaz, E., Campos, D.: Overview of the TREC 2020 deep learning track. CoRR abs/2102.07662 (2021). https://arxiv.org/abs/2102.07662

14. Craswell, N., Mitra, B., Yilmaz, E., Campos, D., Voorhees, E.M.: Overview of the TREC 2019 deep learning track. CoRR abs/2003.07820 (2020). https://arxiv.org/abs/2003.07820

15. Cronen-Townsend, S., Zhou, Y., Croft, W.B.: Predicting query performance. In: Järvelin, K., Beaulieu, M., Baeza-Yates, R.A., Myaeng, S. (eds.) SIGIR 2002: Proceedings of the 25th Annual International ACM SIGIR Conference on Research and Development in Information Retrieval, 11–15 August 2002, Tampere, Finland, pp. 299–306. ACM (2002). https://doi.org/10.1145/564376.564429

16. Cronen-Townsend, S., Zhou, Y., Croft, W.B.: Precision prediction based on ranked list coherence. Inf. Retr. **9**(6), 723–755 (2006)

17. Cummins, R., Jose, J.M., O'Riordan, C.: Improved query performance prediction using standard deviation. In: Ma, W., Nie, J., Baeza-Yates, R., Chua, T., Croft, W.B. (eds.) Proceeding of the 34th International ACM SIGIR Conference on Research and Development in Information Retrieval, SIGIR 2011, Beijing, China, 25–29 July 2011, pp. 1089–1090. ACM (2011). https://doi.org/10.1145/2009916.2010063

18. Déjean, S., Ionescu, R.T., Mothe, J., Ullah, M.Z.: Forward and backward feature selection for query performance prediction. In: Hung, C., Cerný, T., Shin, D., Bechini, A. (eds.) SAC 2020: The 35th ACM/SIGAPP Symposium on Applied Computing, online event, [Brno, Czech Republic], March 30 - April 3, 2020, pp. 690–697. ACM (2020). https://doi.org/10.1145/3341105.3373904

19. Devlin, J., Chang, M., Lee, K., Toutanova, K.: BERT: pre-training of deep bidirectional transformers for language understanding. In: Burstein, J., Doran, C., Solorio, T. (eds.) Proceedings of the 2019 Conference of the North American Chapter of the Association for Computational Linguistics: Human Language Technologies, NAACL-HLT 2019, Minneapolis, MN, USA, 2–7 June 2019, Volume 1 (Long and Short Papers), pp. 4171–4186. Association for Computational Linguistics (2019). https://doi.org/10.18653/v1/n19-1423

20. Diaz, F.: Performance prediction using spatial autocorrelation. In: Kraaij, W., de Vries, A.P., Clarke, C.L.A., Fuhr, N., Kando, N. (eds.) SIGIR 2007: Proceedings of the 30th Annual International ACM SIGIR Conference on Research and Development in Information Retrieval, Amsterdam, The Netherlands, 23–27 July 2007, pp. 583–590. ACM (2007). https://doi.org/10.1145/1277741.1277841

21. Faggioli, G., Zendel, O., Culpepper, J.S., Ferro, N., Scholer, F.: An enhanced evaluation framework for query performance prediction. In: Hiemstra, D., Moens, M.-F., Mothe, J., Perego, R., Potthast, M., Sebastiani, F. (eds.) ECIR 2021. LNCS, vol. 12656, pp. 115–129. Springer, Cham (2021). https://doi.org/10.1007/978-3-030-72113-8_8

22. Hashemi, H., Zamani, H., Croft, W.B.: Performance prediction for non-factoid question answering. In: Fang, Y., Zhang, Y., Allan, J., Balog, K., Carterette, B., Guo, J. (eds.) Proceedings of the 2019 ACM SIGIR International Conference on Theory of Information Retrieval, ICTIR 2019, Santa Clara, CA, USA, 2–5 October 2019, pp. 55–58. ACM (2019). https://doi.org/10.1145/3341981.3344249

23. He, B., Ounis, I.: Query performance prediction. Inf. Syst. **31**(7), 585–594 (2006)

24. He, J., Larson, M., de Rijke, M.: Using coherence-based measures to predict query difficulty. In: Macdonald, C., Ounis, I., Plachouras, V., Ruthven, I., White, R.W. (eds.) ECIR 2008. LNCS, vol. 4956, pp. 689–694. Springer, Heidelberg (2008). https://doi.org/10.1007/978-3-540-78646-7_80

25. Khodabakhsh, M., Bagheri, E.: Semantics-enabled query performance prediction for ad hoc table retrieval. Inf. Process. Manag. **58**(1), 102399 (2021)

26. Kingma, D.P., Ba, J.: Adam: a method for stochastic optimization. In: 3rd International Conference on Learning Representations, pp. 1–15 (2015)

27. Krikon, E., Carmel, D., Kurland, O.: Predicting the performance of passage retrieval for question answering. In: Chen, X., Lebanon, G., Wang, H., Zaki, M.J. (eds.) 21st ACM International Conference on Information and Knowledge Management, CIKM 2012, Maui, HI, USA, October 29 - 02 November 2012, pp. 2451–2454. ACM (2012). https://doi.org/10.1145/2396761.2398664

28. Kurland, O., Shtok, A., Carmel, D., Hummel, S.: A unified framework for post-retrieval query-performance prediction. In: Amati, G., Crestani, F. (eds.) ICTIR 2011. LNCS, vol. 6931, pp. 15–26. Springer, Heidelberg (2011). https://doi.org/10.1007/978-3-642-23318-0_4

29. Mothe, J., Tanguy, L.: Linguistic features to predict query difficulty. In: SIGIR 2005 (2005)

30. Nguyen, T., et al.: MS MARCO: A human generated machine reading comprehension dataset. CoRR abs/1611.09268 (2016). http://arxiv.org/abs/1611.09268

31. Nogueira, R., Cho, K.: Passage re-ranking with BERT. CoRR abs/1901.04085 (2019)

32. Pang, L., Xu, J., Ai, Q., Lan, Y., Cheng, X., Wen, J.: Setrank: learning a permutation-invariant ranking model for information retrieval. In: SIGIR, pp. 499–508. ACM (2020)

33. Pasumarthi, R.K., Wang, X., Bendersky, M., Najork, M.: Self-attentive document interaction networks for permutation equivariant ranking. CoRR abs/1910.09676 (2019)

34. Pérez-Iglesias, J., Araujo, L.: Standard deviation as a query hardness estimator. In: Chavez, E., Lonardi, S. (eds.) SPIRE 2010. LNCS, vol. 6393, pp. 207–212. Springer, Heidelberg (2010). https://doi.org/10.1007/978-3-642-16321-0_21

35. Raiber, F., Kurland, O.: Query-performance prediction: setting the expectations straight. In: Geva, S., Trotman, A., Bruza, P., Clarke, C.L.A., Järvelin, K. (eds.) The 37th International ACM SIGIR Conference on Research and Development in Information Retrieval, SIGIR 2014, Gold Coast, QLD, Australia - 06–11 July 2014, pp. 13–22. ACM (2014). https://doi.org/10.1145/2600428.2609581

36. Raviv, H., Kurland, O., Carmel, D.: Query performance prediction for entity retrieval. In: Geva, S., Trotman, A., Bruza, P., Clarke, C.L.A., Järvelin, K. (eds.) The 37th International ACM SIGIR Conference on Research and Development in Information Retrieval, SIGIR 2014, Gold Coast, QLD, Australia - 06–11 July 2014, pp. 1099–1102. ACM (2014). https://doi.org/10.1145/2600428.2609519

37. google research: GitHub - google-research/bert: TensorFlow code and pre-trained models for BERT. https://github.com/google-research/bert

38. Roitman, H.: An enhanced approach to query performance prediction using reference lists. In: Kando, N., Sakai, T., Joho, H., Li, H., de Vries, A.P., White, R.W. (eds.) Proceedings of the 40th International ACM SIGIR Conference on Research and Development in Information Retrieval, Shinjuku, Tokyo, Japan, 7–11 August 2017, pp. 869–872. ACM (2017). https://doi.org/10.1145/3077136.3080665

39. Roitman, H.: ICTIR tutorial: Modern query performance prediction: Theory and practice. In: Balog, K., Setty, V., Lioma, C., Liu, Y., Zhang, M., Berberich, K. (eds.) ICTIR 2020: The 2020 ACM SIGIR International Conference on the Theory of Information Retrieval, Virtual Event, Norway, 14–17 September 2020, pp. 195–196. ACM (2020). https://dl.acm.org/doi/10.1145/3409256.3409813

40. Roitman, H., Erera, S., Shalom, O.S., Weiner, B.: Enhanced mean retrieval score estimation for query performance prediction. In: Kamps, J., Kanoulas, E., de Rijke, M., Fang, H., Yilmaz, E. (eds.) Proceedings of the ACM SIGIR International Conference on Theory of Information Retrieval, ICTIR 2017, Amsterdam, The Netherlands, 1–4 October 2017, pp. 35–42. ACM (2017). https://doi.org/10.1145/3121050.3121051

41. Roitman, H., Erera, S., Weiner, B.: Robust standard deviation estimation for query performance prediction. In: Kamps, J., Kanoulas, E., de Rijke, M., Fang, H., Yilmaz, E. (eds.) Proceedings of the ACM SIGIR International Conference on Theory of Information Retrieval, ICTIR 2017, Amsterdam, The Netherlands, 1–4 October 2017, pp. 245–248. ACM (2017). https://doi.org/10.1145/3121050.3121087

42. Roitman, H., Kurland, O.: Query performance prediction for pseudo-feedback-based retrieval. In: Piwowarski, B., Chevalier, M., Gaussier, É., Maarek, Y., Nie, J., Scholer, F. (eds.) Proceedings of the 42nd International ACM SIGIR Conference on Research and Development in Information Retrieval, SIGIR 2019, Paris, France, 21–25 July 2019, pp. 1261–1264. ACM (2019). https://doi.org/10.1145/3331184.3331369

43. Roitman, H., Mass, Y., Feigenblat, G., Shraga, R.: Query performance prediction for multifield document retrieval. In: Balog, K., Setty, V., Lioma, C., Liu, Y., Zhang, M., Berberich, K. (eds.) ICTIR 2020: The 2020 ACM SIGIR International Conference on the Theory of Information Retrieval, Virtual Event, Norway, 14–17 September 2020, pp. 49–52. ACM (2020). https://dl.acm.org/doi/10.1145/3409256.3409821

44. Shtok, A., Kurland, O., Carmel, D.: Predicting query performance by query-drift estimation. In: Azzopardi, L., et al. (eds.) ICTIR 2009. LNCS, vol. 5766, pp. 305–312. Springer, Heidelberg (2009). https://doi.org/10.1007/978-3-642-04417-5_30

45. Shtok, A., Kurland, O., Carmel, D.: Using statistical decision theory and relevance models for query-performance prediction. In: Crestani, F., Marchand-Maillet, S., Chen, H., Efthimiadis, E.N., Savoy, J. (eds.) Proceeding of the 33rd International ACM SIGIR Conference on Research and Development in Information Retrieval, SIGIR 2010, Geneva, Switzerland, 19–23 July 2010, pp. 259–266. ACM (2010). https://doi.org/10.1145/1835449.1835494

46. Shtok, A., Kurland, O., Carmel, D.: Query performance prediction using reference lists. ACM Trans. Inf. Syst. 34(4), 19:1–19:34 (2016). https://doi.org/10.1145/2926790

47. Tao, Y., Wu, S.: Query performance prediction by considering score magnitude and variance together. In: Li, J., Wang, X.S., Garofalakis, M.N., Soboroff, I., Suel, T., Wang, M. (eds.) Proceedings of the 23rd ACM International Conference on Conference on Information and Knowledge Management, CIKM 2014, Shanghai, China, 3–7 November 2014, pp. 1891–1894. ACM (2014). https://doi.org/10.1145/2661829.2661906

48. Vinay, V., Cox, I.J., Milic-Frayling, N., Wood, K.R.: On ranking the effectiveness of searches. In: Efthimiadis, E.N., Dumais, S.T., Hawking, D., Järvelin, K. (eds.) SIGIR 2006: Proceedings of the 29th Annual International ACM SIGIR Conference on Research and Development in Information Retrieval, Seattle, Washington, USA, 6–11 August 2006, pp. 398–404. ACM (2006). https://doi.org/10.1145/1148170.1148239

49. Voorhees, E.M.: Overview of the TREC 2004 robust track. In: Proceedings of the Thirteenth Text REtrieval Conference. NIST Special Publication, vol. 500–261, pp. 1–10. National Institute of Standards and Technology (2004)

50. Yang, P., Fang, H., Lin, J.: Anserini: enabling the use of lucene for information retrieval research. In: Kando, N., Sakai, T., Joho, H., Li, H., de Vries, A.P., White, R.W. (eds.) Proceedings of the 40th International ACM SIGIR Conference on Research and Development in Information Retrieval, Shinjuku, Tokyo, Japan, 7–11 August 2017, pp. 1253–1256. ACM (2017). https://doi.org/10.1145/3077136.3080721

51. Zamani, H., Croft, W.B., Culpepper, J.S.: Neural query performance prediction using weak supervision from multiple signals. In: Collins-Thompson, K., Mei, Q., Davison, B.D., Liu, Y., Yilmaz, E. (eds.) The 41st International ACM SIGIR Conference on Research & Development in Information Retrieval, SIGIR 2018, Ann Arbor, MI, USA, 08–12 July 2018, pp. 105–114. ACM (2018). https://doi.org/10.1145/3209978.3210041

52. Zendel, O., Culpepper, J.S., Scholer, F.: Is query performance prediction with multiple query variations harder than topic performance prediction? In: Diaz, F., Shah, C., Suel, T., Castells, P., Jones, R., Sakai, T. (eds.) SIGIR 2021: The 44th International ACM SIGIR Conference on Research and Development in Information Retrieval, Virtual Event, Canada, 11–15 July 2021, pp. 1713–1717. ACM (2021). https://doi.org/10.1145/3404835.3463039

53. Zendel, O., Shtok, A., Raiber, F., Kurland, O., Culpepper, J.S.: Information needs, queries, and query performance prediction. In: Piwowarski, B., Chevalier, M., Gaussier, É., Maarek, Y., Nie, J., Scholer, F. (eds.) Proceedings of the 42nd International ACM SIGIR Conference on Research and Development in Information Retrieval, SIGIR 2019, Paris, France, 21–25 July 2019, pp. 395–404. ACM (2019). https://doi.org/10.1145/3331184.3331253

54. Zhou, Y., Croft, W.B.: Ranking robustness: a novel framework to predict query performance. In: Yu, P.S., Tsotras, V.J., Fox, E.A., Liu, B. (eds.) Proceedings of the 2006 ACM CIKM International Conference on Information and Knowledge Management, Arlington, Virginia, USA, November 6–11, 2006, pp. 567–574. ACM (2006). https://doi.org/10.1145/1183614.1183696

55. Zhou, Y., Croft, W.B.: Query performance prediction in web search environments. In: Kraaij, W., de Vries, A.P., Clarke, C.L.A., Fuhr, N., Kando, N. (eds.) SIGIR 2007: Proceedings of the 30th Annual International ACM SIGIR Conference on Research and Development in Information Retrieval, Amsterdam, The Netherlands, 23–27 July 2007, pp. 543–550. ACM (2007). https://doi.org/10.1145/1277741.1277835

How Can Graph Neural Networks Help Document Retrieval: A Case Study on CORD19 with Concept Map Generation

Hejie Cui[iD], Jiaying Lu[iD], Yao Ge[iD], and Carl Yang[✉][iD]

Department of Computer Science, Emory University, Atlanta, Georgia
{hejie.cui,jiaying.lu,yao.ge,j.carlyang}@emory.edu

Abstract. Graph neural networks (GNNs), as a group of powerful tools for representation learning on irregular data, have manifested superiority in various downstream tasks. With unstructured texts represented as concept maps, GNNs can be exploited for tasks like document retrieval. Intrigued by how can GNNs help document retrieval, we conduct an empirical study on a large-scale multi-discipline dataset CORD-19. Results show that instead of the complex structure-oriented GNNs such as GINs and GATs, our proposed semantics-oriented graph functions achieve better and more stable performance based on the BM25 retrieved candidates. Our insights in this case study can serve as a guideline for future work to develop effective GNNs with appropriate semantics-oriented inductive biases for textual reasoning tasks like document retrieval and classification. All code for this case study is available at https://github.com/HennyJie/GNN-DocRetrieval.

Keywords: Document retrieval · Graph neural networks · Concept maps · Graph representation learning · Textual reasoning.

1 Introduction

Concept map, which models texts as a graph with words/phrases as vertices and relations between them as edges, has been studied to improve information retrieval tasks previously [10,14,46]. Recently, graph neural networks (GNNs) attract tremendous attention due to their superior power established both in theory and through experiments [6,12,16,20,32]. Empowered by the structured document representation of concept maps, it is intriguing to apply powerful GNNs for tasks like document classification [38] and retrieval [45]. Take Fig. 1 as an example. Towards the query about "violent crimes in society", a proper GNN might be able to highlight query-relevant concept of "crime" and its connection to "robbery" and "citizen", thus ranking the document as highly relevant. On the other hand, for another document about precaution, the GNN can capture concepts like "n95 mask" and "vaccine", together with their connections to "prevention", thus ranking it as not so relevant.

M. Hagen et al. (Eds.): ECIR 2022, LNCS 13186, pp. 75–83, 2022.
https://doi.org/10.1007/978-3-030-99739-7_9

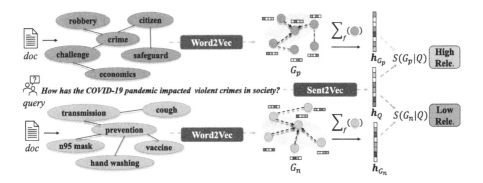

Fig. 1. An overview of GNN-based document retrieval.

Present work. In this work, we explore how GNNs can help document retrieval with generated concept maps. The core contributions are three-fold:

- We use constituency parsing to construct semantically rich concept maps from documents and design quality evaluation for them towards document retrieval.
- We investigate two types of graph models for document retrieval: the structure-oriented complex GNNs and our proposed semantics-oriented graph functions.
- By comparing the retrieval results from different graph models, we provide insights towards GNN model design for textual retrieval, with the hope to prompt more discussions on the emerging areas such as IR with GNNs.

2   GNNs for Document Retrieval

2.1   Overview

In this section, we describe the process of GNN-based document retrieval. As is shown in Fig. 1, concept maps $G = \{V, E\}$ are first constructed for documents. Each node $v_i \in V$ is a concept (usually a word or phrase) in the document, associated with a frequency f_i and an initial feature vector a_i from the pretrained model. The edges in E denote the interactions between concepts. GNNs are then applied to each individual concept map, where node representation $h_i \in \mathbb{R}^d$ is updated through neighborhood transformation and aggregation. The graph-level embedding $h_G \in \mathbb{R}^d$ is summarized over all nodes with a read-out function.

For the training of GNN models, the widely-used triplet loss in retrieval tasks [22,37,42] is adopted. Given a triplet (Q, G_p, G_n) composed by a relevant document G_p (denoted as positive) and an irrelevant document G_n (denoted as negative) to the query Q, the loss function is defined as:

$$L(Q, G_p, G_n) = \max\left\{S(G_n \mid Q) - S(G_p \mid Q) + margin, 0\right\}. \tag{1}$$

The relevance score $S(G \mid Q)$ is calculated as $\frac{h_G \cdot h_Q}{\|h_G\| \|h_Q\|}$, where h_G is the learned graph representation from GNN models and h_Q is the query representation from a pretrained model. In the training process, the embeddings of relevant documents are pulled towards the query representation, whereas those of the irrelevant ones are pushed away. For retrieval in the testing phrase, documents are ranked according to the learned relevance score $S(G \mid Q)$.

2.2   Concept Maps and Their Generation

Concept map generation, which aims to distill structured information hidden under unstructured text and represent it with a graph, has been studied extensively in literature [3,39,40,45]. Since entities and events often convey rich semantics, they are widely used to represent core information of documents [5,18,21]. However, according to our pilot trials, existing concept map construction methods based on name entity recognition (NER) or relation extraction (RE) often suffer from limited nodes and sparse edges. Moreover, these techniques rely on significant amounts of training data and predefined entities and relation types, which restricts the semantic richness of the generated concept maps [34].

To increase node/edge coverage, we propose to identify entities and events by POS-tagging and constituency parsing [23]. Compared to concept maps derived from NER or RE, our graphs can identify more sufficient phrases as nodes and connect them with denser edges, since pos-tagging and parsing are robust to domain shift [26,43]. The identified phrases are filtered via articles removing and lemmas replacing, and then merged by the same mentions. To capture the interactions (edges in graphs) among extracted nodes, we follow the common practice in phrase graph construction [17,27,31] that uses the sliding window technique to capture node co-occurrence. The window size is selected through grid search. Our proposed constituency parsing approach for concept map generation alleviates the limited vocabulary problem of existing NER-based methods, thus bolstering the semantic richness of the concept maps for retrieval.

2.3   GNN-based Concept Map Representation Learning

Structure-Oriented Complex GNNs. Various GNNs have been proposed for graph representation learning [12,16,32,36]. The discriminative power of complex GNNs mainly stems from the 1-WL test for graph isomorphism, which exhaustively capture possible graph structures so as to differentiate non-isomorphic graphs [36]. To investigate the effectiveness of structured-oriented GNNs towards document retrieval, we adopt two state-of-the-art ones, Graph isomorphism network (GIN) [36] and Graph attention network (GAT) [32], as representatives.

Semantics-Oriented Permutation-Invariant Graph Functions. The advantage of complex GNNs in modelling interactions may become insignificant for semantically important task. In contrast, we propose the following series of graph functions oriented from semantics perspectives.

Table 1. The similarity of different concept map pairs.

Pair Type	# Pairs	NCR (%)	NCR+ (%)	ECR (%)	ECR+ (%)
Pos-Pos	762,084	4.96	19.19	0.60	0.78
Pos-Neg	1,518,617	4.12	11.75	0.39	0.52
(*t-score*)	–	(*187.041*)	(*487.078*)	(*83.569*)	(*105.034*)
Pos-BM	140,640	3.80	14.98	0.37	0.43
(*t-score*)	–	(*126.977*)	(*108.808*)	(*35.870*)	(*56.981*)

- **N-Pool**: independently process each single node v_i in the concept map by multi-layer perceptions and then apply a read-out function to aggregate all node embeddings $\boldsymbol{a}_i$ into the graph embedding $\boldsymbol{h}_G$, i.e.,

$$\boldsymbol{h}_G = \text{READOUT}\Big(\{\text{MLP}(\boldsymbol{a}_i) \mid v_i \in V\}\Big). \tag{2}$$

- **E-Pool**: for each edge $e_{ij} = (v_i, v_j)$ in the concept map, the edge embedding is obtained by concatenating the projected node embedding $\boldsymbol{a}_i$ and $\boldsymbol{a}_j$ on its two ends to encode first-order interactions, i.e.,

$$\boldsymbol{h}_G = \text{READOUT}\Big(\{cat\,(\text{MLP}(\boldsymbol{a}_i), \text{MLP}(\boldsymbol{a}_j)) \mid e_{ij} \in E\}\Big). \tag{3}$$

- **RW-Pool**: for each sampled random walk $p_i = (v_1, v_2, \ldots, v_m)$ that encode higher-order interactions among concepts ($m = 2, 3, 4$ in our experiments), the embedding is computed by the sum of all node embeddings on it, i.e.,

$$\boldsymbol{h}_G = \text{READOUT}\Big(\{sum\,(\text{MLP}(\boldsymbol{a}_1), \text{MLP}(\boldsymbol{a}_2), \ldots, \text{MLP}(\boldsymbol{a}_m)) \mid p_i \in P\}\Big). \tag{4}$$

All of the three proposed graph functions are easier to train and generalize. They preserve the *message passing* mechanism of complex GNNs [11], which is essentially *permutation invariant* [15, 24, 25], meaning that the results of GNNs are not influenced by the orders of nodes or edges in the graph; while focusing on the basic semantic units and different level of interactions between them.

3 Experiments

3.1 Experimental Setup

Dataset. We adopt a large scale multi-discipline dataset from the TREC-COVID[1] challenge [29] based on the CORD-19[2] collection [33]. The raw data includes a corpus of 192,509 documents from broad research areas, 50 queries about the pandemic that interest people, and 46,167 query-document relevance labels.

---

[1] https://ir.nist.gov/covidSubmit/.
[2] https://github.com/allenai/cord19.

Experimental Settings and Metrics. We follow the common two-step practice for the large-scale document retrieval task [7,19,28]. The initial retrieval is performed on the whole corpus with full texts through BM25 [30], a traditional yet widely-used baseline. In the second stage, we further conduct re-ranking on the top 100 candidates using different graph models. The node features and query embeddings are initialized with pretrained models from [4,44]. NDCG@20 is adopted as the main evaluation metric for retrieval, which is used for the competition leader board. Besides NDCG@K, we also provide Precision@K and Recall@K (K=10, 20 for all metrics).

3.2 Evaluation of Concept Maps

We empirically evaluate the quality of concept maps generated from Sect. 2.2. The purpose is to validate that information in concept maps can indicate query-document relevance, and provide additional discriminative signals based on the initial candidates. Three types of pairs are constructed: a Pos-Pos pair consists of two documents both relevant to a query; a Pos-Neg pair consists of a relevant and an irrelevant one; and a Pos-BM pair consists of a relevant one and a top-20 one from BM25. Given a graph pair G_i and G_j, their similarity is calculated via four measures: the node coincidence rate (NCR) defined as $\frac{|V_i \cap V_j|}{|V_i \cup V_j|}$; NCR+ defined as NCR weighted by the tf-idf score [1] of each node; the edge coincidence rate (ECR) where an edge is coincident when its two ends are contained in both graphs; and ECR+ defined as ECR weighted by the tf-idf scores of both ends.

It is shown in Table 1 that Pos-Neg pairs are less similar than Pos-Pos under all measures, indicating that concept maps can effectively reflect document semantics. Moreover, Pos-BM pairs are not close to Pos-Pos and even further away than Pos-Neg. This is because the labeled "irrelevant" documents are actually hard negative ones difficult to distinguish. Such results indicate the potential for improving sketchy candidates with concept maps. Besides, student's t-Test [13] is performed, where standard critical values of (Pos-Pos, Pos-Neg) and (Pos-Pos, Pos-BM) under 95% confidence are 1.6440 and 1.6450, respectively. The calculated *t-scores* shown in Table 1 strongly support the significance of differences.

3.3 Retrieval Performance Results

In this study, we focus on the performance improvement of GNN models based on sketchy candidates. Therefore, two widely-used and simple models, the forementioned BM25 and Anserini[3], are adopted as baselines, instead of the heavier language models such as BERT-based [8,9,41] and learning to rank (LTR)-based [2,35] ones. The retrieval performance are shown in Table 2. All the values are reported as the averaged results of five runs under the best settings.

---

[3] https://git.uwaterloo.ca/jimmylin/covidex-trec-covid-runs/-/tree/master/round5, whichisrecognizedbythecompetitionorganizersasabaselineresult.

Table 2. The retrieval performance results of different models.

.5 Type	.5 Methods	Precision (%)		Recall (%)		NDCG (%)	
		$k = 10$	$k = 20$	$k = 10$	$k = 20$	$k = 10$	$k = 20$
Traditional	BM25	55.20	49.00	1.36	2.39	51.37	45.91
	Anserini	54.00	49.60	1.22	2.25	47.09	43.82
Structure-Oriented	GIN	35.24	34.36	0.77	1.50	30.59	29.91
	GAT	46.48	43.26	1.08	2.00	42.24	39.49
Semantics-Oriented	N-Pool	58.24	52.20	1.38	2.41	53.38	48.80
	E-Pool	59.60	53.88	1.40	2.49	56.11	51.16
	RW-Pool	**59.84**	**53.92**	**1.42**	**2.53**	**56.19**	**51.41**

For the structure-oriented GIN and GAT, different read-out functions including mean, sum, max and a novel proposed tf-idf (i.e., weight the nodes using the tf-idf scores) are experimented, and tf-idf achieves the best performance. It is shown that GIN constantly fails to distinguish relevant documents while GAT is relatively better. However, they both fail to improve the baselines. This performance deviation may arise from the major inductive bias on complex structures, which makes limited contribution to document retrieval and is easily misled by noises. In contrast, our three proposed semantics-oriented graph functions yield significant and consistent improvements over both baselines and structure-oriented GNNs. Notably, E-Pool and RW-Pool improve the document retrieval from the initial candidates of BM25 by 11.4% and 12.0% on NDCG@20, respectively. Such results demonstrate the potential of designing semantics-oriented GNNs for textual reasoning tasks such as classification, retrieval, etc.

3.4 Stability and Efficiency

We further examine the stability and efficiency of different models across runs. As is shown in Fig. 2(a), GIN and GAT are less consistent, indicating the diffi-

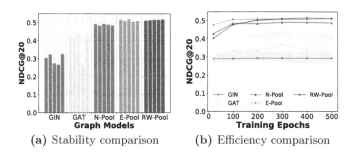

(a) Stability comparison   (b) Efficiency comparison

Fig. 2. Stability and efficiency comparison of different graph models.

culty in training over-complex models. The training efficiency in Fig. 2(b) shows that GIN can hardly improve during training, while GAT fluctuates a lot and suffers from overfitting. In contrast, our proposed semantics-oriented functions perform more stable in Fig. 2(a), and improve efficiently during training in Fig. 2(b), demonstrating their abilities to model the concepts and interactions important for the retrieval task. Among the three graph functions, E-Pool and RW-Pool are consistently better than N-Pool, revealing the utility of simple graph structures. Moreover, RW-Pool converges slower but achieves better and more stable results in the end, indicating the potential advantage of higher-order interactions.

4 Conclusion

In this paper, we investigate how can GNNs help document retrieval through a case study. Concept maps with rich semantics are generated from unstructured texts with constituency parsing. Two types of GNNs, structure-oriented complex models and our proposed semantics-oriented graph functions are experimented and the latter achieves consistently better and stable results, demonstrating the importance of semantic units as well as their simple interactions in GNN design for textual reasoning tasks like retrieval. In the future, more textual datasets such as news, journalism and downstream tasks can be included for validation. Other types of semantics-oriented graph functions can also be designed based on our permutation-invariant schema, such as graphlet based-pooling.

References

1. Baeza-Yates, R., Ribeiro-Neto, B., et al.: Modern Information Retrieval, vol. 463 (1999)
2. Burges, C.J.C., et al.: Learning to rank using gradient descent. In: ICML (2005)
3. Chen, N., Kinshuk, Wei, C., Chen, H.: Mining e-learning domain concept map from academic articles. Comput. Educ. **50**(5), 1009–1021 (2008)
4. Chen, Q., Peng, Y., Lu, Z.: Biosentvec: creating sentence embeddings for biomedical texts. In: ICHI, pp. 1–5 (2019)
5. Christensen, J., Mausam, Soderland, S., Etzioni, O.: Towards coherent multi-document summarization. In: NAACL, pp. 1163–1173 (2013)
6. Cui, H., Lu, Z., Li, P., Yang, C.: On positional and structural node features for graph neural networks on non-attributed graphs. CoRR abs/2107.01495 (2021)
7. Dang, V., Bendersky, M., Croft, W.B.: Two-stage learning to rank for information retrieval. In: Serdyukov, P., et al. (eds.) ECIR 2013. LNCS, vol. 7814, pp. 423–434. Springer, Heidelberg (2013). https://doi.org/10.1007/978-3-642-36973-5_36
8. Deshmukh, A.A., Sethi, U.: IR-BERT: leveraging BERT for semantic search in background linking for news articles. CoRR abs/2007.12603 (2020)
9. Devlin, J., Chang, M., Lee, K., Toutanova, K.: BERT: pre-training of deep bidirectional transformers for language understanding. In: NAACL (2019)
10. Farhi, S.H., Boughaci, D.: Graph based model for information retrieval using a stochastic local search. Pattern Recognit. Lett. **105**, 234–239 (2018)
11. Gilmer, J., Schoenholz, S.S., Riley, P.F., Vinyals, O., Dahl, G.E.: Neural message passing for quantum chemistry. In: ICML, pp. 1263–1272 (2017)

12. Hamilton, W., Ying, Z., Leskovec, J.: Inductive representation learning on large graphs. In: NeurIPS (2017)
13. Hogg, R.V., McKean, J., et al.: Introduction to Mathematical Statistics (2005)
14. Kamphuis, C.: Graph databases for information retrieval. In: Jose, J.M., et al. (eds.) ECIR 2020. LNCS, vol. 12036, pp. 608–612. Springer, Cham (2020). https://doi.org/10.1007/978-3-030-45442-5_79
15. Keriven, N., Peyré, G.: Universal invariant and equivariant graph neural networks. In: NeurIPS (2019)
16. Kipf, T.N., Welling, M.: Semi-supervised classification with graph convolutional networks. In: ICLR (2017)
17. Krallinger, M., Padron, M., Valencia, A.: A sentence sliding window approach to extract protein annotations from biomedical articles. BMC Bioinform. **6**, 1–12 (2005)
18. Li, M., et al.: Connecting the dots: event graph schema induction with path language modeling. In: EMNLP, pp. 684–695 (2020)
19. Liu, T.Y.: Learning to Rank for Information Retrieval, pp. 181–191 (2011). https://doi.org/10.1007/978-3-642-14267-3_14
20. Liu, Z., et al.: Geniepath: graph neural networks with adaptive receptive paths. In: AAAI, vol. 33, no. 1, pp. 4424–4431 (2019)
21. Lu, J., Choi, J.D.: Evaluation of unsupervised entity and event salience estimation. In: FLAIRS (2021)
22. Manmatha, R., Wu, C., Smola, A.J., Krähenbühl, P.: Sampling matters in deep embedding learning. In: ICCV, pp. 2840–2848 (2017)
23. Manning, C., Surdeanu, M., Bauer, J., Finkel, J., Bethard, S., McClosky, D.: The Stanford CoreNLP natural language processing toolkit. In: ACL, pp. 55–60 (2014)
24. Maron, H., Ben-Hamu, H., Shamir, N., Lipman, Y.: Invariant and equivariant graph networks. In: ICLR (2019)
25. Maron, H., Fetaya, E., Segol, N., Lipman, Y.: On the universality of invariant networks. In: ICML, pp. 4363–4371 (2019)
26. McClosky, D., Charniak, E., Johnson, M.: Automatic domain adaptation for parsing. In: NAACL Linguistics, pp. 28–36 (2010)
27. Mihalcea, R., Tarau, P.: Textrank: bringing order into text. In: EMNLP, pp. 404–411 (2004)
28. Nogueira, R., Cho, K.: Passage re-ranking with bert. arXiv preprint arXiv:1901.04085 (2019)
29. Roberts, K., et al.: Searching for scientific evidence in a pandemic: an overview of TREC-COVID. J. Biomed. Inform. **121**, 103865 (2021)
30. Robertson, S., Walker, S., Jones, S., Hancock-Beaulieu, M., Gatford, M.: Okapi at trec-3. In: TREC (1994)
31. Rose, S., Engel, D., Cramer, N., Cowley, W.: Automatic keyword extraction from individual documents. Text Min. Appl. Theory **1**, 1–20 (2010)
32. Veličković, P., Cucurull, G., Casanova, A., Romero, A., Lio, P., Bengio, Y.: Graph attention networks. In: ICLR (2018)
33. Wang, L.L., Lo, K., Chandrasekhar, Y., et al.: CORD-19: the COVID-19 open research dataset. In: Proceedings of the 1st Workshop on NLP for COVID-19 at ACL (2020)
34. Wang, X., Yang, C., Guan, R.: A comparative study for biomedical named entity recognition. Int. J. Mach. Learn. Cybern. **9**(3), 373–382 (2015). https://doi.org/10.1007/s13042-015-0426-6
35. Wu, Q., Burges, C.J.C., Svore, K.M., Gao, J.: Adapting boosting for information retrieval measures. Inf. Retr. **13**, 254–270 (2010)

36. Xu, K., Hu, W., Leskovec, J., Jegelka, S.: How powerful are graph neural networks? In: ICLR (2019)
37. Yang, C., et al.: Multisage: empowering GCN with contextualized multi-embeddings on web-scale multipartite networks. In: KDD, pp. 2434–2443 (2020)
38. Yang, C., Zhang, J., Wang, H., Li, B., Han, J.: Neural concept map generation for effective document classification with interpretable structured summarization. In: SIGIR, pp. 1629–1632 (2020)
39. Yang, C., et al.: Relation learning on social networks with multi-modal graph edge variational autoencoders. In: WSDM, pp. 699–707 (2020)
40. Yang, C., Zhuang, P., Shi, W., Luu, A., Li, P.: Conditional structure generation through graph variational generative adversarial nets. In: NeurIPS (2019)
41. Yilmaz, Z.A., Wang, S., Yang, W., Zhang, H., Lin, J.: Applying BERT to document retrieval with birch. In: EMNLP, pp. 19–24 (2019)
42. Ying, R., He, R., Chen, K., Eksombatchai, P., Hamilton, W.L., Leskovec, J.: Graph convolutional neural networks for web-scale recommender systems. In: KDD, pp. 974–983 (2018)
43. Yu, J., El-karef, M., Bohnet, B.: Domain adaptation for dependency parsing via self-training. In: Proceedings of the 14th International Conference on Parsing Technologies, pp. 1–10 (2015)
44. Zhang, Y., Chen, Q., Yang, Z., Lin, H., Lu, Z.: Biowordvec, improving biomedical word embeddings with subword information and mesh. Sci. Data **6**, 1–9 (2019)
45. Zhang, Y., Zhang, J., Cui, Z., Wu, S., Wang, L.: A graph-based relevance matching model for ad-hoc retrieval. In: AAAI (2021)
46. Zhang, Z., Wang, L., Xie, X., Pan, H.: A graph based document retrieval method. In: CSCWD, pp. 426–432 (2018)

Leveraging Content-Style Item Representation for Visual Recommendation

Yashar Deldjoo[1], Tommaso Di Noia[1], Daniele Malitesta[1(✉)],
and Felice Antonio Merra[2]

[1] Politecnico di Bari, Bari, Italy
{yashar.deldjoo,tommaso.dinoia,daniele.malitesta}@poliba.it
[2] Amazon Science Berlin, Berlin, Germany
felmerra@amazon.de

Abstract. When customers' choices may depend on the visual appearance of products (e.g., fashion), visually-aware recommender systems (VRSs) have been shown to provide more accurate preference predictions than pure collaborative models. To refine recommendations, recent VRSs have tried to recognize the influence of each item's visual characteristic on users' preferences, for example, through attention mechanisms. Such visual characteristics may come in the form of content-level item metadata (e.g., image tags) and reviews, which are not always and easily accessible, or image regions-of-interest (e.g., the collar of a shirt), which miss items' style. To address these limitations, we propose a pipeline for visual recommendation, built upon the adoption of those features that can be easily extracted from item images and represent the item content on a stylistic level (i.e., color, shape, and category of a fashion product). Then, we inject such features into a VRS that exploits attention mechanisms to uncover users' personalized importance for each content-style item feature and a neural architecture to model non-linear patterns within user-item interactions. We show that our solution can reach a competitive accuracy and beyond-accuracy trade-off compared with other baselines on two fashion datasets. Code and datasets are available at: https://github.com/sisinflab/Content-Style-VRSs.

Keywords: Visual recommendation · Attention · Collaborative filtering

1 Introduction and Related Work

Recommender systems (RSs) help users in their decision-making process by guiding them in a personalized fashion to a small subset of interesting products or services amongst massive corpora. In applications where visual factors are at

Authors are listed in alphabetical order.

F. A. Merra—Work performed while at Politecnico di Bari, Italy.

play (e.g., fashion [22], food [14], or tourism [33]), customers' choices are highly dependent on the visual product appearance that attracts attention, enhances emotions, and shapes their first impression about products. By incorporating this source of information when modeling users' preference, visually-aware recommender systems (VRSs) have found success in extending the expressive power of pure collaborative recommender models [10,12,13,17,18].

Recommendation can hugely benefit from items' side information [4]. To this date, several works have leveraged the high-level representational power of convolutional neural networks (CNNs) to extract item visual features, where the adopted CNN may be either pretrained on different datasets and tasks, e.g., [3, 11,18,26,29], or trained end-to-end in the downstream recommendation task, e.g., [23,38]. While the former family of VRSs builds upon a more convenient way of visually representing items (i.e., reusing the knowledge of pretrained models), such representations are not entirely in line with correctly providing users' visual preference estimation. That is, CNN-extracted features cannot capture what each user enjoys about a product picture since she might be more attracted by the color and shape of a specific bag, but these features do not necessarily match what the pretrained CNN learned when classifying the product image as a bag.

Recently, there have been a few attempts trying to uncover user's personalized visual attitude towards finer-grained item characteristics, e.g., [7–9,21]. These solutions disentangle product images at *(i)* content-level, by adopting item metadata and/or reviews [9,31], *(ii)* region-level, by pointing the user's interest towards parts of the image [8,36] or video frames [7], and *(iii)* both content- and region-level [21]. It is worth mentioning that most of these approaches [7,8,21,36] exploit attention mechanisms to weight the importance of the content or the region in driving the user's decisions.

Despite their superior performance, we recognize practical and conceptual limitations in adopting both content- and region-level item features, especially in the fashion domain. The former rely on additional side information (e.g., image tags or reviews), which could be not-easily and rarely accessible, as well as time-consuming to collect, while the latter ignore stylistic characteristics (e.g., color or texture) that can be impactful on the user's decision process [41].

Driven by these motivations, we propose a pipeline for visual recommendation, which involves a set of visual features, i.e., color, shape, and category of a fashion product, whose extraction is straightforward and always possible, describing items' content on a stylistic level. We use them as inputs to an attention- and neural-based visual recommender system, with the following purposes:

- We disentangle the visual item representations on the stylistic content level (i.e., color, shape, and category) by making the attention mechanisms weight the importance of each feature on the user's visual preference and making the neural architecture catch non-linearities in user/item interactions.
- We reach a reasonable compromise between accuracy and beyond-accuracy performance, which we further justify through an ablation study to investigate the importance of attention (in all its configurations) on the recommendation

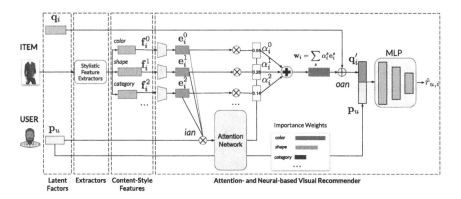

Fig. 1. Our proposed pipeline for visual recommendation, involving content-style item features, attention mechanisms, and a neural architecture.

performance. Notice that no ablation is performed on the content-style input features, as we learn to weight their contribution through the end-to-end attention network training procedure.

2 Method

In the following, we present our visual recommendation pipeline (Fig. 1).

Preliminaries. We indicate with $\mathcal{U}$ and $\mathcal{I}$ the sets of users and items. Then, we adopt $\mathbf{R}$ as the user/item interaction matrix, where $r_{ui} \in \mathbf{R}$ is 1 for an interaction, 0 otherwise. As in latent factor models such as matrix factorization (MF) [25], we use $\mathbf{p}_u \in \mathbb{R}^{1 \times h}$ and $\mathbf{q}_i \in \mathbb{R}^{1 \times h}$ as user and item latent factors, respectively, where $h << |\mathcal{U}|, |\mathcal{I}|$. Finally, we denote with $\mathbf{f}_i \in \mathbb{R}^{1 \times v}$ the visual feature for item image i, usually the fully-connected layer activation of a pre-trained convolutional neural network (CNN).

Content-Style Features. Let $\mathcal{S}$ be the set of content-style features to characterize item images. Even if we adopt $\mathcal{S} = \{\text{color}, \text{shape}, \text{category}\}$, for the sake of generality, we indicate with $\mathbf{f}_i^s \in \mathbb{R}^{1 \times v_s}$ the s-th content-style feature of item i. Since all $\mathbf{f}_i^s$ do not necessarily belong to the same latent space, we project them into a common latent space $\mathbb{R}^{1 \times h}$, i.e., the same as the one of $\mathbf{p}_u$ and $\mathbf{q}_i$. Thus, for each $s \in \mathcal{S}$, we build an encoder function $enc_s : \mathbb{R}^{1 \times v_s} \mapsto \mathbb{R}^{1 \times h}$, and encode the s-th content-style feature of item i as:

$$\mathbf{e}_i^s = enc_s(\mathbf{f}_i^s) \tag{1}$$

where $\mathbf{e}_i^s \in \mathbb{R}^{1 \times h}$, and enc_s is either trainable, e.g., a multi-layer perceptron (MLP), or handcrafted, e.g., principal-component analysis (PCA). In this work, we use an MLP-based encoder for the color feature, a CNN-based encoder for the shape, and PCA for the category.

Attention Network. We seek to produce recommendations conditioned on the visual preference of user u towards each content-style item characteristic. That is, the model is supposed to assign different importance weights to each encoded feature $\mathbf{e}_i^s$ based on the predicted user's visual preference $(\hat{r}_{u,i})$. Inspired by previous works [7,8,21,36], we use attention. Let $ian(\cdot)$ be the function to aggregate the inputs to the attention network $\mathbf{p}_u$ and $\mathbf{e}_i^s$, e.g., element-wise multiplication. Given a user-item pair (u,i), the network produces an attention weight vector $\mathbf{a}_{u,i} = [a_{u,i}^0, a_{u,i}^1, \ldots, a_{u,i}^{|\mathcal{S}|-1}] \in \mathbb{R}^{1 \times |\mathcal{S}|}$, where $a_{u,i}^s$ is calculated as:

$$a_{u,i}^s = \boldsymbol{\omega}_2(\boldsymbol{\omega}_1 ian(\mathbf{p}_u, \mathbf{e}_i^s) + \mathbf{b}_1) + \mathbf{b}_2 = \boldsymbol{\omega}_2(\boldsymbol{\omega}_1(\mathbf{p}_u \odot \mathbf{e}_i^s) + \mathbf{b}_1) + \mathbf{b}_2 \qquad (2)$$

where $\odot$ is the Hadamard product (element-wise multiplication), while $\boldsymbol{\omega}_*$ and $\mathbf{b}_*$ are the matrices and biases for each attention layer, i.e., the network is implemented as a 2-layers MLP. Then, we normalize $\mathbf{a}_{u,i}$ through the temperature-smoothed *softmax* function [20], so that $\sum_s a_{u,i}^s = 1$, getting the normalized weight vector $\boldsymbol{\alpha}_{u,i} = [\alpha_{u,i}^0, \alpha_{u,i}^1, \ldots, \alpha_{u,i}^{|\mathcal{S}|-1}]$. We leverage the attention values to produce a unique and weighted stylistic representation for item i, conditioned on user u:

$$\mathbf{w}_i = \sum_{s \in \mathcal{S}} \alpha_{u,i}^s \mathbf{e}_i^s \qquad (3)$$

Finally, let $oan(\cdot)$ be the function to aggregate the latent factor $\mathbf{q}_i$ and the output of the attention network $\mathbf{w}_i$ into a unique representation for item i, e.g., through addition. We calculate the final item representation $\mathbf{q}_i'$ as:

$$\mathbf{q}_i' = oan(\mathbf{q}_i, \mathbf{w}_i) = \mathbf{q}_i + \mathbf{w}_i \qquad (4)$$

Neural Inference. To capture non-linearities in user/item interactions, we adopt an MLP to run the prediction. Let $concat(\cdot)$ be the concatenation function and $out(\cdot)$ be a trainable MLP, we predict rating $\hat{r}_{u,i}$ for user u and item i as:

$$\hat{r}_{u,i} = out(concat(\mathbf{p}_u, \mathbf{q}_i')) \qquad (5)$$

Objective Function and Training. We use Bayesian personalized ranking (BPR) [32]. Given a set of triples $\mathcal{T}$ (user u, positive item p, negative item n), we seek to optimize the following objective function:

$$\underset{\Theta}{\arg\min} \sum_{(u,p,n) \in \mathcal{T}} -\ln(sigmoid(\hat{r}_{u,p} - \hat{r}_{u,n})) + \lambda \|\Theta\|^2 \qquad (6)$$

where Θ and λ are the set of trainable weights and the regularization term, respectively. We build $\mathcal{T}$ from the training set by picking, for each randomly sampled (u,p) pair, a negative item n for u (i.e., not-interacted by u). Moreover, we adopt mini-batch Adam [24] as optimizing algorithm.

3 Experiments

Datasets. We use two popular categories from the Amazon dataset [17,28], i.e., Boys & Girls and Men. After having downloaded the available item images, we filter out the items and the users with less than 5 interactions [17,18]. Boys & Girls counts 1,425 users, 5,019 items, and 9,213 interactions (sparsity is 0.00129), while Men counts 16,278 users, 31,750 items, and 113,106 interactions (sparsity is 0.00022). In both cases, we have, on average, > 6 interactions per user.

Feature Extraction and Encoding. Since we address a fashion recommendation task, we extract color, shape/texture, and fashion category from item images [34,41]. Unlike previous works, we leverage such features because they are easy to extract and always accessible and represent the content of item images at a stylistic level. We extract the **color** information through the 8-bin RGB color histogram, the **shape/texture** as done in [34], and the **fashion category** from a pretrained ResNet50 [6,11,15,37], where "category" refers to the classification task on which the CNN is pretrained. As for the features encoding, we use a trainable MLP and CNN for color (a vector) and shape (an image), respectively. Conversely, following [30], we adopt PCA to compress the fashion category feature, also to level it out to the color and shape features that do not benefit from a pretrained feature extractor.

Baselines. We compare our approach with pure collaborative and visual-based approaches, i.e., BPRMF [32] and NeuMF [19] for the former, and VBPR [18], DeepStyle [26], DVBPR [23], ACF [7], and VNPR [30] for the latter.

Evaluation and Reproducibility. We put, for each user, the last interaction into the test set and the second-to-last into the validation one (i.e., temporal leave-one-out). Then, we measure the model accuracy with the hit ratio ($HR@k$, the validation metric) and the normalized discounted cumulative gain ($nDCG@k$) as performed in related works [7,19,39]. We also measure the fraction of items covered in the catalog ($iCov@k$), the expected free discovery ($EFD@k$) [35], and the diversity with the 1's complement of the Gini index ($Gini@k$) [16]. For the implementation, we used the framework Elliot [1,2].

3.1   Results

What are the Accuracy and Beyond-Accuracy Recommendation Performance? Table 1 reports the accuracy and beyond-accuracy metrics on top-20 recommendation lists. On Amazon Boys & Girls, our solution and DeepStyle are the best and second-best models on accuracy and beyond-accuracy measures, respectively (e.g., 0.03860 vs. 0.03719 for the HR). In addition, our approach outperforms all the other baselines on novelty and diversity, covering a broader fraction of the catalog (e.g., $iCov \simeq 90\%$). As for Amazon Men, the proposed approach is still consistently the most accurate model, even beating BPRMF, whose accuracy performance is superior to all other visual baselines. Considering that BPRMF covers only the 0.6% of the item catalog, it follows that its superior performance on accuracy comes from recommending the most popular items [5,27,40]. Given that, we maintain the competitiveness of our solution, being the best on the accuracy, but also covering about 29% of the item catalog and supporting the discovery of new products (e.g., $EFD = 0.01242$ is

Table 1. Accuracy and beyond-accuracy metrics on top-20 recommendation lists.

Model	HR	nDCG	iCov	EFD	Gini
Amazon Boys & Girls—configuration file					
BPRMF	.01474	.00508	.68181	.00719	.28245
NeuMF	.02386	.00999	.00638	.01206	.00406
VBPR	.03018	.01287	.71030	.02049	.30532
DeepStyle	.03719	.01543	.85017	.02624	.44770
DVBPR	.00491	.00211	.00438	.00341	.00379
ACF	.01544	.00482	.70731	.00754	.40978
VNPR	.01053	.00429	.51584	.00739	.13664
Ours	**.03860**	**.01610**	**.89878**	**.02747**	**.49747**
Amazon Men—configuration file					
BPRMF	.01947	.00713	.00605	.00982	.00982
NeuMF	.01333	.00444	.00076	.00633	.00060
VBPR	.01554	.00588	.59351	.01042	.17935
DeepStyle	.01634	.00654	.84397	.01245	.33314
DVBPR	.00123	.00036	.00088	.00069	.00065
ACF	.01548	.00729	.19380	.01147	.02956
VNPR	.00528	.00203	.59443	.00429	.16139
Ours	**.02021**	**.00750**	.28995	.01242	.06451

Table 2. Ablation study on different configurations of attention, *ian*, and *oan*.

Components		Boys & Girls		Men	
$ian(\cdot)$	$oan(\cdot)$	HR	iCov	HR	iCov
No Attention		.01263	.01136	.01462	.02208
Add	Add	.02316	.00757	.02083	.00076
Add	Mult	.02246	.00458	.00768	.00079
Concat	Add	.01404	.00518	**.02113**	.00076
Concat	Mult	.02456	.00458	.00891	.00085
Mult	*Add*	**.03860**	**.89878**	.02021	**.28995**
Mult	Mult	.02807	.00478	.01370	.01647

the second to best value). That is, the proposed method shows a competitive performance trade-off on accuracy and beyond-accuracy metrics.

How performance is affected by different configurations of attention, *ian* , and *oan*? Following [8,21], we feed the attention network by exploring three aggregations for the inputs of the attention network (*ian*), i.e., element-wise multiplication/addition and concatenation, and two aggregations for the output of the attention network (*oan*), i.e., element-wise addition/multiplication. Table 2 reports the HR, i.e., the validation metric, and the $iCov$, i.e., a beyond-accuracy metric. No ablation study is run on the content-style features, as their relative influence on recommendation is learned during the training. First, we observe that attention mechanisms, i.e., all rows but *No Attention*, lead to better-tailored recommendations. Second, despite the {Concat, Add} choice reaches the highest accuracy on Men, the {Mult, Add} combination we used in this work is the most competitive on both accuracy and beyond-accuracy metrics.

4 Conclusion and Future Work

Unlike previous works, we argue that in visual recommendation scenarios (e.g., fashion), items should be represented by easy-to-extract and always accessible visual characteristics, aiming to describe their content from a stylistic perspective (e.g., color and shape). In this work, we disentangled these features via attention to assign users' personalized importance weights to each content-style feature. Results confirmed that our solution could reach a competitive accuracy and beyond-accuracy trade-off against other baselines, and an ablation study justified the adopted architectural choices. We plan to extend the content-style features for other visual recommendation domains, such as food and social media. Another area where item content visual features can be beneficial is in improving accessibility to extremely long-tail items (distant tails), for which traditional CF or hybrid approaches are not helpful due to the scarcity of interaction data.

Acknowledgment. The authors acknowledge partial support of the projects: CTE Matera, ERP4.0, SECURE SAFE APULIA, Servizi Locali 2.0.

References

1. Anelli, V.W., et al.: Elliot: a comprehensive and rigorous framework for reproducible recommender systems evaluation. In: SIGIR, pp. 2405–2414. ACM (2021)
2. Anelli, V.W., et al.: V-elliot: design, evaluate and tune visual recommender systems. In: RecSys, pp. 768–771. ACM (2021)
3. Anelli, V.W., Deldjoo, Y., Di Noia, T., Malitesta, D., Merra, F.A.: A study of defensive methods to protect visual recommendation against adversarial manipulation of images. In: SIGIR, pp. 1094–1103. ACM (2021)
4. Anelli, V.W., Di Noia, T., Di Sciascio, E., Ferrara, A., Mancino, A.C.M.: Sparse feature factorization for recommender systems with knowledge graphs. In: RecSys, pp. 154–165. ACM (2021)
5. Boratto, L., Fenu, G., Marras, M.: Connecting user and item perspectives in popularity debiasing for collaborative recommendation. Inf. Process. Manag. **58**(1), 102387 (2021)
6. Chen, J., Ngo, C., Feng, F., Chua, T.: Deep understanding of cooking procedure for cross-modal recipe retrieval. In: ACM Multimedia, pp. 1020–1028. ACM (2018)
7. Chen, J., Zhang, H., He, X., Nie, L., Liu, W., Chua, T.: Attentive collaborative filtering: multimedia recommendation with item- and component-level attention. In: SIGIR, pp. 335–344. ACM (2017)
8. Chen, X., et al.: Personalized fashion recommendation with visual explanations based on multimodal attention network: towards visually explainable recommendation. In: SIGIR, pp. 765–774. ACM (2019)
9. Cheng, Z., Chang, X., Zhu, L., Kanjirathinkal, R.C., Kankanhalli, M.S.: MMALFM: explainable recommendation by leveraging reviews and images. ACM Trans. Inf. Syst. **37**(2), 16:1–16:28 (2019)
10. Chong, X., Li, Q., Leung, H., Men, Q., Chao, X.: Hierarchical visual-aware minimax ranking based on co-purchase data for personalized recommendation. In: WWW, pp. 2563–2569. ACM/IW3C2 (2020)
11. Deldjoo, Y., Di Noia, T., Malitesta, D., Merra, F.A.: A study on the relative importance of convolutional neural networks in visually-aware recommender systems. In: CVPR Workshops, pp. 3961–3967. Computer Vision Foundation/IEEE (2021)
12. Deldjoo, Y., Schedl, M., Cremonesi, P., Pasi, G.: Recommender systems leveraging multimedia content. ACM Comput. Surv. (CSUR) **53**(5), 1–38 (2020)
13. Deldjoo, Y., Schedl, M., Hidasi, B., He, X., Wei, Y.: Multimedia recommender systems: algorithms and challenges. In: Recommender Systems Handbook. Springer, US (2022)
14. Elsweiler, D., Trattner, C., Harvey, M.: Exploiting food choice biases for healthier recipe recommendation. In: SIGIR, pp. 575–584. ACM (2017)

15. Gao, X., et al.: Hierarchical attention network for visually-aware food recommendation. IEEE Trans. Multim. **22**(6), 1647–1659 (2020)
16. Gunawardana, A., Shani, G.: Evaluating recommender systems. In: Ricci, F., Rokach, L., Shapira, B. (eds.) Recommender Systems Handbook, pp. 265–308. Springer, Boston, MA (2015). https://doi.org/10.1007/978-1-4899-7637-6_8
17. He, R., McAuley, J.J.: Ups and downs: Modeling the visual evolution of fashion trends with one-class collaborative filtering. In: WWW, pp. 507–517. ACM (2016)
18. He, R., McAuley, J.J.: VBPR: visual Bayesian personalized ranking from implicit feedback. In: AAAI, pp. 144–150. AAAI Press (2016)
19. He, X., Liao, L., Zhang, H., Nie, L., Hu, X., Chua, T.: Neural collaborative filtering. In: WWW, pp. 173–182. ACM (2017)
20. Hinton, G.E., Vinyals, O., Dean, J.: Distilling the knowledge in a neural network. CoRR abs/1503.02531 (2015)
21. Hou, M., Wu, L., Chen, E., Li, Z., Zheng, V.W., Liu, Q.: Explainable fashion recommendation: a semantic attribute region guided approach. In: IJCAI, pp. 4681–4688. ijcai.org (2019)
22. Hu, Y., Yi, X., Davis, L.S.: Collaborative fashion recommendation: a functional tensor factorization approach. In: ACM Multimedia, pp. 129–138. ACM (2015)
23. Kang, W., Fang, C., Wang, Z., McAuley, J.J.: Visually-aware fashion recommendation and design with generative image models. In: ICDM, pp. 207–216. IEEE Computer Society (2017)
24. Kingma, D.P., Ba, J.: Adam: a method for stochastic optimization. In: ICLR (Poster) (2015)
25. Koren, Y., Bell, R.M., Volinsky, C.: Matrix factorization techniques for recommender systems. Computer **42**(8), 30–37 (2009)
26. Liu, Q., Wu, S., Wang, L.: Deepstyle: learning user preferences for visual recommendation. In: SIGIR, pp. 841–844. ACM (2017)
27. Mansoury, M., Abdollahpouri, H., Pechenizkiy, M., Mobasher, B., Burke, R.: Feedback loop and bias amplification in recommender systems. In: CIKM, pp. 2145–2148. ACM (2020)
28. McAuley, J.J., Targett, C., Shi, Q., van den Hengel, A.: Image-based recommendations on styles and substitutes. In: SIGIR. ACM (2015)
29. Meng, L., Feng, F., He, X., Gao, X., Chua, T.: Heterogeneous fusion of semantic and collaborative information for visually-aware food recommendation. In: ACM Multimedia, pp. 3460–3468. ACM (2020)
30. Niu, W., Caverlee, J., Lu, H.: Neural personalized ranking for image recommendation. In: WSDM, pp. 423–431. ACM (2018)
31. Packer, C., McAuley, J.J., Ramisa, A.: Visually-aware personalized recommendation using interpretable image representations. CoRR abs/1806.09820 (2018)
32. Rendle, S., Freudenthaler, C., Gantner, Z., Schmidt-Thieme, L.: BPR: Bayesian personalized ranking from implicit feedback. In: UAI, pp. 452–461. AUAI Press (2009)
33. Sertkan, M., Neidhardt, J., Werthner, H.: Pictoure - A picture-based tourism recommender. In: RecSys, pp. 597–599. ACM (2020)
34. Tangseng, P., Okatani, T.: Toward explainable fashion recommendation. In: WACV, pp. 2142–2151. IEEE (2020)
35. Vargas, S.: Novelty and diversity enhancement and evaluation in recommender systems and information retrieval. In: SIGIR, p. 1281. ACM (2014)
36. Wu, Q., Zhao, P., Cui, Z.: Visual and textual jointly enhanced interpretable fashion recommendation. IEEE Access **8**, 68736–68746 (2020)

37. Yang, X., et al.: Interpretable fashion matching with rich attributes. In: SIGIR, pp. 775–784. ACM (2019)
38. Yin, R., Li, K., Lu, J., Zhang, G.: Enhancing fashion recommendation with visual compatibility relationship. In: WWW, pp. 3434–3440. ACM (2019)
39. Zhang, Y., Zhu, Z., He, Y., Caverlee, J.: Content-collaborative disentanglement representation learning for enhanced recommendation. In: RecSys, pp. 43–52. ACM (2020)
40. Zhu, Z., Wang, J., Caverlee, J.: Measuring and mitigating item under-recommendation bias in personalized ranking systems. In: SIGIR, pp. 449–458. ACM (2020)
41. Zou, Q., Zhang, Z., Wang, Q., Li, Q., Chen, L., Wang, S.: Who leads the clothing fashion: Style, color, or texture? A computational study. CoRR abs/1608.07444 (2016)

Does Structure Matter? Leveraging Data-to-Text Generation for Answering Complex Information Needs

Hanane Djeddal[1(✉)], Thomas Gerald[1], Laure Soulier[1],
Karen Pinel-Sauvagnat[2], and Lynda Tamine[2]

[1] CNRS-ISIR, Sorbonne University, Paris, France
{djeddal,gerald,soulier}@isir.upmc.fr
[2] Université Paul Sabatier, IRIT, Toulouse, France
{sauvagnat,tamine}@irit.fr

Abstract. In this work, our aim is to provide a structured answer in natural language to a complex information need. Particularly, we envision using generative models from the perspective of data-to-text generation. We propose the use of a content selection and planning pipeline which aims at structuring the answer by generating intermediate plans. The experimental evaluation is performed using the TREC Complex Answer Retrieval (CAR) dataset. We evaluate both the generated answer and its corresponding structure and show the effectiveness of planning-based models in comparison to a text-to-text model.

Keywords: Answer generation · Complex search tasks · Data-to-text generation

1 Introduction

Complex search tasks (e.g., exploratory) involve open-ended and multifaceted queries that require information retrieval (IR) systems to aggregate information over multiple unstructured documents [19]. To address these requirements, most interactive IR methods adopt the dynamic multi-turn retrieval approach by designing session-based predictive models relying on Markov models [21], query-flow graphs [7] for relevance prediction and sequence-to-sequence models for query suggestion [1,15]. One drawback of those approaches remains in the iterative querying process, requesting users to visit different contents to complete their information need. Moreover, while there is a gradual shift today towards new interaction paradigms through natural-sounding answers [5,16], most approaches still rely on a ranked list of documents as the main form of answer.

We envision here solving complex search tasks triggered by open-ended queries, by considering single-turn (vs. multi-turn) interaction with users and providing natural language generated answers (vs. a ranked list of documents).

© The Author(s), under exclusive license to Springer Nature Switzerland AG 2022
M. Hagen et al. (Eds.): ECIR 2022, LNCS 13186, pp. 93–101, 2022.
https://doi.org/10.1007/978-3-030-99739-7_11

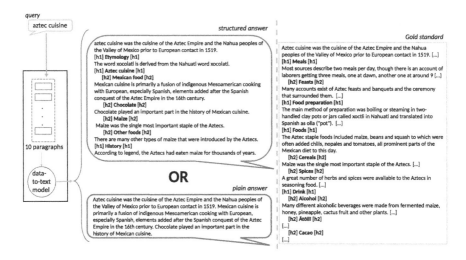

Fig. 1. Example of a query from the CAR dataset [6] and variants of outputs (structured or plain answers) obtained using a sequential DTT planning-based model.

We focus on the upstream part of the search process, once a ranked list of candidate documents has been identified in response to a complex information need. In a close line of research, open-domain QA attempt to retrieve and reason over multiple seed passages either to extract [2,4] or to generate in a natural language form [14,16,18] answers to open-domain questions. Most open-domain QA approaches adopt the "Retriever Reader" framework: the retriever ranks candidate passages, then the reader constructs the response based on the most relevant retrieved candidate [9,17,24]. Compared with open-domain QA, answer generation to open-ended queries has two main specific issues: 1) all the documents provided by the reader potentially contribute both as evidence and source to generate the answer leading to difficulties in discriminating between relevance and salience of the spans; 2) while most QA problems target a single-span answer [22] included in one document, open-ended queries are characterized by multiple facets [19,20] that could target a multiple-span answer.

Our objective is to generate an answer that covers the multiple facets of an open-ended query using as input, an initial ranked list of documents. We basically assume that the list of documents cover the different query facets. A naive approach would be to exploit text-to-text models [11,12]. However, we believe that answering multi-faceted queries would require the modelling of structure prior to generating the answer's content [5] To fit with this requirement, we adopt a data-to-text (DTT) generation approach [10] that introduces the notion of structure by guiding the generation with an intermediary plan aiming at determining *what to say* on the basis of the input data. This intermediary step, called *content selection/planning*, reinforces the factualness and the coverage of the generated text since: 1) it organizes the data structure in a latent form to better fit with the generated output, and 2) it provides a structure to the generated answer based on the elements of the initial data. Figure 1 presents an example of a query from TREC Complex Answer Retrieval (CAR) dataset [6] and the two variants of answers (*plain answers, structured answers*) generated by our proposed model trained respectively on two different train-test datasets (See Sect. 3). To sum up, given a ranked list of documents relevant to a complex information need, this work investigates the potential of content selection and planning DTT generation models for single-turn answer generation.

2   A Data-to-text Approach for Answer Generation

We introduce here the model used for generating natural language answers to open-ended queries formulated by users while completing complex search tasks. The designed model is driven by the intuition that the response should be guided by a structure to cover most of the query facets. This prior is modeled through a hierarchical plan which corresponds to a textual object relating the structure of the response with multiple-level titles (titles, subtitles, etc.).

More formally, we consider a document collection $\mathcal{D}$ and a set $Q \times A \times P$ of query-answer-plan triplets, where $q \in Q$ refer to queries, answers $a \in A$ the final response in natural language provided to the user and plans $p \in P$ represent the hierarchical structure of answers a. All documents d, queries q, answers a are represented by sets of tokens. For modeling the structure of plans p, we use $p = \{h_1, ..., h_i, ..., h_{|p|}\}$ where h_i represents a line in the plan expressing a heading (title, subtitles, etc.). The h_i are modeled as sets of tokens.

Given a query q and a document collection $\mathcal{D}$, our objective is to generate an answer a. To do so, we follow the "Retriever Generator" framework in which: 1) a ranking model $\mathcal{M}_{ret}$ retrieves a ranked list $\mathcal{D}_q$ of documents in response to query q, where $\mathcal{D}_q = \{d_q^1, ..., d_q^n\}$ and 2) a text generation model $\mathcal{M}_{gen}$ generates answer a given the retrieved list $\mathcal{D}_q$ and query q. As outlined earlier, the challenges of our task mainly rely on aggregating information over the ranked list of documents and generating a structured answer in natural language. Thus, we fix the retrieval model $\mathcal{M}_{ret}$ and focus on the generation model $\mathcal{M}_{gen}$. The latter exploits the DTT generation model based on content selection and planning [10]. To generate the intermediary plan p and the answer a, we rely on two successive encoder-decoders (based on T5 [12] as the building-box model):

- The **planning encoder-decoder** encodes each document $d_q \in \mathcal{D}_q$ concatenated with the query q and decodes a plan p. The training of such network is guided by the auto-regressive generation loss:

$$\mathcal{L}_{planning}(q, p) = P(p|q, \mathcal{D}_q) = \prod_{j=1}^{|p|} \prod_{k=1}^{|h_j|} P(h_{jk}|h_{j,<k}, q, \mathcal{D}_q) \qquad (1)$$

where j and k point out resp. to the heading h_j and the kth token h_{jk} in heading h_j. $h_{j,<k}$ corresponds to the token sequence in heading h_j before the kth token.

- The **content generation encoder-decoder** encodes each heading h_p in the plan p (generated by the planning encoder-decoder) concatenated with the embedding of the document list $\mathcal{D}_q$. The latter is obtained by the planning encoder-decoder since the T5 model provides embeddings for both documents independently and the set of documents. After the encoding, the network then decodes an answer a. The training is also guided by the auto-regressive generation loss:

$$\mathcal{L}_{answer}(q, a, p) = P(a|q, p, \mathcal{D}_q) = \prod_{k=1}^{|a|} P(a_k|a_{<k}, q, p, \mathcal{D}_q) \tag{2}$$

where a_k and $a_{<k}$ resp. express the kth token in answer a and the token sequence of answer a before the kth token. The final loss is a combination of both losses:

$$\mathcal{L} = \sum_{\{q,a,p\}\in Q\times A\times P} \mathcal{L}_{planning}(q, p) + \mathcal{L}_{answer}(q, a, p) \tag{3}$$

3   Evaluation Setup

Dataset. We selected the TREC CAR (Complex Answer Retrieval) 2017 corpus [6]. This dataset includes: (1) queries - denoting complex search tasks with multiple facets, (2) plans - expressing the different expected facets, and (3) paragraphs extracted from English Wikipedia - corresponding to texts associated with plan sections. The TREC CAR task consists of retrieving the paragraphs associated to each plan section to build a structured answer combining both plan sections and paragraphs. We used these *structured answers* as the final objective of our generation model given the queries; and the plans as the structure prior. Due to the structure prior constraint, we removed in the training set answers without any plans. To compare the models abilities to generate *structured answers*, we also evaluate a new form of expected answer (*plain answers*) where structure is not taken into account. For this aim, we built a new dataset upon the initial TREC CAR dataset but only considering the paragraphs (without plans). Thus, we obtain two versions of datasets (for *structured answers* and *plain answers*) which both follow the original split of the TREC CAR dataset[1]. Second, for computational reasons, we reduced the number of entries in our training set by considering only a half of Fold 0. Also, due to the memory constraints of generation models and the length of Wikipedia articles, we reduced the document size by only keeping the first sentence of paragraphs. Some statistics of the original dataset and our two adapted datasets are given in Table 1.

---

[1] The large train set for training, and the Y1 benchmark test set for testing.

Model Variants And Baselines. We implement two versions of our model[2]:

- **Planning-seq**: a sequential model where the planning module (Eq. 1) and the content generation module (Eq. 2) are trained separately. At inference, both modules are used sequentially.
- **Planning-e2e**: the end-to-end version of our model (Eq. 3). The content module is fed with the output embeddings of the planning module, and document tokens.

Besides, we compare our models with two baselines: 1) the **T5** model [12] which is fine-tuned on each dataset, and 2) **Ext**, an extractive method where we extract, for each sentence in the ground truth, a sentence in the input supporting documents that maximizes the F1 score of BERTScore [23]. All models consider for each topic a set of 10 relevant paragraphs ranked using BM25 as input.

Metrics. To evaluate the quality of the generation, we consider three well-known metrics: 1) the ROUGE-L mid metric (Rouge-P, Rouge-R, Rouge-F) [8] measuring the exact match between the generated and the reference texts, 2) the BERTScore [23] (the F1 score is reported) which computes similarity between the generated and the gold reference text embeddings, 3) the QuestEval [13]

Table 1. Statistics on the TREC CAR 2017 dataset and its adaptation for experiments.

	Original dataset		Structured answers		Plain answers	
	Train	Test	Train	Test	Train	Test
#answers	598 308	132	46 224	132	46 224	132
#tokens/answers	1376.48	5456.94	609.31	1724.63	449.21	1409.79
#headings/plan	6.10	17.69	6.22	17.69	–	–

Table 2. Effectiveness of the answer generation. In bold are the highest metric value among the generation models (T5, Planning-seq, Planning-e2e).

		# tokens	Rouge-P	Rouge-R	Rouge-F	BERTScore	QuestEval
structured answers	EXT	898.22	36.50	26.99	29.86	85.50	41.99
	T5	126.25	**76.19**	08.41	14.25	**84.95**	39.06
	Planning-seq	181.39	62.94	09.57	15.36	84.44	37.47
	Planning-e2e	203.48	63.4	**10.21**	**16.09**	84.91	**39.31**
plain answers	EXT	885.35	34.35	26.73	28.99	86.30	42.34
	T5	110.62	**78.05**	09.24	15.48	85.51	39.89
	Planning-seq	163.58	65.73	10.34	16.27	84.29	38.46
	Planning-e2e	126.91	75.92	**10.34**	**17.05**	**85.67**	**40.78**

[2] Code available at https://github.com/hanane-djeddal/Complex-Answer-Generation/.

framework which relies on question answering models to assess whether a summary contains all the source information: if the same questions are asked to the generated and the reference texts, the produced answers should be consistent.

To evaluate the model's ability to generate structure (namely the plans), we use the METEOR score [3] capturing how well-ordered the output words are.

4   Results

We perform the experimental evaluations w.r.t. two objectives. First, we measure the effectiveness of the generated answers. Second, we provide a thorough analysis of the generated plans.

Answer Generation Effectiveness. Table 2 reports the results of the different settings and models used for generating answers. It is worth of recall that the EXT baseline does not address the generation task and is built from the ground truth leading to provide high value trends. With this in mind, we can outline that:

- Planning-based generation models are competitive regarding the T5 generation baseline: our models allow to generate longer answers (avg. 200 tokens), thus increasing the recall metric (Rouge-R). The smaller precision (76.19 for T5, up to 63.4 for our models) does not hinder the semantic content of the answer (see BERTScore and QuestEval values which are very close to the EXT metrics). This suggests that our models are able to generate answers with the adequate content, even if noisy at some points.
- One can see the general trend towards higher metrics for all models in the *plain answers* setting compared to the *structured answers* setting (e.g. Rouge-P reaching up to 78.05 vs. 76.15) over all models. The *plain answers* setting is less difficult since the expected answer is not structured (only composed of paragraphs); evaluation metrics are higher since the gold reference is not based on both plans and paragraphs (as in the *structured answers* setting). In the *plain answers* setting, our models are most effective (with an advantage for Planning-e2e with, for instance 17.05 Rouge-F vs. 15.48 for T5). Even if the *plain answers* setting does not expect plans in the final answer, our models generate an intermediary plan that guides the answer generation. In contrast, T5 directly generates the answer. This reinforces our intuition about the importance of structure prior for generating an answer to a complex information need.
- Our end-to-end model seems more effective than the sequential one (e.g., resp. 40.78 vs. 38.46 for the QuestEval metric), suggesting the relevance of guiding the learning of the planning encoder-decoder by the answer generation task.

Analyses of the Generated Plans. To get a deeper understanding of our model behavior regarding the structure prior, we analyze the plans generated by the different encoder-decoders: the intermediate one provided by the planning encoder-decoder and the final one included in the final answer after the generation

Table 3. Analysis of the intermediate and final plans (resp. noted IP and FP) in our sequential and end-to-end planning-based models for the *structured answers* setting.

		#tokens	#heading	depth	Rouge-P	Rouge-R	Rouge-F	BERTScore	Meteor
T5	FP	1.41	2.24	1.14	**39.89**	04.69	07.69	77.40	3.24
Planning-seq	IP	1.83	**4.42**	**1.45**	31.20	**8.29**	**11.51**	**81.25**	**5.97**
	FP	**1.88**	4.11	**1.45**	31.31	7.93	11.03	80.49	5.55
Planning-e2e	IP	1.57	3.37	1.15	35.15	07.34	11.12	81.21	5.51
	FP	1.64	3.27	1.16	34.79	06.38	09.78	80.70	4.71

encoder-decoder (we simply extracted headings of the structured answer - red and blue lines in Fig. 1). We report in Table 3 the different evaluation metrics presented in Sect. 4 to measure the quality of plans and add some plan statistics (the average number of tokens for each plan section - #token; the number of generated plan sections by query -#heading; the mean depth of plan sections i.e. i of h_i -*depth*). Comparison of intermediate and final plans obtained by our models with the final one generated by T5 highlights that: 1) our plans are longer and more complex (more tokens by plan section - up to 1.83 in average, more and deeper headings - up to 4/5 headings in average), 2) our plans generally cover more facets (higher recall), in correct order (higher Meteor) with a better relevant semantics (higher BERTScore). The lowest precision (up to 35.15 vs. 39.89 for the T5) might be explained by the plan sizes. Moreover, the comparison of intermediate vs. final plans underlines a general trend towards lower quality of plans in the final step (e.g., 11.51 vs. 11.03 for Planning-seq in terms of Rouge-F). But the previous discussion on answer effectiveness, and the higher performance of our models regarding T5) suggests that there is a balance to reach between raw text and plan generation and that the structure prior is however highly beneficial for generating a good answer.

5 Conclusion

Traditionally, IR approaches solving complex information needs focused on leveraging multi-turn interactions to provide optimal rankings of candidate documents at each turn. In this paper we have suggested alternative retrieval models that do not rely on the interactive updating of queries and document rankings as answers. We suggest that data-to-text generation is an alternative way to generate in a single-turn, a natural language and structured answer. Experimental evaluation of a planning-based DTT model using the TREC CAR dataset shows the potential of our intuition. We believe that our work opens up novel research areas regarding answer generation and explanation in conversational IR systems.

Acknowledgements. We would like to thank projects ANR JCJC SESAMS (ANR-18- CE23-0001) and ANR COST (ANR-18-CE23-0016) for supporting this work. This work was performed using HPC resources from GENCI-IDRIS (Grant 2021–101681).

References

1. Ahmad, W.U., Chang, K., Wang, H.: Context attentive document ranking and query suggestion. In: Proceedings of the 42nd International ACM SIGIR, SIGIR 2019, pp. 385–394 (2019)
2. Asai, A., Hashimoto, K., Hajishirzi, H., Socher, R., Xiong, C.: Learning to retrieve reasoning paths over wikipedia graph for question answering. In: International Conference on Learning Representations (2020)
3. Banerjee, S., Lavie, A.: Meteor: An automatic metric for mt evaluation with improved correlation with human judgments. In: Proceedings of the ACL Workshop on Intrinsic and Extrinsic Evaluation Measures for Machine Translation and/or Summarization, pp. 65–72 (2005)
4. Chen, D., Fisch, A., Weston, J., Bordes, A.: Reading wikipedia to answer open-domain questions. ACL (2017)
5. Culpepper, J.S., Diaz, F., Smucker, M.D.: Research frontiers in information retrieval: Report from the third strategic workshop on information retrieval in lorne (SWIRL 2018). SIGIR Forum **1**, 34–90 (2018)
6. Dietz, L., Verma, M., Radlinski, F., Craswell, N.: Trec complex answer retrieval overview. In: TREC (2017)
7. Hassan Awadallah, A., White, R.W., Pantel, P., Dumais, S.T., Wang, Y.M.: Supporting complex search tasks. In: CIKM 2014 Proceedings of the 23rd ACM International Conference on Conference on Information and Knowledge Management, New York, NY, USA, pp. 829–838 (2014)
8. Lin, C.Y.: Rouge: A package for automatic evaluation of summaries. In: Text Summarization Branches Out, pp. 74–81 (2004)
9. Min, S., Zhong, V., Zettlemoyer, L., Hajishirzi, H.: Multi-hop reading comprehension through question decomposition and rescoring. In: ACL (2019)
10. Puduppully, R., Dong, L., Lapata, M.: Data-to-text generation with content selection and planning. In: The Thirty-Third AAAI Conference on Artificial Intelligence, AAAI 2019, pp. 6908–6915 (2019)
11. Radford, A., Narasimhan, K.: Improving language understanding by generative pre-training (2018)
12. Raffel, C., et al.: Exploring the limits of transfer learning with a unified text-to-text transformer. J. Mach. Learn. Res. **21**(140), 1–67 (2020)
13. Scialom, T., et al.: Questeval: Summarization asks for fact-based evaluation. arXiv (2021)
14. Song, L., Wang, Z., Hamza, W., Zhang, Y., Gildea, D.: Leveraging context information for natural question generation. In: Walker, M.A., Ji, H., Stent, A. (eds.) Proceedings of the 2018 Conference of the North American Chapter of the Association for Computational Linguistics: Human Language Technologies, pp. 569–574. Association for Computational Linguistics (2018)
15. Sordoni, A., Bengio, Y., Vahabi, H., Lioma, C., Grue Simonsen, J., Nie, J.Y.: A hierarchical recurrent encoder-decoder for generative context-aware query suggestion. In: Proceedings of the 24th ACM International on Conference on Information and Knowledge Management. Association for Computing Machinery, CIKM 2015, pp. 553–562 (2015)
16. Tan, C., Wei, F., Yang, N., Du, B., Lv, W., Zhou, M.: S-net: From answer extraction to answer synthesis for machine reading comprehension. In: AAAI (2018)

17. Tu, M., Wang, G., Huang, J., Tang, Y., He, X., Zhou, B.: Multi-hop reading comprehension across multiple documents by reasoning over heterogeneous graphs. In: Proceedings of the 57th Annual Meeting of the Association for Computational Linguistics ACL (2019)
18. Wang, Y., et al.: Multi-passage machine reading comprehension with cross-passage answer verification. CoRR (2018)
19. White, R.W., Roth, R.A.: Exploratory Search: Beyond the Query-Response Paradigm. Morgan & Claypool Publishers, [San Rafael, Calif.] (2009)
20. Wildemuth, B.M., Freund, L.: Assigning search tasks designed to elicit exploratory search behaviors. In: HCIR 2012, Proceedings of the Symposium on Human-Computer Interaction and Information Retrieval. ACM, pp. 1–10 (2012)
21. Yang, H., Guan, D., Zhang, S.: The query change model: Modeling session search as a Markov decision process. ACM Trans. Inf. Syst. **33**(4), 20:1–20:33 (2015)
22. Yang, Z., et al.: HotpotQA: a dataset for diverse, explainable multi-hop question answering. In: Proceedings of the 2018 Conference on Empirical Methods in Natural Language Processing ACL, pp. 2369–2380 (2018)
23. Zhang, T., Kishore, V., Wu, F., Weinberger, K.Q., Artzi, Y.: Bertscore: Evaluating text generation with bert. arXiv preprint (2019)
24. Zhang, X., et al.: Answer complex questions: path ranker is all you need. In: SIGIR 2021, pp. 449–458. ACM (2021)

Temporal Event Reasoning Using Multi-source Auxiliary Learning Objectives

Xin Dong[1(✉)], Tanay Kumar Saha[2], Ke Zhang[2], Joel Tetreault[2], Alejandro Jaimes[2], and Gerard de Melo[1,3]

[1] Rutgers University, New Jersey, USA
xd48@rutgers.edu, gdm@demelo.org
[2] Dataminr Inc., New York, USA
{tsaha,kzhang,jtetreault,ajaimes}@dataminr.com
[3] Hasso Plattner Institute/University of Potsdam, Potsdam, Germany

Abstract. Temporal event reasoning is vital in modern information-driven applications operating on news articles, social media, financial reports, etc. Recent works train deep neural nets to infer temporal events and relations from text. We improve upon the state-of-the-art by proposing an approach that injects additional temporal knowledge into the pretrained model from two sources: (*i*) part-of-speech tagging and (*ii*) question constraints. Auxiliary learning objectives allow us to incorporate this temporal information into the training process. Our experiments show that these types of multi-source auxiliary learning objectives lead to better temporal reasoning. Our model improves over the state-of-the-art model on the TORQUE question answering benchmark by 1.1% and on the MATRES relation extraction benchmark by 2.8% in F1 score.

Keywords: Temporal event reasoning · Auxiliary learning · Question answering

1 Introduction

Temporal event reasoning is a crucial yet under-explored aspect of interpreting text in modern information systems, enabling people to infer the timeline of narrated events. Past work has often cast this as a Relation Extraction task [2,12,13] that involves predicting temporal relationships between two events mentioned in a given piece of text, such as BEFORE or AFTER. Another recently proposed task is that of reading comprehension about temporal relations [11]. Given an input text, the system answers temporal questions pertaining to some event. Compared with the aforementioned temporal relationship prediction task, the advantage of such a Question Answering (QA) problem formulation is that questions can encode a richer, more diverse range of complex temporal relationships

Work done as a Research Intern at Dataminr, Inc.

and phenomena, such as overlap, uncertainty, negation, hypotheticals, and repetition, to name a few. For instance, we may ask a challenging question incorporating negation such as "What has not happened after investigators made good progress?"

Table 1. Excerpts from input passages with different verb POS tags.

Example	POS tag	Temporal information
People have **predicted** his demise so many times...	VBN: verb, past participle	event has happened
Security Council **passed a** resolution ...	VBD: verb, past tense	event happened

Table 2. Question Answering samples from TORQUE [11].

Passage: They were traveling in an up-armored high-mobility, multi-purpose, wheeled vehicle when this occurred. Those injured were evacuated by air to a nearby forward operating base for treatment.	
Questions	**Answers**
What events have already finished?	traveling, occurred, evacuated
What will happen in the future?	No answer.
What events happened during their travel?	occurred, evacuated
What events have begun but has not finished?	treatment
What happened after it occurred?	evacuated, treatment
What happened before the injured were treated?	traveling, occurred, evacuated

Auxiliary learning is a common means of improving the performance on a primary task of interest [6, 8, 15]. In our work, we propose two auxiliary tasks to acquire better temporal reasoning abilities: (i) part-of-speech (POS) tagging, and (ii) question constraints. POS tagging as an auxiliary task is able to ensure a better understanding of tense-related information within a sentence. For example, as shown in Table 1, the word "predicted" in "People have predicted his demise so many times ..." is labeled as VBN (past participle), while "passed" is labeled as VBD (past tense) in "Security Council passed a resolution ...". Being able to capture such distinctions enables the model to more accurately distinguish what happened from what *has* (perhaps more recently) happened.

The second auxiliary task, question constraints, can be viewed as a self-supervised task and is induced based on a temporal question answering dataset. As shown in Table 2, for a given text passage, the dataset provides a set of questions, and different questions tend to call for different answers. For example, the set of answers to "What events have already finished?" and "What will happen in the future?" should typically be disjoint. Hence, we explore the value of question constraint rules between pairs of questions for a passage. We induce such rules automatically based on their answer overlap, and subsequently enforce them by training the model with the auxiliary classification task of identifying the kind of answer overlap.

We propose a novel multi-source auxiliary learning objective that incorporates the two auxiliary tasks to improve the performance in two temporal event reasoning tasks. Our method achieves a new state-of-the-art performance on the TORQUE [11] dataset (QA setup), improving over previous work by 0.8 F1 points (absolute). Having fine-tuned the model on this QA setup so as to learn complex temporal cues, we further demonstrate the generalizability of our approach by showing that the fine-tuned encoder can then be further fine-tuned to improve the top performance on MATRES [12] (Relation Extraction setup) by 2.3 F1 points. Finally, we show that our approach is particularly performant in a low-resource setting, yielding absolute improvements of up to 19.5%.

2   Related Work

Temporal Question Answering. Great strides have been made with new architectures and new self-supervised objectives to improve over vanilla BERT [3]. However, while models such as RoBERTa [10] and AlBERT [7] enable a better understanding of predicates and arguments for conventional QA tasks, our experiments show that they fail to yield substantial gains on temporal QA. Recently, Han et al. [5] presented a temporal-related language model with new self-supervised objectives for improved Temporal QA. In contrast to our approach, this method requires pre-defined event and temporal lexicons.

Temporal Relation Extraction. Compared with temporal QA, temporal relation (TempRel) extraction is widely studied in temporal event reasoning. Many TempRel datasets have been collected, such as TB-Dense [2], RED [13], and MATRES [12], and a variety of models target this task. For instance, Han et al. (2019) [4] present a joint event and temporal relation extraction model. Wang et al. (2020) [16] enforce logical constraints within and across temporal relations via differentiable learning objectives. Zhou et al. (2020) [18] incorporate probabilistic soft logic regularization and global inference.

Auxiliary Learning. There is a long history of research on multi-task learning [14], e.g., the Multi-Task Deep Neural Network (MT-DNN) seeks to learn representations across diverse natural language understanding tasks [9]. In auxiliary learning, there is a single primary task, and the role of the auxiliary tasks is to improve the performance and generalizability of this primary task. Trinh et al. (2018) [15] propose a method for better capturing long term dependencies in RNNs with an extra unsupervised auxiliary loss. Xu et al. (2021) [17] propose multi-task recurrent modular networks for any multi-task recurrent models.

3   Method

Following standard practice when training a deep network on multiple tasks [9], our model consists of a shared encoder and several task-specific classifiers on top of it. There is one such classifier for the primary task as well as two further ones for our proposed auxiliary tasks. This architecture allows the shared encoder to jointly learn from each of the tasks.

Shared Encoder. The encoder is from a pre-trained contextual representation model, denoted as $f_{se}(\cdot; \theta_{se})$. Given an input text sequence $\mathbf{s}$ consisting of T tokens $[\mathbf{x}_1, \mathbf{x}_2, ..., \mathbf{x}_T]$, this encoder infers a contextual hidden representation $\mathbf{h_t} \in \mathbb{R}^d$ d of dimensionality d for each input token $\mathbf{x}_t$.

Primary Task. Our primary task-specific classification module $f_p(\cdot; \theta_p)$ is responsible for the question answering task. It is applied for fine-tuning on top of the pre-trained model $f_{se}(\cdot; \theta_{se})$ and consists of a fully-connected layer with softmax activation to map $\mathbf{h_t} \in \mathbb{R}^d$ into $\mathbb{R}^{|\mathcal{Y}_p|}$. Here, $\mathcal{Y}_p$ is defined as a set of binary output class labels denoting whether a given token is deemed a valid answer in response to the question.

Auxiliary Tasks. The model is additionally trained on two auxiliary tasks.

1. *POS tagging.* Our auxiliary POS tagging classification module $f_{pos}(\cdot; \theta_{pos})$ draws its input from the shared encoder $f_{se}(\cdot; \theta_{se})$. It then applies a linear mapping $\mathbf{h_t} \in \mathbb{R}^d$ into $\mathbb{R}^{|\mathcal{Y}_{pos}|}$ followed by a softmax activation to predict a distribution over the set of POS tag classes $\mathcal{Y}_{pos}$.

2. *Question Constraint Classification (Question CC).* For a given passage p from our primary QA task, we have a corresponding question set $Q = \{(q_i, a_i) \mid i \in \{1, ..., n\}, a_i \neq \emptyset\}$, where n is the number of questions and a_i is the answer set for question q_i. From this, we can obtain a set of question pairs $\mathcal{C} = \{\langle q_i, q_j \rangle \mid i < j; i, j \in \{1, ..., n\}\}$ and a set of answer pairs $\mathcal{A} = \{\langle a_i, a_j \rangle \mid i < j; i, j \in \{1, ..., n\}\}$. We consider the overlap of answers between two questions to acquire a constraint label for the question pair. In particular, the constraint label is chosen from a set of five relations $\mathcal{Y}_{qc} = \{$ EQUAL, SUBSET, SUPERSET, DISJOINT, OVERLAP$\}$, based on the corresponding conditions $(a_i = a_j)$, $(a_i \subset a_j)$, $(a_i \supset a_j)$, $(a_i \cap a_j = \emptyset)$, and $(a_i \cap a_j \neq \emptyset; a_i \cap a_j \neq a_i; a_i \cap a_j \neq a_j)$. To predict such labels, our model incorporates a question classification module $f_{qc}(\cdot; \theta_{qc})$ consisting of a fully-connected layer mapping $\mathbf{h_0} \in \mathbb{R}^d$ into $\mathbb{R}^{|\mathcal{Y}_{qc}|}$ with softmax activation.

Auxiliary Learning Objectives. To inject the temporal knowledge into the primary QA training, we jointly learn the primary task along with the two auxiliary tasks. Hence, the overall loss function becomes

$$\mathcal{L} = \mathcal{L}_p + \lambda_1 \mathcal{L}_{pos} + \lambda_2 \mathcal{L}_{qc}, \tag{1}$$

where $\mathcal{L}_p, \mathcal{L}_{pos}, \mathcal{L}_{qc}$ are the QA loss, POS tagging loss, and question constraint classification loss, respectively, and λ_1, λ_2 are coefficients to control the influence of each auxiliary task loss term.

4   Experiments

4.1   Experimental Setup

Tasks and Datasets. For evaluation, we use TORQUE [11], a reading comprehension dataset of temporal ordering questions and answers. It provides 3.2k passages (~50 tokens/passage), 24.9k events (7.9 events/passage), and 21.2k user-provided questions. For end-to-end training, the task is modeled as a binary classification problem that requires predicting for each token in the passage whether it is an answer. We also investigate pretraining on TORQUE to then improve on MATRES [12], a temporal relation (TempRel) extraction benchmark, consisting of 275 documents with entity relationships labeled as BEFORE, AFTER, EQUAL, or VAGUE. Regarding metrics, TORQUE is evaluated in terms of F1 score, Exact Match (EM), and Consistency (C). The latter is defined as the percentage of contrast groups for which a model's predictions have F1 $\leq$ 80% for all questions in a group. The contrast groups provided by TORQUE consist of questions with contrasting changes to the temporal keywords, e.g., "What happened *after* the snow started?" versus "What happened *before* the snow started?". For MATRES, we report standard micro-averaged F1 scores.

Table 3. Hyper-parameter settings.

Parameter	TORQUE	MATRES
Max. sequence length	180	220
Batch size	12	10
Learning rate	1×10^{-5}	5×10^{-6}
# of training epochs	10	5
λ_1	0.001	–
λ_2	0.001	–

Model Details. For POS tagging as the auxiliary task, we invoke NLTK [1] to obtain POS tags on the TORQUE training set. The size of the POS tag inventory is 36. For question constraint classification, the number of question pairs extracted from the training set for the five labels defined in Sect. 3 are 4,307, 11,610, 6,181, 42,928, and 7,146, respectively. We adopt RoBERTa-Large [10] as the pre-trained encoder. To further evaluate the effectiveness of auxiliary learning, we use models fine-tuned on TORQUE first to evaluate on MATRES. We tune the hyper-parameters based on the respective development sets and list their values in Table 3. On TORQUE, as for the original baseline, we report average results over 3 random seeds, while on MATRES, we consider averages over 5 runs.

4.2   Results and Analysis

Table 4. Results from TORQUE experiments.

Method	F1	EM	C
RoBERTa-Large [11]	75.2	51.1	34.5
RoBERTa-Large			
+ Question CC	75.7	**51.3**	<u>36.2</u>
+ POS Tagging	<u>75.8</u>	50.7	35.6
+ POS Tagging + Question CC	**76.0**	<u>51.2</u>	**36.7**

TORQUE (Question Answering Setup). The current SOTA method on TORQUE is RoBERTa-Large [11]. Table 4 compares our approach against this baseline to evaluate the effectiveness of auxiliary learning. We first evaluate on RoBERTa-Large with either POS tagging or Question CC as the auxiliary task. Compared with RoBERTa-Large, we observe that adding Question CC improves the Consistency score, while POS tagging in particular improves the F1 score. This shows that our answer constraints lead to a better understanding of the differences between questions, while the POS tagging auxiliary task enables the model to better capture subtle differences. Our full method outperforms RoBERTa-Large across all three metrics, demonstrating that our multi-source auxiliary learning objective is effective for our primary QA task.

Table 5. Results on TORQUE with different ratios of training data.

Ratio	30%			50%			100%		
Method	F1	EM	C	F1	**EM**	C	**F1**	**EM**	**C**
RoBERTa-Large	57.3	37.9	20.1	73.3	46.3	32.0	75.2	51.1	34.5
Our approach	68.5	39.4	25.1	74.3	48.5	34.5	76.0	51.2	36.7
Improvement (%)	19.5%	4.0%	24.8%	1.4%	4.8%	7.8%	1.1%	0.2%	6.4%

Influence of Amount of Training Data for TORQUE. To assess the effectiveness of our method with limited amounts of training data on TORQUE, we compare our full multi-source auxiliary learning approach with RoBERTa-Large using different ratios of training data. As shown in Table 5, our method yields significant improvements over RoBERTa-Large in terms of F1 and C scores, especially with 30% of training data, which suggests that our auxiliary tasks are particularly fruitful when training data is scarce, although this also means that less supervision is available for POS tagging and question constraint induction Table 6.

Table 6. Results on MATRES dataset.

Method	F1
Want et al. [16]	78.8
RoBERTa-Large	80.1
+ TORQUE	80.6
+ TORQUE (Question CC)	80.4
+ TORQUE (POS Tagging)	80.7
+ TORQUE (POS Tagging + Question CC)	**81.1**

MATRES (Relation Extraction Setup). As TORQUE provides more complex temporal information, we assess to what extent we can transfer the knowledge learned on it to the MATRES relation extraction task, so as to evaluate the generalizability of our auxiliary learning. As baselines, in addition to RoBERTa-Large, we consider Wang et al. [16], which incorporates temporal logic constraints among events into the training loss function. Our model is fine-tuned on TORQUE first and then further fine-tuned on MATRES. This outperforms the baselines, showing that MATRES can benefit from the auxiliary information provided by training on TORQUE first. In this regard, compared to versions with just one additional auxiliary task, our full auxiliary learning model proves the most effective at acquiring an understanding of temporal relationships.

5   Conclusion

We propose a method to inject additional temporal information with multi-source auxiliary learning objectives into pre-trained models for temporal event reasoning. In particular, we consider part-of-speech prediction and question answer constraint classification as additional objectives, and investigate how pretraining on question answering can benefit temporal relation extraction. Our experiments show that we achieve state-of-the-art results on TORQUE as well as on MATRES, and that our auxiliary learning method is particularly useful in low-resource settings.

References

1. Bird, S., Klein, E., Loper, E.: Natural Language Processing with Python. O'Reilly Media (2009)
2. Chambers, N., Cassidy, T., McDowell, B., Bethard, S.: Dense event ordering with a multi-pass architecture. Trans. Assoc. Comput. Linguist. **2**, 273–284 (2014)

3. Devlin, J., Chang, M.W., Lee, K., Toutanova, K.: BERT: pre-training of deep bidirectional transformers for language understanding. In: Proceedings of the 2019 Conference of the North American Chapter of the Association for Computational Linguistics: Human Language Technologies, Volume 1 (Long and Short Papers), pp. 4171–4186. Association for Computational Linguistics, Minneapolis, Minnesota, June 2019. https://doi.org/10.18653/v1/N19-1423, https://www.aclweb.org/anthology/N19-1423

4. Han, R., Ning, Q., Peng, N.: Joint event and temporal relation extraction with shared representations and structured prediction. In: Proceedings of the 2019 Conference on Empirical Methods in Natural Language Processing and the 9th International Joint Conference on Natural Language Processing (EMNLP-IJCNLP), pp. 434–444. Association for Computational Linguistics, Hong Kong, China (2019). https://doi.org/10.18653/v1/D19-1041, https://www.aclweb.org/anthology/D19-1041

5. Han, R., Ren, X., Peng, N.: Deer: A data efficient language model for event temporal reasoning. arXiv preprint arXiv:2012.15283 (2020)

6. Jaderberg, M., et al.: Reinforcement learning with unsupervised auxiliary tasks. In: 5th International Conference on Learning Representations, ICLR 2017, Toulon, France, 24–26 April 2017, Conference Track Proceedings. OpenReview.net (2017). https://openreview.net/forum?id=SJ6yPD5xg

7. Lan, Z., Chen, M., Goodman, S., Gimpel, K., Sharma, P., Soricut, R.: Albert: a lite bert for self-supervised learning of language representations. In: International Conference on Learning Representations (2020). https://openreview.net/forum?id=H1eA7AEtvS

8. Liu, S., Davison, A., Johns, E.: Self-supervised generalisation with meta auxiliary learning. In: Wallach, H., Larochelle, H., Beygelzimer, A., d'Alché-Buc, F., Fox, E., Garnett, R. (eds.) Advances in Neural Information Processing Systems, vol. 32. Curran Associates, Inc. (2019). https://proceedings.neurips.cc/paper/2019/file/92262bf907af914b95a0fc33c3f33bf6-Paper.pdf

9. Liu, X., He, P., Chen, W., Gao, J.: Multi-task deep neural networks for natural language understanding. In: Proceedings of the 57th Annual Meeting of the Association for Computational Linguistics, pp. 4487–4496. Association for Computational Linguistics, Florence, Italy, July 2019. https://doi.org/10.18653/v1/P19-1441, https://www.aclweb.org/anthology/P19-1441

10. Liu, Y., et al.: Roberta: A robustly optimized bert pretraining approach. arXiv preprint arXiv:1907.11692 (2019)

11. Ning, Q., Wu, H., Han, R., Peng, N., Gardner, M., Roth, D.: TORQUE: a reading comprehension dataset of temporal ordering questions. In: Proceedings of the 2020 Conference on Empirical Methods in Natural Language Processing (EMNLP), pp. 1158–1172. Association for Computational Linguistics, Online, November 2020. https://doi.org/10.18653/v1/2020.emnlp-main.88, https://www.aclweb.org/anthology/2020.emnlp-main.88

12. Ning, Q., Wu, H., Roth, D.: A multi-axis annotation scheme for event temporal relations. In: Proceedings of the 56th Annual Meeting of the Association for Computational Linguistics (Volume 1: Long Papers), pp. 1318–1328. Association for Computational Linguistics, Melbourne, Australia, July 2018. https://doi.org/10.18653/v1/P18-1122, https://www.aclweb.org/anthology/P18-1122

13. O'Gorman, T., Wright-Bettner, K., Palmer, M.: Richer event description: integrating event coreference with temporal, causal and bridging annotation. In: Proceedings of the 2nd Workshop on Computing News Storylines (CNS 2016), pp. 47–56 (2016)
14. Ruder, S.: An overview of multi-task learning in deep neural networks. arXiv preprint arXiv:1706.05098 (2017)
15. Trinh, T., Dai, A., Luong, T., Le, Q.: Learning longer-term dependencies in RNNS with auxiliary losses. In: International Conference on Machine Learning, pp. 4965–4974. PMLR (2018)
16. Wang, H., Chen, M., Zhang, H., Roth, D.: Joint constrained learning for event-event relation extraction. In: Proceedings of the 2020 Conference on Empirical Methods in Natural Language Processing (EMNLP), pp. 696–706. Association for Computational Linguistics, Online, November 2020. https://doi.org/10.18653/v1/2020.emnlp-main.51, https://www.aclweb.org/anthology/2020.emnlp-main.51
17. Xu, D., et al.: Multi-task recurrent modular networks. In: AAAI, vol. 35, no. 12, pp. 10496–10504 (2021)
18. Zhou, Y., et al.: Clinical temporal relation extraction with probabilistic soft logic regularization and global inference. arXiv e-prints pp. arXiv-2012 (2020)

Enhanced Sentence Meta-Embeddings
for Textual Understanding

Sourav Dutta[(⊠)] and Haytham Assem

Huawei Research Centre, Dublin, Ireland
{sourav.dutta2,haytham.assem}@huawei.com

Abstract. Sentence embeddings provide vector representations for sentences and short texts, enabling the capture of contextual and semantic meaning for different applications. However, the diversity of sentence embedding techniques poses a challenge, in terms of choosing the model best suited for the downstream task. As such, *meta-embeddings* study different techniques for combining embeddings from multiple sources. In this paper, we propose *CINCE*, a *principled meta-embedding framework* for aggregating various semantic information, captured by different embeddings techniques, via multiple *component analysis* strategies. Experiments on *SentEval* benchmark exhibit improved performance for semantic understanding and text classification, compared to existing approaches.

Keywords: Sentence meta-embedding · Independent component analysis · Canonical correlation analysis · Semantic understanding

1 Introduction and Background

Distributed *word embeddings* like word2vec [32], GloVe [36], and fastText [5] have shown to efficiently capture the generic semantic meaning of words as well as the relationships among them. Pre-trained language models like BERT [14] and XLM [27] provide "dynamic" word embeddings modelling the different meanings or senses of words depending on the context of use. On the other hand, *sentence embeddings* provide dense vector representations capturing the overall contextual and semantic meanings of sentences and short texts. With the success of sentence embeddings in several downstream natural language understanding tasks like semantic content similarity, sentiment analysis, question answering, and text classification [2,15,38], several sentence embedding techniques from word embeddings were proposed – by intelligent combination of word embeddings [30], using [CLS] token or pooling methods on language models [14,29], and other strategies based on advanced learning architectures. The use of specialized learning networks, like LSTM networks, encoder-decoder, and Siamese networks, were proposed for state-of-the-art sentence embeddings frameworks such as InferSent [12], LASER [2], USE [6], EMU [18], SBERT [38], and DuEAM [16] to name a few.

Motivation. Such embeddings form a core component in several modern day natural language understanding (NLU) and information retrieval (IR) applications. For example, text classification (e.g., spam detection or user comment sentiment analysis), FAQ

© The Author(s), under exclusive license to Springer Nature Switzerland AG 2022
M. Hagen et al. (Eds.): ECIR 2022, LNCS 13186, pp. 111–119, 2022.
https://doi.org/10.1007/978-3-030-99739-7_13

retrieval [3], and parallel sentence identification have necessitated richer semantic representation of texts. The diversity of different sentence embedding techniques captures different aspects of semantics, and demonstrates varying degrees of success on various classes of tasks. This poses a unique challenge of choosing the appropriate model, the performance of which depends on the nature of the downstream task and dataset characteristics. Thus, along with new sentence embedding algorithms, methods to better combine sentence embeddings from multiple pre-trained architectures – known as *"meta-embedding"* – have become an interesting area of research.

State-of-the-Art. Word-level meta-embeddings had been studied in the context of simple operations like concatenation and averaging [9,46], and have been to shown to perform quite well in several word analogy tasks. Further, the use of transformation functions like Singular Value Decomposition (SVD) [46], Auto-encoding [4], and adversarial learning [28] have been proposed for obtained effective word meta-embeddings. Learning word meta-embedding in supervised setting [33] and by decoupling information from different representations [8] have been shown to provide improvements in several NLP tasks. However, the simple techniques of concatenation and averaging have, in general, demonstrated the best performance and robustness across various tasks [9].

Sentence meta-embeddings have been constructed from such word-level meta-embeddings via self-attention [24] or ensemble methods [17]. However, similar to word meta-embedding, simple concatenation and averaging of sentence embeddings from different pre-trained models (like SBERT, USE, etc.) have shown to be surprisingly good [34,37]. Correlation based aggregation of multiple sentence embeddings have reported improved performance in capturing semantic similarity [37], compared to concatenation, SVD and Auto-encoding.

Contribution. This paper proposes *Canonicalized Independent Component based Embeddings* (CINCE), a principled and effective novel sentence meta-embedding framework. We amalgamate different aspects of semantic information captured by diverse embedding techniques, by combining different component analysis techniques – enabling the learning of complementary linguistic and semantic cues. Specifically, *CINCE* leverages a pipeline of principal component analysis (PCA), independent component analysis (ICA), and canonical correlation analysis (CCA) for creating improved sentence meta-embeddings. Empirical results on *SentEval* benchmark tasks demonstrate the effectiveness of our proposed framework for semantic similarity and classification tasks.

1.1 Preliminary Concepts

Principal Component Analysis (PCA) computes an orthogonal linear transformation to obtain the *principal components* of a data collection, i.e., identifies the coordinate axes that optimally describes the variance of the data points [35]. PCA uses singular value decomposition (SVD) to identify important dimensions and in dimensionality reduction [19]. Kernel PCA (KPCA) [39] proposes non-linear transformations to be performed in high-dimensional space for improved performance and de-noising [31].

`Independent Component Analysis` (ICA) is used for *blind-source separation* (BSS), where a multi-variate signal is decomposed into its additive sub-components, assumed to be non-Gaussian and statistically independent. Mathematically, given an observation vector x to be a mixture of several underlying signals, s, ICA computes the *mixing matrix*, A, such that $x = As$ [10,22]. Several algorithms have been designed to solve the BSS problem [21], and non-linear variants have also been proposed [41].

`Canonical Correlation Analysis` (CCA) explores the relationships between two multivariate sets of variables to compute linear combinations of variables having the maximum correlation [20,25]. In general, CCA provides an efficient way to gain insight into which dimensions are correlated and the amount of shared variance – enabling a unified view of the observations. Generalized CCA (GCCA) involves extension to multiple sets of observations [23], and finds numerous applications in machine learning [42]. Neural architecture based CCA approaches have also been studied in the literature [1].

2   *CINCE* Framework

Text representations have been shown to capture various surface-level (like sentence length), syntactic (like hierarchical structure), and semantic (like tense) information [13], albeit in varying degrees. However, interpretability of the exact linguistic information encoded by such high-dimensional encoding is an active area of study [40]. As such, existing solutions like concatenation and averaging for meta-embeddings, although effective, seems ad-hoc without proper intuition (in our opinion).

The proposed *CINCE* framework assumes sentence embedding techniques to capture different linguistic information or signals, and the corresponding representation as a combination of the semantic cues thus captured. In this setting, the working of our meta-embedding framework aims to extract and combine the diverse linguistic and semantic information embedded in the different text representations, enabling better understanding of natural language texts.

Consider $S = \{s_1, s_2, \cdots, s_n\}$ to be a collection of input sentences, and $\mathcal{E} = \{e_1, e_2, \cdots, e_m\}$ to be different sentence embeddings techniques. *Sentence meta-embedding* ($\mathcal{M}$) then learns an embedding function, $\mathcal{F}$, for sentence representation by combining embeddings from the different techniques, that is, $\mathcal{M}(S, \mathcal{E}) = \{\mathcal{F}_{e_1, e_2, \cdots, e_m}(s_i)\}$. The working of the proposed *CINCE* framework hinges on 3 sequential modules, using *component analysis* approaches as follows.

I. Principal Component Analysis (PCA): Sentence embeddings obtained from the different techniques, $\mathcal{E}_i(S) = \{e_i(s_1), e_i(s_2), \cdots, e_i(s_n)\}$, are independently provided as input to this module. To capture the most important encoding dimensions within each embedding strategy, we apply PCA and transform the input data onto the orthogonal component axes. Observe, the axes (formed by linear combinations of the input dimensions) intuitively represent the linguistic characteristics (and their combinations) that are important for semantic understanding of the text. Thus, projecting the sentence embeddings on the PCA components, enables us to capture deeper information from the input text, as well as reducing noise. Specifically, we apply Kernel PCA (KPCA) with a cosine kernel and the number of dimensions is chosen such that 99% of the initial cumulative energy is preserved (to reduce information loss). Intuitively, this might capture the dominant aspects from the embedding techniques, and the cosine kernel is chosen since text similarity is pre-dominantly computed using cosine similarity measure.

II. Independent Component Analysis (ICA): Assuming the transformed sentence representations (obtained above) to inherently be a mixture of the different linguistic signals, we next apply ICA to extract the underlying independent source signals (capturing different semantic aspects). Specifically, to approximate the optimal underlying signals, we apply *FastICA* [21] with logcosh (shown to be fast and efficient in modelling the optimization function) on the PCA-transformed embeddings from each of the input embedding techniques.

III. Canonical Correlation Analysis (CCA): Finally, the different linguistic signals obtained above (from the embedding techniques) are amalgamated into a single vector representation by quantifying the association between them. To this end, we use Generalized CCA (GCCA) (similar to [37]) to perform orthogonal linear combinations that best explain the encoding variability both within and across the embedding techniques. This provides an holistic view and enables the meta-embedding to capture the prominent linguistic aspects and semantic signals across the embeddings, as discussed in [26].

Overall, the operations of the *CINCE* framework for sentence meta-embedding function, $\mathcal{F}$, can be summarized as the following transformations: $\mathcal{F} : \mathcal{E}' = \{KPCA(\mathcal{E}_i)\} \rightarrow \mathcal{E}'' = \{ICA(\mathcal{E}'_i)\} \rightarrow \mathcal{E}''' = GCCA(\{\mathcal{E}''\})$. Hence, $\mathcal{M}(\mathcal{S}, \mathcal{E}) = \mathcal{E}''' = \{\mathcal{F}(\mathcal{E}_i)\}$ provides an efficient and principled sentence meta-embeddings for *CINCE* capturing diverse linguistic aspects and semantic information from different embeddings sources. In fact, combinations of such techniques have been shown to work well in extracting underlying signals in images [43].

3   Experimental Setup

In this section, we empirically analyze the performance of the *CINCE* meta-embedding framework against competing methodologies.

Embeddings. We consider 3 state-of-the-art sentence embedding techniques as:

(i) *SBERT* [38] – provides text representations using Siamese network based teacher-student to learn from a fine-tuned transformer based Siamese architecture. We use the *distiluse-base-cased* instance obtained from www.sbert.net;

(ii) *LASER* [2] – computes sentence embeddings based on a trained Bi-LSTM encoder network, obtained from github.com/facebookresearch/LASER; and

(iii) *USE* [6] – a transformer based architecture trained on skip-thought and NLI tasks for sentence embedding generation. We use the package made available via SpaCy library, from http://spacy.io/universe/project/spacy-universal-sentence-encoder.

We considered specialized architectures for learning effective sentence representations as baselines, while ParaNMT [44] used in [37] involves averaging on word-level

Table 1. Mean Pearson's and Spearman's Correlation (ρ) scores achieved by the approaches on the STS tasks of SentEval benchmark. (Best results are marked in **bold**, second-best results are underlined, and $*$ indicates statistically significant results using paired bootstrap resampling).

Benchmark /	dim	STS-12	STS-13	STS-14	STS-15	STS-16	STS-B
Approaches		*(mean Pearson's $r \times 100$ / Spearman's $\rho \times 100$)*					
SBERT	512	68.84/66.98	68.97/70.00	75.77/73.27	81.30/82.05	78.65/80.02	81.83/81.91
USE	512	69.58/68.15	69.16/70.35	76.56/73.51	80.69/81.58	78.81/79.73	81.58/81.52
LASER	1024	62.90/62.30	48.69/51.61	67.83/67.04	75.03/75.38	71.78/72.32	78.15/78.11
Average	1024	70.99/69.06	68.74/69.43	77.27/74.78	81.97/82.60	79.77/80.78	82.80/<u>82.66</u>
Concatenation	2048	<u>71.29</u>/**69.38**	<u>68.85</u>/69.76	<u>77.65</u>/75.10	<u>82.56</u>/**83.17**	<u>80.08</u>/**81.10**	<u>83.63</u>/**83.77***
Auto-Encoder	1024	67.65/67.22	61.19/63.99	74.18/72.27	80.63/81.56	78.27/79.62	80.88/81.22
GCCA	1024	68.41/68.09	65.26/66.64	77.32/<u>75.93</u>	79.01/80.00	76.63/80.22	79.52/79.61
CINCE	1024	**71.56**/<u>69.23</u>	**75.35***/**75.48***	**79.26***/**76.50***	**82.66**/<u>82.71</u>	**80.12**/<u>80.92</u>	**83.91**/82.06
GCCA+NMT	1024	72.80/71.60	69.60/69.40	81.70/79.50	84.20/85.50	81.30/83.30	83.90/84.40
CINCE+NMT	1024	74.11/72.92	77.22/76.27	82.96/81.03	86.00/86.98	84.54/83.28	86.34/86.10

embeddings (hence not considered). However, for completeness, we also provide comparison with [37] using ParaNMT (instead of LASER).

Baselines. We evaluate *CINCE* against the following 5 meta-embedding strategies: (1) *Independent*: different sentence embedding methods are considered independently; (2) *Average*: embeddings obtained from the different techniques are averaged (with 0 padding for dimension matching); (3) *Concatenation*: sentence embeddings obtained from the 3 methods are concatenated; (4) *Auto-Encoder*: meta-embeddings obtained by training auto-encoder to minimize reconstruction loss [4]. We use the same setup as that of [37]; and (5) *GCCA*: canonical correlation analysis based meta-embeddings of [37].

Note that, although higher dimensional embeddings (for concatenation) might have an advantage over lower dimensional ones [45], no dimensional reduction (that might potentially affect the quality of embeddings) have been performed for the concatenation baseline, and we report the best results obtained.

Dataset. To adjudge the quality of meta-embeddings obtained, we perform experiments on unsupervised as well as supervised tasks of the *SentEval* benchmark [11] (from github.com/facebookresearch/SentEval). Specifically, we evaluate on semantic similarity correlation score (for unsupervised STS tasks) and 10-fold classification accuracy (on review, question, opinion and sentiment classification).

Setup. For the unsupervised STS tasks, our framework is trained on sentences from the Billion Word Corpus (using only the news.en-00001-of-00100 file) [7], similar to the setup of [37]. For the supervised tasks, we used the training dataset (without the classification labels) to learn meta-embeddings on the 3 embeddings methods mentioned above. Thus, our meta-embedding learning setup is an unsupervised approach in principle. For classification, we use a dense network with 100 hidden layers and 0.2 dropout, trained for 10 epochs with 64 batch size, Adam optimizer and ReLu activation. The embedding dimension for *CINCE* was fixed at 1024, while other parameters (for component analysis) were kept at default value, so as not to overly study the effect of

Table 2. SentEval classification task accuracy. (Best results in **bold** and second-best underlined).

Tasks/Methods	CR	MR	MPQA	TREC	SUBJ	SST-5
SBERT	82.22	74.25	86.97	91.60	91.67	44.75
USE	79.66	71.50	86.44	92.40	90.91	42.53
LASER	82.01	74.50	87.95	92.40	91.94	45.52
Average	81.11	74.82	88.40	93.40	92.47	44.48
Concatenation	83.21	**76.09**	**88.60**	**95.00**	92.93	46.79
Auto-Encoder	80.76	72.96	87.86	91.40	91.59	44.43
GCCA	78.15	74.68	87.86	94.60	92.65	44.62
CINCE	**83.28**	75.62	**88.60**	**95.00**	**93.04**	**46.92**

parameter tuning. Observe, for learning the meta-embeddings we would need all the input sentences apriori, i.e., a static input set. The use of *online* variant of PCA, ICA, and CCA for handling dynamic input provides an interesting direction of future work.

3.1   Empirical Results

This section presents the obtained results on the SentEval benchmark tasks by the different meta-embedding approaches, and are shown in Table 1 and Table 2.

STS Tasks: The unsupervised Semantic Textual Similarity (STS) task involves computing the degree of semantic equivalence (on a similarity scale of 0 to 5) between paired snippets of text. From Table 1, we observe that *CINCE* achieves the *best Pearson's correlation score* across **all** the STS datasets, while for Spearman's score it is either the best or second best with comparable results. In fact, the performance improvements on the STS-13 and STS-14 datasets is *statistically significant.* We also observe that the *concatenation* strategy is a simple yet effective approach (shown in literature [9]), however, concatenating 10–15 different embedding techniques (with 10K dimension size) is technically infeasible, while our framework would be able to handle such scenarios.

Note that, higher dimensional embeddings (as in this case of concatenation) might have an advantage over low-dimensional ones [45]. Hence, if compared with other strategies that produce the same meta-embedding dimension size (like GCCA or average), *CINCE* is seen to produce a significant performance improvement of around **4%** Pearson's score, across all the benchmarking tasks. For completeness, *CINCE* is also seen to perform better than GCCA technique of [37], with the same setup as of the authors (i.e., using USE, SBERT, and ParaNMT embeddings) (bottom of Table 1).

Classification Tasks: On the supervised classification tasks (e.g., review classification, subjectivity and sentiment classification) of SentEval, we similarly observe *CINCE* to perform slightly better (in 5 out of 6 datasets) than the other existing approaches, as shown in Table 2. On average, we report comparable results to the concatenation based baseline, albeit with lower dimensional embeddings that might enhance real-time performance in enterprise production settings.

Overall, we observe that our principled approach of capturing various semantic information from the different embeddings techniques, generates better sentence meta-embeddings for improved semantic similarity understanding and text classification tasks. We see that the proposed *CINCE* framework obtains higher average Pearson's as well as Spearman's correlation (ρ) score on STS benchmark and better accuracy on classification tasks, when compared to the existing techniques. Note, meta-embedding from *CINCE* (KPCA + ICA + GCCA) is agnostic to the domain, as long as different sentence embeddings are available.

4   Conclusion

This paper proposed *CINCE*, a novel and principled framework for extracting sentence meta-embeddings. We show how our technique combines diverse semantic information, from different sentence embeddings, captured via a sequence of component analysis techniques. Experiments on supervised and unsupervised benchmark tasks showcased the efficacy of our framework in better capturing textual semantic and contextual similarity.

References

1. Andrew, G., Arora, R., Bilmes, J., Livescu, K.: Deep canonical correlation analysis. In: International Conference on Machine Learning (ICML), pp. 1247–1255 (2013)
2. Artetxe, M., Schwenk, H.: Massively multilingual sentence embeddings for zero-shot cross-lingual transfer and beyond. Trans. Assoc. Comput. Linguist. **7**, 597–610 (2019)
3. Assem, H., Dutta, S., Burgin, E.: DTAFA: decoupled training architecture for efficient FAQ retrieval. In: Annual Meeting of the Special Interest Group on Discourse and Dialogue (SIGDIAL), pp. 423–430 (2021)
4. Bao, C., Bollegala, D.: Learning word meta-embeddings by autoencoding. In: Conference on Computational Linguistics (COLING), pp. 1650–1661 (2018)
5. Bojanowski, P., Grave, E., Joulin, A., Mikolov, T.: Enriching word vectors with subword information. Trans. Assoc. Comput. Linguist. **5**, 135–146 (2017)
6. Cer, D., et al.: Universal Sentence Encoder, arXiv preprint arXiv:1803.11175 (2018)
7. Chelba, C., et al.: One billion word benchmark for measuring progress in statistical language modeling. In: Conference of the International Speech Communication Association (INTERSPEECH), pp. 2635–2639 (2014)
8. Chen, W., Sheng, M., Mao, J., Sheng, W.: Investigating word meta-embeddings by disentangling common and individual information. IEEE Access **8**, 11692–11699 (2020)
9. Coates, J.N., Bollegala, D.: Frustratingly easy meta-embedding - computing meta-embeddings by averaging source word embeddings. In: North American Chapter of the Association for Computational Linguistics (NAACL-HLT), pp. 194–198 (2018)
10. Comon, P.: Independent component analysis - a new concept? Signal Process. **36**, 287–314 (1994)
11. Conneau, A., Kiela, D.: SentEval: an evaluation toolkit for universal sentence representations. In: International Conference on Language Resources and Evaluation (LREC), pp. 1699–1704 (2018)
12. Conneau, A., Kiela, D., Schwenk, H., Barrault, L., Borders, A.: Supervised learning of universal sentence representations from natural language inference data. In: Empirical Methods in Natural Language Processing (EMNLP), pp. 670–680 (2017)

13. Conneau, A., Kruszewski, G., Lample, G., Barrault, L., Baroni, M.: What you can cram into a single $&!#* vector: probing sentence embeddings for linguistic properties. In: Annual Meeting of the Association for Computational Linguistics (ACL), pp. 2126–2136 (2018)

14. Devlin, J., Chang, M., Lee, K., Toutanova, K.: BERT: pre-training of deep bidirectional transformers for language understanding. In: North American Chapter of the Association for Computational Linguistics (NAACL-HLT), pp. 4171–4186 (2019)

15. Dutta, S.: "Alignment is All You Need": analyzing cross-lingual text similarity for domain-specific applications. In: The Web Conference - Workshop on Cross-lingual Event-centric Open Analytics, pp. 85–92 (2021)

16. Goswami, K., Dutta, S., Assem, H., Fransen, T., McCrae, J.: Cross-lingual sentence embedding using multi-task learning. In: Empirical Methods in Natural Language Processing (EMNLP) (2021)

17. Hazem, A., Hernandez, N.: Meta-embedding sentence representation for textual similarity. In: Recent Advances in Natural Language Processing (RANLP), pp. 465–473 (2019)

18. Hirota, W., Suhara, Y., Golshan, B., Tan, W.C.: EMU: enhancing multilingual sentence embeddings with semantic specialization. In: AAAI Conference on Artificial Intelligence (AAAI), pp. 7935–7943 (2020)

19. Hotelling, H.: Analysis of a complex of statistical variables into principal components. J. Educ. Psychol. **24**(417–441), 498–520 (1933)

20. Hotelling, H.: Relations between two sets of variates. Biometrika **28**(3–4), 321–377 (1936)

21. Hyvärinen, A., Oja, E.: Independent component analysis: algorithms and applications. Neural Netw. **13**(4–5), 411–430 (2000)

22. Jutten, C., Hérault, J.: Blind separation of sources, Part I: an adaptive algorithm based on neuromimetic architecture. Signal Process. **24**, 1–10 (1991)

23. Kettenring, J.R.: Canonical analysis of several sets of variables. Biometrika **58**, 433–451 (1971)

24. Kiela, D., Wang, C., Cho, K.: Dynamic meta-embeddings for improved sentence representations. In: Empirical Methods in Natural Language Processing (EMNLP), pp. 1466–1477 (2018)

25. Knapp, T.R.: Canonical correlation analysis: a general parametric significance-testing system. Psychol. Bull. **85**(2), 410–416 (1978)

26. Kornblith, S., Norouzi, M., Lee, H., Hinton, G.: Similarity of neural network representations revisited. In: International Conference on Machine Learning (ICML), pp. 3519–3529 (2019)

27. Lample, G., Conneau, A.: Cross-lingual language model pretraining. In: Conference on Neural Information Processing Systems (NIPS), pp. 1–11 (2019)

28. Lange, L., Adel, H., Strötgen, J., Klakow, D.: Adversarial learning of feature-based meta-embeddings (2020). arXiv:2010.12305

29. Li, B., Zhou, H., He, J., Wang, M., Yang, Y., Li, L.: On the sentence embeddings from pre-trained language models. In: Empirical Methods in Natural Language Processing (EMNLP), pp. 9119–9130 (2020)

30. Lin, Z., et al.: A structured self-attentive sentence embedding. In: International Conference on Learning Representations (ICLR), pp. 1–15 (2017)

31. Mika, S., Schölkopf, B., Smola, A., Müller, K.R., Scholz, M., Rätsch, G.: Kernel PCA and de-noising in feature spaces. In: Conference on Neural Information Processing Systems (NIPS), pp. 536–542 (1998)

32. Mikolov, T., Sutskever, I., Chen, K., Corrado, G.S., Dean, J.: Distributed representations of words and phrases and their compositionality. In: Conference on Neural Information Processing Systems (NIPS), pp. 3111–3119 (2013)

33. O'Neill, J., Bollegala, D.: Meta-embedding as auxiliary task regularization. In: European Conference on Artificial Intelligence (ECAI), pp. 2124–2131 (2020)

34. Patel, R.N., Burgin, E., Assem, H., Dutta, S.: Efficient multi-lingual sentence classification framework with sentence meta encoders. In: IEEE International Conference on Big Data (2021)
35. Pearson, K.: On lines and planes of closest fit to systems of points in space. Phil. Mag. **2**(11), 559–572 (1901)
36. Pennington, J., Socher, R., Manning, C.D.: GloVe: global vectors for word representation. In: Empirical Methods in Natural Language Processing (EMNLP), pp. 1532–1543 (2014)
37. Poerner, N., Waltinger, U., Schütze, H.: Sentence meta-embeddings for unsupervised semantic textual similarity. In: Annual Meeting of the Association for Computational Linguistics (ACL), pp. 7027–7034 (2020)
38. Reimers, N., Gurevych, I.: Sentence-BERT: sentence embeddings using Siamese BERT-networks. In: Empirical Methods in Natural Language Processing (EMNLP), pp. 3982–3992 (2019)
39. Schölkopf, B.: Nonlinear component analysis as a kernel eigenvalue problem. Neural Comput. **10**(5), 1299–1319 (1998)
40. Senel, L.K., Utlu, I., Yucesoy, V., Koc, A., Cukur, T.: Semantic structure and interpretability of word embeddings. IEEE/ACM Trans. Audio Speech Lang. Process. **26**(10), 1769–1779 (2018)
41. Shi, X.: Nonlinear ICA, pp. 97–112. Springer, Heidelberg (2011). https://doi.org/10.1007/978-3-642-11347-5_5
42. Sørensen, M., Kanatsoulis, C.I., Sidiropoulos, N.D.: Generalized canonical correlation analysis: a subspace intersection approach. arXiv:2003.11205 (2020)
43. Tsatsishvilil, V., Cong, F., Toiviainen, P., Ristaniemil, T.: Combining PCA and multiset CCA for dimension reduction when group ICA is applied to decompose naturalistic fMRI data. In: International Joint Conference on Neural Networks (IJCNN), pp. 1–6 (2015)
44. Wieting, J., Gimpel, K.: ParaNMT-50M: pushing the limits of paraphrastic sentence embeddings with millions of machine translations. In: Annual Meeting of the Association for Computational Linguistics (ACL), pp. 451–462 (2018)
45. Wieting, J., Kiela, D.: No Training required: exploring random encoders for sentence classification. In: International Conference on Learning Representations (ICLR), pp. 1–16 (2019)
46. Yin, W., Schütze, H.: Learning word meta-embeddings. In: Annual Meeting of the Association for Computational Linguistics (ACL), pp. 1351–1360 (2016)

Match Your Words! A Study of Lexical Matching in Neural Information Retrieval

Thibault Formal[1,2(✉)], Benjamin Piwowarski[2,3], and Stéphane Clinchant[1]

[1] Naver Labs Europe, Meylan, France
{thibault.formal,stephane.clinchant}@naverlabs.com
[2] Sorbonne Université, Institute for Intelligent Systems and Robotics, UMR 7222,
Paris, France
benjamin@piwowarski.fr
[3] CNRS, Paris, France

Abstract. Neural Information Retrieval models hold the promise to replace lexical matching models, e.g. BM25, in modern search engines. While their capabilities have fully shone on in-domain datasets like MS MARCO, they have recently been challenged on out-of-domain zero-shot settings (BEIR benchmark), questioning their actual generalization capabilities compared to bag-of-words approaches. Particularly, we wonder if these shortcomings could (partly) be the consequence of the inability of neural IR models to perform lexical matching off-the-shelf. In this work, we propose a measure of discrepancy between the lexical matching performed by any (neural) model and an "ideal" one. Based on this, we study the behavior of different state-of-the-art neural IR models, focusing on whether they are able to perform lexical matching *when it's actually useful*, i.e. for important terms. Overall, we show that neural IR models fail to properly generalize term importance on out-of-domain collections or terms almost unseen during training.

Keywords: Neural Information Retrieval · BERT · Lexical matching

1 Introduction

Over the last two years, the effectiveness of neural IR systems has risen substantially. Neural retrievers based on pre-trained Language Models like BERT [4] – whether dense or sparse – hold the promise to replace lexical matching models (e.g. BM25) for first-stage ranking in modern search engines. Despite this success, little is known regarding their actual inner working in the IR setting. Previous works scrutinizing BERT-based ranking models either relied on axiomatic approaches adapted to neural models [1,17], controlled experiments [11], or direct investigation of the learned representations [7,9] or attention [19]. This line of work has shown – among other findings – that these models, which rely on contextualized semantic matching, are actually still quite sensitive to lexical match and term statistics in documents/collections [7,9]. However, these observations are based on specifically tailored approaches that cannot directly be applied to any given model. To generalize these findings, we introduce instead an intuitive black box approach: we propose to "count" query terms appearing in top documents retrieved by various state-of-the-art neural systems, in order to compare their ability to perform *lexical matching*.

Furthermore, previous studies have been conducted on the MS MARCO dataset, on which models have been trained. The BEIR benchmark [18] has shown that the only systems improving the overall performance over BM25 in the zero-shot setting have (somehow) a lexical bias, e.g. models like doc2query-T5 [13] or ColBERT [10]. Therefore, we also propose to study the extent to which neural IR models are able to *generalize* lexical matching, for query terms that either have not been seen in the training set or with different collection statistics (e.g. common in the training set but rare on an out-of-domain evaluation set).

In this work, we first develop indicators that help measuring to what extent a lexical match is "important" for the user (user relevance) or for the model (system relevance). By comparing both values – i.e. computing the difference between the user and the system, we can look at the following research questions: **(RQ1).** To what extent neural retrievers perform accurate lexical matching (Sect. 3.1)? **(RQ2).** Do they generalize term matching to unseen query terms (Sect. 3.1)? **(RQ3).** Do they generalize term matching to new collections (Sect. 3.2)?

2 Methodology

Our analysis rationale is the following: the more a term is important for a query (w.r.t. relevant documents), the more frequent the term should be retrieved by the system in top retrieved documents. Therefore, we first need to define what it means for a term to be *important for lexical matching*, and how to accurately measure frequency in top documents. Roughly speaking, we are interested in the models ability to retrieve documents containing query terms, *when they are deemed important*. Note that we are not interested in expansion mechanisms in our analysis since they are more related to semantic matching.

Intuitively, term importance w.r.t. relevance can be measured by the extent to which a term allows to distinguish relevant from non-relevant documents in a collection of documents. It is thus natural to use the Robertson-Sparck Jones (RSJ) weight [14,20]. The RSJ weights have been shown, if estimated correctly, to order documents in the optimal order w.r.t. the Probability Ranking Principle [15]. For a given user information need U, the user RSJ_U weight for term t is defined as follows (the conditioning on query q is implicit):

$$\text{RSJ}_{t,U} = \log \frac{p(t|R)p(\neg t|\neg R)}{p(\neg t|R)p(t|\neg R)} \tag{1}$$

where $P(t|R)$ is the probability that term t occurs in a relevant document. $\text{RSJ}_{t,U}$ is thus high when a term, for a document to be relevant, is both *necessary* $(p(.|R))$ and *sufficient* $(p(.|\neg R))$. Intuitively, it is low for e.g. stopwords, as they have equal *odds* to appear in relevant and irrelevant documents. The above weight can be estimated using the set of relevant documents and collection statistics.

We now want to compute the same weight, when relevance is defined by the *system* (and not the *user*). In other words, we would like to measure how much a model "retrieves" term t. One way to proceed is to suppose that top-K documents are *relevant from the point of view of the system*, for a suitable K. While a more accurate definition of system relevance could be used, we found out in our preliminary analysis that results were not very sensitive to the choice of K. We hence define the system RSJ_S weight for term t as:

$$\text{RSJ}_{t,S} = \log \frac{p(t|\text{top-}K)p(\neg t|\neg\text{top-}K)}{p(\neg t|\text{top-}K)p(t|\neg\text{top-}K)} \tag{2}$$

Intuitively, it gives us a mean to properly count occurrences of query terms in retrieved documents – taking into account collection statistics. It is estimated similarly to Eq. 1. Once RSJ_U and RSJ_S have been computed, we can look at the difference between both, i.e. $\Delta\text{RSJ}_t = \text{RSJ}_{t,S} - \text{RSJ}_{t,U}$. If $\Delta\text{RSJ}_t > 0$ (resp. $\Delta\text{RSJ}_t < 0$), it means that the model overestimates (resp. underestimates) the importance of term t when considering its document ordering. In other words, the model retrieves "too much" (resp. "too few") this term. Please note that a high correlation between RSJ_S and RSJ_U **is not** indicative of the absolute performance of a model, as RSJ_U is neither a perfect model nor performance measure. However, we argue that it can still indicate partly the performance of the model w.r.t. lexical matching, especially for terms whose RSJ_U are high.

3   Experiments

We conducted experiments by analyzing models trained on MS MARCO [12], using public model parameters when available (indicated by $\star$). We evaluated models on the *in-domain* TREC Deep Learning 2019–2020 datasets [2,3] (97 queries in total), and two *out-of-domain* datasets from the BEIR [18] benchmark (TREC-COVID (bio-medical) and FiQA-2018 (financial), with respectively 50 and 648 test queries). For all our experiments, we measure the system relevance by using top-$K = 100$. For the term-level analysis, we keep stopwords, and use standard tokenization and Porter stemming. We solely focus on first-stage retrievers (and not re-rankers), for which lexical matching might be more critical. We thus compare various state-of-the-art models (based on the BEIR benchmark), considering different types of approaches (sparse and dense). We include two lexical models, the standard BM25 [16] and doc2query-T5 ($\star$) [13]; SPLADE ($\star$) [5,6], an expansion-based sparse approach; ColBERT [10], an interaction-based architecture; two dense retrievers, TAS-B ($\star$) [8] and a standard Bi-encoder trained with contrastive loss and in-batch negatives.

3.1   Lexical Match in Neural IR

In Fig. 1, we plot the relationship between the user weight and ΔRSJ, for each term in the test queries appearing at least 10 times in the training queries (left,

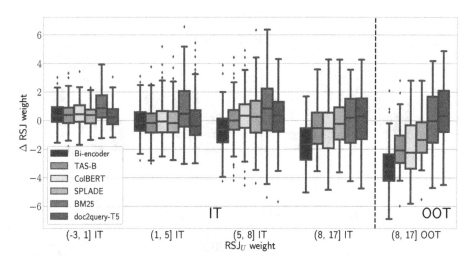

Fig. 1. ΔRSJ with respect to user RSJ$_U$ (x-axis, binned), splitting according to query terms seen during training (IT, left) or not (OOT, right). We consider that terms appearing in less than 10 training queries are OOT, leading to 499 and 42 terms in TREC queries, for IT and OOT respectively. Note that due to the fact that OOT terms are also generally rare in the collection, their RSJ$_U$ is always > 8, hence the single bin.

IT for In-Training). We first note that lexical-based models tend to overestimate the importance of query terms (ΔRSJ > 0). The second observation is that models are roughly similar in their estimations for low user RSJ$_U$ weights (below 5). Then, there is a clear distinction between the bi-encoder and other neural models (both dense and sparse): we can see that it retrieves less documents, on average, containing precisely the important query terms. Comparing dense and sparse/interaction models overall – by considering the average ΔRSJ over terms – we observe that, interestingly, dense models underestimate RSJ$_U$ ($\overline{\Delta\text{RSJ}}$ = −0.07 for TAS-B and −0.26 for the bi-encoder) while sparse/interaction slightly overestimate it ($\overline{\Delta\text{RSJ}}$ = 0.03 for ColBERT and SPLADE). Note again, as mentioned in Sect. 2, that the measure is not necessarily indicative of performance: for instance, TAS-B performs better than BM25 on TREC, suggesting that the model is better for semantic search. To illustrate the above, let us consider a query from the TREC DL set: "does (-1.12) legionella (14.85) pneumophila (13.12) cause (4.34) pneumonia (8.34)" (terms with associated RSJ$_U$). BM25 is able to correctly estimate importance for legionella (RSJ$_S$ = 15.08) contrary to neural approaches which tend to under-estimate it (RSJ$_S$ = 10.63, 13.42, 13.65 for the bi-encoder, SPLADE and ColBERT respectively).

We now shift our attention to the behavior of models for query words that are *(almost) not* in the training set. In Fig. 1, we show the distribution of ΔRSJ for terms appearing in less than 10 training queries (out of > 500k) (right, OOT for Out-Of-Training). Comparing with ΔRSJ for terms in the training set, we

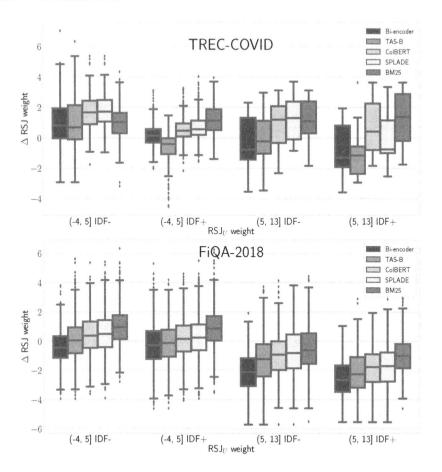

Fig. 2. ΔRSJ with respect to RSJ_U (x-axis, binned) in the zero-shot setting. IDF-includes 108 and 933 terms, while IDF+ includes 112 and 428 terms for respectively TREC-COVID and FiQA-2018. Note that bins are not similar compared to Fig. 1, as RSJ weights have different distributions on BEIR datasets.

can see that all neural models are affected somehow, showing that lexical match does not fully generalize to "new" terms. For the $(8, 17]$ bin, and for every model (except BM25), the difference in mean between IT/OOT is significant, based on a t-test with $p = 0.01$.

Finally, we also looked at the relationship between IT/OOT and model performance. More precisely, for terms in the $(8, 17]$ bin, we computed the mean ndcg@10 for queries containing at least one term either in IT or OOT (respectively 55 and 37 queries out of the 97, with 9 queries in both sets). We found that BM25 and doc2query-T5 performance increased by 0.1 and 0.02 respectively, while for all neural models the performance decreased (≈ 0 for TAS-B, -0.11 for SPLADE, -0.27 for the bi-encoder and -0.38 for ColBERT). The fact that BM25 performance increased is likely due to the fact that the mean IDF increased (from 7.3 to 10.9), i.e. important terms are more discriminative in the OOT query set. With this in mind, the decrease of all neural models might suggest that a potential reason for the relative performance decrease (w.r.t. BM25) is due to a worse estimate of high RSJ_U.

3.2 Lexical Match and Zero-Shot Transfer Learning

We now analyze whether term importance can generalize to the zero-shot setting[1]. We distinguish two categories of words, namely those that occurred 5 times more in the target collection than in MS MARCO (IDF+), or those for which term statistics were more preserved (IDF-), allowing us to split query terms in sets of roughly equal size. Since term importance is related to the collection frequency (albeit loosely), we can compare ΔRSJ in those two settings. Figure 2 shows the ΔRSJ with respect to RSJ_U for the TREC-COVID and FiQA-2018 collections from the BEIR benchmark [18].

We can first observe that neural models underestimate RSJ_U for terms that are more frequent in the target collection than in the training one (IDF+). It might indicate that models have learned a dataset-specific term importance – confirming the results obtained in the previous section on out-of-training terms. When comparing dense and sparse/interaction models overall – by considering the average ΔRSJ over terms – we observe than dense models underestimate even more RSJ_U than on in-domain ($\overline{\Delta\text{RSJ}} = -0.17$ for TAS-B and -0.38 for the bi-encoder) while sparse/interaction seem to overestimate ($\overline{\Delta\text{RSJ}} = 0.18$ for ColBERT and 0.30 for SPLADE), but however to a lesser extent than BM25 ($\overline{\Delta\text{RSJ}} = 0.83$). Finally, we observed that when transferring, all the models have a higher ΔRSJ variance compared to their trained version on MS MARCO: in all cases, the standard deviation (when normalized by BM25 one) is around 0.8 for MS MARCO, but around 1.1 for TREC-COVID and FiQA-2018. This further strengthens our point on the issue of generalizing lexical matching to out-of-domain collections.

4 Conclusion

In this work, we analyzed how different neural IR models predict the importance of lexical matching for query terms. We proposed to use the Robertson-Sparck Jones (RSJ) weight as an appropriate measure to compare term importance w.r.t. the user and system relevance. We introduce a black box approach that enables a systematic comparison of different models w.r.t. term matching. We have also investigated the behavior of lexical matching in the zero-shot setting. Overall, we have shown that lexical matching properties are heavily influenced by the presence of the term in the training collection. The rarer the term, the harder it is to find documents containing that term for most neural models. Furthermore, this phenomenon is amplified if term statistics change across collections.

---

[1] We excluded doc2query-T5 from the analysis, due to the high computation cost for obtaining the expanded collections.

References

1. Camara, A., Hauff, C.: Diagnosing BERT with Retrieval Heuristics. In: ECIR. p. 14 (2020), zSCC: NoCitationData[s0]
2. Craswell, N., Mitra, B., Yilmaz, E., Campos, D.: Overview of the trec 2020 deep learning track (2021)
3. Craswell, N., Mitra, B., Yilmaz, E., Campos, D., Voorhees, E.M.: Overview of the trec 2019 deep learning track (2020)
4. Devlin, J., Chang, M., Lee, K., Toutanova, K.: BERT: pre-training of deep bidirectional transformers for language understanding. In: Burstein, J., Doran, C., Solorio, T. (eds.) Proceedings of the 2019 Conference of the North American Chapter of the Association for Computational Linguistics: Human Language Technologies, NAACL-HLT 2019, Minneapolis, MN, USA, June 2–7, 2019, Volume 1 (Long and Short Papers). pp. 4171–4186. Association for Computational Linguistics (2019). 10.18653/v1/n19-1423, https://doi.org/10.18653/v1/n19-1423
5. Formal, T., Lassance, C., Piwowarski, B., Clinchant, S.: SPLADE v2: sparse lexical and expansion model for information retrieval. arXiv:2109.10086 [cs], September 2021. http://arxiv.org/abs/2109.10086, arXiv: 2109.10086
6. Formal, T., Piwowarski, B., Clinchant, S.: Splade: sparse lexical and expansion model for first stage ranking. In: Proceedings of the 44th International ACM SIGIR Conference on Research and Development in Information Retrieval, SIGIR 2021, pp. 2288–2292. Association for Computing Machinery, New York (2021). https://doi.org/10.1145/3404835.3463098, https://doi.org/10.1145/3404835.3463098
7. Formal, T., Piwowarski, B., Clinchant, S.: A white box analysis of ColBERT. In: Hiemstra, D., Moens, M.-F., Mothe, J., Perego, R., Potthast, M., Sebastiani, F. (eds.) ECIR 2021. LNCS, vol. 12657, pp. 257–263. Springer, Cham (2021). https://doi.org/10.1007/978-3-030-72240-1_23
8. Hofstätter, S., Lin, S.C., Yang, J.H., Lin, J., Hanbury, A.: Efficiently Teaching an Effective Dense Retriever with Balanced Topic Aware Sampling. In: SIGIR, July 2021
9. Jiang, Z., Tang, R., Xin, J., Lin, J.: How does BERT rerank passages? an attribution analysis with information bottlenecks. In: EMNLP Workshop, Black Box NLP, p. 14 (2021)
10. Khattab, O., Zaharia, M.: ColBERT: Efficient and Effective Passage Search via Contextualized Late Interaction over BERT. arXiv:2004.12832 [cs], April 2020. http://arxiv.org/abs/2004.12832, arXiv: 2004.12832
11. MacAvaney, S., Feldman, S., Goharian, N., Downey, D., Cohan, A.: ABNIRML: analyzing the behavior of neural IR models. arXiv:2011.00696 [cs], Nov 2020. http://arxiv.org/abs/2011.00696, zSCC: 0000000 arXiv: 2011.00696
12. Nguyen, T., Rosenberg, M., Song, X., Gao, J., Tiwary, S., Majumder, R., Deng, L.: Ms marco: a human generated machine reading comprehension dataset. CoRR abs/1611.09268 (2016). http://dblp.uni-trier.de/db/journals/corr/corr1611.html#NguyenRSGTMD16
13. Nogueira, R., Lin, J.: From doc2query to docTTTTTquery, p. 3, zSCC: 0000004
14. Robertson, S.E., Jones, K.S.: Relevance weighting of search terms. J. Am. Soc. Inf. Sci. **27**(3), 129–146 (1976). 10/dvgb84, https://onlinelibrary.wiley.com/doi/abs/10.1002/asi.4630270302,_eprint: https://onlinelibrary.wiley.com/doi/pdf/10.1002/asi.4630270302
15. Robertson, S.E.: The probability ranking principle in IR. J. Documentation **33**(4), 294–304 (1977). 10/ckqfpm, https://doi.org/10.1108/eb026647, publisher: MCB UP Ltd

16. Robertson, S.E., Zaragoza, H.: The Probabilistic Relevance Framework: BM25 and Beyond. Foundations and Trends in Information Retrieval (2009)
17. Sciavolino, C., Zhong, Z., Lee, J., Chen, D.: Simple entity-centric questions challenge dense retrievers. In: Empirical Methods in Natural Language Processing (EMNLP) (2021)
18. Thakur, N., Reimers, N., Rücklé, A., Srivastava, A., Gurevych, I.: BEIR: a heterogenous benchmark for zero-shot evaluation of information retrieval models. arXiv:2104.08663 [cs], September 2021. http://arxiv.org/abs/2104.08663
19. Yates, A., Nogueira, R., Lin, J.: Pretrained transformers for text ranking: Bert and beyond. In: Proceedings of the 14th ACM International Conference on Web Search and Data Mining, WSDM 2021, pp. 1154–1156. Association for Computing Machinery, New York (2021). https://doi.org/10.1145/3437963.3441667, https://doi.org/10.1145/3437963.3441667
20. Yu, C.T., Salton, G.: Precision weighting - an effective automatic indexing method. J. ACM **23**(1), 76–88 (1976). 10/d3fgsz, https://doi.org/10.1145/321921.321930

CARES: CAuse Recognition for Emotion in Suicide Notes

Soumitra Ghosh[1], Swarup Roy[2], Asif Ekbal[1]([✉]),
and Pushpak Bhattacharyya[1]

[1] Indian Institute of Technology Patna, Patna, India
asif@iitp.ac.in, pb@cse.iitb.ac.in
[2] National Institute of Technology, Durgapur, India

Abstract. Inspired by recent advances in emotion-cause extraction in texts and its potential in research on computational studies in suicide motives and tendencies and mental health, we address the problem of *cause identification* and *cause extraction* for emotion in suicide notes. We introduce an emotion-cause annotated suicide corpus of 5769 sentences by labeling the benchmark CEASE-v2.0 dataset (4932 sentences) with causal spans for existing annotated emotions. Furthermore, we expand the utility of the existing dataset by adding emotion and emotion cause annotations for an additional 837 sentences collected from 67 non-English suicide notes (Hindi, Bangla, Telugu). Our proposed approaches to emotion-cause identification and extraction are based on pre-trained transformer-based models that attain performance figures of 83.20% accuracy and 0.76 Ratcliff-Obershelp similarity, respectively. The findings suggest that existing computational methods can be adapted to address these challenging tasks, opening up new research areas.

Keywords: Emotion cause · Suicide notes · XLM-R · BERT · SpanBERT

1 Introduction

Suicide continues to be one of the major causes of death across the world. Suicide rates have risen by 60% globally in the previous 45 years[1]. Suicide is currently one of the top three causes of mortality for those aged 15 to 44. Both men and women commit suicide at higher rates throughout Europe, notably in Eastern Europe. India and China account for about 30% of all suicides globally[2]. By writing a suicide note, a person communicates sentiments that would otherwise lie covert and fester. Suicide notes may be used as both an explanation and a therapeutic tool to help family members comprehend the suicide [10].

---

[1] https://www.who.int/news/item/17-06-2021-one-in-100-deaths-is-by-suicide.
[2] https://www.befrienders.org/suicide-statistics.

M. Hagen et al. (Eds.): ECIR 2022, LNCS 13186, pp. 128–136, 2022.
https://doi.org/10.1007/978-3-030-99739-7_15

In this work, we primarily aim at achieving two objectives:

- *Produce a gold standard corpus annotated with causal spans for emotion annotated sentences in suicide notes.*
- *Develop a benchmark setup for emotion cause recognition in suicide notes, specifically, cause identification and cause extraction.*

Given the limits of automated approaches in suicidal research, this study will aid the research community by introducing a publicly available emotion cause annotated corpus of 5769 sentences from suicide notes (whose availability is otherwise scarce). Additionally, the attempt to make the available benchmark CEASE-v2.0 dataset [8] multilingual, by adding 17% new sentences from 67 non-English suicide notes (Hindi, Bangla and Telugu) will increase the utility of the CEASE-v2.0 dataset. Lastly, the proposed evaluation methods on the dataset can serve as solid baselines for future research on this dataset and related topics.

2 Background

The Emotion Cause Extraction (ECE) task aims to identify the possible causes of certain emotions from text. The basic premise is that such phrases are good descriptors of the underlying causes of the expressed emotions [19]. When compared to emotion classification, this is a far more challenging problem. In the early 1960s, several scholars [1,20] began analyzing the content of suicide notes to investigate the various reasons for suicide. The socioeconomic and psychological causes of suicides were examined in [17].

Emotion cause analysis has received greater attention in recent years in the sentiment analysis and text mining fields [6,16]. The work in [9] introduced an emotion cause annotated dataset built from SINA[3] news for event-driven emotion cause extraction. Deep learning has lately piqued the interest of the ECE community. The authors in [2] proposed a joint neural network-based method for emotion extraction and emotion cause extraction, capturing mutual benefits across these two emotion analysis tasks. In [21], a two-step technique was designed to solve the related job of emotion cause pair extraction (ECPE). Emotion identification and cause extraction was conducted first by a multitask framework, followed by emotion-cause pairing and filtering. To tackle the emotion cause pair extraction challenge, [3] suggested a graph neural network-based solution. Recently, [14] introduced the task of recognizing emotion cause in conversations (RECCON) and presented the emotion cause annotated RECCON dataset.

Despite numerous studies on ECE in other fields, no such study on suicide research utilizing computational techniques exists to our knowledge. Our effort aims at bridging this gap by performing ECE in suicide notes. We consider the CEASE-v2.0 [8] suicide notes corpus, which is an improved version of the CEASE dataset provided in the introductory paper [7] by the same authors.

---

[3] http://news.sina.com.cn/society/.

3   Corpus Development

Following the methods for data collection and annotation in [7,8], we introduce multilingual CARES_CEASE-v2.0 corpus with emotion cause annotations.

Data Collection: We collected 67 suicide notes, from various Indian news websites such as 'patrika.com', 'jkstudenttimes.com', etc., of three most popularly spoken languages in India, Hindi (46 notes), Bengali (15 notes) and Telugu (6 notes). After proper anonymization, each note was sentence tokenized and the resultant bag of sentences were shuffled for reconstruction of the actual notes.

Annotations: Three annotators (two undergraduate students and one doctoral researcher of computer science discipline) with sufficient subject knowledge and experience on construction of supervised corpora, were engaged to manually digitize the notes (wherever direct transcripts were not available) and annotate the newly added 837 multilingual sentences with fine-grained emotion labels and then perform the causal span marking task at sentence-level. Each sentence is tagged with at most 3 emotions[4] from a set of 15 fine-grained emotion labels as described in [8]. A Fleiss-Kappa [18] score of 0.65 was attained among the three annotators, which is 0.06 points better than the work in [8] signifying that the annotations are of significantly good quality.

For each emotion (E) annotated sentence (S) in CARES_CEASE-v2.0[5], annotators were instructed to extract the causal span, $C(S)$, that adequately represented the source of the emotion E. We consider the span-level aggregation method discussed in [9] to mark the final causal span for a sentence S. If there was no specific $C(S)$ for E in S, the annotators labelled the sentence as 'no cause'. Based on prior studies on span extraction, we use the macro-F1 [15] metric to measure the inter-rater agreement and attain 0.8294 F1-score, which depicts that the annotations are of substantially good quality. Our annotations process differs from [14] primarily because suicide notes are not set in a conversational setting, and thus no conversational context is available.

A sample of annotated sample is shown below:

Sentence 1: "Mom, Dad, I could not become your good son, forgive me."
Emotion: *forgiveness*; Cause: *I could not become your good son*
Sentence 2 *(Transliterated Hindi)*: "ab jine ki ichaa nhi ho rhi hai."
(English Translation): "no longer want to live."
Emotion: *hopelessness*; Cause: *no cause*

Corpus Statistics: The average sentence length is 13.31 words. The longest note has 80 sentences totaling 1467 words and the shortest one contains 13 words. Table 1 shows some data about the newly collected non-English suicide notes.

---

[4] *forgiveness, happiness_peacefulness, love, pride, hopefulness, thankfulness, blame, anger, fear, abuse, sorrow, hopelessness, guilt, information, instructions.*

[5] Dataset available at https://www.iitp.ac.in/ ai-nlp-ml/resources.html#CARES.

Table 1. Distribution of the collected notes across various attributes. NA: Not available

Gender		Marital status		Age		Data type	
Category	Count	Category	Count	Interval	Count	Category	Type
				10–20	16		
Male	32	Married	15	21–30	16	Non-code-mixed	41
Female	33	Unmarried	43	31–40	3	Code-mixed	22
NA	2	NA	9	50–60	2	Transliterated	4
				NA	30		

4 Cause Recognition for Emotion in Suicide Notes

We address the task of cause recognition for emotion in suicide notes as two independent sub-tasks: (A) *Emotion Cause Identification* (whether cause is present or not), (B) *Emotion Cause Extraction* (extract the causal span). Figure 1 shows the overall framework for the cause recognition setup.

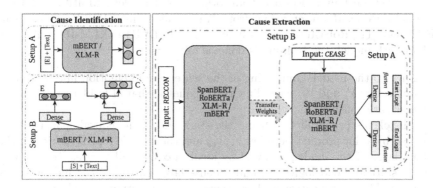

Fig. 1. Architecture of the setup for cause recognition for emotion in suicide notes.

4.1 Emotion Cause Identification

Given a sentence, s, and its associated most prominent emotion, e, the task is to identify whether s is causal or not. We propose a multitask solution (Setup B) to address the problem where the input is formed as: $< [CLS], p, [SEP], s >$, where, p indicates the associated polarity (positive, negative, neutral) of s where s is a sequence of words in the input sentence. Emotions are mapped to their associated polarity labels following the weak labelling scheme discussed in [8]. We set the max length of the input sequence as 30. The output is an emotion label from the 15-emotion tag set and a label to indicate *causal* or *non-causal*.

The input sequence is passed through a pre-trained transformer module followed by two task-specific dense layers. The softmax emotion output is fed as additional features to the task-specific causal features to enhance the output of the cause identification task, which is also generated through a softmax activation function. The model is trained using a unified loss function as shown below:

$$\lambda = \alpha * \lambda_E + \beta * \lambda_C \tag{1}$$

λ_E and λ_C are the categorical crossentropy losses for the emotion identification and cause identification task and α and β are the loss weights for the tasks, respectively. As an ablation experiment, we develop a single-task setup (Setup A) following the work in [14] where the input sequence contains the emotion label in place of polarity and outputs whether S is 'causal' or 'non-causal'.

4.2   Emotion Cause Extraction

We formulate the emotion cause extraction task for any non-neutral sentence as follows: Given a phrase s with an emotion label e, determine the causal span $c(s)$ in s that is relevant to emotion e. The input sequence (I) is formed as $< [CLS], s, [SEP], e >$. We set the max length of the input sequence as 30. We finetune four pretrained transformer-based models, Multilingual Bidirectional Encoder Representations from Transformers (mBERT) Base [5], Span-BERT Base [11], RoBERTa Base [13], and XLM-RoBERTa (XLM-R) [4] models, and for each model, the training is done as follows:

1. I is fed to the model, comprising of the sentence and emotion information.
2. Two vectors V_s and V_{cs} with dimensions equal to that of hidden states in the models are considered.
3. Each token's probability of being the start/end of the causal span is determined by a dot product between V_s/V_{cs} and the token's representation in the final layer of the model, followed by a softmax over all tokens. To compute span start and end logits, we add a dense layer on top of the hidden-states output. The sparse categorical crossentropy loss function is used in this case.
4. The model is then finetuned, allowing V_s and V_{cs} to learn along the way.

The CARES_CEASE-v2.0 dataset is used in setup A for training and testing. To train setup B, we use a transfer learning (TL) method, first training the models on the RECCON [14] dataset (only causal sentences with non-conversational context: 3613 sentences) and then finetuning on our dataset. We save the best model as per the 'Exact Match' metric [14] score on the validation set.

5   Experimental Setting

The pre-trained models are sourced from the open-source libraries[6] huggingface transformers (RoBERTa, XLM-R, and SpanBERT) and tensorflowhub

---

[6] https://huggingface.co/model and https://tfhub.dev/google/collections/bert/1.

(mBERT). To avoid overfitting, we use ReLU activations for the dense layers (100 neurons each) and apply dropouts of 0.5 to all dense layer outputs. For the *cause identification* task, we divided the CARES CEASE-v2.0 dataset (5769 sentences) into train, test, and validation sets in the ratio 7:2:1. Only the causal utterances (1479 sentences) from the dataset are used in the *cause extraction* task. We divided the CARES CEASE-v2.0 dataset in a 7:2:1 ratio for train, test, and validation purposes. The experiments are run on an NVIDIA GeForce RTX 2080 Ti GPU and the models are optimized using Adam [12] optimizer with learning rates of 3e−5 (for mBERT and SpanBERT) and 5e−5 (for RoBERTa and XLM-R). We kept the batch size as 8 to fully utilize the GPU.

6   Results and Discussion

We observe from Table 2 that for the *cause identification* task, the multitask mBERT-based system (Setup B) performs (83.20% accuracy) better than the XLM-R variant in most of the metrics. Despite being trained on highly skewed emotion data, it achieves commendable results on the emotion identification task (overall weighted-F1 of 74.48%). For the *cause extraction* task, Table 3 shows that the TL approach has proved to be effective for all the experimented models with an increase in scores for the various metrics. In the TL scenario (Setup B), the RoBERTa and XLM-R models both achieved overall top scores for three of the six metrics evaluated, with a joint highest score of 0.76 for the ROS metric. This shows the efficacy of the RoBERTa-based models compared to SpanBERT and mBERT when dealing with multi-lingual code-mixed data.

Analysis: Empirical investigation shows that learning the cause identification task together with the emotion identification task simultaneously increases performance relative to learning the task separately, regardless of the differences in pre-trained encoders. For the causal extraction task, the *Full Match (FM)* and *Partial Match (PM)* measures give a quantitative estimation of the model's performance. For quantitative evaluation, we employed an edit distance-based, token-based and sequence-based measure in the form of *Hamming distance (HD)*, *Jaccard Similarity (JS)* and *Ratcliff-Obershelp Similarity (ROS)* metrics, respectively. Manual analysis of some predicted samples with the gold annotations makes us believe that the ROS measure, based on the longest sequence matching approach, suits the training objective of causal span extraction and better estimates a model's performance from a qualitative standpoint. Although trained for span extraction tasks, we also notice that the SpanBERT model performs poorly on multilingual code-mixed data because it is exclusively trained on non-code-mixed English data. With a ROS score of 0.76, the cross-lingual XLM-R model adapts well to our multilingual data as well as the cause extraction task.

Table 2. Results for the cause identification task. Values in bold are the maximum scores (%) attained for a metric. A: accuracy, m-F1: macro-F1, w-F1: weighted-F1

| Models | CEASE-v2.0 | | | | | | CARES_CEASE-v2.0 | | | | | |
| | Setup A | | Setup B | | | | Setup A | | Setup B | | | |
	A^C	$m - F1^C$	A^C	$m - F1^C$	A^E	$w - F1^E$	A^C	$m - F1^C$	A^C	$m - F1^C$	A^E	$w - F1^E$
mBERT	81.56	78.29	82.47	80.05	70.11	70.90	81.73	80.41	**83.20**	**81.89**	75.67	**74.48**
XLM-R	80.45	79.44	80.95	79.42	70.92	60.53	80.87	79.94	81.55	79.67	**76.36**	72.81

Table 3. Results for the cause extraction task on the two setups. Values in bold are the overall maximum scores attained for a particular metric.

| | | CEASE-v2.0 | | | | | CARES_CEASE-v2.0 | | | | |
		FM (%)	PM (%)	HD	JS	ROS	FM (%)	PM (%)	HD	JS	ROS
Setup A	SpanBERT	23.98	24.66	0.40	0.55	0.68	31.17	17.62	0.49	0.66	0.76
	RoBERTa	34.12	21.96	0.48	0.63	0.73	28.73	19.51	0.42	0.58	0.69
	XLM-R	35.81	20.61	0.45	0.65	0.74	31.98	23.58	0.45	0.64	0.74
	mBERT	31.42	**33.11**	0.49	0.65	0.75	29.00	26.29	0.48	0.62	0.73
Setup B	SpanBERT	36.49	28.72	0.51	0.65	0.75	28.18	29.00	0.45	0.62	0.73
	RoBERTa	**38.51**	29.05	0.50	0.65	0.74	34.42	23.04	0.49	**0.67**	**0.76**
	XLM-R	34.80	26.35	0.49	0.65	**0.76**	35.23	21.41	**0.52**	0.66	**0.76**
	mBERT	33.78	26.01	0.50	0.64	0.74	29.54	28.73	0.48	0.61	0.73

7   Conclusion and Future Work

This study focuses on addressing the task of emotion cause recognition in suicide notes by extending the size of an existing standard suicide notes dataset with multilingual data and providing gold standard emotion cause annotations on the same. Empirical results indicates that no one model can be declared to be the best at performing any of the specific sub-tasks since their performance varies with the many assessment criteria used in this study.

Future efforts might focus on extracting multiple causes from sentences and develop efficient ways to model discourse relations among the causes.

Acknowledgement. Soumitra Ghosh acknowledges the partial support from the project titled 'Development of CDAC Digital Forensic Centre with AI based Knowledge Support Tools' supported by MeitY, Gov. of India and Gov. of Bihar (project #: P-264).

Ethical Implications. We followed the policies of using the original data and did not violate any copyright issues. The study was deemed exempt by our Institutional Review Board. The codes and data will be made available for research purposes only, after filling and signing an appropriate data compliance form.

References

1. Capstick, A.: Recognition of emotional disturbance and the prevention of suicide. BMJ **1**(5180), 1179 (1960)
2. Chen, Y., Hou, W., Cheng, X., Li, S.: Joint learning for emotion classification and emotion cause detection. In: Riloff, E., Chiang, D., Hockenmaier, J., Tsujii, J. (eds.) Proceedings of the 2018 Conference on Empirical Methods in Natural Language Processing, Brussels, Belgium, October 31 - November 4, 2018, pp. 646–651. Association for Computational Linguistics (2018). https://doi.org/10.18653/v1/d18-1066. https://doi.org/10.18653/v1/d18-1066
3. Chen, Y., Hou, W., Li, S., Wu, C., Zhang, X.: End-to-end dblp:journals/jmlr/srivastavahkss14emotion-cause pair extraction with graph convolutional network. In: Proceedings of the 28th International Conference on Computational Linguistics, pp. 198–207. International Committee on Computational Linguistics, Barcelona, Spain (Online), December 2020. https://doi.org/10.18653/v1/2020.coling-main.17. https://aclanthology.org/2020.coling-main.17
4. Conneau, A., et al.: Unsupervised cross-lingual representation learning at scale. In: Proceedings of the 58th Annual Meeting of the Association for Computational Linguistics, pp. 8440–8451. Association for Computational Linguistics, Online, July 2020. https://doi.org/10.18653/v1/2020.acl-main.747. https://aclanthology.org/2020.acl-main.747
5. Devlin, J., Chang, M.W., Lee, K., Toutanova, K.: BERT: pre-training of deep bidirectional transformers for language understanding. In: Proceedings of the 2019 Conference of the North American Chapter of the Association for Computational Linguistics: Human Language Technologies, Volume 1 (Long and Short Papers), pp. 4171–4186. Association for Computational Linguistics, Minneapolis, Minnesota, June 2019. https://doi.org/10.18653/v1/N19-1423. https://aclanthology.org/N19-1423
6. Ghazi, D., Inkpen, D., Szpakowicz, S.: Detecting emotion stimuli in emotion-bearing sentences. In: Gelbukh, A. (ed.) Computational Linguistics and Intelligent Text Processing. LNCS, vol. 9042, pp. 152–165. Springer, Cham (2015). https://doi.org/10.1007/978-3-319-18117-2_12
7. Ghosh, S., Ekbal, A., Bhattacharyya, P.: Cease, a corpus of emotion annotated suicide notes in English. In: Calzolari, N., et al. (eds.) Proceedings of The 12th Language Resources and Evaluation Conference, LREC 2020, Marseille, France, 11–16 May 2020, pp. 1618–1626. European Language Resources Association (2020). https://aclanthology.org/2020.lrec-1.201/
8. Ghosh, S., Ekbal, A., Bhattacharyya, P.: A multitask framework to detect depression, sentiment and multi-label emotion from suicide notes. Cognitive Computation, pp. 1–20 (2021)
9. Gui, L., Wu, D., Xu, R., Lu, Q., Zhou, Y.: Event-driven emotion cause extraction with corpus construction. In: Proceedings of the 2016 Conference on Empirical Methods in Natural Language Processing, pp. 1639–1649. Association for Computational Linguistics, Austin, November 2016. https://doi.org/10.18653/v1/D16-1170. https://aclanthology.org/D16-1170
10. Ho, T., Yip, P.S., Chiu, C., Halliday, P.: Suicide notes: what do they tell us? Acta Psychiatr. Scand. **98**(6), 467–473 (1998)
11. Joshi, M., Chen, D., Liu, Y., Weld, D.S., Zettlemoyer, L., Levy, O.: SpanBERT: improving pre-training by representing and predicting spans. Trans. Assoc. Comput. Linguist. **8**, 64–77 (2020)

12. Kingma, D.P., Ba, J.: Adam: a method for stochastic optimization. In: Bengio, Y., LeCun, Y. (eds.) 3rd International Conference on Learning Representations, ICLR 2015, San Diego, CA, USA, 7–9 May, 2015, Conference Track Proceedings (2015). http://arxiv.org/abs/1412.6980

13. Liu, Y., et al.: Roberta: a robustly optimized BERT pretraining approach. CoRR abs/1907.11692 (2019). http://arxiv.org/abs/1907.11692

14. Poria, S., et al.: Recognizing emotion cause in conversations. Cognitive Computation, pp. 1–16 (2021)

15. Rajpurkar, P., Zhang, J., Lopyrev, K., Liang, P.: SQuAD: 100,000+ questions for machine comprehension of text. In: Proceedings of the 2016 Conference on Empirical Methods in Natural Language Processing, pp. 2383–2392. Association for Computational Linguistics, Austin, Texas, November 2016. https://doi.org/10.18653/v1/D16-1264. https://aclanthology.org/D16-1264

16. Russo, I., Caselli, T., Rubino, F., Boldrini, E., Martínez-Barco, P.: EMOCause: an easy-adaptable approach to extract emotion cause contexts. In: Proceedings of the 2nd Workshop on Computational Approaches to Subjectivity and Sentiment Analysis (WASSA 2.011), pp. 153–160. Association for Computational Linguistics, Portland, Oregon, June 2011. https://aclanthology.org/W11-1720

17. Shneidman, E.S., Farberow, N.L.: A socio-psychological investigation of suicide. In: Perspectives in Personality Research, pp. 270–293. Springer (1960)

18. Spitzer, R.L., Cohen, J., Fleiss, J.L., Endicott, J.: Quantification of agreement in psychiatric diagnosis: a new approach. Arch. Gen. Psychiatry **17**(1), 83–87 (1967)

19. Talmy, L.: Toward a Cognitive Semantics, vol. 2. MIT Press (2000)

20. Wagner, F.: Suicide notes. Danish Med. J. **7**, 62–64 (1960)

21. Xia, R., Ding, Z.: Emotion-cause pair extraction: a new task to emotion analysis in texts. In: Proceedings of the 57th Annual Meeting of the Association for Computational Linguistics, pp. 1003–1012. Association for Computational Linguistics, Florence, Italy, July 2019. https://doi.org/10.18653/v1/P19-1096. https://aclanthology.org/P19-1096

Identifying Suitable Tasks for Inductive Transfer Through the Analysis of Feature Attributions

Alexander J. Hepburn$^{(\boxtimes)}$ⓘ and Richard McCreadieⓘ

University of Glasgow, University Avenue, Glasgow G12 8QQ, Scotland
a.hepburn.1@research.gla.ac.uk, richard.mccreadie@glasgow.ac.uk

Abstract. Transfer learning approaches have shown to significantly improve performance on downstream tasks. However, it is common for prior works to only report where transfer learning was beneficial, ignoring the significant trial-and-error required to find effective settings for transfer. Indeed, not all task combinations lead to performance benefits, and brute-force searching rapidly becomes computationally infeasible. Hence the question arises, *can we predict whether transfer between two tasks will be beneficial without actually performing the experiment?* In this paper, we leverage explainability techniques to effectively predict whether task pairs will be complementary, through comparison of neural network activation between single-task models. In this way, we can avoid grid-searches over all task and hyperparameter combinations, dramatically reducing the time needed to find effective task pairs. Our results show that, through this approach, it is possible to reduce training time by up to 83.5% at a cost of only 0.034 reduction in positive-class F1 on the TREC-IS 2020-A dataset.

Keywords: Explainability · Transfer learning · Classification

1 Introduction

Transfer learning is a method of optimisation where models trained on one task are repurposed for another downstream task. The intuition behind this approach is clear; as human beings, we often apply knowledge learned from previous experience when learning a new, related skill. Hence, transfer learning aims to mimic this biological behaviour by exploiting the relatedness between tasks.

However, there remains an ever-present question that researchers have long strived to answer, *Why is pretraining useful for my task?* More specifically, *What information encoded in a pretrained model is transferrable for my task?* If, hypothetically, we are capable of approximating, prior to training, which auxiliary tasks will be useful in practice, we are then able to avoid the often laborious process of trial-and-error over all task and parameter combinations. Hence, we propose a solution which leverages recent research in explainability to identify

© The Author(s), under exclusive license to Springer Nature Switzerland AG 2022
M. Hagen et al. (Eds.): ECIR 2022, LNCS 13186, pp. 137–143, 2022.
https://doi.org/10.1007/978-3-030-99739-7_16

the properties that characterise particular tasks and by extension, the properties which make these tasks related.

Through the evaluation of 803 models, we calculate the per-document term activity for each task and use these to predict the performance outputs of each combined task pair. We show that there exists correlation between strongly-attributed shared terms between pairs of single tasks and their combined performance output, and that, by ranking each task pair by their performance, we can reduce the time it takes to find the best-performing model by up to 83.5% (with a cost of only 0.034 reduction in positive-class F1).

2   Improving Performance Through Inductive Transfer

The concept of *inductive transfer*, introduced by Pan and Yang [9], can be considered a method of transfer learning wherein the source and target tasks are different, the goal of which is to leverage domain information in the source task—encoded in the training signals as an inductive bias—to be transferred to a downstream, target task.

However, the necessary conditions for what constitutes a suitable auxiliary task for use in pretraining is unclear. Mou et al. [7] note that the difficulty in transferability in this domain lies in the discreteness of word tokens and their embeddings. Similar to this work, Bingel and Søgaard [1] identified beneficial task relations for multi-task learning and found that performance gains were predictable from the dataset characteristics. While ground has been covered in understanding and quantifying the relationship between pairs of tasks, what constitutes task relatedness remains an open question. To this end, we first demonstrate the efficacy of transfer learning as a method of improving classifier performance. We utilise the dataset provided by the TREC Incident Streams Track (TREC-IS) which features a number of multi-label classification tasks wherein each label is representative of some information need (known as *information types*) to end users of automated crisis and disaster systems. More importantly, these labels exhibit some level of conceptual relatedness, and as such, is an appropriate framework for this investigation. The track features 25 labels which manual assessors may ascribe to each document, however, to limit the number of models trained, we use the track's **Task 2** formulation, which restricts the number of information types to 12^1.

We experiment with transfer learning across these information types, that is to say, we train a particular classifier on one, *source* task and then use the resulting model as a pretrained baseline for tuning another downstream *target* task, using a pretrained BERT transformer model as defined by Devlin et al. [3] as the base model for our experiments.

Table 1 shows the single- and multi-task model results from previous experiments, containing each task's baseline performance (omitting 4 tasks which showed no performance change) and their respective best-performing auxiliary

---

[1] More information on metrics and tasks can be found at http://trecis.org.

Table 1. Information type categorisation performance with and without inductive transfer from a source task. Metrics are micro-averaged across events and range from 0 to 1, higher is better.

Target	Model	Inductive transfer (source)				Target parameters			Evaluation scores	
		Transfer-From	LR	#E	B#	LR	#E	B#	Positive F1	Accuracy
New Sub Event	BERT→Target	None	-	-	-	2e−05	4	16	0.0258	0.9604
	BERT→Source→Target	Other (Best)	1e−05	2	32	2e−05	2	32	0.0578	0.9432
First Party Observation	BERT→Target	None	-	-	-	2e−05	4	32	0.0259	0.9646
	BERT→Source→Target	Move People (Best)	1e−05	2	32	1e−05	1	32	0.1142	0.9538
Service Available	BERT→Target	None	-	-	-	3e−05	3	16	0.0944	0.9821
	BERT→Source→Target	Other (Best)	1e−05	1	32	1e−05	1	32	0.1095	0.9783
Move People	BERT→Target	None	-	-	-	2e−05	3	32	0.1964	0.9835
	BERT→Source→Target	Other (Best)	1e−05	1	32	1e−05	2	32	0.2423	0.9853
Emerging Threats	BERT→Target	None	-	-	-	3e−05	2	32	0.2329	0.8323
	BERT→Source→Target	Location (Best)	1e−05	2	32	1e-05	1	32	0.2612	0.8135
Multimedia Share	BERT→Target	None	-	-	-	2e−05	3	32	0.4356	0.6760
	BERT→Source→Target	Other (Best)	1e−05	2	32	2e−05	1	32	0.4709	0.6422
Location	BERT→Target	None	-	-	-	3e−05	2	16	0.5904	0.6939
	BERT→Source→Target	Multimedia Share (Best)	1e−05	1	32	1e−05	1	32	0.6178	0.7196
Other	BERT→Target	None	-	-	-	5e−05	4	16	0.6831	0.5638
	BERT→Source→Target	Multimedia Share (Best)	2e-05	1	32	1e−05	2	32	0.6853	0.7187
AVERAGE	BERT→Target	None	-			Varies			0.2856	0.8321
	BERT→Source→Target	Varies	Varies			Varies			0.3199	0.8443

task when used as a prior. With the exception of those omitted tasks, we observed performance increases across the board , as can be seen from comparing the BERT→Target and BERT→Source→Target rows for each task in the above table. However, obtaining these improvements was not a trivial process. We found that performance increases were highly dependent on the target information type and that the effectiveness of transfer was highly sensitive to changes in model hyperparameters. Moreover, there were no easily discernible patterns that we could use as heuristics to speed up the process of finding the best model, leading to an exhaustive grid-search over all task and parameter combinations, calling into question the practicality of such an approach in production. Hence, if we are to realise these performance gains, a cheaper approach to finding effective pairs of tasks is needed.

3   Optimising Transfer Learning with Explainability

As the complexity of deep neural models grows exponentially, there is an increasing need for methods to enable a deeper understanding of the latent patterns of a neural model, such as when trying to understand cases where that model has failed. In order to understand this behaviour, we must explore methods of explaining the inner working of language models.

Explainability is a field focused on model understanding and the predictive transparency of machine learning-based systems. A number of explainability techniques take the form of gradient-based approaches [8, 10]. One such gradient-based approach, known as *attribution*-based explanations, allow us to assess what the dominant features were that contributed to a particular prediction. Various algorithms can assign an importance score to each given input feature

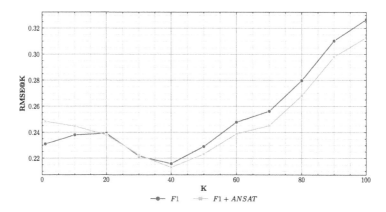

Fig. 1. RMSE@k results from our XGBoost regression model. RMSE metrics range from 0 to ∞, lower is better.

and effectively summarise and visualise these scores in a human-readable manner. Attribution-based explainability has become especially popular in the literature [5,6,12–14], however, research into explainability for transfer learning is sparse.

In this work, we investigate: 1) whether there exists correlation between the shared, *important* linguistic properties of a pair of tasks and their combined performance output; 2) whether we can compute this relationship prior to training these combined models; and 3) whether we can, as a result, reduce the time taken to produce high-performance models. As such, we divide the remainder of this paper into the following research questions:

RQ1. Does there exist some degree of correlation between the shared, *active* terms between pairs of tasks and their combined performance output?

RQ2. Can we leverage this knowledge, prior to training, to reduce the overall runtime required to produce effective models?

To this end, we compute the *conductance* of latent features in the context of each document. Introduced by Dhamdhere et al. [4,11] the conductance of a hidden unit can be described as the flow of attributions via said unit. By computing the conductance, we are able to quantify the bearing each individual input feature has on a particular prediction (with respect to a given input sequence).

For each BERT→Target model and each document in our test set, we calculate the effect any individual feature (term) had on the prediction output of its document using conductance. The conductance c of each term x_i within a document is scored $\{c_{x_i} \in \mathbb{R} : -1 \leq c_{x_i} \leq 1\}$ wherein $c_{x_i} \in [-1, 0)$ represents conductance scores that attribute towards the negative class and $c_{x_i} \in (0, 1]$ attribute towards our target class. We eliminate negatively attributed terms in order to capture the most *active* terms that represent our target class. We determine activity by testing against a range of thresholds for term activity (TAT), beginning from the mean of positively-attributed conductance scores, 0.05, and

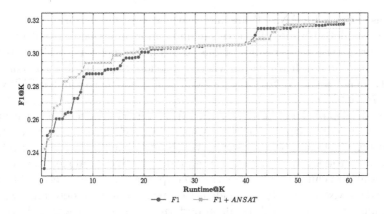

Fig. 2. Performance vs. Runtime results from XGBRegressor. F1 metrics range from 0 to 1, higher is better. Runtime is reported in hours.

increasing to a reasonable upper bound at 0.05 increments. As such, we decided to test the set of conductance thresholds: $[0.05, 0.7] \cap 0.05\mathbb{Z}$. We then averaged the total number of active terms across documents and only consider those terms which are above said thresholds. Our calculations result in the following formulation, Average Number of Shared Active Terms ($ANSAT$), which provides a quantified comparison metric between each pair of models:

Definition 1. *Let $\mathcal{M}$ represent a neural model with layers $l \in L$ and $\mathcal{D}$ represent the collection of positive-class documents (with respect to task sets A, B, and AB) containing words $w \in d \in D$, and with conductance threshold TAT then:*

$$ANSAT(M_A, M_B, D, TAT)$$
$$= \sum_{d \in D} \frac{\left(\sum_{w \in d} \begin{cases} 1, & if \left(\frac{\sum_{l \in L_{M_A}} conduct(w,l)}{|L_{M_A}|} \geq TAT \right) AND \left(\frac{\sum_{l \in L_{M_B}} conduct(w,l)}{|L_{M_B}|} \geq TAT \right) \\ 0, & otherwise \end{cases} \right)}{|D|}$$

$$(1)$$

Through this formulation, we can estimate the pretraining similarity between two tasks (A and B) via their underlying datasets (positive-class documents only) D_A and D_B, as well as the intersection of both, D_{AB}. We then use these estimates to predict the effectiveness of a combined model M_{AB} created via transfer learning, i.e. BERT→Source(A)→Target(B). In particular, we train an XGBoost [2] regression model (XGBRegressor) to produce a prediction of the effectiveness of M_{AB}, given various feature combinations. We use this model to predict the combined performance of M_{AB} combinations for each target task B given a set of source tasks A∈S, using Positive F1 as our target metric.

To answer RQ1, we compare the performance predicted by our XGBoost model when using only individual model effectiveness (M_A and M_B F1-scores) as features vs. those same features + the ANSAT similarity estimations. If active terms as defined by ANSAT are indicative of transfer performance then the XGBoost model with these features should be more effective than the one without. Figure 1 shows the results of our experiment, reporting RMSE at ranks 10–100 with different feature sets, where $F1$ denotes M_A and M_B F1-scores and $ANSAT$ denotes the ANSAT scores for D_A, D_B and D_{AB} (under TAT values $[0.05, 0.7] \cap 0.05\mathbb{Z}$).

Near the top of the ranking (K = 5, 10), we observe the feature set using F1 only to marginally outperform F1 + ANSAT by 1.76% and 2.76%, respectively. At ranks K = 40 and above, however, we observe that the inclusion of ANSAT results in considerably lower error than using F1 scores alone. Indeed, from these results we can conclude that the overlap of active terms between tasks as measured by ANSAT is valuable evidence when attempting to determine whether the combination of tasks will result in performance gains, answering RQ1.

To answer RQ2, we consider the potential real-world benefits of such performance prediction models when used to reduce task-pair training time. For this experiment, we assume you have a certain budget to train K task-pair combinations and check their performance. For a task, the more combinations you try, the more likely you will find a good combination. As our XGBoost models are predicting which combinations will work well together, we can use this to determine the order of combinations to try, where the goal is to find the best performing combination for each task as early as possible, such that we can end the search early. Figure 2 reports the Positive F1 performance of the best performing model for different depths K, where the x-axis is a conversion of K into the number of hours needed to train that many models for all tasks (Runtime@K).

From the collection of 803 models used as the dataset for our regression model, our best, average performance (F1) was 0.3199, which took 60.6 h to train. By utilising our regression model, we are able to achieve an F1-score of 0.3003 (only 6.12% worse than our best-performing F1 model), at only 30 h or 50.5% less training time, using the $F1 + ANSAT$ feature space. If we were to accept a 0.034 or 10.78% reduction in F1-score, we can further reduce our time to 10 h or a 83.5% reduction in training time. We note that at lower ranks of K, we observe a consistent increase in performance when including ANSAT in our feature space. At ranks 7, and 10, we observe 8.71%, and 8.01% performance increases, respectively, when including ANSAT alongside F1. Considering these results, there is clearly significant scope for improving performance by leveraging attribution-based techniques, answering RQ2.

4   Conclusions and Future Work

In this work, we presented an approach for estimating the suitability for pairs of tasks to be used in transfer learning by comparing their shared, active terms. It is clear that there exists some correlation between term activity and performance, as highlighted by our results. By predicting the projected performance output of each task pair, we managed to achieve up to 83.5% reduction in training time (for only a 0.034 or 10.78% reduction in F1). However, while we have demonstrated the value of using conductance to estimate combined model performance pre-training, there is clearly more work needed to increase the accuracy of these estimations, and hence further reduce the space of models that need to be searched. As such, for future work, we propose further analysis into the quantifiable properties that constitute related tasks which could further improve inductive transfer between such tasks.

References

1. Bingel, J., Søgaard, A.: Identifying beneficial task relations for multi-task learning in deep neural networks. arXiv:1702.08303 [cs], February 2017
2. Chen, T., Guestrin, C.: Xgboost: A scalable tree boosting system. CoRR abs/1603.02754 (2016). http://arxiv.org/abs/1603.02754
3. Devlin, J., Chang, M.W., Lee, K., Toutanova, K.: Bert: pre-training of deep bidirectional transformers for language understanding. In: NAACL-HLT (2019)
4. Dhamdhere, K., Sundararajan, M., Yan, Q.: How important is a neuron? CoRR abs/1805.12233 (2018). http://arxiv.org/abs/1805.12233
5. Ismail, A.A., Gunady, M.K., Bravo, H.C., Feizi, S.: Benchmarking deep learning interpretability in time series predictions. CoRR abs/2010.13924 (2020). https://arxiv.org/abs/2010.13924
6. Liu, N., Ge, Y., Li, L., Hu, X., Chen, R., Choi, S.: Explainable recommender systems via resolving learning representations. CoRR abs/2008.09316 (2020). https://arxiv.org/abs/2008.09316
7. Mou, L., Meng, Z., Yan, R., Li, G., Xu, Y., Zhang, L., Jin, Z.: How Transferable are Neural Networks in NLP Applications? arXiv:1603.06111 [cs], October 2016
8. Mundhenk, T.N., Chen, B.Y., Friedland, G.: Efficient saliency maps for explainable AI. CoRR abs/1911.11293 (2019). http://arxiv.org/abs/1911.11293
9. Pan, S.J., Yang, Q.: A survey on transfer learning. IEEE Trans. Knowl. Data Eng. **22**, 1345–1359 (2010)
10. Ribeiro, M.T., Singh, S., Guestrin, C.: "why should I trust you?": Explaining the predictions of any classifier. CoRR abs/1602.04938 (2016). http://arxiv.org/abs/1602.04938
11. Sundararajan, M., Taly, A., Yan, Q.: Axiomatic attribution for deep networks. CoRR abs/1703.01365 (2017). http://arxiv.org/abs/1703.01365
12. Wu, Z., Kao, B., Wu, T.H., Yin, P., Liu, Q.: Perq: predicting, explaining, and rectifying failed questions in kb-qa systems. In: Proceedings of the 13th International Conference on Web Search and Data Mining, WSDM 2020, pp. 663–671. Association for Computing Machinery, New York (2020). https://doi.org/10.1145/3336191.3371782. https://doi.org/10.1145/3336191.3371782
13. Zeiler, M.D., Fergus, R.: Visualizing and understanding convolutional networks. CoRR abs/1311.2901 (2013). http://arxiv.org/abs/1311.2901
14. Zhang, Z., Rudra, K., Anand, A.: Explain and predict, and then predict again. CoRR abs/2101.04109 (2021). https://arxiv.org/abs/2101.04109

Establishing Strong Baselines For TripClick Health Retrieval

Sebastian Hofstätter[✉], Sophia Althammer, Mete Sertkan, and Allan Hanbury

TU Wien, Vienna, Austria
{sebastian.hofstatter,sophia.althammer,mete.sertkan,
allan.hanbury}@tuwien.ac.at

Abstract. We present strong Transformer-based re-ranking and dense retrieval baselines for the recently released TripClick health ad-hoc retrieval collection. We improve the – originally too noisy – training data with a simple negative sampling policy. We achieve large gains over BM25 in the re-ranking task of TripClick, which were not achieved with the original baselines. Furthermore, we study the impact of different domain-specific pre-trained models on TripClick. Finally, we show that dense retrieval outperforms BM25 by considerable margins, even with simple training procedures.

1 Introduction

The latest neural network advances in Information Retrieval (IR) – specifically the ad-hoc passage retrieval task – are driven by available training data, especially the large web-search-based MSMARCO collection [1]. Here, neural approaches lead to enormous effectiveness gains over traditional techniques [8,13,20,24]. A valid concern is the generalizability and applicability of the developed techniques to other domains and settings [14,16,31,34].

The newly released TripClick collection [27] with large-scale click log data from the *Trip Database*, a health search engine, provides us with the opportunity to re-test previously developed techniques on this new ad-hoc retrieval task: keyword search in the health domain with large training and evaluation sets. TripClick provides three different test sets (Head, Torso, Tail), grouped by their query frequency, so we can analyze model performance for different slices of the overall query distribution.

This study conducts a range of controlled ad-hoc retrieval experiments using pre-trained Transformer [32] models with various state-of-the-art retrieval architectures on the TripClick collection. We aim to reproduce effectiveness gains achieved on MSMARCO in the click-based health ad-hoc retrieval setting. Typically, neural ranking models are trained with a triple of one query, a relevant and a non-relevant passage. As part of our evaluation study, we discovered a flaw in the provided neural training data of TripClick: The original negative sampling strategy included non-clicked results, which led to inadequate training. Therefore, we re-created the training data with an improved negative sampling

M. Hagen et al. (Eds.): ECIR 2022, LNCS 13186, pp. 144–152, 2022.
https://doi.org/10.1007/978-3-030-99739-7_17

strategy, based solely on BM25 negatives, with better results than published baselines.

As the TripClick collection was released only recently, we are the first to study a wide-ranging number of BERT-style ranking architectures and answer the fundamental question:

RQ1. How do established ranking models perform on re-ranking TripClick?

In the re-ranking setting, where the neural models score a set of 200 candidates produced by BM25, we observe large effectiveness gains for $BERT_{CAT}$, ColBERT, and TK for every one of the three frequency-based query splits. $BERT_{CAT}$ improves over BM25 on Head by 100%, on Torso by 66% and Tail still by 50%.

We compare the general BERT-Base & DistilBERT with the domain-specific SciBERT & PubMedBERT models to answer:

RQ2. Which BERT-style pre-trained checkpoint performs best on TripClick?

Although the general domain models show good effectiveness results, they are outperformed by the domain-specific pre-training approaches. Here, PubMed-BERT slightly outperforms SciBERT on re-ranking with $BERT_{CAT}$ & ColBERT. An ensemble of all domain-specific models with $BERT_{CAT}$ again outperforms all previous approaches and sets new state-of-the-art results for TripClick.

Finally, we study the concept of retrieving passages directly from a nearest neighbor vector index, also referred to as dense retrieval, and answer:

RQ3. How well does dense retrieval work on TripClick?

Dense retrieval outperforms BM25 considerably for initial candidate retrieval, both in top-10 precision results and for all recall cutoffs, except top-1000. In contrast to re-ranking, SciBERT outperforms PuBMedBERT on dense retrieval results.

We publish our source code as well as the improved training triples at: https://github.com/sebastian-hofstaetter/tripclick.

2   Background

We describe the collection, the BERT-style pre-training instances, ranking architectures, and training procedures we use below.

2.1   TripClick Collection

TripClick contains 1.5 million passages (with an average length of 259 words), 680 thousand click-based training queries (with an average of 4.4 words), and $3,525$ test queries. The TripClick collection includes three test sets with $1,175$ queries each grouped by their frequency and called Head, Torso, and Tail queries.

For the Head queries a DCTR [3] click model was employed to created relevance signals, the other two sets use raw clicks.

In comparison to the widely analyzed MSMARCO collection [10], TripClick is yet to be fully understood. This includes the quality of the click labels and the effect of various filtering mechanisms of the professional search production UI, that are not part of the released data.[1]

2.2   Re-ranking and Retrieval Models

We study multiple architectures with different aspects on the efficiency vs. effectiveness tradeoff scale. Here, we give a brief overview, for more detailed comparisons see Hofstätter et al. [8].

BERT$_{CAT}$ – Concatenated Scoring. The base re-ranking model BERT$_{CAT}$ [20,24,40] concatenates query and passage sequences with special tokens and computes a score by reducing the pooled CLS representation with a single linear layer. It represents one of the current state-of-the art models in terms of effectiveness, however it exhibits many drawbacks in terms of efficiency [9,39].

ColBERT. The ColBERT model [13] delays the interactions between every query and document representation after BERT. The interactions in the ColBERT model are aggregated with a max-pooling per query term and sum of query-term scores. The aggregation only requires simple dot product computations, however the storage cost of pre-computing passage representations is very high as it depends on the total number of terms in the collection.

TK (Transformer-Kernel). The Transformer-Kernel model [12] is not based on BERT pre-training, but rather uses shallow and independently computed Transformers followed by a set of RBF kernels to count match signals in a term-by-term match matrix, for very efficient re-ranking.

BERT$_{DOT}$ – Dense Retrieva. The BERT$_{DOT}$ model matches a single CLS vector of the query with a single CLS vector of a passage [17,18,39], independently computed. This decomposition of interactions to a single dot-product allows us to pre-compute every contextualized passage representation and employ a nearest neighbor index for dense retrieval, without a traditional first stage.

2.3   Pre-trained BERT Instances

The 12-layer BERT-Base model [5] (and the 6-layer distilled version DistilBERT [29]) and its vocabulary are based on the Books Corpus and English Wikipedia articles. The SciBERT model [2] uses the identical architecture to the BERT-Base model, but the vocabulary and the weights are pre-trained on Semantic

---

[1] The TripDatabase allows users to use different ranking schemes, such as popularity, source quality and pure relevance, as well as filtering results by facets. Unfortunately, this information is not available in the public dataset.

Scholar articles (with 82% articles from the broad biomedical domain). Similarly the PubMedBERT model [7] and its vocabulary are trained on PubMed articles using the same architecture as the BERT model.

2.4 Related Studies

At the time of writing, this is the first paper evaluating on the novel TripClick collection. However many other tasks have been set up before in the biomedical retrieval domain, such as BioASQ [23], TREC Precision Medicine tracks [6,28] or the timely created TREC-COVID [22,30,33] (which is based on CORD-19 [35], a collection of scientific articles concerned with the coronavirus pandemic).

For TREC-COVID, MacAvaney et al. [19] train a neural re-ranking model on a subset of the MS MARCO dataset containing only medical terms (Med-MARCO) and demonstrate its domain-focused effectiveness on a transfer to TREC-COVID. Xiong et al. [38] and Lima et al. [15] explore medical domain specific BERT representations for the retrieval from the TREC-COVID corpus and show that using SciBERT for dense retrieval outperforms the BM25 baseline by a large margin. Wang et al. [36] explore continuous active learning for the retrieval task from the COVID-19 corpus, this method is also studied for retrieval in the precision medicine track [4,28]. Reddy et al. [26] demonstrate synthetic training for question answering of COVID-19 related questions.

Many of these related works are concerned with overcoming the lack of large training data on previous medical collections. Now with TripClick we have a large-scale medical retrieval dataset. In this paper we jumpstart work on this collection, by showcasing the effectiveness of neural ranking approaches on TripClick.

3 Experiment Design

n our experiment setup, we largely follow Hofstätter et al. [8], except where noted otherwise. Mainly we rely on PyTorch [25] and HuggingFace Transformer [37] libraries as foundation for our neural training and evaluation methods. For TK, we follow Rekabsaz et al. [27] and utilize a PubMed-trained 400 dimensional word embedding as starting point [21]. For validation and testing we utilize the data splits outlined in TripClick by Rekabsaz et al. [27].

3.1 Training Data Generation

The TripClick dataset conveniently comes with a set of pre-generated training triples for neural training. Nevertheless, we found this training set to produce less than optimal results and the trained BERT models show no robustness against increased re-ranking depth. This phenomena of having to tune the best re-ranking depth for effectiveness, rather than efficiency, has been studied as part of early non-BERT re-rankers [11]. With the advent of Transformer-based re-rankers, this technique became obsolete [12].

Table 1. Effectiveness results for the three frequency-binned TripClick query sets. *The nDCG & MRR cutoff is at rank 10.*

Model	BERT Instance	Head (DCTR) nDCG	MRR	Torso (RAW) nDCG	MRR	Tail (RAW) nDCG	MRR
Original Baselines							
1 BM25	–	.140	.276	.206	.283	.267	.258
2 ConvKNRM	–	.198	.420	.243	.347	.271	.265
3 TK	–	.208	.434	.272	.381	.295	.280
Our Re-Ranking (BM25 Top-200)							
4 TK	–	.232	.472	.300	.390	.345	.319
5 ColBERT	SciBERT	.270	.556	.326	.426	.374	.347
6	PubMedBERT-Abstract	.278	.557	.340	.431	.387	.361
7	DistilBERT	.272	.556	.333	.427	.381	.355
8	BERT-Base	.287	.579	.349	.453	.396	.366
9 BERT$_{CAT}$	SciBERT	.294	.595	.360	.459	.408	.377
10	PubMedBERT-Full	.298	.582	.365	.462	.412	.381
11	PubMedBERT-Abstract	.296	.587	.359	.456	.409	.380
12	*Ensemble (Lines: 9,10,11)*	**.303**	**.601**	**.370**	**.472**	**.420**	**.392**

In the TripClick dataset, the clicked results are considered as positives samples for training. However, we discovered a flaw in the published negative sampling procedure, that non-clicked results – ranked above the clicked ones – are included as negative sampled passages. We hypothesize this leads to many false negatives in the training set, confusing the models during training. We confirm this thesis by observing our training telemetry data, showing low pairwise training accuracy as well as a lack of clear distinction in the scoring margins of the BERT$_{CAT}$ models. For all results presented in this study we generate new training data with the following simple procedure:

1. We generate 500 BM25 candidates for every training query
2. For every pair of query - relevant (clicked) passage in the training set we randomly sample, without replacement, up to 20 negative candidates from the candidates created in 1.
   - We remove candidates present in the relevant pool, regardless of relevance grade.
   - We discard positional information (we expect position bias to be in the training data – a potential for future work).
3. After shuffling the training triples we save 10 million triples for training

Our new training set gave us a 45–50% improvement on MRR@10 (from .41 to .6) and nDCG@10 (from .21 to .30) for the HEAD validation queries using the same PubMedBERT$_{CAT}$ model and setup. The models are now also robust against increasing the re-ranking depth.

4   TripClick Effectiveness Results

In this section, we present the results for our research questions, first for re-ranking and then for dense retrieval.

4.1   Re-ranking

We present the original baselines, as well as our re-ranking results for all three frequency-based TripClick query sets in Table 1. All neural models re-rank the top-200 results of BM25. While the original baselines do improve the frequent Head queries by up to 6 points nDCG@10 (TK-L3 vs. BM25-L1); they hardly improve the Tail queries with only 1–3 points difference in nDCG@10 (CK-L2 & TK-L3 vs. BM25-L1). This is a pressing issue, as those queries make up 83% of all Trip searches [27].

Turning to our results in Table 1, to answer **RQ1** *How do established ranking models perform on re-ranking TripClick?* We can see that our training approach for TK (Line 4) strongly outperforms the original TK (L3), especially on the Tail queries. This is followed by ColBERT (L5 & 6) and BERT$_{CAT}$ (L7 to L12) which both improve strongly over the previous model. This trend directly follows previous observations of effectiveness improvements per model architecture on MSMARCO [8,13].

To understand if there is a clear benefit of the BERT model choice we study: **RQ2** *Which BERT-style pre-trained checkpoint performs best on TripClick?* We find that although the general domain models show good effectiveness results (L7 & 8), they are outperformed by the domain-specific pre-training approaches (L9 to L11). Here, PubMedBERT (L5 + L10 & 11) slightly outperforms SciBERT (L9 + L10 & 11) on re-ranking with BERT$_{CAT}$ & ColBERT. An ensemble of all domain-specific models with BERT$_{CAT}$ (L12) again outperforms all previous approaches, and sets new state-of-the-art results for TripClick.

4.2   Dense Retrieval

To answer **RQ3** *How well does dense retrieval work on TripClick?* we present our results in Table 2. Dense retrieval with BERT$_{DOT}$ (L13 to L15) outperforms BM25 (L1) considerably for initial candidate retrieval, both in terms of top-10 precision results, as well as for all recall cutoffs, except top-1000. We also provided the judgement coverage for the top-10 results, and surprisingly, the coverage for dense retrieval increases compared to BM25. Future annotation campaigns should explore the robustness of these click-based evaluation results.

Table 2. BERT$_{DOT}$ dense retrieval effectiveness results for the HEAD TripClick query set. *J@10 indicates the ratio of judged results at cutoff 10.*

Model	BERT Instance	Head (DCTR)					
		J@10	nDCG@10	MRR@10	R@100	R@200	R@1K
Original Baselines							
1 BM25	–	31%	.140	.276	.499	.621	**.834**
Retrieval (Full Collection Nearest Neighbor)							
13	DistilBERT	39%	.236	.512	.550	.648	.813
14 BERT$_{DOT}$	SciBERT	41%	**.243**	**.530**	.562	.640	.793
15	PubMedBERT	40%	.235	.509	**.582**	**.673**	.828

5 Conclusion

Test collection diversity is a fundamental requirement of IR research. Ideally, we as a community develop methods that work on the largest possible set of problem settings. However, neural models require large training sets, which restricted most of the foundational research to the public MSMARCO and other web search collections. Now, with TripClick we have a another large-scale collection available. In this paper we show that in contrast to the original baselines, neural models perform very well on TripClick – both in the re-ranking task and the full collection retrieval with nearest neighbor search. We make our techniques openly available to the community to foster diverse neural information retrieval research.

References

1. Bajaj, P., et al.: MS MARCO: a human generated MAchine Reading COmprehension dataset. In: Proceedings of NIPS (2016)
2. Beltagy, I., Lo, K., Cohan, A.: SciBERT: a pretrained language model for scientific text. In: Proceedings of EMNLP-IJCNLP (2019)
3. Chuklin, A., Markov, I., de Rijke, M.: Click Models for Web Search. Morgan & Claypool, San Rafael (2015)
4. Cormack, G., Grossman, M.: Technology-assisted review in empirical medicine: waterloo participation in clef ehealth 2018. In CLEF (Working Notes) (2018)
5. Devlin, J., Chang, M., Lee, K., Toutanova, K.: Bert: pre-training of deep bidirectional transformers for language understanding. In: Proceedings of NAACL (2019)
6. Fernández-Pichel, M., Losada, D., Pichel, J.C., Elsweiler, D.: Citius at the trec 2020 health misinformation track (2020)
7. Gu, Y., et al.: Domain-specific language model pretraining for biomedical natural language processing (2020)
8. Hofstätter, S., Althammer, S., Schröder, M., Sertkan, M., Hanbury, A.: Improving efficient neural ranking models with cross-architecture knowledge distillation. arXiv preprint2010.02666 (2020)

9. Hofstätter, S., Hanbury, A.: Let's measure run time! Extending the IR replicability infrastructure to include performance aspects. In: Proceedings of OSIRRC (2019)

10. Hofstätter, S., Lipani, A., Althammer, S., Zlabinger, M., Hanbury, A.: Mitigating the position bias of transformer models in passage re-ranking. In: Proceedings of ECIR (2021)

11. Hofstätter, S., Rekabsaz, N., Eickhoff, C., Hanbury, A.: On the effect of low-frequency terms on neural-IR models. In: Proceedings of SIGIR (2019)

12. Hofstätter, S., Zlabinger, M., Hanbury, A.: Interpretable & Time-Budget-Constrained Contextualization for Re-Ranking. In: Proceedings of ECAI (2020)

13. Khattab, O., Zaharia, M.: Colbert: efficient and effective passage search via contextualized late interaction over Bert. In: Proceedings of SIGIR (2020)

14. Li, M., Li, M., Xiong, K., Lin, J.: Multi-task dense retrieval via model uncertainty fusion for open-domain question answering. In: Findings of EMNLP (2021)

15. Lima, L.C., et al.: Denmark's participation in the search engine TREC COVID-19 challenge: lessons learned about searching for precise biomedical scientific information on COVID-19. arXiv preprint2011.12684 (2020)

16. Lin, J.: A proposed conceptual framework for a representational approach to information retrieval. arXiv preprint2110.01529 (2021)

17. Lu, W., Jiao, J., Zhang, R.: Twinbert: distilling knowledge to twin-structured Bert models for efficient retrieval. arXiv preprint arXiv:2002.06275 (2020)

18. Luan, Y., Eisenstein, J., Toutanova, K., Collins, M.: Sparse, dense, and attentional representations for text retrieval. arXiv preprint arXiv:2005.00181 (2020)

19. MacAvaney, S., Cohan, A., Goharian, N.: Sledge: a simple yet effective baseline for COVID-19 scientific knowledge search. arXiv preprint2005.02365 (2020)

20. MacAvaney, S., Yates, A., Cohan, A., Goharian, N.: CEDR: contextualized embeddings for document ranking. In: Proceedings of SIGIR (2019)

21. McDonald, R., Brokos, G.-I., Androutsopoulos, I.: Deep relevance ranking using enhanced document-query interactions. arXiv preprint1809.01682 (2018)

22. Möller, T., Reina, A., Jayakumar, R., Pietsch, M.: COVID-QA: a question answering dataset for COVID-19. In: Proceedings of the 1st Workshop on NLP for COVID-19 at ACL 2020, Online, July 2020. Association for Computational Linguistics (2020)

23. Nentidis, A., et al.: Overview of BioASQ 2020: the Eighth BioASQ challenge on large-scale biomedical semantic indexing and question answering, pp. 194–214, September 2020

24. Nogueira, R., Cho, K.: Passage re-ranking with bert. arXiv preprint arXiv:1901.04085 (2019)

25. Paszke, A., et al.: Automatic differentiation in PYTORCH. In: Proceedings of NIPS-W (2017)

26. Reddy, R.G., et al.: End-to-end QA on COVID-19: domain adaptation with synthetic training. arXiv preprint2012.01414 (2020)

27. Rekabsaz, N., Lesota, O., Schedl, M., Brassey, J., Eickhoff, C.: Tripclick: the log files of a large health web search engine. arXiv preprint2103.07901 (2021)

28. Roberts, K., et al.:. Overview of the TREC 2019 precision medicine track. The ... text REtrieval Conference: TREC. Text REtrieval Conference, 26 (2019)

29. Sanh, V., Debut, L., Chaumond, J., Wolf, T.: Distilbert, a distilled version of Bert: smaller, faster, cheaper and lighter. arXiv preprint arXiv:1910.01108 (2019)

30. Tang, R., et al.: Rapidly bootstrapping a question answering dataset for COVID-19. CoRR, abs/2004.11339 (2020)

31. Thakur, N., Reimers, N., Rücklé, A., Srivastava, A., Gurevych, I.: Beir: a heterogenous benchmark for zero-shot evaluation of information retrieval models. arXiv preprint arXiv:2104.08663 4 2021
32. Vaswani, A., Shazeer, N., Parmar, N., Uszkoreit, J., et al.: Attention is all you need. In: Proceedings of NIPS (2017)
33. Voorhees, E., et al.: TREC-COVID: constructing a pandemic information retrieval test collection. ArXiv, abs/2005.04474 (2020)
34. Wang, K., Reimers, N., Gurevych, I.: TSDAE: using transformer-based sequential denoising auto-encoderfor unsupervised sentence embedding learning. arXiv preprint arXiv:2104.06979, April 2021
35. Wang, L.L., et al.: CORD-19: the COVID-19 open research dataset. arXiv preprint2004.10706 (2020)
36. Wang, X.J., Grossman, M.R., Hyun, S.G.: Participation in TREC 2020 COVID track using continuous active learning. arXiv preprint2011.01453 (2020)
37. Wolf, T., et al.: Huggingface's transformers: state-of-the-art natural language processing. ArXiv, pages arXiv-1910 (2019)
38. Xiong, C., et al.: CMT in TREC-COVID round 2: mitigating the generalization gaps from web to special domain search. arXiv preprint2011.01580 (2020)
39. Xiong, L., et al.: Approximate nearest neighbor negative contrastive learning for dense text retrieval. arXiv preprint arXiv:2007.00808 (2020)
40. Yilmaz, Z.A., Yang, W., Zhang, H., Lin,J.: Cross-domain modeling of sentence-level evidence for document retrieval. In: Proceedings of EMNLP-IJCNLP (2019)

Less is Less: When are Snippets Insufficient for Human vs Machine Relevance Estimation?

Gabriella Kazai[1]([✉]), Bhaskar Mitra[2], Anlei Dong[3], Nick Craswell[4],
and Linjun Yang[4]

[1] Microsoft, Cambridge, UK
[2] Microsoft, Montréal, Canada
[3] Microsoft, Silicon Valley, USA
[4] Microsoft, Redmond, WA, USA
{gkazai,bmitra,anldong,nickcr,linjya}@microsoft.com

Abstract. Traditional information retrieval (IR) ranking models process the full text of documents. Newer models based on Transformers, however, would incur a high computational cost when processing long texts, so typically use only snippets from the document instead. The model's input based on a document's URL, title, and snippet (UTS) is akin to the summaries that appear on a search engine results page (SERP) to help searchers decide which result to click. This raises questions about when such summaries are sufficient for relevance estimation by the ranking model or the human assessor, and whether humans and machines benefit from the document's full text in similar ways. To answer these questions, we study human and neural model based relevance assessments on 12k query-documents sampled from Bing's search logs. We compare changes in the relevance assessments when only the document summaries and when the full text is also exposed to assessors, studying a range of query and document properties, e.g., query type, snippet length. Our findings show that the full text is beneficial for humans and a BERT model for similar query and document types, e.g., tail, long queries. A closer look, however, reveals that humans and machines respond to the additional input in very different ways. Adding the full text can also hurt the ranker's performance, e.g., for navigational queries.

Keywords: Relevance estimation · Crowdsourcing · Neural IR

1 Introduction

In adhoc retrieval, ranking models typically process text from the URL, title and body of the documents. While the URL and title are short, the body may include thousands of terms. Recently, Transformer-based ranking models have demonstrated significant improvements in retrieval effectiveness (Lin et al. 2020), but are notoriously memory and compute intensive. Their training and inference cost grows prohibitively with long input. A common solution is to estimate document

© The Author(s), under exclusive license to Springer Nature Switzerland AG 2022
M. Hagen et al. (Eds.): ECIR 2022, LNCS 13186, pp. 153–162, 2022.
https://doi.org/10.1007/978-3-030-99739-7_18

relevance based only on sub-parts of the document, e.g., query-biased snippets. Such approaches are motivated by the *scope hypothesis* (Robertson et al. 2009), which states that the relevance of a document can be inferred by considering only its most relevant parts. Several neural approaches, *e.g.*, Hofstätter et al. (2021); Yan et al. (2019), have operationalized this hypothesis in their model design. Document summaries based on URL, title and query-biased snippet (UTS) are also typically presented on SERPs to searchers. While the model uses UTS to estimate relevance when ranking, the human searcher uses UTS to estimate relevance when deciding whether to click a result. These scenarios motivate us to study when snippets are sufficient replacements of the full body text for relevance estimation by humans and machines. Concretely, by collecting human relevance assessments and relevance rankings from a machine-learned model both for UTS only and UTS plus body text inputs, we study whether humans and machines benefit from the document's full text under similar conditions and in similar ways, or if humans and machines respond to the additional input differently.

2   Related Work

Automatic document summarization dates as far back as the foundational work by Luhn (1958) and Edmundson (1964). In the context of search, several early user studies (Tombros and Sanderson, 1998; Sanderson, 1998; White et al. 2003) demonstrated the usefulness of query-biased snippets for assessing document relevance. Demeester et al. (2012, 2013) studied how well the document's relevance can be predicted based on the snippet alone in federated search. Unlike these prior works, our goal is to study the differences in human and machine relevance assessments when only document summaries or when also the body texts are inspected. Past studies have also employed diverse measures of snippet quality based on manual assessment (Kaisser et al. 2008), eye-tracking studies (Lagun and Agichtein, 2012; Cutrell and Guan, 2007), view-port analysis (Lagun and Agichtein, 2011), historical clickthrough data (Clarke et al. 2007; Yue et al. 2010), and A/B testing (Savenkov et al. 2011), but did not try to understand when and why human and model assessments differ.

The application of passage-based document views for adhoc document ranking have been explored in the context of traditional retrieval methods (Bendersky and Kurland, 2008; Salton et al. 1993), but gained more attention recently (Nogueira and Cho 2019; Yan et al. 2020; Hofstätter et al. 2020, 2021; Li et al. 2020) in the context of Transformer-based (Vaswani et al. 2017) neural ranking models. While these models typically evaluate several passages per document, single query-biased summaries can be applied under stricter efficiency concerns. Our work helps to understand the feasibility of estimating document relevance based on just the UTS information.

Finally, our work is similar to Bolotova et al. (2020) in the sense that we too study humans and a BERT model, but while Bolotova et al. (2020) focused on attention, we study changes in relevance estimation due to input change.

3   Experiment Design

To answer our research questions, we collect both human and neural model based relevance assessments in two conditions: 1) when the human/machine assessor is only shown the query-biased summary, made up of the URL, title and snippet (UTS), and 2) when the body text is also exposed (UTSB). We use snippets returned by Bing's API.

We collect relevance assessments from humans via a Human Intelligent Task (HIT) with multiple judging steps, ensuring that the same person labels both conditions. First, we ask assessors to estimate a search result's relevance to the query based on its UTS information alone (UTS label). We then show assessors the web page and ask them to re-assess its relevance (UTSB label). Both labels use a five point scale. Next, we ask if seeing the web page led to a revised assessment ('Revised'; this is auto-filled), if it helped to confirm the UTS based estimate ('Confirmed') or if the page did not provide further help in the assessment ('Not needed'). Finally, assessors are asked to highlight parts of the body text that explain why the body text provided additional benefit over the UTS. Figure 1 shows the final HIT state. We use UHRS, an internal crowdsourcing platform, to collect judgments from trusted, quality monitored, long-term judges and pay

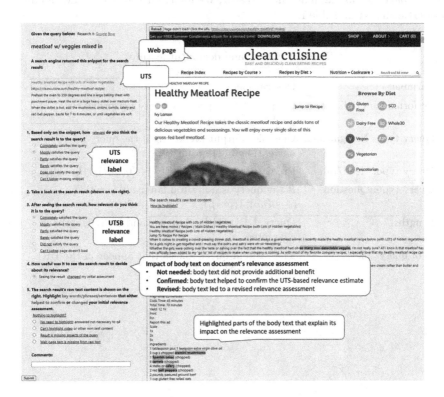

Fig. 1. Human intelligent task to collect UTS and UTSB labels from assessors

them their standard hourly rate. We obtain an inter-assessor agreement rate of
0.44 for the UTS and 0.53 for the UTSB labels (Krippendorff α).

For our Neural Ranker based relevance estimation, we follow the state-of-the-
art neural ranking approach (Nogueira and Cho 2019) and train a UTS and a
UTSB ranker, starting with a pretrained BERT-style (Devlin et al. 2019) model.
The model inputs comprise sentence A, which is the query, and sentence B, which
is either UTS or UTSB, respectively. Query and UTS have an expected length of
less than 128 tokens, so we use an input sequence length of 512 tokens in all our
experiments, truncating the input if it is longer. This allows the UTSB model
to see significantly more document text than is seen from snippet alone, and
allows us to observe systematic differences between UTS and UTSB. We use the
[CLS] vector as input to a single layer neural network to obtain the probability
of the document being relevant. We refer to the probability prediction values as
UTS and UTSB ranking scores and use the ranking orders they impose to study
whether neural models benefit from the body text.

For our dataset, we sample 1k queries at random from Bing's search logs, then
for each query, we scrape the Bing SERP and collect a total of 12k query-URL
pairs. We collect human labels for every query-URL and run ranking experiments with our dataset as the test set. For our investigation of when the body
text impacts a human/machine assessor, we focus on the query and document
properties listed in Table 1.

Table 1. Query and document features

Variable	Description
Performance predictor	Output of a proprietary query performance prediction model ($\in [0,1]$)
Query type: Navigational	Classifier output predicting if the query is navigational (1) or not (0)
Query type: Head/tail	Predicted query popularity ($\in [0(tail), 1(head)]$)
Query type: Question	If the query is a natural language question ($\in [0(no), 1(yes)]$)
Lengths	Query, URL, Title, Snippet, and Body lengths in characters
% of query tokens	The ratio of query tokens that appear in the URL, Title, Snippet, Body

Table 2. The UTSB model's performance improvement over the UTS model, measured using RBP (on a 100 point scale) and either the UTS or UTSB human labels as ground-truth (GT).

	ΔRBP@3	ΔRBP@10
UTS label GT	0.165	0.071
UTSB label GT	0.797	0.587
% improved/degraded	33/31	45/43

Table 3. Reasons when human assessors could not highlight parts of the body text to explain why it was beneficial over the UTS

	UTS > UTSB	UTS < UTSB
Missing term	76%	12%
Other	20%	48%
Video	4%	40%

4   Results and Discussions

Impact of Body Text on Human Assessors: We stipulate that UTS alone is insufficient in cases when human assessors either revised their initial assessment upon seeing the body text ('Revised') or when the body text was needed to confirm their UTS label ('Confirmed'). Overall, assessors indicated that UTS alone was insufficient (body text was beneficial) in 48% of the cases. Of these, 'Revised' made up 59% and 'Confirmed' the other 41%. When assessors revised their ratings, they initially overestimated the document's relevance in 54% of cases (UTS > UTSB) and underestimated it in 46% of cases (UTS < UTSB). The higher ratio of overestimates could hint at possible SEO manipulation methods succeeding or assessors exhibiting confirmation bias with UTS. Using statistical analysis (t-test) to compare the sample means of the query document properties (Table 1) across cases where the body text benefited judges or not, we found that the body text was helpful for predictably poor performing, long, not-navigational, tail and question type queries (all stat. sig. $p < 0.01$).

Impact of Body Text on Neural Ranker: We assume that UTS is insufficient when the UTSB model outperforms the UTS model. We calculate the two models' performance using RBP (Moffat and Zobel, 2008) with both the human UTS and UTSB labels as ground-truths. As it can be seen in Table 2, the UTSB model outperforms the UTS model (ΔRBP > 0), where the benefit from body text is more evident at the top ranks (ΔRBP@3 > ΔRBP@10). We also see that the ranker learns to make better use of the body text when the training labels also consider the body text (2nd row). Looking at the ratio of queries where the UTSB model outperforms the UTS model (3rd row), we see that there is room for improvement: the percentage of queries that benefit from the body text is just higher than those that body text degrades. Differences in the sample means of the query document properties (Table 1) for the improved and degraded queries reveals that improved queries are long, tail, not-navigational and of question type, while degraded queries are short, head and navigational, and the documents long (all stat. sig. $p < 0.01$).

Explanation of Body Text's Impact: We make use of the interpretML framework[1] and train two Explainable Boosting Machine (EBM) glassbox regression models (tree-based, cyclic gradient boosting Generalized Additive Models) (Lou et al. 2013). For each query-URL pair input, we use the properties listed in Table 1 as features and construct the target labels as follows:

- Δ**Label**: Target label for the EBM model used to explain human assessors' reaction to seeing the body text, mapped as –1 if UTS > UTSB (UTS label overestimated document relevance), 0 if UTS = UTSB, and 1 if UTS < UTSB (UTS underestimated).
- Δ**Rank**: To model the neural rankers' reaction we opt to use the ranking position (rp) since the UTS and UTSB scores are not directly comparable (different trained models) and use –1 if UTSrp < UTSBrp (UTSB model's relevance estimation decreased compared to UTS), 0 if UTS rp = UTSB rp, and 1 if UTSrp > UTSBrp (UTSB's estimate increased compared to UTS).

Table 4. The EBM models' top 5 feature importance scores for human and machine assessors, explaining the delta observed in the human assessors' UTS and UTSB labels (ΔLabel) and the neural models' UTS and UTSB based rankings (ΔRank), respectively.

Δ**Label** (UTSB label - UTS label)	Δ**Rank** (UTS rp - UTSB rp)
Question (0.2825)	%QueryWords in Tokenized Body (0.2858)
Body length (0.2434)	Snippet length (0.2831)
Performance predictor (0.2418)	Title length (0.2478)
%QueryWords in Tokenized Title (0.2218)	Body length (0.1658)
Query length (0.2141)	%QueryWords in Tokenized Snippet (0.1459)

Table 4 shows the EBM models' top 5 feature importance scores for human and machine assessors, telling us which of the query and document properties explain the delta observed in the human assessors' UTS and UTSB labels (ΔLabel) and the neural models' UTS and UTSB based rankings (ΔRank), respectively. We can see that a change in labels or rankings is explained by very different factors: body length is the only common factor in the top 5. The top explanation of change in the humans' UTS vs UTSB assessments is whether the query is phrased as a question, while the top reason for the ranker is the ratio of query tokens that are present in the body text.

To examine how ΔLabel and ΔRank change with each feature, in Fig. 2, we plot EBM's learnt per-feature functions. Each plot shows how a given feature contributes to the model's prediction. For example, the Query length plot shows that for short queries, human assessors (blue line) are more likely to underestimate ($y > 0$) the document's relevance based on UTS alone, while for

---

[1] https://interpret.ml/docs/ebm.html.

long queries, they tend to overestimate $(y < 0)$. The neural model (orange line) shows a similar but more subtle trend: for short queries, body text increases the ranker's relevance estimate over the UTS estimate, while for long queries the predicted relevance decreases with body text. The Question plot shows that humans tend to underestimate the document's relevance when the query is more likely to be a question. This indicates that document summaries fail to convince searchers that the document answers their question. The ranker's predicted relevance, however, decreases with body text for question type queries. Looking at the Snippet length plot, we see that the neural model is more likely to decrease its estimate of the document's relevance with body text when snippets are short, but increase it for long snippets. This suggests that when snippets include more context, the ranker is more likely to see these as evidence of irrelevance, which is diminished when body text is added. Snippet length has the opposite impact on humans: the longer the snippet, the more likely they overestimate the result's relevance. Overall, we see very little similarities (parallel trends) in the human vs

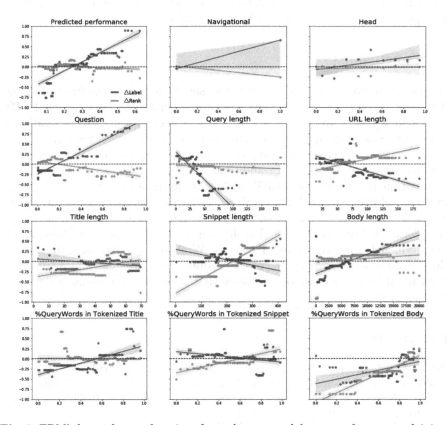

Fig. 2. EBM's learnt feature functions for each query and document feature, explaining the Δ changes: $y > 0$ means that 'seeing' the body text led to an increase in the relevance estimate compared to UTS. (Color figure online)

ranker feature plots, indicating that humans and machines react to body text in fundamentally different ways. Human assessors are more likely to overestimate relevance from UTS for long, tail, and not-navigational queries, and underestimate when the query is head, navigational or a question. They also overestimate for long snippets and short documents, and underestimate for long documents and short snippet. Unlike humans, the neural model results in more near-flat plots: the most impact is seen for document (rather than query) properties, e.g., Snippet length and ratio of query tokens in the snippet and body.

Additional Considerations: When assessors revised their relevance assessment but were unable to highlight parts of the body text to explain the change (in 72% of overestimates and 22% of underestimates), they were asked to indicate a reason. Table 3 shows that the absence of query terms in the document was the main reason for overestimates without highlighted text (76%). This suggests that informing users of missing query terms on the SERP is a helpful strategy. On the other hand, a major reason when assessors underestimated a document was when video (or other non-textual content) was present on the page (40%) - an aspect that was not considered by the neural model.

5   Conclusions

We studied when human and machine assessors benefit from the full text of the document to estimate its relevance. We showed that both humans and BERT style models benefit from the body text in similar cases (long, not navigational, tail and question type queries), but that full text impacts their relevance assessments in different ways (e.g., full text increases humans' relevance estimates but decreases the ranker's). In addition, we observe differences in the properties of queries where the BERT model's performance improves or degrades with the full text, e.g., performance degrades for navigational queries (ΔRBP@3 of –1.07). This indicates that more work is necessary on BERT style models when considering full text as input or that different types of queries (e.g., head v tail) require models to be optimized differently. While our findings are a function of the query-biased summaries, the observed differences in human and model reactions to additional information indicate that different mechanisms are needed for human vs machine inputs.

References

Bendersky, M., Kurland, O.: Utilizing passage-based language models for document retrieval. In: Macdonald, C., Ounis, I., Plachouras, V., Ruthven, I., White, R.W. (eds.) ECIR 2008. LNCS, vol. 4956, pp. 162–174. Springer, Heidelberg (2008). https://doi.org/10.1007/978-3-540-78646-7_17

Bolotova, V., Blinov, V., Zheng, Y., Croft, W.B., Scholer, F., Sanderson, M.: Do people and neural nets pay attention to the same words: studying eye-tracking data for non-factoid QA evaluation. In: Proceedings of CIKM, pp. 85–94 (2020). https://doi.org/10.1145/3340531.3412043

Clarke, C.L., Agichtein, E., Dumais, S., White, R.W.: The influence of caption features on clickthrough patterns in web search. In: Proceedings of SIGIR, pp. 135–142. ACM (2007)

Cutrell, E., Guan, Z.: What are you looking for? an eye-tracking study of information usage in web search. In: Proceedings of the SIGCHI conference on Human factors in computing systems, pp. 407–416 (2007)

Demeester, T., Nguyen, D., Trieschnigg, D., Develder, C., Hiemstra, D.: What snippets say about pages in federated web search. In: Hou, Y., Nie, J.-Y., Sun, L., Wang, B., Zhang, P. (eds.) AIRS 2012. LNCS, vol. 7675, pp. 250–261. Springer, Heidelberg (2012). https://doi.org/10.1007/978-3-642-35341-3_21

Demeester, T., Nguyen, D., Trieschnigg, D., Develder, C., Hiemstra, D.: Snippet-based relevance predictions for federated web search. In: Serdyukov, P., et al. (eds.) ECIR 2013. LNCS, vol. 7814, pp. 697–700. Springer, Heidelberg (2013). https://doi.org/10.1007/978-3-642-36973-5_63

Devlin, J., Chang, M.W., Lee, K., Toutanova, K.: BERT: pre-training of deep bidirectional transformers for language understanding. In: Proceedings of the 2019 Conference of the North American Chapter of the Association for Computational Linguistics: Human Language Technologies, Volume 1 (Long and Short Papers), pp. 4171–4186, Association for Computational Linguistics, Minneapolis, Minnesota, June 2019. https://doi.org/10.18653/v1/N19-1423, https://www.aclweb.org/anthology/N19-1423

Edmundson, H.: Problems in automatic abstracting. Commun. ACM 7(4), 259–263 (1964)

Hofstätter, S., Mitra, B., Zamani, H., Craswell, N., Hanbury, A.: Intra-document cascading: learning to select passages for neural document ranking. In: Proceedings of SIGIR, ACM. ACM (2021)

Hofstätter, S., Zamani, H., Mitra, B., Craswell, N., Hanbury, A.: Local self-attention over long text for efficient document retrieval. In: Proceedings of SIGIR. ACM (2020)

Kaisser, M., Hearst, M.A., Lowe, J.B.: Improving search results quality by customizing summary lengths. In: Proceedings of ACL-08: HLT, pp. 701–709 (2008)

Lagun, D., Agichtein, E.: Viewser: Enabling large-scale remote user studies of web search examination and interaction. In: Proceedings of SIGIR, pp. 365–374. ACM (2011)

Lagun, D., Agichtein, E.: Re-examining search result snippet examination time for relevance estimation. In: Proceedings of SIGIR, pp. 1141–1142. ACM (2012)

Li, C., Yates, A., MacAvaney, S., He, B., Sun, Y.: Parade: Passage representation aggregation for document reranking. arXiv preprint arXiv:2008.09093 (2020)

Lin, J., Nogueira, R., Yates, A.: Pretrained transformers for text ranking: Bert and beyond. arXiv preprint arXiv:2010.06467 (2020)

Lou, Y., Caruana, R., Gehrke, J., Hooker, G.: Accurate intelligible models with pairwise interactions. In: Proceedings of the 19th ACM SIGKDD International Conference on Knowledge Discovery and Data Mining, pp. 623–631, KDD 2013. ACM (2013). https://doi.org/10.1145/2487575.2487579

Luhn, H.P.: The automatic creation of literature abstracts. IBM J. Res. Dev. 2(2), 159–165 (1958)

Moffat, A., Zobel, J.: Rank-biased precision for measurement of retrieval effectiveness. ACM Trans. Inf. Syst. 27(1) (2008). https://doi.org/10.1145/1416950.1416952

Nogueira, R., Cho, K.: Passage re-ranking with BERT. CoRR abs/1901.04085 (2019). http://arxiv.org/abs/1901.04085

Robertson, S., Zaragoza, H., et al.: The probabilistic relevance framework: Bm25 and beyond. Found. Trends® Inf. Retrieval 3(4), 333–389 (2009)

Salton, G., Allan, J., Buckley, C.: Approaches to passage retrieval in full text information systems. In: Proceedings of SIGIR. ACM (1993)

Sanderson, M.: Accurate user directed summarization from existing tools. In: Proceedings of CIKM, pp. 45–51. ACM (1998). https://doi.org/10.1145/288627.288640

Savenkov, D., Braslavski, P., Lebedev, M.: Search snippet evaluation at yandex: lessons learned and future directions. In: Forner, P., Gonzalo, J., Kekäläinen, J., Lalmas, M., de Rijke, M. (eds.) CLEF 2011. LNCS, vol. 6941, pp. 14–25. Springer, Heidelberg (2011). https://doi.org/10.1007/978-3-642-23708-9_4

Tombros, A., Sanderson, M.: Advantages of query biased summaries in information retrieval. In: Proceedings of SIGIR, pp. 2–10. ACM (1998)

Vaswani, A., et al.: Attention is all you need. In: Proceedings of NeurIPS (2017)

White, R.W., Jose, J.M., Ruthven, I.: A task-oriented study on the influencing effects of query-biased summarisation in web searching. Inf. Process. Manag. **39**(5), 707–733 (2003)

Yan, M., et al.: IDST at TREC 2019 deep learning track: Deep cascade ranking with generation-based document expansion and pre-trained language modeling. In: TREC (2019)

Yan, M., et al.: IDST at TREC 2019 deep learning track: Deep cascade ranking with generation-based document expansion and pre-trained language modeling. In: TREC (2020)

Yue, Y., Patel, R., Roehrig, H.: Beyond position bias: examining result attractiveness as a source of presentation bias in clickthrough data. In: Proceedings of the 19th International Conference on World Wide Web, pp. 1011–1018. ACM (2010)

Leveraging Transformer Self Attention Encoder for Crisis Event Detection in Short Texts

Pantelis Kyriakidis$^{(\boxtimes)}$, Despoina Chatzakou, Theodora Tsikrika, Stefanos Vrochidis, and Ioannis Kompatsiaris

Information Technologies Institute, Centre for Research and Technology Hellas, Thessaloniki, Greece
{pantelisk,dchatzakou,theodora.tsikrika,stefanos,ikom}@iti.gr

Abstract. Analyzing content generated on social media has proven to be a powerful tool for early detection of crisis-related events. Such an analysis may allow for timely action, mitigating or even preventing altogether the effects of a crisis. However, the high noise levels in short texts present in microblogging platforms, combined with the limited publicly available datasets have rendered the task difficult. Here, we propose deep learning models based on a transformer self-attention encoder, which is capable of detecting event-related parts in a text, while also minimizing potential noise levels. Our models' efficacy is shown by experimenting with CrisisLexT26, achieving up to 81.6% f1-score and 92.7% AUC.

Keywords: Self attention · Multihead attention · Crisis event detection

1 Introduction

Over the years, many methods have been introduced in an effort to effectively detect crisis events from online textual content, and thus keep relevant stakeholders and the community at large informed about these events and their aftermath. Crisis event detection methods were initially utilizing handcrafted feature engineering to enrich their models with semantic and linguistic knowledge [18,23,24]. Nevertheless, the handcrafted features are not able to capture the multilevel correlations between words and tend to be overspecialized to the designed domain (e.g. applied on a single language), time consuming, and prone to error propagation. However, over the last decade, Convolutional Neural Networks (CNNs) [22] and Recurrent Neural Networks (RNNs) [17] proved their ability to capture semantic information [20,34] and, thereafter, pioneering works introduced CNNs [8,32], RNNs [31], and even hybrid approaches [13] to the event detection task. Graph Convolutional Networks (GCNs) which enable the convolution of words that are dependent on each other by using the syntactic representation of a text have also been employed [30]. Finally, attention mechanisms have been

© The Author(s), under exclusive license to Springer Nature Switzerland AG 2022
M. Hagen et al. (Eds.): ECIR 2022, LNCS 13186, pp. 163–171, 2022.
https://doi.org/10.1007/978-3-030-99739-7_19

applied, either in combination with GCNs [36] or in a more straightforward way [26], to combine attention vectors for words and entities with the original word embeddings before being forwarded to a perceptron layer. However all these studies focused on the sentence level of large and well organized documents by identifying event trigger words.

Another research area for event detection has been the categorization of short social media posts by their informativeness during a crisis. A widely used dataset in this context is the publicly available CrisisLexT26 [33], which contains Twitter posts from 26 different crisis events. CNNs have been extensively studied for this task (e.g. [5,6,29]), with one of the most well known architectures for sentence classification, the Multi-Channel CNN [20], achieving noteworthy results [4].

However, little emphasis has been placed thus far on handling short, noisy (informal language, syntactic errors, unordered sumarizations, etc.) texts with only a few studies utilizing attention mechanisms in this direction [7,19]. To this end, this paper proposes a novel way of dealing with such texts in the context of the crisis-related event detection. Our method is based on a state-of-the-art self attention encoder [35], which is utilized as a denoiser for the text before it is further forwarded to other types of layers; to the best of our knowledge, no prior work investigates the effect of the self-attention encoder in the event detection task. We also propose three models modifying to some extent the way in which the self attention is utilized, while experimenting with different neural architectures. To be comparable with the state of the art, the Multi-Channel CNN [4] is used as baseline, as well as a variation thereof. Finally, we open-source our implementations for reproducibility and extensibility purposes [1].

2   Methodology

To handle short and noisy texts, we base our models on a self attention method. We assume that the attention will be immune to any temporal inconsistency in the text and be able to reinforce dependencies between relevant words. Before presenting the proposed architectures, we first describe their key components.

Language Modeling. To create a mathematical representation of the input words we use an embedding of the language into a high dimensional euclidean space where neighboring words are also semantically close; input texts are modeled in a D-dimensional vector based on the words' vectors representations. We chose a 300-dimensional Word2Vec model [28] (pretrained on Google news) so as to be comparable with the chosen baseline [4]. Furthermore, to give sequential information to each word embedding, other works (e.g. [35]) add positional encoding to the embedding vector [15]. However, we argue that in short texts (e.g. Twitter posts), positional information would be unnecessary in the embedding layer, as these texts tend to be very unorganized and high in noise.

Self Attention Encoder. Overall, the Self Attention Encoder [35] consists of a block of 2 sub-layers. The first is a *multi-head self attention* and the second is a *position-wise feed-forward* layer, i.e. a fully connected layer with shared parameters over the sequence, applied in each position. Each layer output is added with a residual connection [16] from the previous layer output followed by Layer Normalization [3] aiming to a more stable and better regularized network.

Intuitively, the attention vector of a query word sums to 1 and spans the "attention" of the query to the most important words (keys); we use as queries (Q), as keys (K), and as values (V) the same input sequence, where every word is a query, a key, and a value, and their dimension is d. The attention mechanism that is applied at the core of each head is a scaled dot product attention (SDPA) which is a normalized version of the simple dot product attention and adheres to the following equations: (1): $E = \frac{(QK^T)}{\sqrt{d}}$; the attention scores are calculated, (2): $A = softmax(E)$; the attention is distributed to every key based on the attention scores, and (3): $C = AV$; the output context vector is calculated as an attention weighted sum of all values.

Multi-head Attention. This is the layer where the attentions are calculated. In order to attend in more than one ways, queries, keys, and values are projected h times through learned projection matrices, where h is the number of attention heads. These h tuples of Q, K, V are forwarded to each head's SDPA. Outputs are concatenated, and once again projected through a learned weight matrix. In the original work, these projections are linear [35]. However, we found that in our case non-linear projection with the Rectifier Linear Unit (ReLU) as an activation function boosted the performance.

GRU and CNN. GRUs are well known neural architectures capable of capturing sequential information [9]. Specifically, an update and reset gate is used to decide what information should be passed to the output. CNNs [22] capture salient features from chunks of information where each chunk is an n-gram of words from the input text where the convolution operation is performed on.

2.1 Proposed Neural Network Architectures

In this section, we introduce the overall design of the three proposed architectures. All models are optimized with ADAM optimizer [21], learning rate of 0.001, and with dropout of 0.5 before the output, for regularization purposes.

Stacked-Self Attention Encoders (Stacked-SAE). The first proposed architecture consists of a stack of 4 (experimentally chosen) self-attention encoders. It is the deepest of the architectures and the most complex one. The output is aggregated with Global Average Pooling [25] and finally projected onto the output layer, i.e. a fully connected layer with softmax as activation function. We expect that a deeper architecture will be able to create better representations and capture more complex patterns.

Attention Denoised Parallel GRUs (AD-PGRU). This and the next architecture use only one self-attention encoder as a feature extraction mechanism. As a result, the attention weighted output added to the input with the residual connection is expected to reinforce the important words eliminating significant part of the noise. Afterwards and inspired by [20], the signal is passed to a Parallel GRU architecture composed of 3 units that reduces the sequence to a single vector in the end and concatenates all before forwarding to the output. Each unit is expected to learn a different sequential representation of the input.

Attention Denoised Multi-channel CNN (AD-MCNN). The final architecture replaces the parallel GRU of the previous design, with the model proposed in [20] for sentence classification, i.e. three parallel CNN layers operating under different kernel sizes, so as to capture different N-gram combinations from the text, and a max-over-time pooling operation [10]. For the parameterization, we followed the settings proposed in [4], where the aforementioned architecture was repurposed for the event detection task (experimenting on CrisisLexT26).

3 Dataset and Experimental Setup

Ground Truth. For experimentation purposes, the CrisisLexT26 [33] dataset is used; it is publicly available and widely used in related work. It contains 26 different crisis events from 2012 and 2013 and consists of ≈28k tweets (≈1k posts per event). The labels for each tweet concern its: (i) *Informativeness*, whether it is related to the specific crisis or not, (ii) *Information Source*, e.g. government and NGO, and (iii) *Information Type*, e.g. affected individuals.

Experimental Setups. Two experimental setups were designed: (i) *Binary classification:* the focus is on the *Informativeness* category and the objective is to detect the relatedness of a post to a crisis event; and (ii) *Multi-class classification:* with the focus being on the *Information Type* category (7 types overall).

Since the class distribution is highly imbalanced we decided to train our models both for imbalanced and balanced dataset setups. For the balanced setup, we performed oversampling of the minority class using the pretrained BERT model [12] tailored to the Masked Language Model task [2].

We divided the dataset in a stratified fashion with a 0.8–0.2 train-test split, and derived a 10 run average for each experiment creating stochasticity by alternating the global network seed, while all the other randomized parameters, such as the dataset split seed, are constant. This is to ensure that all randomness comes from network weights initialization.

4 Experimental Results

Next, we evaluate the performance of the proposed architectures and compare the results to the baseline. We implemented two baseline architectures: (i) **Multi-channel CNN (MCNN)** [4], which has shown the best performance so far in the CrisisLexT26 dataset; and (ii) **MCNN-MA**: a recently proposed architecture for Sentiment Analysis [14] that uses MCNN and multi-head attention afterwards, which we adapted to our task with suitable parameterization.

Table 1. Experimental results. ("*": statistically significant over baselines)

Binary classification

	Imbalanced				Balanced			
	Precision	Recall	F1-score	AUC	Precision	Recall	F1-score	AUC
MCNN	**0.841**	0.772	0.800	0.921	0.798	0.793	0.793	0.918
MCNN-MA	0.772	0.771	0.770	0.888	0.691	0.768	0.711	0.871
Stacked-SAE	0.821	0.799*	0.808*	0.915	**0.809**	0.784	0.793	0.910
AD-PGRU	0.835	0.799*	0.814*	**0.927***	0.808	0.802	**0.803***	0.921
AD-MCNN	0.834	**0.802***	**0.816***	0.925*	0.804	**0.805**	0.802*	**0.923***

Multiclass classification

	Imbalanced				Balanced			
	Precision	Recall	F1-score	AUC	Precision	Recall	F1-score	AUC
MCNN	**0.671**	0.627	0.640	0.913	0.624	**0.648**	0.632	**0.910**
MCNN-MA	0.616	0.589	0.598	0.883	0.561	0.577	0.563	0.873
Stacked-SAE	0.644	0.630*	0.633	0.906	0.622	0.636	0.626	0.900
AD-PGRU	0.648	0.637*	0.638	0.910	**0.627**	0.640	0.630	0.909
AD-MCNN	0.656	**0.644***	**0.647***	0.914	0.627	0.648	0.633	0.910

For the **binary classification** (Table 1), we observe that the proposed architectures outperform the baselines in terms of F1 and AUC, both for the balanced and imbalanced datasets, with AD-MCNN being arguably the best performing. This confirms the hypothesis that using the self-attention encoder as a denoiser has a positive impact on the overall performance. When MCNN is used as a feature extractor on the embeddings, the convolution window limits the interactions to only neighboring words. So if attention is placed after the CNNs (MCNN-MA), the original signal is altered. On the contrary, the use of attention before MCNN resolves this issue, since it is neither restricted by the distance between words, nor by the words located between them, as an RNN would be. Class-specific results (not reported due to lack of space) on the imbalanced setup, indicate a substantial improvement on the minority class' recall (≈ 0.1 increase) for AD-MCNN. MCNN shows better—though not statistically significant—precision because it lacks the ability to effectively separate the two classes (predicts fewer non-events); a problem somewhat addressed in the balanced setup, where the precision is balanced in conjunction with an optimized recall value. As for the **multiclass classification** (Table 1) the overall results follow a similar behavior with the binary one. Although we see comparable performance on the balanced setup, we argue that the AD-MCNN would widen the difference with MCNN if more samples per class were available (on average $\approx 2.8k$ original samples).

In Sect. 1 we claimed that our attention mechanism is utilized as a denoiser for the text before it is further forwarded to other types of layers. To support our argument we provide an illustrative example of the attention scoring of a tweet: *"RT @user: texas: massive explosion u/d - local hospitals notified. every available*

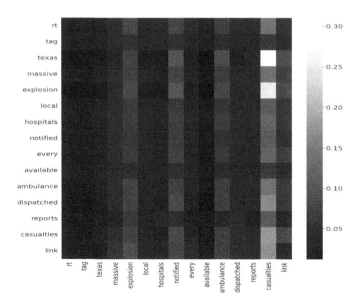

Fig. 1. Attention visualization: highest scores for combinations of location ("texas") and type of incident ("explosion") with an important consequence ("casualties").

ambulance dispatched. reports of casualties. HTTP: ...". Figure 1 depicts the scores of a single (due to space limitation) attention head, when applied to the above example and indicates that combinations of highly relevant words are being matched up with higher scores, while non-important combinations exhibit low attention scores, resulting in the claimed denoising behavior.

Finally, in Sect. 2 we argued that positional encoding might not be fit for this specific domain given the informal text used in Twitter, the brevity of the messages, the unordered use of hashtags, etc. Using the AD-MCNN method with the use of positional encoding, we validated this hypothesis, observing 0.01 and 0.004 performance decrease in F1-score and AUC, respectively.

Complexity. In terms of model size, the attention denoiser adds 361.200 parameters to be learned to the 461.954 of the MCNN. It also adds 4.378.851 FLOPS to the 23.930.923 of MCNN. We should mention that the complexity will vary depending on the task as the layer uses the input text shape internally. Also, contrary to an RNN layer (where the computation is serialized), the Transformer can process the input sequence in parallel making it notably faster.

5   Conclusions and Future Work

This work proposed three effective models for crisis-related event detection while combating the noise inherent in short social media posts. Our hypothesis was that self attention would act as a denoiser, enhancing important features i.e., every vector in the sequence is enhanced with context from other directly related word vectors. We validated our hypothesis by building and evaluating three attention enhanced models that improved performance against strong baselines. In the future, we intend to further evaluate our models, especially when more data is available, as well as to evaluate the impact of more recent language models for word embeddings (e.g. [12,27]), especially multilingual ones [11].

Acknowledgements. This research has received funding from the European Union's H2020 research and innovation programme as part of the INFINITY (GA No 883293) and AIDA (GA No 883596) projects.

References

1. Crisis-event-detection-in-short-texts - implementation. https://github.com/M4D-MKLab-ITI/Crisis-Event-Detection-in-Short-Texts
2. Bert for masked lm (2020). shorturl.at/drRV4
3. Ba, J.L., Kiros, J.R., Hinton, G.E.: Layer normalization. Stat **1050**, 21 (2016)
4. Burel, G., Alani, H.: Crisis event extraction service (crees)-automatic detection and classification of crisis-related content on social media (2018)
5. Burel, G., Saif, H., Alani, H.: Semantic wide and deep learning for detecting crisis-information categories on social media. In: d'Amato, C., et al. (eds.) ISWC 2017. LNCS, vol. 10587, pp. 138–155. Springer, Cham (2017). https://doi.org/10.1007/978-3-319-68288-4_9
6. Caragea, C., Silvescu, A., Tapia, A.H.: Identifying informative messages in disaster events using convolutional neural networks. In: International Conference on Information Systems for Crisis Response and Management, pp. 137–147 (2016)
7. Chen, J., Hu, Y., Liu, J., Xiao, Y., Jiang, H.: Deep short text classification with knowledge powered attention. In: Proceedings of the AAAI Conference on Artificial Intelligence, vol. 33, pp. 6252–6259 (2019)
8. Chen, Y., Xu, L., Liu, K., Zeng, D., Zhao, J.: Event extraction via dynamic multi-pooling convolutional neural networks. In: Proceedings of the 53rd Annual Meeting of the Association for Computational Linguistics and the 7th International Joint Conference on Natural Language Processing (Volume 1: Long Papers), pp. 167–176 (2015)
9. Cho, K., Van Merriënboer, B., Gulcehre, C., Bahdanau, D., Bougares, F., Schwenk, H., Bengio, Y.: Learning phrase representations using rnn encoder-decoder for statistical machine translation. arXiv preprint arXiv:1406.1078 (2014)
10. Collobert, R., Weston, J., Bottou, L., Karlen, M., Kavukcuoglu, K., Kuksa, P.: Natural language processing (almost) from scratch. J. Mach. Learn. Res. **12**(ARTICLE), 2493–2537 (2011)
11. Conneau, A., et al.: Unsupervised cross-lingual representation learning at scale. In: Proceedings of the 58th Annual Meeting of the Association for Computational Linguistics, pp. 8440–8451 (2020)
12. Devlin, J., Chang, M.W., Lee, K., Toutanova, K.: Bert: Pre-training of deep bidirectional transformers for language understanding. In: Proceedings of the 2019 Conference of the North American Chapter of the Association for Computational Linguistics: Human Language Technologies, Volume 1 (Long and Short Papers), pp. 4171–4186 (2019)

13. Feng, X., Qin, B., Liu, T.: A language-independent neural network for event detection. Sci. China Inf. Sci. **61**(9), 1–12 (2018). https://doi.org/10.1007/s11432-017-9359-x

14. Feng, Y., Cheng, Y.: Short text sentiment analysis based on multi-channel CNN with multi-head attention mechanism. IEEE Access **9**, 19854–19863 (2021)

15. Gehring, J., Auli, M., Grangier, D., Yarats, D., Dauphin, Y.N.: Convolutional sequence to sequence learning. In: International Conference on Machine Learning, pp. 1243–1252. PMLR (2017)

16. He, K., Zhang, X., Ren, S., Sun, J.: Deep residual learning for image recognition. In: Proceedings of the IEEE Conference on Computer Vision and Pattern Recognition, pp. 770–778 (2016)

17. Hochreiter, S., Schmidhuber, J.: Long short-term memory. Neural Comput. **9**(8), 1735–1780 (1997)

18. Hong, Y., Zhang, J., Ma, B., Yao, J., Zhou, G., Zhu, Q.: Using cross-entity inference to improve event extraction. In: Proceedings of the 49th annual meeting of the association for computational linguistics: human language technologies, pp. 1127–1136 (2011)

19. Kabir, M.Y., Madria, S.: A deep learning approach for tweet classification and rescue scheduling for effective disaster management. In: Proceedings of the 27th ACM SIGSPATIAL International Conference on Advances in Geographic Information Systems, pp. 269–278 (2019)

20. Kim, Y.: Convolutional neural networks for sentence classification. In: Proceedings of the 2014 Conference on Empirical Methods in Natural Language Processing (EMNLP), pp. 1746–1751. Association for Computational Linguistics, Doha, Qatar, October 2014. https://doi.org/10.3115/v1/D14-1181, https://www.aclweb.org/anthology/D14-1181

21. Kingma, D.P., Ba, J.: Adam: a method for stochastic optimization. In: 3rd International Conference on Learning Representations, ICLR 2015 (2015)

22. LeCun, Y., Bottou, L., Bengio, Y., Haffner, P.: Gradient-based learning applied to document recognition. Proc. IEEE **86**(11), 2278–2324 (1998)

23. Li, Q., Ji, H., Hong, Y., Li, S.: Constructing information networks using one single model. In: Proceedings of the 2014 Conference on Empirical Methods in Natural Language Processing (EMNLP), pp. 1846–1851 (2014)

24. Li, Q., Ji, H., Huang, L.: Joint event extraction via structured prediction with global features. In: Proceedings of the 51st Annual Meeting of the Association for Computational Linguistics (Volume 1: Long Papers), pp. 73–82 (2013)

25. Lin, M., Chen, Q., Yan, S.: Network in network. arXiv preprint arXiv:1312.4400 (2013)

26. Liu, S., Chen, Y., Liu, K., Zhao, J.: Exploiting argument information to improve event detection via supervised attention mechanisms. In: Proceedings of the 55th Annual Meeting of the Association for Computational Linguistics (Volume 1: Long Papers), pp. 1789–1798 (2017)

27. Liu, Y., et al.: Roberta: A robustly optimized Bert pretraining approach. arXiv preprint arXiv:1907.11692 (2019)

28. Mikolov, T., Chen, K., Corrado, G., Dean, J.: Efficient estimation of word representations in vector space. arXiv preprint arXiv:1301.3781 (2013)

29. Nguyen, D., Al Mannai, K.A., Joty, S., Sajjad, H., Imran, M., Mitra, P.: Robust classification of crisis-related data on social networks using convolutional neural networks. In: Proceedings of the International AAAI Conference on Web and Social Media, vol. 11 (2017)

30. Nguyen, T., Grishman, R.: Graph convolutional networks with argument-aware pooling for event detection. In: Proceedings of the AAAI Conference on Artificial Intelligence, vol. 32 (2018)
31. Nguyen, T.H., Cho, K., Grishman, R.: Joint event extraction via recurrent neural networks. In: Proceedings of the 2016 Conference of the North American Chapter of the Association for Computational Linguistics: Human Language Technologies, pp. 300–309 (2016)
32. Nguyen, T.H., Grishman, R.: Event detection and domain adaptation with convolutional neural networks. In: Proceedings of the 53rd Annual Meeting of the Association for Computational Linguistics and the 7th International Joint Conference on Natural Language Processing (Volume 2: Short Papers), pp. 365–371 (2015)
33. Olteanu, A., Vieweg, S., Castillo, C.: What to expect when the unexpected happens: Social media communications across crises. In: Proceedings of the 18th ACM Conference on Computer Supported Cooperative Work & Social Computing (2015)
34. Tai, K.S., Socher, R., Manning, C.D.: Improved semantic representations from tree-structured long short-term memory networks. In: Proceedings of the 53rd Annual Meeting of the Association for Computational Linguistics and the 7th International Joint Conference on Natural Language Processing (Volume 1: Long Papers), pp. 1556–1566 (2015)
35. Vaswani, A., et al.: Attention is all you need. In: Proceedings of the 31st International Conference on Neural Information Processing Systems, pp. 6000–6010 (2017)
36. Yan, H., Jin, X., Meng, X., Guo, J., Cheng, X.: Event detection with multi-order graph convolution and aggregated attention. In: Proceedings of the 2019 Conference on Empirical Methods in Natural Language Processing and the 9th International Joint Conference on Natural Language Processing (EMNLP-IJCNLP), pp. 5770–5774 (2019)

What Drives Readership? An Online Study on User Interface Types and Popularity Bias Mitigation in News Article Recommendations

Emanuel Lacic[1], Leon Fadljevic[1], Franz Weissenboeck[2], Stefanie Lindstaedt[1], and Dominik Kowald[1(✉)]

[1] Know-Center GmbH, Graz, Austria
{elacic,lfadljevic,slind,dkowald}@know-center.at
[2] Die Presse Verlags-GmbH, Vienna, Austria
franz.weissenboeck@diepresse.com

Abstract. Personalized news recommender systems support readers in finding the right and relevant articles in online news platforms. In this paper, we discuss the introduction of personalized, content-based news recommendations on DiePresse, a popular Austrian online news platform, focusing on two specific aspects: (i) user interface type, and (ii) popularity bias mitigation. Therefore, we conducted a two-weeks online study that started in October 2020, in which we analyzed the impact of recommendations on two user groups, i.e., anonymous and subscribed users, and three user interface types, i.e., on a desktop, mobile and tablet device. With respect to user interface types, we find that the probability of a recommendation to be seen is the highest for desktop devices, while the probability of interacting with recommendations is the highest for mobile devices. With respect to popularity bias mitigation, we find that personalized, content-based news recommendations can lead to a more balanced distribution of news articles' readership popularity in the case of anonymous users. Apart from that, we find that significant events (e.g., the COVID-19 lockdown announcement in Austria and the Vienna terror attack) influence the general consumption behavior of popular articles for both, anonymous and subscribed users.

Keywords: News recommendation · User interface · Popularity bias

1 Introduction

Similar to domains such as social networks or social tagging systems [14,17,21], the personalization of online content has become one of the key drivers for news portals to increase user engagement and convince readers to become paying subscribers [8,9,22]. A natural way for news portals to do this, is to provide their users with articles that are fresh and popular. This is typically achieved via simple most-popular news recommendations, especially since this approach has been

M. Hagen et al. (Eds.): ECIR 2022, LNCS 13186, pp. 172–179, 2022.
https://doi.org/10.1007/978-3-030-99739-7_20

shown to provide accurate recommendations in offline evaluation settings [11]. However, such an approach could amplify popularity bias with respect to users' news consumption. This means that the equal representation of non-popular, but informative content in the recommendation lists is put into question, since articles from the "long tail" do not have the same chance of being represented and served to the user [1]. Since nowadays, readers tend to consume news content on smaller user interface types (e.g., mobile devices) [10,20], the impact of popularity bias may even get amplified due to the reduced number of recommendations that can be shown [12].

In this paper, we therefore discuss the introduction of personalized, content-based news articles on DiePresse, a popular Austrian news platform, focusing on two aspects: (i) user interface type, and (ii) popularity bias mitigation. To do so, we performed a two-weeks online study that started in October 2020, in which we compared the impact of recommendations with respect to different user groups, i.e., anonymous (cold-start [18]) and subscribed (logged-in and paying) users, as well as different user interface types, i.e., desktop, mobile and tablet devices (see Sect. 2). Specifically, we address two research questions:

RQ1: How does the user interface type impact the performance of news recommendations?

RQ2: Can we mitigate popularity bias by introducing personalized, content-based news recommendations?

We investigate RQ1 in Sect. 3.1 and RQ2 in Sect. 3.2. Additionally, we discuss the impact of two significant events, i.e., (i) the COVID-19 lockdown announcement in Austria, and (ii) the Vienna terror attack, on the consumption behavior of users. We hope that our findings will help other news platform providers assessing the impact of introducing personalized recommendations.

2 Experimental Setup

In order to answer our two research questions, we performed a two-weeks online user study, which started on the 27th of October 2020 and ended on the 9th of November 2020. Here, we focused on three user interface types, i.e., desktop, mobile and tablet devices, as well as investigated two user groups, i.e., anonymous and subscribed users. About 89% of the traffic (i.e., $2,371,451$ user interactions) was produced by the 1,182,912 anonymous users, where a majority of them (i.e., 77.3%) read news articles on a mobile device. Interestingly, the 15,910 subscribed users exhibited a more focused reading behavior and only interacted with a small subset of all articles that were read during our online study (i.e., around 18.7% out of $17,372$ articles). Within the two-weeks period, two significant events happened: (i) the COVID-19 lockdown announcement in Austria on the 31st of October 2020, and (ii) the Vienna terror attack on the 2nd of November 2020. The articles related to these events were the most popular ones in our study.

Calculation of Recommendations. We follow a content-based approach to recommend news articles to users [19]. Therefore, we represent each news article using a 25-dimensional topic vector calculated using Latent Dirichlet Allocation (LDA) [3]. Each user was also represented by a 25-dimensional topic vector, where the user's topic weights are calculated as the mean of the news articles' topic weights read by the user. In case of subscribed users, the read articles consist of the entire user history and in case of anonymous users, the read articles consist of the articles read in the current session. Next, these topic vectors are used to match users and news articles using Cosine similarity in order to find top-n news article recommendations for a given user. For our study, we set $n = 6$ recommended articles. For this step, only news articles are taken into account that have been published within the last 48 h. Additionally, editors had the possibility to also include older (but relevant) articles into this recommendation pool (e.g., a more general article describing COVID-19 measurements).

In total, we experimented with four variants of our content-based recommendation approach: (i) recommendations only including articles of the last 48 h, (ii) recommendations also including the editors' choices, and (iii) and (iv) recommendations, where we also included a collaborative component by mixing the user's topic vector with the topic vectors of similar users for the variants (i) and (ii), respectively. Additionally, we also tested a most-popular approach, since this algorithm was already present in DiePresse before the user study started. However, we did not find any significant differences between these five approaches with respect to recommendation accuracy in our two-weeks study and therefore, we did not distinguish between the approaches and report the results for all calculated recommendations in the remainder of this paper.

3 Results

3.1 RQ1: User Interface Type

Most studies focus on improving the accuracy of the recommendation algorithms, but recent research has shown that this has only a partial effect on the final user experience [13]. The user interface is namely a key factor that impacts the usability, acceptance and selection behavior within a recommender system [6]. Additionally, in news platforms, we can see a trend that shifts from classical desktop devices to mobile ones. Moreover, users are biased towards clicking on higher ranked results (i.e., position bias) [4]. When evaluating personalized news recommendations, it becomes even more important to understand the user acceptance of recommendations for smaller user interface types, where it is much harder for the user to see all recommended options due to the limited size. In our study, we therefore investigate to what extent the user interface type impacts the performance of news recommendations (RQ1). As mentioned, we differentiate between three different user interface types, i.e., interacting with articles on a (i) desktop, (ii) mobile, and (iii) tablet device. In order to measure the acceptance of recommendations shown via the chosen user interface type, we use the following two evaluation metrics [9]:

Table 1. RQ1: Acceptance of recommended articles with respect to user interface type.

Metric	Desktop	Mobile	Tablet
RSR: Recommendation-Seen-Ratio (%)	**26.88**	17.55	26.71
CTR: Click-Through-Rate (%)	10.53	**13.40**	11.37

Recommendation-Seen-Ratio (RSR) is defined as the ratio between the number of times the user actually saw recommendations (i.e., scrolled to the corresponding recommendation section in the user interface) and the number of recommendations that were generated for a user.

Click-Through-Rate (CTR) is measured by the ratio between the number of actually clicked recommendations and the number of seen recommendations.

As shown in Table 1, the smaller user interface size of a mobile device heavily impacts the probability of a user to actually see the list of recommended articles. This may be due to the fact that reaching the position where the recommendations are displayed is harder in comparison to a larger desktop or tablet device, where the recommendation section can be reached without scrolling. Interestingly enough, once a user has seen the list of recommended articles, users who use a mobile device exhibit a much higher CTR. Again, we hypothesize that if a user has put more effort into reaching the list of recommended articles, the user is more likely to accept the recommendation and interact with it.

When looking at Fig. 1, we can see a consistent trend during the two weeks of our study regarding the user interface types for both the RSR and CTR measures. However, notable differences are the fluctuations of the evaluation measures for the two significant events that happened during the study period. For instance, the positive peak in the RSR and the negative peak in CTR that can be spotted around the 31st of October was caused by the COVID-19 lockdown announcement in Austria. For the smaller user interfaces (i.e., mobile and tablet devices) this actually increased the likelihood of the recommendation to be seen since users have invested more energy in engaging with the content of the news articles. On the contrary, we saw a drop in the CTR, which was mostly caused by anonymous users since the content-based, personalized recommendations did not provide articles that they expected at that moment (i.e., popular ones solely related to the event). Another key event can be spotted on the 2nd of November, the day the Vienna terror attack happened. This was by far the most read article with a lot of attack-specific information during the period of the online study. Across all three user interface types, this has caused a drop in the likelihood of a recommendation to be seen at all. Interestingly enough, the CTR in this case does not seem to be influenced. We investigated this in more detail and noticed that a smaller drop was only noticeable for the relatively small number of subscribed users using a mobile device and thus, this does not influence the results shown in Fig. 1. The differences between all interface types shown in Table 1 and Fig. 1 are statistically significant according to a Kruskal-Wallis followed by a Dunn test except for mobile vs. tablet device with respect to CTR.

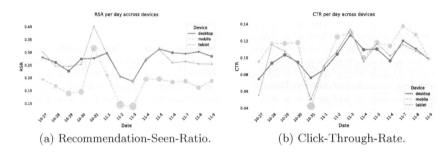

(a) Recommendation-Seen-Ratio.   (b) Click-Through-Rate.

Fig. 1. RQ1: Acceptance of recommended articles for the two weeks of our study with respect to (a) RSR, and (b) CTR. The size of the dots represent the number of reading events on a specific day for a specific user interface type.

3.2   RQ2: Mitigating Popularity Bias

Many recommender systems are affected by popularity bias, which leads to an overrepresentation of popular items in the recommendation lists. One potential issue of this is that unpopular items (i.e., so-called long-tail items) are recommended rarely [15,16]. The news article domain is an example where ignoring popularity bias could have a significant societal effect. For example, a potentially controversial news article could easily impose a narrow ideology to a large population of readers [7]. This effect could even be strengthened by providing unpersonalized, most-popular news recommendations as it is currently done by many online news platforms (including DiePresse) since these popularity-based approaches are easy to implement and also provide good offline recommendation performance [9,10]. We hypothesize that the introduction of personalized, content-based recommendations (see Sect. 2) could lead to more balanced recommendation lists in contrast to most-popular recommendations. This way also long-tail news articles are recommended and thus, popularity bias could be mitigated. Additionally, we believe that this effect differs between different user groups and thus, we distinguish between anonymous and subscribed users.

We measure popularity bias in news article consumption by means of the skewness [2] of the article popularity distribution, i.e., the distribution of the number of reads per article. Skewness measures the asymmetry of a probability distribution, and thus a high, positive skewness value depicts a right-tailed distribution, which indicates biased news consumption with respect to article popularity. On the contrary, a small skewness value depicts a more balanced popularity distribution with respect to head and tail, and thus indicates that also non-popular articles are read. As another measure, we calculate the kurtosis of the popularity distribution, which measures the "tailedness" of a distribution. Again, higher values indicate a higher tendency for popularity bias. For both metrics, we hypothesize that the values at the end of our two-weeks study are smaller than at the beginning, which would indicate that the personalized recommendations helped to mitigate popularity bias.

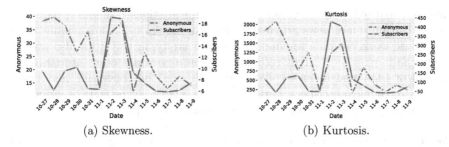

(a) Skewness.   (b) Kurtosis.

Fig. 2. RQ2: Impact of personalized, content-based recommendations on the popularity bias in news article consumption measured by (a) skewness and (b) kurtosis based on the number of article reads for each day.

The plots in Fig. 2 show the results addressing RQ2. For both metrics, i.e., skewness and kurtosis, we see a large gap between anonymous users and subscribers at the beginning of the study (i.e., 27th of October 2020), where only most-popular recommendations were shown to the users. While anonymous users have mainly read popular articles, subscribers were also interested in unpopular articles. This makes sense since subscribed users typically visit news portals for consuming articles within their area of interest, which will also include articles from the long-tail, while anonymous users typically visit news portals for getting a quick overview of recent events, which will mainly include popular articles. Based on this, a most-popular recommendation approach does not impact subscribers as much as it impacts anonymous users.

However, when looking at the last day of the study (i.e., 9th of November 2020), there is a considerably lower difference between anonymous and subscribed users anymore. We also see that the values at the beginning and at the end of the study are nearly the same in case of subscribed users, which shows that these users are not prone to popularity bias, and thus also personalized recommendations do not affect their reading behavior in this respect. With respect to RQ2, we find that the introduction of personalized recommendations can help to mitigate popularity bias in case of anonymous users. Furthermore, we see two significant peaks in the distributions that are in line with the COVID-19 lockdown announcement in Austria and the Vienna terror attack. Hence, in case of significant events also subscribed users are prone to popularity bias.

4   Conclusion

In this paper, we discussed the introduction of personalized, content-based news recommendations on DiePresse, a popular Austrian news platform, focusing on two specific aspects: user interface type (RQ1), and popularity bias mitigation (RQ2). With respect to RQ1, we find that the probability of recommendations to be seen is the highest for desktop devices, while the probability of clicking the recommendations is the highest for mobile devices. With respect to RQ2, we find that personalized, content-based news recommendations result in a more balanced distribution of news articles' readership popularity for anonymous users. For future work, we plan to conduct a longer study, in which we also want to study the impact of different recommendation algorithms (e.g., use BERT [5] instead of LDA and include collaborative filtering) on converting anonymous users into paying subscribers. Furthermore, we plan to investigate other evaluation metrics, such as recommendation diversity, serendipity and novelty.

Acknowledgements. This work was funded by the H2020 projects TRUSTS (GA: 871481), TRIPLE (GA: 863420), and the FFG COMET program. The authors want to thank Aliz Budapest for supporting the study execution.

References

1. Abdollahpouri, H.: Popularity bias in ranking and recommendation. In: Proceedings of the 2019 AAAI/ACM Conference on AI, Ethics, and Society, pp. 529–530 (2019)
2. Bellogín, A., Castells, P., Cantador, I.: Statistical biases in information retrieval metrics for recommender systems. Inf. Retrieval J. **20**(6), 606–634 (2017)
3. Blei, D.M., Ng, A.Y., Jordan, M.I.: Latent Dirichlet allocation. J. Mach. Learn. Res. **3**, 993–1022 (2003)
4. Craswell, N., Zoeter, O., Taylor, M., Ramsey, B.: An experimental comparison of click position-bias models. In: Proceedings of the 2008 International Conference on Web Search and Data Mining, pp. 87–94 (2008)
5. Devlin, J., Chang, M.W., Lee, K., Toutanova, K.: Bert: pre-training of deep bidirectional transformers for language understanding. arXiv preprint arXiv:1810.04805 (2018)
6. Felfernig, A., Burke, R., Pu, P.: Preface to the special issue on user interfaces for recommender systems. User Modeling User-Adapted Interact. **22**(4–5), 313 (2012)
7. Flaxman, S., Goel, S., Rao, J.M.: Filter bubbles, echo chambers, and online news consumption. Public Opinion Quart. **80**(S1), 298–320 (2016)
8. Garcin, F., Faltings, B.: Pen recsys: a personalized news recommender systems framework. In: Proceedings of the 2013 International News Recommender Systems Workshop and Challenge, NRS 2013, pp. 3–9. Association for Computing Machinery, New York (2013). https://doi.org/10.1145/2516641.2516642
9. Garcin, F., Faltings, B., Donatsch, O., Alazzawi, A., Bruttin, C., Huber, A.: Offline and online evaluation of news recommender systems at swissinfo.ch. In: Proceedings of the 8th ACM Conference on Recommender Systems, RecSys 2014, pp. 169–176 (2014). https://doi.org/10.1145/2645710.2645745
10. Karimi, M., Jannach, D., Jugovac, M.: News recommender systems – survey and roads ahead. Inf. Process. Manage. **54**(6), 1203–1227 (2018). https://doi.org/10.1016/j.ipm.2018.04.008.https://www.sciencedirect.com/science/article/pii/S030645731730153X
11. Kille, B., et al.: Overview of clef newsreel 2015: News recommendation evaluation lab. In: Working Notes of CLEF 2015 - Conference and Labs of the Evaluation forum (2015)
12. Kim, J., Thomas, P., Sankaranarayana, R., Gedeon, T., Yoon, H.J.: Eye-tracking analysis of user behavior and performance in web search on large and small screens. J. Assoc. Inf. Sci. Technol. **66**(3), 526–544 (2015)

13. Knijnenburg, B.P., Willemsen, M.C., Gantner, Z., Soncu, H., Newell, C.: Explaining the user experience of recommender systems. User Modeling User-Adapted Interac. **22**(4), 441–504 (2012)
14. Kowald, D., Dennerlein, S.M., Theiler, D., Walk, S., Trattner, C.: The social semantic server a framework to provide services on social semantic network data. In: Proceedings of the I-SEMANTICS 2013 Posters & Demonstrations Track co-located with 9th International Conference on Semantic Systems (I-SEMANTICS 2013), pp. 50–54 (2013)
15. Kowald, D., Muellner, P., Zangerle, E., Bauer, C., Schedl, M., Lex, E.: Support the underground: characteristics of beyond-mainstream music listeners. EPJ Data Sci. **10**(1), 1–26 (2021). https://doi.org/10.1140/epjds/s13688-021-00268-9
16. Kowald, D., Schedl, M., Lex, E.: The unfairness of popularity bias in music recommendation: a reproducibility study. Adv. Inf. Retrieval **12036**, 35 (2020)
17. Lacic, E., Kowald, D., Seitlinger, P., Trattner, C., Parra, D.: Recommending items in social tagging systems using tag and time information. In: Proceedings of the 1st International Workshop on Social Personalisation (SP'2014) co-located with the 25th ACM Conference on Hypertext and Social Media (2014)
18. Lacic, E., Kowald, D., Traub, M., Luzhnica, G., Simon, J.P., Lex, E.: Tackling cold-start users in recommender systems with indoor positioning systems. In: Poster Proceedings of the 9th ACM Conference on Recommender Systems. Association of Computing Machinery (2015)
19. Lops, P., De Gemmis, M., Semeraro, G.: Content-based recommender systems: state of the art and trends. In: Recommender Systems Handbook, pp. 73–105 (2011)
20. Newman, N., Fletcher, R., Levy, D., Nielsen, R.K.: The Reuters Institute digital news report 2016. Reuters Institute for the Study of Journalism (2015)
21. Seitlinger, P., Kowald, D., Kopeinik, S., Hasani-Mavriqi, I., Lex, E., Ley, T.: Attention please! a hybrid resource recommender mimicking attention-interpretation dynamics. In: Proceedings of the 24th International Conference on World Wide Web, pp. 339–345 (2015)
22. de Souza Pereira Moreira, G., Ferreira, F., da Cunha, A.M.: News session-based recommendations using deep neural networks. In: Proceedings of the 3rd Workshop on Deep Learning for Recommender Systems, DLRS 2018, pp. 15–23. Association for Computing Machinery, New York (2018). https://doi.org/10.1145/3270323.3270328

GameOfThronesQA: Answer-Aware Question-Answer Pairs for TV Series

Aritra Kumar Lahiri and Qinmin Vivian Hu[✉]

Ryerson University, Toronto, Canada
{aritra.lahiri,vivian}@ryerson.ca

Abstract. In this paper, we offer a corpus of question answer pairs related to the TV series generated from paragraph contexts. The data set called **GameofThronesQA V1.0** contains 5237 unique question answer pairs from the Game Of Thrones TV series across the eight seasons. In particular, we provide a pipeline approach for answer aware question generation, where the answers are extracted based on the named entities from the TV series. This is different to the traditional methods which generate questions first and find the relevant answers later. Furthermore, we provide a comparative analysis of the generated corpus with the benchmark datasets such as SQuAD, TriviaQA, WikiQA and TweetQA. The snapshot of the dataset is provided as an appendix for review purpose and will be released to public later.

Keywords: Datasets · NLP tasks · Question answer generation · Information extraction · Named entity

1 Introduction

This paper describes **a TV series story-line and its characters details** in the form of question answer pairs. The initial motivation is to provide users easier and faster information retrieval tasks as well as the benchmark datasets catered around TV series related information in OTT platforms like Netflix, Amazon Prime, Hulu, HBO. Formally, we offer a dataset called **GameOfThronesQA V1.0** which so far contains 5237 unique question answer pairs from the Game Of Thrones TV series across the eight seasons. For this dataset we have used the Game Of Thrones Wiki Web page [5] as the raw data source.

Figure 1 shows the examples of our QA pairs. The given paragraph extracted from GameOfThrones Wiki [5] as the input which we call the source text. Based on this input, we apply our approach to achieve the highlighted answer spans and obtain the **answers** centering the named entities extracted by [4]. After that, we generate the **questions** based on the answers, where BERT [3] is used as a core for sequence generation. This is basically how we generate the answer-aware question-answer pairs. It is worth to point out that our answer aware approach is different to the traditional question generation approaches [1,2], since the

M. Hagen et al. (Eds.): ECIR 2022, LNCS 13186, pp. 180–189, 2022.
https://doi.org/10.1007/978-3-030-99739-7_21

researchers traditionally generate the questions first via the given text or the extracted entities, then find the answers based on those questions.

In our approach, we first split the text into sentences, then for each sentence that has answers, we highlight the sentence with Highlight tokens. Now for the target text we concatenate the answers in that sentence with Separation tokens. Next we solve the problem of creating unique questions when we have the extracted answers. Here we consider about the theory of classification/clustering by adopting nine factoid based question categories as {What, Who, Which, Why, How, If, Whose, When, Where} These categories are embedded in the proposed answer aware question generation model (AAQG), which is our pipeline way to make sure the generated question is well paired with the given answer. Our experimental analysis will show the QA pair distribution over nine categories. Back to the examples in Fig. 1, we extract seven entities with three confidence levels, and generate six unique question answer pairs are generated based on the given passage.

Our experimental results on this version show the qualitative analysis of our dataset and the comparative analysis of our dataset with TriviaQA [13], TweetQA [12], WikiQA [9] and SQuAD [18].

Cersei marries Robert Baratheon. After the end of the civil war which ended the reign of House Targaryen, Cersei was married to King Robert Baratheon, a political marriage Robert agreed to in thanks for her father's last-minute alliance. She was only 19 when she married Robert and became Queen. At first, Cersei was infatuated with the handsome Robert, but their love began to wane when he called her Lyanna on their wedding night. She nevertheless bore Robert a son who died shortly after his birth. Cersei was devastated by the loss of her son and refused to have the body taken from her, forcing Robert to hold her while they took her son from her. In time, Cersei's feelings for Robert turned to hatred and she returned to her brother as a result. Her three children, Joffrey, Myrcella and Tommen, were officially Robert's, but in reality were the products of her incestuous relationship with Jaime.

Question:
Who was Cersei married to after the end of the civil war?
Answer: King Robert Baratheon

Question:
How old was Cersei when she married Robert?
Answer: 19

Question:
Who did Robert call Cersei on their wedding night?
Answer: Lyanna

Question: Who did Cersei have a son who died shortly after his birth?
Answer: Robert

Question: What did Cersei's feelings for Robert turn to in time?
Answer: hatred

Question: Who were Cersei's three children?
Answer: Joffrey, Myrcella and Tommen

Fig. 1. Examples: question answer pairs

2   Answer-Aware Question-Answer Pairs

We present how we generate the answer-aware question-answer pairs as: (1) how answer spans are selected from the named entities; (2) how the questions are generated based on answers, including a sequence-level classification, a span level prediction, and a token-level prediction; and (3) how to output the question answer pairs.

Mathematically, let (1) I be the context paragraph input; (2) $S = (s_1, s_2, \ldots, s_m)$ be the split sentence set of I; (3) $E = (e_{11}, e_{12}, \ldots, e_{1n_1}, \ldots, e_{m1}, e_{m2}, \ldots, e_{m_m})$ be the named entity set from S; Since in our answer span generation, we will need to know the exact position of extracted answer spans, the input for our candidate answer span set of our QA pairs is denoted by (4) $A = (I, S, E)$

Sentence: Arya was born and raised at Winterfell.
Entities: Arya, Winterfell
Prepend token: Winterfell [SEP] Arya was born and raised at Winterfell. Here the input is processed as - answer: Winterfell context: Arya was born and raised at Winterfell.
Highlight token: Arya was born and raised at [HL]Winterfell[HL]

Fig. 2. An example for answer span selection

2.1   Answer Span Selection

Our idea is to find an answer span among the extracted entities and select the answer span with special highlight tokens.

Taking the sentence "Arya was born and raised at Winterfell" as a simple example as shown in Fig. 2, the named entity algorithm [4] extracts the entities "Arya" and "Winterfell" first. We then define two tokens to select answer span as:

- **Unique Prepend format** is noted as the [SEP] token, to identify an answer span among extracted entities and is placed before the sentence containing the answer span.
- **Unique Highlight format** is noted as the [HL] token, to highlight the context entity that gives the location of the answer span in the context and avoids duplicates issues while performing named entity extraction.

2.2   Answer Aware Question Generation (AAQG)

For the Question Generation part, we propose our **Answer Aware Question Generation (AAQG)** Model, which generates the question based on the answer span selection. The model works on the principle of sequential question generation by fine-tuning adapted BERT-HLSQG [2] model by *Chan et al.*. The

AAQG model is trained by using SQuAD V2 dataset that contains 81K records split into train, test and dev as 81577, 8964, 8964.

Given the context paragraph **input** $\mathbf{I} = [i_1, ..., i_{|I|}]$ and the selected answer phrase $\mathbf{A} = [a_1, ..., a_{|A|}]$, we have the question-answer pairs as our **output** to be represented as:

$$QA = [A : A_1 ....An, Q : Q_1 ....Q_n] \tag{1}$$

Then, we formulate new context I', with **Highlight Tokens [HL]** as the input sequence to the AAQG model.

$$I' = [i_1, i_2, ..., [HL], a_1, ..., a_{|A|}, [HL], ..., i_{|I|}] \tag{2}$$

Given the above I', the input sequence S to the AAQG model is denoted by

$$S_i = ([CLS]; I_0; [SEP]; Q_1, ..., Q_i; [MASK]) \tag{3}$$

Here I_0 is the first input context and $Q_1,...,Q_i$ denotes the questions generated from AAQG model for the input sequence S.

In order to generate Q_i in Eq. 1, we apply a fine-tuned BERT-HLSQG [2] to calculate the label probabilities as:

$$Q_i = argmax_w P_r(w|S_i) \tag{4}$$

where $P_r(w|S_i) \in softmax(h_{[MASK]}W_{AAQG} + b_{AAQG})$. Note that we take the final hidden state for the last token [**MASK**] in the input sequence and connect it to the next connected layer $\mathbf{W_{AAQG}}$.

2.3   Question Answer Pair Output

We provide a task based pipeline abstraction for generating question answer pairs. It can be interpreted by a simple method where a context paragraph is passed as an input argument to the pipeline and the output is generated based on the task selected. We currently implemented three tasks for the pipeline to support which includes the following - i) **qg** - single question generation , ii) **multi-qa-qg** for multiple QA pairs generation and iii) **e2e-qg** for end to end QA pair generation. The pipeline is mimicked similar to the Transformers pipeline [28] for easy inference.

3   Empirical Study

3.1   Settings

We follow the PyTorch implementation of BERT for initial setup with SQuAd v2 dataset with 81K records. The hyper-parameters used for our experiments include batch size 32 for both training and validation set of the data, gradient accumulation steps as 8 and we train across 10 epochs. We also apply Adam Optimizer with an initial learning rate of 1e−4. We split the training set (80%), a development set (10%), and a test set (10%). We report results on the 10% test set, to directly compare the state-of-the art results on the QG tasks. Overall, with this AAQG model we generate around 5237 unique QA pairs.

Table 1. Comparison: TriviaQA Vs GameofThronesQA

Dataset	Metric	
	QA- exact match	QA - F1
TriviaQA-Wiki	80.86	84.50
TriviaQA-Web	82.99	87.18
GameOfThronesQA	81.8	91.1

3.2   Evaluation Metrics

We utilize the evaluation scripts released by Sharma et al. [7]. The package includes BLEU 1, BLEU 2, BLEU 3, BLEU 4 [24], METEOR (Denkowski and Lavie, 2014) [15] and ROUGE (Lin, 2004) [17] evaluation scripts.

3.3   TriviaQA vs. GameofThronesQA

The TriviaQA dataset [13] by Joshi et al., is released in 2017 and contains questions and answers crawled from 14 different trivia and quiz-league websites. As an extraction based dataset similar to our GameofThronesQA dataset, we compare our results with this dataset as here answering questions requires more cross-sentence reasoning compared to earlier datasets in the same category. Trivia QA Wiki and Trivia QA Web datasets utilize **QA-Exact Match** and **QA-F1** to compute their QA pair similarity quotient. Therefore, we compare their current state of the art leaderboard scores with our dataset and observe that our dataset performs almost similar to the **QA-Exact Match** and **QA-F1** scores of Trivia QA datasets (which is shown in Table 1).

3.4   TweetQA vs. GameofThronesQA

The TweetQA dataset [12], published in 2019 by Xiong et al., is considered as the first major large-scale dataset that revolves around social media content. This dataset contains 13,757 triplets of tweet, questions and answers collected from 10898 news articles in CNN and NBC. As shown in Table 2 we illustrate the factoid based question category distributions between our dataset and TweetQA where the difference in the "Who" based question percentage is attributed to the fact that our data contains more named entities as answers.

Table 2. Factoid question category distribution: TweetQA vs GameOfThronesQA

Dataset	Distribution percentage							
	What	Who	How	Where	Why	Which	When	Others
TweetQA	42.33	29.36	7.79	7.00	2.61	2.43	2.16	6.32
GameOfThronesQA	36	41	4	7	5	1	1	5

3.5 WikiQA vs. GameOfThronesQA

The WikiQA dataset [9] was released by Microsoft in 2015 and it is one of the earliest datasets in QA research. The document collection is based on Wikipedia as each question is associated with sentences extracted from Wikipedia articles. The dataset includes 3,047 questions and 29,258 sentences, out of which 1,473 were labeled as correct answers for their respective questions. As a retrieval based dataset compared to our extraction based dataset, the distribution of answer classes are illustrated in Table 3.

Table 3. Generated Answer Classes: GameOfThronesQA VS WikiQA

Class	WIKI QA	GameOfThrones QA
Location	373(11)	884 (14.8)
Human	494 (16)	989 (17.3)
Numeric	658 (22)	234 (3)
Abbreviation	16 (1)	2 (0.1)
Entity	419 (14)	3605 (60.74)
Description	1087 (36)	221 (3.7)

3.6 SQuAD vs. GameOfThronesQA

Stanford Question Answering Datasets (SQuAD) [18] was published latest in 2018 by Rajpurkar et al. The document collection for this dataset is based on passages extracted from Wikipedia articles. The first version of the SQuAD dataset had 107,702 questions and the second version added 53,775 new questions.

As shown in Fig. 3, there are predominantly more questions related to the "What" type in SQuAD whereas our dataset consists mostly "Who" based question. This can be attributed to the fact that our GameofThronesQA is a more character oriented dataset, because the questions were generated using named entities as answers primarily. On the other hand, the SQuAD training dataset is primarily built on Reading Comprehension which explains why "What" based questions requires more concept based knowledge.

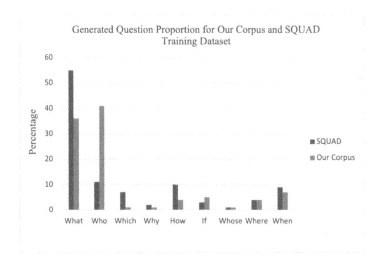

Fig. 3. Distribution of factoid question categories

4   Limitation and Conclusions

We provide a question answering data set called GameofThroneQA whose content comes from the Game of Thrones Wiki Web page [5]. We investigate our data set with other state-of-the-art question answering sets such as TriviaQA, TweetQA, WikiQA and SQuAD. Then we find that our data set (1) has good match and F1 scores on the QA pairs, (2) has a different focus on human entities, which reflects in the "who" and "what" questions.

We are looking to experiment with the long answer span extraction, since the current answers are primarily based on named entities. Also, we have long term goals of broadening the context of our dataset from one TV series to multiple ones based on genre similarity, which can be useful for model training purposes as well as tasks such as the information retrieval tasks (e.g. plot precise positioning), summarization tasks (e.g. the story-line), recommendation tasks (e.g. precise recommendation for shows in OTT platforms).

Acknowledgments. This study was supported in part by the Discovery and CREATE grants from the Natural Sciences and Engineering Research Council (NSERC) of Canada.

5   Appendix

Some sample QA pairs in our corpus is given below for users to review and get an understanding of our generated dataset. The generated 5237 QA pairs are all unique, although there may be some answers which are same for different questions.

```
{
  "answer": "Sandor",
  "question": "Who convinces Arya to abandon her quest for vengeance?"
},
{
  "answer": "Arya's younger brother Bran",
  "question": "Who is chosen as king in Westeros?"
},
{
  "answer": "Arya decides to leave Westeros and sail west to discover what lies beyond where the maps of the known world end.",
  "question": "What does Arya decide to leave Westeros after bidding farewell to her siblings?"
},
{
  "answer": "Arya Stark",
  "question": "Who is the youngest daughter of Lady Catelyn and Lord Ned Stark?"
},
{
  "answer": "Lady Catelyn and Lord Ned Stark",
  "question": "Who are Arya Stark's daughters?"
},
{
  "answer": "Winterfell",
  "question": "Where was Arya born and raised?"
},
{
  "answer": "Jon Snow",
  "question": "Who is Arya's bastard half-brother?"
},
{
  "answer": "Sansa",
  "question": "Who is Arya's older sister?"
},
{
  "answer": "Arya",
  "question": "Who rejects the notion that she must become a lady and marry for influence and power?"
},

{
  "answer": "Hand of the King",
  "question": "What does Robert name Arya's father?"
},
{
  "answer": "Nymeria",
  "question": "What is Arya's name after a great warrior-queen of Essos?"
},
{
  "answer": "a sword",
  "question": "What does Arya receive as a gift from Jon?"
},
{
  "answer": "Needle",
  "question": "What does Arya name as a play on words that she may now enjoy doing needlework?"
},
{
  "answer": "Mycah",
  "question": "Who is the son of the butcher in the King's retinue?"
},
{
  "answer": "Sansa and Joffrey",
  "question": "Who spot Nymeria fighting?"
},
{
  "answer": "Mycah",
  "question": "Who does Nymeria cut when he threatens Arya?"
},
{
  "answer": "Joffrey",
  "question": "Who threatens Arya with a sword?"
},
{
  "answer": "rocks",
  "question": "What does Arya drive Nymeria off with so she won't be punished?"
},
```

References

1. Du, X., Cardie, C.: Harvesting paragraph-level question-answer Pairs from Wikipedia. In: Association for Computational Linguistics (ACL) (2018)
2. Chan, Y.-H., Fan, Y.-C.: A recurrent BERT-based model for question generation. In: Proceedings of the Second Workshop on Machine Reading for Question Answering, pp. 154–162, Hong Kong, China, 4 November 2019. (ACL) (2019)

3. Devlin, J., et al.: BERT: pre-training of deep bidirectional transformers for language understanding. arXiv preprint arXiv:1810.04805 (2018)
4. Lee, J., et al.: BioBERT: a pre-trained biomedical language representation model for biomedical text mining, Bioinformatics, vol. 36(4), pp. 1234–1240, BioBERT: a pre-trained biomedical language representation model for biomedical text mining, 15 February 2020
5. https://gameofthrones.fandom.com/wiki/
6. Duan, N., Tang, D., Chen, P., Zhou, M.: Question generation for question answering. In: Proceedings of the 2017 Conference on Empirical Methods in Natural Language Processing, pp. 866–874 (2017). http://www.aclweb.org/anthology/D13-1160
7. Indurthi, S.R., et al.: Generating natural language question-answer pairs from a knowledge graph using a RNN based question generation model. In: Proceedings of the 15th Conference of the European Chapter of the Association for Computational Linguistics, vol. 1, Long Papers (2017)
8. Kwiatkowski, T., et al.: Natural questions: a benchmark for question answering research. Trans. Assoc. Comput. Linguist. **7**, 453–466 (2019)
9. Yang, Y., Yih, W.-T., Meek, C.: WIKIQA: a challenge dataset for open-domain question answering. In: Proceedings of the 2015 Conference on Empirical Methods in Natural Language Processing (2015)
10. Cambazoglu, B.B., et al.: A Review of Public Datasets in Question Answering Research (2020)
11. Raffel, C., et al.: Exploring the limits of transfer learning with a unified text-to-text transformer. arXiv preprint arXiv:1910.10683 (2019)
12. Xiong, W., et al.: TWEETQA: a social media focused question answering dataset. arXiv preprint arXiv:1907.06292 (2019)
13. Joshi, M., et al.: TriviaQA: a large scale distantly supervised challenge dataset for reading comprehension. arXiv preprint arXiv:1705.03551 (2017)
14. Bordes, A., Usunier, N., Chopra, S., Weston, J.: Large-scale simple question answering with memory networks. arXiv preprint arXiv:1506.02075 (2015)
15. Banerjee, S., Lavie, A.: METEOR: an automatic metric for MT evaluation with improved correlation with human judgments. In: Proceedings of the ACL workshop on intrinsic and extrinsic evaluation measures for machine translation and/or summarization (2005)
16. Chen, D., Fisch, A., Weston, J., Bordes, A.: Reading Wikipedia to answer open domain questions. In: Proceedings of the 55th Annual Meeting of the Association for Computational Linguistics, vol. 1: Long Papers. Association for Computational Linguistics, pp. 1870–1879 (2017). https://doi.org/10.18653/v1/P17-1171
17. Lin, C.-Y.: Rouge: a package for automatic evaluation of summaries. Text summarization branches out (2004)
18. Rajpurkar, P., Zhang, J., Lopyrev, K., Liang, P.: Squad: 100,000+ questions for machine comprehension of text. In: Proceedings of the 2016 Conference on Empirical Methods in Natural Language Processing (EMNLP). Association for Computational Linguistics, Austin, Texas, pp. 2383–2392 (2016). https://aclweb.org/anthology/D16-1264
19. Serban, I.V., et al.: Generating factoid questions with recurrent neural networks: the 30 m factoid question-answer corpus. In: Proceedings of the 54th Annual Meeting of the Association for Computational Linguistics (vol. 1: Long Papers). Association for Computational Linguistics, Berlin, Germany, pp. 588–598 (2016). http://www.aclweb.org/anthology/P16-1056

20. Srivastava, N., Hinton, G., Krizhevsky, A., Sutskever, I., Salakhutdinov, R.: Dropout: a simple way to prevent neural networks from overfitting. J. Mach. Learn. Res. **15**(1), 1929–1958 (2014)
21. Wang, S.: R3: Reinforced ranker-reader for open-domain question answering (2018)
22. Yao, X., Bouma, G., Zhang, Y.: Semantics-based question generation and implementation. Dialog. Discourse **3**(2), 11–42 (2012)
23. Zhou, Q., Yang, N., Wei, F., Tan, C., Bao, H., Zhou, M.: Neural question generation from text: a preliminary study. arXiv preprint arXiv:1704.01792 (2017)
24. Papineni, K., Roukos, S., Ward, T., Zhu, W.-J.: BLEU: a method for automatic evaluation of machine translation. In: Proceedings of 40th Annual Meeting of the Association for Computational Linguistics. Association for Computational Linguistics, Philadelphia, Pennsylvania, USA, pp. 311–318 (2002). https://doi.org/10.3115/1073083.1073135
25. Winograd, T.: Understanding natural language. Cogn. Psychol. **3**(1), 1–191 (1972)
26. Ryu, P.-M., Jang, M.-G., Kim, H.-K.: Open domain question answering using Wikipedia-based knowledge model. Inf. Process. Manage. **50**(5), 683–692 (2014)
27. Hermann, K.M., Kocisky, T., Grefenstette, E., Espeholt, L., Kay, W., Suleyman, M., Blunsom, P.: Teaching machines to read and comprehend. In: Advances in Neural Information Processing Systems (NIPS) (2015)
28. https://huggingface.co/t5-base

Leveraging Customer Reviews
for E-commerce Query Generation

Yen-Chieh Lien[1], Rongting Zhang[2], F. Maxwell Harper[2],
Vanessa Murdock[2], and Chia-Jung Lee[2(✉)]

[1] University of Massachusetts Amherst, Amherst, USA
ylien@cs.umass.edu
[2] Amazon, Seattle, USA
{rongtz,fmh,vmurdock,cjlee}@amazon.com

Abstract. Customer reviews are an effective source of information about what people deem important in products (e.g. "strong zipper" for tents). These crowd-created descriptors not only highlight key product attributes, but can also complement seller-provided product descriptions. Motivated by this, we propose to leverage customer reviews to generate queries pertinent to target products in an e-commerce setting. While there has been work on automatic query generation, it often relied on proprietary user search data to generate query-document training pairs for learning supervised models. We take a different view and focus on leveraging reviews without training on search logs, making reproduction more viable by the public. Our method adopts an ensemble of the statistical properties of review terms and a zero-shot neural model trained on adapted external corpus to synthesize queries. Compared to competitive baselines, we show that the generated queries based on our method both better align with actual customer queries and can benefit retrieval effectiveness.

Keywords: Query generation · Reviews · Weak learning · Zero-shot

1 Introduction

Customer reviews contain diverse descriptions about how people reflect the properties, pros and cons of the products that they have experienced. For example, properties such as "for underwater photos" or "for kayaking recording" were mentioned in reviews for action cameras, as well as "compact" or "strong zipper" for tents. These descriptors not only paint a rich picture of what people deem important, but also can complement and uncover shopping considerations that may be absent in seller-provided product descriptions. Motivated by this, our work investigates ways to generate queries that surface key properties about the target products using reviews.

Y.-C. Lien—Work done while an intern at Amazon.

M. Hagen et al. (Eds.): ECIR 2022, LNCS 13186, pp. 190–198, 2022.
https://doi.org/10.1007/978-3-030-99739-7_22

Previous work on automatic query generation often relied on human labels or logs of queries and engaged documents (or items) [16–20] to form relevance signals for training generative models. Despite the reported effectiveness, the cost of acquiring high quality human labels is high, whereas the access to search logs is often only limited to site owners. As we approach the problem using reviews, it brings an advantage of not requiring any private, proprietary user data, making reproduction more viable by the public in general. Meanwhile, generation based on reviews is favorable as the outcome may likewise produce human-readable language patterns, potentially facilitating people-facing experiences such as related search recommendation.

We propose a simple yet effective ensemble method for query generation. Our approach starts with building a candidate set of "query-worthy" terms from reviews. To begin, we first leverage syntactic and statistical signals to build up a set of terms from reviews that are most distinguishable for a given product. A second set of candidate terms is obtained through a zero-shot sequence-to-sequence model trained according to adapted external relevance signals. Our ensemble method then devises a statistics-based scoring function to rank the combined set of all candidates, from which a query can be formulated by providing a desired query length.

Our evaluation examines two crucial aspects of query quality. To quantify how readable the queries are, we take the human-submitted queries from logs as ground truth to evaluate how close the generated queries are to them for each product. Moreover, we investigate whether the generated queries can benefit retrieval tasks, similar to prior studies [6,7,15]. We collect pairs of product descriptions and generated queries, both of which can be derived from public sources, to train a deep neural retrieval model. During inference, we take human-submitted queries on the corresponding product to benchmark the retrieval effectiveness. Compared with the competitive alternatives YAKE [1,2] and Doc2Query [6], our approach shows significantly higher similarity with human-submitted queries and benefits retrieval performance across multiple product types.

2   Related Work

Related search recommendation (or query suggestion) helps people automatically discover related queries pertinent to their search journeys. With the advances in deep encoder-decoder models [9,12], query generation [6,16,17,19,20] sits at the core of many recent recommendation algorithms. Sordoni et al. [17] proposed hierarchical RNNs [24] to generate next queries based on observed queries in a session. Doc2Query [6] adapted T5 [12] to generate queries according to input documents. Ahmad et al. [20] jointly optimized two companion ranking tasks, document ranking and query suggestion, by RNNs. Our approach differs in that we do not require in-domain logs of query-document relations for supervision.

Studies also showed that generated queries can be used for enhancing retrieval effectiveness [6,7,15]. Doc2Query [6] leveraged the generated queries to enrich and expand document representations. Liang et al. [7] proposed to synthesize query-document relations based on MSMARCO [8] and Wikipedia for training large-scale neural retrieval models. Ma et al. [15] explored a similar zero-shot learning method for a different task of synthetic question generation, while Puri et al. [21] improve QA performance by incorporating synthetic questions. Our work resembles the zero-shot setup but differs in how we adapt external corpus particularly for e-commerce query generation.

Customer reviews have been adopted as a useful resource for summarization [22] and product question answering. Approaches to PQA [10,11,13,14] often take in reviews as input, conditioned on which answers are generated for user questions. Deng et al. [11] jointly learned answer generation and opinion mining tasks, and required both a reference answer and its opinion type during training phase. While our work also depends on reviews as input, we focus on synthesizing the most relevant queries without requiring ground-truth labels.

3   Method

Our approach involves a candidate generation phrase to identify key terms from reviews, and a selection phrase that employs an unsupervised scoring function to rank and aggregate the term candidates into queries.

3.1   Statistics-Based Approach

We started with a pilot study to characterize the opportunity of whether and how reviews could be useful for query generation. We found that a subset of terms in reviews resemble that of search queries, which are primarily composed of combinations of nouns, adjectives and participles to reflect critical semantics. For example, given a headphone, the actual queries that had led to purchases may contain nouns such as "earbuds" or "headset" to denote product types, adjectives such as "wireless" or "comfortable" to reflect desired properties, and participles such as "running" or "sleeping" to emphasize use cases.

Inspired by this, we first leverage part-of-speech analysis to scope down reviews to the three types of POS-tags. From this set, we then rely on conventional tf-idf corpus statistics to mine distinguishing terms salient in a product type but not generic across the entire catalog. Specifically, an importance score $I_t^D = \frac{p(t,R_D)}{p(t,R_G)}$ is used to estimate the salience of a term t in a product type D by contrasting its density in review set R_D to generic reviews R_G, where $p(t,R) = \frac{freq(t,R)}{\Sigma_{r \in R}|r|}$. Beyond unigrams, we also consider if the relative frequency of bigram phrases containing the unigrams $\frac{freq([t,t'],R_D)}{freq(t,R_D)}$ is above some threshold; in this case, bigrams will replace unigrams and become the candidates. We apply I_t^D to each review sentence, and collect top scored terms or phrases as candidates.

A straightforward way to form queries is to directly use the candidates as-is. We additionally consider an alternative which trains a seq2seq model using the candidates as weak supervision (i.e. encode review sentences to fit the candidates). By doing so, we anticipate the terms decoded during inference can generalize more broadly compared to a direct application. The two methods are referred to as Stats-base and Stats-s2s respectively.

3.2   Zero-Shot Generation Based on Adapted External Corpus

Recent findings [7,15] suggest that zero-shot domain adaptation can deliver high effectiveness given the knowledge embedded in large-scale language models via pre-training tasks. With this, we propose to rely on fine-tuning T5 [12] on MSMARCO query-passage pairs to capture the notion of generic relevance, and apply the trained model to e-commerce reviews to identify terms that are more probable to be adopted in queries.

This idea has been experimented by Nogueira et al. [6], where their Doc2Query approach focused on generating queries as document expansion for improving retrieval performance. Different from [6], our objective is to generate queries that are not only beneficial to retrieval but also similar to actual queries in terms of syntactic forms. Thus, a direct application of Doc2Query on MSMARCO creates a gap in our case since MSMARCO "queries" predominantly follow a natural-language question style, resulting in generated queries of similar forms[1]. To tighten the loop, we propose to apply POS-tag analysis to MSMARCO queries and retain only terms that satisfy the selected POS-tags (i.e. nouns, adjectives and participles). For example, an original query "what does physical medicine do" is first transformed into "physical medicine" as pre-processing. After the adaptation, we conduct T5 seq2seq model training and apply it in a zero-shot fashion to generate salient terms based on input review sentences.

3.3   Ensemble Approach to Query Generation

For a product p in the product type D, we employ both statistical and zero-shot approaches on its reviews to construct candidates for generating queries, which we denote as C_p. To select representative terms from the set, we devise a scoring function $S_t = freq(t, C_p) \cdot log(\frac{|\{p' \in D\}|}{|\{p'|p' \in D, t \in C_{p'}\}|})$ to rank all candidates, where higher ranked terms are more distinguishable for a specific product based on the tf-idf intuition. Given a desired query length n, we formulate the pseudo queries for a product by selecting all possible $\binom{k}{n}$ combinations from the top-k scored terms in the C_p set[2]. A final post-processing step removes any redundant words after stemming from the queries and adds product types if not already included.

---

[1] Original Doc2Query is unsuitable since question-style queries are rare in e-commerce.

[2] Our experiment sets $k = 3$ and $n = 1, 2, 3$ per its popularity in generic search queries.

4   Experiments

Our evaluation set is composed of products from three different product types, together with the actual queries[3] that were submitted by people who purchased those products on `Amazon.com`. As shown in Table 1, we consider *headphones*, *tents* and *conditioners* to evaluate our method across diverse product types, for which people tend to behave and shop differently with variances reflected in search queries. The query vocabulary size for conditioners, for instance, is about thrice the size of tents, with headphones sitting in-between the two.

As our approach disregards the actual queries for supervision, we primarily consider competitive baselines that do not involve using query logs. In particular, we compare to the unsupervised approach YAKE [1,2] which reportedly outperforms a variety of seminal key word extraction approaches, including RAKE [4], TextRank [3] and SingleRank [5] methods. In addition, we leverage the zero-shot Doc2Query model on adapted corpus as our baseline to reflect the absence of e-commerce logs. For generation, we initialize separate Huggingface **T5-base** [12] weights with conditional generation head and fine-tune for Stats-s2s and Doc2Query models respectively. Training is conducted on review sentences broken down by NLTK. For retrieval, we fine-tune a Sentence-Transformer [23] **ms-marco-TinyBERT**[4] pre-trained with MSMARCO data, which was shown to be effective for semantics matching. Our experiments use a standard AdamW optimizer with learning rate 0.001 and $\beta_1, \beta_2 = (0.9, 0.999)$, and conduct 2 and 4 epochs training on a batch size of 16 respectively for generation and retrieval.

Table 1. Statistics of the three product types used in the experiments. For each product type, the dev and test split respectively contains 500 disjoint products.

	Headphone		Tent		Conditioner	
	Dev	Test	Dev	Test	Dev	Test
# of reviews	23,165	23,623	19,208	18,734	17,055	17,689
# of sentences	102,281	103,771	97,553	97,320	68,691	70,829

4.1   Intrinsic Similarity Evaluation

Constructing readable and human-like queries is desirable since it is practically useful for applications such as related search recommendation. A natural way to reflect readability is to evaluate the similarity between the generated and customer-submitted queries since the latter is created by human. In practice, we

---

[3] Note that we use actual data only for the purpose of evaluation not training.
[4] https://www.sbert.net/docs/pretrained-models/ce-msmarco.html.

consider customer-submitted queries that had led to at least 5 purchases on the corresponding products as ground-truth queries, to which the generated queries are then compared. We use conventional metrics adopted in generative tasks including corpus BLEU and METEOR for evaluation. The results in Table 2 show that our ensemble approach consistently achieves the highest similarity with human-queries across product types, suggesting that the statistical and zero-shot methods could be mutually beneficial.

Table 2. The similarity in BLEU and METEOR between generated queries and real queries. ⋆ stands for p-value < 0.05 in T-test compared to the second best performing method in each column. The bottom shows example generated queries by ensemble.

	Headphone		Tent		Conditioner	
	BLEU	METEOR	BLEU	METEOR	BLEU	METEOR
YAKE	0.1014	0.1371	0.2794	0.2002	0.3143	0.1998
Doc2Query	0.1589	0.1667	0.3684	0.2145	0.4404	0.264
Stats-base	0.1743	0.2001	0.3294	0.2201	0.4048	0.2723
Stats-s2s	0.1838	0.2004	0.321	0.2189	0.3931	0.2641
Ensemble	**0.2106⋆**	**0.2024**	**0.394⋆**	**0.2334⋆**	**0.5047⋆**	**0.2956⋆**
Examples	Noise cancelling headphone		Lightweight tent		Detangling conditioner	
	Truck driver headphone		Alps backpacking tent		Shea moisture conditioner	
	Hearing aids headphone		Air mattresses queen tent		Dry hair conditioner	

4.2 Extrinsic Retrieval Evaluation

We further study how the generated queries can benefit e-commerce retrieval. Our evaluation methodology leverages pairs of generated queries and product descriptions to train a retrieval model and validates its quality based on actual queries. During training, we fine-tune a Sentence-Transformer based on top-3 generated queries of each product. For each query, we prepare its corresponding relevant product description, together with 49 negative product descriptions randomly sampled from the same product type. During inference, instead of generated queries, we use customer-submitted queries to fetch descriptions from the product corpus, and an ideal retrieval model should rank the corresponding product description at the top. We also include BM25 as a common baseline. Table 3 shows that Doc2Query and the ensemble methods are the most effective and are on par in aggregate, with some variance in different product types. Stats-s2s slightly outperforms Stats-base overall, which may hint a potential for better generalization.

Table 3. The retrieval effectiveness for queries generated by baselines and our method.

	Headphone			Tent			Conditioner		
	MRR	P@1	P@10	MRR	P@1	P@10	MRR	P@1	P@10
BM25	0.28	0.19	0.06	0.43	0.29	0.11	0.56	0.47	0.14
YAKE	0.23	0.11	0.07	0.46	0.34	0.11	0.54	0.43	0.14
Doc2Query	0.28	0.18	**0.08**	**0.49**	**0.40**	0.12	0.58	**0.49**	0.15
Stats-base	0.28	0.16	0.07	0.44	0.29	0.12	0.54	0.42	0.15
Stats-s2s	0.27	0.17	0.07	0.44	0.32	0.12	0.56	0.46	**0.16**
Ensemble	**0.29**	**0.20**	0.07	0.46	0.33	**0.13**	**0.59**	0.48	0.15

5 Conclusion

This paper connected salient review descriptors with zero-shot generative models for e-commerce query generation, without requiring human labels or search logs. The empirical results showed that the ensemble queries both better resemble customer-submitted queries and benefit training effective rankers. Besides MSMARCO, our future plan seeks to incorporate other publicly available resources such as community question-answering threads to generalize the notion of relevance. It is worth to consider ways to combine weak labels with few strong labels and dive deep into the impact of employing different hyper-parameters. A user study that characterizes the extent to which the generated queries can reflect people's purchase intent will further help qualitative understanding.

References

1. Campos, R., Mangaravite, V., Pasquali, A., Jorge, A.M., Nunes, C., Jatowt, A.: A text feature based automatic keyword extraction method for single documents. In: Pasi, G., Piwowarski, B., Azzopardi, L., Hanbury, A. (eds.) ECIR 2018. LNCS, vol. 10772, pp. 684–691. Springer, Cham (2018). https://doi.org/10.1007/978-3-319-76941-7_63
2. Campos, R., Mangaravite, V., Pasquali, A., Jorge, A., Nunes, C., Jatowt, A.: YAKE! Keyword extraction from single documents using multiple local features. Inf. Sci. **509**, 257–289 (2020)
3. Mihalcea, R., Tarau, P.: TextRank: bringing order into texts. In: Proceedings of the 2004 Conference on Empirical Methods in Natural Language Processing, pp. 404–411 (2004)
4. Mining, T., Rose, S., Engel, D., Cramer, N., Cowley, W.: Automatic keyword extraction from individual documents. In: Text Mining: Theory and Applications, vol. 1, pp. 1–20 (2010)
5. Wan, X., Xiao, J.: Single document keyphrase extraction using neighborhood knowledge. In: Proceedings of the 23rd AAAI Conference on Artificial Intelligence, pp. 855–860 (2008)
6. Nogueira, R., Yang, W., Lin, J.J., Cho, K.: Document expansion by query prediction. ArXiv (2019)

7. Liang, D., et al.: Embedding-based Zero-shot retrieval through query generation. ArXiv (2020)
8. Bajaj, P., et a.: A human generated MAchine Reading COmprehension dataset. ArXiv, MS MARCO (2016)
9. Lewis, M., et al.: BART: denoising sequence-to-sequence pre-training for natural language generation, translation, and comprehension. In: Proceedings of the 58th Annual Meeting of the Association for Computational Linguistics, pp. 7871–7880 (2020)
10. Liu, Y., Lee, K.-Y.: E-commerce query-based generation based on user review. ArXiv (2020)
11. Deng, Y., Zhang, W., Lam, W.: Opinion-aware answer generation for review-driven question answering in e-commerce. In: Proceedings of the 29th ACM International Conference on Information and Knowledge Management, pp. 255–264 (2020)
12. Raffel, C., et al.: Exploring the limits of transfer learning with a unified text-to-text transformer. J. Mach. Learn. Res. (JMLR) **21**, 1–67 (2020)
13. Chen, S., Li, C., Ji, F., Zhou, W., Chen, H.: Review-driven answer generation for product-related questions in e-commerce. In: Proceedings of the 12th ACM International Web Search and Data Mining Conference, pp. 411–419 (2019)
14. Gao, S., Ren, Z., Zhao, Y., Zhao, D., Yin, D., Yan, R.: Product-aware answer generation in e-commerce question-answering. In: Proceedings of the 12th ACM International Web Search and Data Mining Conference, pp. 429–437 (2019)
15. Ma, J., Korotkov, I., Yang, Y., Hall, K., McDonald, R.: Zero-shot neural passage retrieval via domain-targeted synthetic question generation. In: Proceedings of the 16th Conference of the European Chapter of the Association for Computational Linguistics, pp. 1075–1088 (2021)
16. Chen, R.-C., Lee, C.-J.: Incorporating behavioral hypotheses for query generation. In: Proceedings of the 2020 Conference on Empirical Methods in Natural Language Processing, pp. 3105–3110, (2020)
17. Sordoni, A., Bengio, Y., Vahabi, H., Lioma, C., Simonsen, J.G., Nie, J.Y.: A hierarchical recurrent encoderdecoder for generative context-aware query suggestion. In: Proceedings of the 24th ACM International Conference on Information and Knowledge Management, pp. 553–562 (2015)
18. Jiang, J.-Y., Wang, W.: RIN: reformulation inference network for context-aware query suggestion. In: Proceedings of the 27th ACM International Conference on Information and Knowledge Management, pp. 197–206 (2018)
19. Kim, K., Lee, K., Hwang, S.-W., Song, Y.-I., Lee, S.: Query generation for multimodal documents. In: Proceedings of the 16th Conference of the European Chapter of the Association for Computational Linguistics, pp. 659–668 (2021)
20. Ahmad, W.U., Chang, K.-W., Wang, H.: Context attentive document ranking and query suggestion. In: Proceedings of the 42nd International ACM SIGIR Conference on Research and Development in Information Retrieval, pp. 385–394 (2019)
21. Puri, R., Spring, R., Patwary, M., Shoeybi, M., Catanzaro, B.: Training question answering models from synthetic data. In: Proceedings of the 2020 Conference on Empirical Methods in Natural Language Processing, pp. 5811–5826 (2020)
22. Zhang, X., et al.: DSGPT: domain-specific generative pre-training of transformers for text generation in e-commerce title and review summarization. In: Proceedings of the 44th International ACM SIGIR Conference on Research and Development in Information Retrieval, pp. 2146–2150 (2021)
23. Reimers, N., Gurevych, I.: Sentence-BERT: sentence embeddings using Siamese BERT-networks. In: Proceedings of the 2019 Conference on Empirical Methods in Natural Language Processing, pp. 3982–3992 (2019)

24. Rumelhart, D.E., Hinton, G.E., Williams, R.J.: Learning internal representations by error propagation. Technical report ICS 8504. Institute for Cognitive Science, University of California, San Diego, California, September 1985

Question Rewriting? Assessing Its Importance for Conversational Question Answering

Gonçalo Raposo$^{(\boxtimes)}$ ⓘ, Rui Ribeiroⓘ, Bruno Martinsⓘ, and Luísa Coheurⓘ

INESC-ID, Instituto Superior Técnico, Universidade de Lisboa, Lisbon, Portugal
{goncalo.cascalho.raposo,rui.m.ribeiro,bruno.g.martins,
luisa.coheur}@tecnico.ulisboa.pt

Abstract. In conversational question answering, systems must correctly interpret the interconnected interactions and generate knowledgeable answers, which may require the retrieval of relevant information from a background repository. Recent approaches to this problem leverage neural language models, although different alternatives can be considered in terms of modules for (a) representing user questions in context, (b) retrieving the relevant background information, and (c) generating the answer. This work presents a conversational question answering system designed specifically for the Search-Oriented Conversational AI (SCAI) shared task, and reports on a detailed analysis of its question rewriting module. In particular, we considered different variations of the question rewriting module to evaluate the influence on the subsequent components, and performed a careful analysis of the results obtained with the best system configuration. Our system achieved the best performance in the shared task and our analysis emphasizes the importance of the conversation context representation for the overall system performance.

Keywords: Conversational question answering · Conversational search · Question rewriting · Transformer-based neural language models

1 Introduction

Conversational question answering extends traditional Question Answering (QA) by involving a sequence of interconnected questions and answers [3]. Systems addressing this problem need to understand an entire conversation flow, often using explicit knowledge from an external datastore to generate a natural and correct answer for the given question. One way of approaching this problem is to

Work supported by national funds through Fundação para a Ciência e a Tecnologia (FCT), under project UIDB/50021/2020; by FEDER, Programa Operacional Regional de Lisboa, Agência Nacional de Inovação (ANI), and CMU Portugal, under project Ref. 045909 (MAIA) and research grant BI|2020/090; and by European Union funds (Multi3Generation COST Action CA18231).

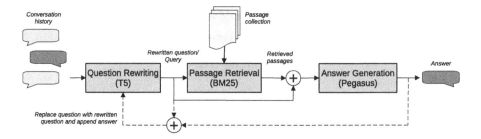

Fig. 1. Proposed conversational question answering system. Question rewriting is performed using T5, passage retrieval using BM25, and answer generation using Pegasus. Dashed lines represent different inputs explored for question rewriting.

divide it into 3 steps (see Fig. 1): initial question rewriting, retrieval of relevant information regarding the question, and final answer generation.

In a conversational scenario, questions may contain acronyms, coreferences, ellipses, and other natural language elements that make it difficult for a system to understand the question. Question rewriting aims to solve this problem by reformulating the question and making it independent of the conversation context [5], which has been shown to improve systems performance [11].

After an initial understanding of the question and its conversational context, the next challenge is the retrieval of relevant information [14] to use explicitly in the answer generation [4]. For this step, the rewritten question is used as a query to an external datastore, and thus the performance of the initial rewriting module can affect the conversational passage retrieval [12].

The last module has the task of generating an answer that incorporates the retrieved information conditioned on the rewritten question. The Question Rewriting in Conversational Context (QReCC) dataset [1] brings these tasks together, supporting the training and evaluation of neural models for conversational QA. Although there are datasets for each individual task (e.g., CANARD for question rewriting [5] and TREC CAsT for passage retrieval [4]), to the best of our knowledge, QReCC is the only dataset that contemplates all these tasks.

This work presents a conversational QA system[1] implemented according to the dataset and task definition of the Search-Oriented Conversational AI (SCAI) QReCC 2021 shared task[2], specifically focusing on the question rewriting module. Participating as team *Rachael*, our system achieved the 1st place in this shared task. Besides evaluating the system performance as a whole, using many variations of the question rewriting module, our work highlights the importance of this module and how much it impacts the performance of subsequent ones.

---

[1] Available at https://github.com/gonced8/rachael-scai.
[2] https://scai.info/scai-qrecc/.

2   Conversational Question Answering

To perform conversational question rewriting, the proposed system uses the model `castorini/t5-base-canard`[3] from the HuggingFace model hub [13]. This consists of a T5 model [8] which was fine-tuned for question rewriting using the CANARD dataset [5]. No further fine-tuning was performed with QReCC data.

In order to incorporate relevant knowledge when answering the questions, our system uses a passage retrieval module built with Pyserini [7], i.e., an easy-to-use Python toolkit that allows searching over a document collection using sparse and dense representations. In our implementation, the retrieval is performed using the BM25 ranking function [10] , with its parameters set to $k_1 = 0.82$ and $b = 0.68$. This function is used to retrieve the top-10 most relevant passages.

Since our system needs to extract the most important information from the retrieved passages, which are often large, we used a Transformer model pre-trained for summarization. We chose the Pegasus model [15], more specifically, the version `google/pegasus-large`[4], which can handle inputs up to 1024 tokens.

We further fine-tuned the Pegasus model for 10 epochs in the task of answer generation, which can be seen as a summarization of the relevant text passages conditioned on the rewritten question. The training instances used the ground truth rewritten question concatenated with the ground truth passages (and additional ones retrieved with BM25), and the ground truth answers as the target.

3   Evaluation

3.1   Experimental Setup

The dataset used for both training and evaluation was the one used in the SCAI QReCC 2021 shared task, which is a slight adaption of the QReCC dataset. The training data contains 11k conversations with 64k question-answer (QA) pairs, while the test data contains 3k conversations with 17k questions-answer pairs. For each QA pair, we have also the corresponding truth rewrites and relevant passages, which are not considered during testing (unless specified otherwise).

To evaluate each module, we used the same automatic metrics as the shared task: ROUGE1-R [6] for question rewriting, Mean Reciprocal Rank (MRR) for passage retrieval, and F1 plus Exact Match (EM) [9] for the model answer evaluation. We additionally used ROUGE-L to assess the answer. When the system performs retrieval without first rewriting the question, we still report (between parentheses) the ROUGE1-R metric comparing the queries and truth rewrites.

3.2   Results

Question Rewriting Input. We first studied different inputs to the question rewriting module in terms of the conversation history. Instead of using the original questions, one could replace them with the corresponding previous model

---

[3] https://huggingface.co/castorini/t5-base-canard.

[4] https://huggingface.co/google/pegasus-large.

rewrites. Moreover, one could use only the questions or also include the answers generated by the model. Regarding the length of the conversation history considered for question rewriting, we use all the most recent interactions that fit in the input size supported by the model.

Table 1. Evaluation of multiple variations of the input used in the question rewriting module: Question (Q), Model Answer (MA), Model Rewritten (MR).

Description	Rewriting Input	Rewriting	Retrieval	Answer		
		ROUGE1-R	MRR	F1	EM	ROUGEL-F1
SCAI baseline: question	–	–	–	0.117	0.000	0.116
SCAI baseline: retrieved	–	(0.571)	0.065	0.067	0.001	0.073
SCAI baseline: GPT-3	–	–	–	0.149	0.001	0.152
No rewriting ($h = 1$)	–	(0.571)	0.061	0.136	0.005	0.143
No rewriting ($h = 7$)	–	(0.571)	0.145	0.155	0.003	0.160
Questions	(Q) + Q	0.673	**0.158**	0.179	**0.011**	0.181
Questions + answers	(Q + MA) + Q	0.681	0.150	0.179	0.010	0.181
Rewritten questions	(MR) + Q	0.676	0.157	0.187	0.010	0.188
Rewritten + answers	(MR + MA) + Q	**0.685**	0.149	**0.189**	0.010	**0.191**
Ground truth rewritten	–	(1)	0.385	0.302	0.028	0.293

The results of our analysis are shown in Table 1, which also includes 3 baselines from the SCAI shared task[5]. The first baseline – question – uses the question as the answer; the second baseline – retrieval – uses the question to retrieve the top-100 most relevant passages using BM25, and selects the one with the highest score; the third baseline – GPT-3 – uses this Transformer Decoder [2] to generate the answer, prompting the model with an example conversation and the current conversation history. Among the baselines, GPT-3 achieved the best performance, which could be expected from this large language model. Moreover, the question baseline achieved better results than the retrieval baseline. This might be caused by the retrieved relevant passage being paragraph-like instead of conversational (thus, significantly different from the ground truth answer) since the performance doubled when we introduced the generation module.

Regarding our results, we observe that the variations without question rewriting had the worst performance, especially when only the last question is considered ($h = 1$). When introducing question rewriting, we explored 4 variations of the question rewriting input, all exhibiting higher scores than without question rewriting. In particular, the highest scores occur in 2 of the variations: when using only the questions, and when using both the model rewritten questions and model answers. The variation without model outputs in the question rewriting should be more resilient to diverging from the conversation topic.

When we used the ground truth rewritten questions instead, the performance of the passage retrieval and answer generation components increased about 1.6 ∼ 2.5×, highlighting the importance of good question rewriting.

---

[5] https://www.tira.io/task/scai-qrecc.

Impact of Question Rewriting. After this initial evaluation, we used the system with the highest F1 score (rewriting using model rewritten questions and answers) to further evaluate the impact of question rewriting. We computed the aforementioned metrics for each QA pair and used the scores to classify the results into different splits reflecting result quality, allowing us to analyze a module's performance when the previous ones succeeded (✓) or failed (✗).

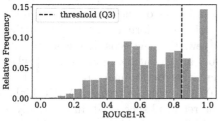
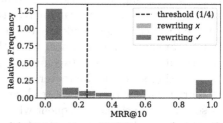

(a) Distribution of ROUGE1-R scores for question rewriting.

(b) Distribution of MRR scores (retrieval) when question rewriting succeeds or fails.

Fig. 2. Analysis of the influence of question rewriting on passage retrieval performance. Relative frequencies refer to the number of QA pairs of each split.

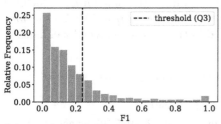
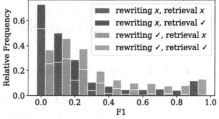

(a) Distribution of F1 scores for the answer generation component.

(b) Distribution of F1 scores when rewriting and retrieval succeed and fail.

Fig. 3. Analysis of the influence of question rewriting and passage retrieval on answer generation performance. Relative frequencies refer to each split.

To classify the performance of the question rewriting module using ROUGE scores, we used the 3rd quartile of the score distribution as a threshold (shown in Fig. 2a), since we are unable to choose a value that corresponds exactly to right/wrong rewriting decisions. As for classifying the passage retrieval using the MRR score, an immediate option would be to classify values greater than 0 as successful. However, although our system retrieves the top-10 most relevant passages, the answer generation model is limited by its maximum input size, which resulted in less important passages being truncated. A preliminary analysis showed us that, in most QA pairs, the model only considered 3–4 passages, and therefore we defined the threshold of a successful retrieval as MRR $\geq 1/4$.

When the question rewriting succeeds (ROUGE1-R $\geq$ Q3), the passage retrieval also exhibits better performance, as seen by MRR scores greater than 0 being more than twice more frequent (see Fig. 2b). Although both splits have many examples where the retrieval fails completely (MRR = 0), they are about twice more frequent when the question rewriting fails. Fig. 3a shows the distribution of F1 scores for answer generation, revealing that 75% of the results have an F1 score lower than 0.25. In turn, Fig. 3b shows 4 splits for when the question rewriting and retrieval modules each succeed or fail. Comparing the stacked bars together, one can analyze the influence of question rewriting in the obtained F1 score. Independently of the retrieval performance, F1 scores higher than 0.2 are much more frequent when the rewriting succeeds than when it fails. In particular, F1 scores between 0.3 and 0.8 are about 2× more frequent when the rewriting succeeds. Moreover, poor rewriting performance results in about 2× more results with an F1 score close to 0. Analyzing in terms of MRR, higher F1 scores are much more frequent when the retrieval succeeded. Interestingly, if the rewriting fails but the retrieval succeeds (less probable, as seen in Fig. 2b), the system is still able to generate answers with a high F1 score.

Error Example. In Table 2, we present a representative error where the system achieves a high ROUGE1-R score in the rewriting module but fails to retrieve the correct passage and to generate a correct answer. The only difference between the model and truth rewritten questions is in the omitted first name *Ryan*, which led the system to retrieve a passage referring to a different person (*Michael Dunn*). Although the first name was not mentioned in the context, maybe by enhancing

Table 2. Example conversation where the retrieval and generation failed.

Context		Q: When was Dunn's death?
		A: Dunn died on August 12, 1955, at the age of 59
Question		What were the circumstances?
Rewriting	Truth	What were the circumstances of Ryan Dunn's death?
ROUGE1-R: 0.889	Model	What were the circumstances of Dunn's death?
Retrieval	Truth	http://web.archive.org/web/20191130012451id_/
		https://en.wikipedia.org/wiki/Ryan_Dunn_p3
MRR: 0	Model	https://frederickleatherman.wordpress.com/2014/02/
		16/racism-is-an-insane-delusion-about-people-of-
		color/?replytocom=257035_p1
Generation	Truth	Ryan Dunn's Porsche 911 GT3 veered off the road,
		struck a tree, and burst into flames in West Goshen
		Township, Chester County, Pennsylvania.
F1: 0.051, EM: 0,	Model	The Florida Department of Law Enforcement
ROUGEL-F1:		concluded that Dunn's death was a homicide caused
0.128		by a single gunshot wound to the chest

the question with information from the previous turn (e.g., the age or day of death) the system could have performed better in the subsequent modules.

4 Conclusions and Future Work

This work presented a conversational QA system composed of 3 modules: question rewriting, passage retrieval, and answer generation. The results obtained from its evaluation on the QReCC dataset show the influence of each individual module in the overall system performance, and emphasize the importance of question rewriting. When the question rewriting succeeded, both the retrieval and answer generation improved – lower scores were up to $2\times$ less frequent while higher scores were also about $2\times$ more frequent. Future work should explore how to better control the question rewriting and its interaction with passage retrieval. Moreover, the impact of question rewriting or the use of other input representations should be validated with different datasets and models. Although our system with automatic question rewriting achieved the 1st place in the SCAI QReCC shared task, significant improvements can perhaps still be achieved with a better rewriting module (e.g., by fine-tuning T5 in the QReCC dataset).

References

1. Anantha, R., Vakulenko, S., Tu, Z., Longpre, S., Pulman, S., Chappidi, S.: Open-domain question answering goes conversational via question rewriting. In: Proceedings of the 2021 Conference of the North American Chapter of the Association for Computational Linguistics: Human Language Technologies, pp. 520–534. Association for Computational Linguistics, Online, June 2021. https://doi.org/10.18653/v1/2021.naacl-main.44, https://aclanthology.org/2021.naacl-main.44
2. Brown, T., et al.: Language models are few-shot learners. In: Larochelle, H., Ranzato, M., Hadsell, R., Balcan, M.F., Lin, H. (eds.) Advances in Neural Information Processing Systems, vol. 33, pp. 1877–1901. Curran Associates, Inc. (2020). https://proceedings.neurips.cc/paper/2020/file/1457c0d6bfcb4967418bfb8ac142f64a-Paper.pdf
3. Choi, E., et al.: QuAC: question answering in context. In: Proceedings of the 2018 Conference on Empirical Methods in Natural Language Processing, pp. 2174–2184. Association for Computational Linguistics, Brussels, Belgium, October–November 2018. https://doi.org/10.18653/v1/D18-1241, https://aclanthology.org/D18-1241
4. Dalton, J., Xiong, C., Callan, J.: CAst 2020: the conversational assistance track overview. In: The Twenty-Ninth Text REtrieval Conference (TREC 2020) Proceedings (2020). https://trec.nist.gov/pubs/trec29/trec2020.html
5. Elgohary, A., Peskov, D., Boyd-Graber, J.: Can you unpack that? Learning to rewrite questions-in-context. In: Proceedings of the 2019 Conference on Empirical Methods in Natural Language Processing and the 9th International Joint Conference on Natural Language Processing, pp. 5918–5924. Association for Computational Linguistics, Hong Kong, China, November 2019. https://doi.org/10.18653/v1/D19-1605, https://aclanthology.org/D19-1605
6. Lin, C.Y.: ROUGE: a package for automatic evaluation of summaries. In: Text Summarization Branches Out, pp. 74–81. Association for Computational Linguistics, Barcelona, Spain, July 2004. https://aclanthology.org/W04-1013

7. Lin, J., Ma, X., Lin, S.C., Yang, J.H., Pradeep, R., Nogueira, R.: Pyserini: a python toolkit for reproducible information retrieval research with sparse and dense representations. In: Proceedings of the 44th International ACM SIGIR Conference on Research and Development in Information Retrieval, pp. 2356–2362. Association for Computing Machinery, New York, NY, USA (2021). https://doi.org/10.1145/3404835.3463238

8. Raffel, C., et al.: Exploring the limits of transfer learning with a unified text-to-text transformer. J. Mach. Learn. Res. **21**(140), 1–67 (2020). http://jmlr.org/papers/v21/20-074.html

9. Rajpurkar, P., Zhang, J., Lopyrev, K., Liang, P.: SQuAD: 100,000+ questions for machine comprehension of text. In: Proceedings of the 2016 Conference on Empirical Methods in Natural Language Processing, pp. 2383–2392. Association for Computational Linguistics, Austin, Texas, November 2016. https://doi.org/10.18653/v1/D16-1264, https://aclanthology.org/D16-1264

10. Robertson, S., Zaragoza, H.: The probabilistic relevance framework: BM25 and beyond. Found. Trends® Inf. Retrieval **3**(4), 333–389 (2009). https://doi.org/10.1561/1500000019

11. Vakulenko, S., Longpre, S., Tu, Z., Anantha, R.: Question rewriting for conversational question answering. In: Proceedings of the 14th ACM International Conference on Web Search and Data Mining, pp. 355–363. Association for Computing Machinery, New York, NY, USA (2021). https://doi.org/10.1145/3437963.3441748

12. Vakulenko, S., Voskarides, N., Tu, Z., Longpre, S.: A comparison of question rewriting methods for conversational passage retrieval, January 2021

13. Wolf, T., et al.: Transformers: state-of-the-art natural language processing. In: Proceedings of the 2020 Conference on Empirical Methods in Natural Language Processing: System Demonstrations, pp. 38–45. Association for Computational Linguistics, Online, October 2020. https://doi.org/10.18653/v1/2020.emnlp-demos.6, https://www.aclweb.org/anthology/2020.emnlp-demos.6

14. Yu, S., Liu, Z., Xiong, C., Feng, T., Liu, Z.: Few-shot conversational dense retrieval. In: Proceedings of the 44th International ACM SIGIR Conference on Research and Development in Information Retrieval, pp. 829–838. Association for Computing Machinery, New York, NY, USA (2021). https://doi.org/10.1145/3404835.3462856

15. Zhang, J., Zhao, Y., Saleh, M., Liu, P.: PEGASUS: pre-training with extracted gap-sentences for abstractive summarization. In: III, H.D., Singh, A. (eds.) Proceedings of the 37th International Conference on Machine Learning. Proceedings of Machine Learning Research, vol. 119, pp. 11328–11339. PMLR, 13–18 July 2020. https://proceedings.mlr.press/v119/zhang20ae.html

How Different are Pre-trained Transformers for Text Ranking?

David Rau[(✉)] and Jaap Kamps

University of Amsterdam, Amsterdam, The Netherlands
{d.m.rau,kamps}@uva.nl

Abstract. In recent years, large pre-trained transformers have led to substantial gains in performance over traditional retrieval models and feedback approaches. However, these results are primarily based on the MS Marco/TREC Deep Learning Track setup, with its very particular setup, and our understanding of why and how these models work better is fragmented at best. We analyze effective BERT-based cross-encoders versus traditional BM25 ranking for the passage retrieval task where the largest gains have been observed, and investigate two main questions. On the one hand, what is similar? To what extent does the neural ranker already encompass the capacity of traditional rankers? Is the gain in performance due to a better ranking of the same documents (prioritizing precision)? On the other hand, what is different? Can it retrieve effectively documents missed by traditional systems (prioritizing recall)? We discover substantial differences in the notion of relevance identifying strengths and weaknesses of BERT that may inspire research for future improvement. Our results contribute to our understanding of (black-box) neural rankers relative to (well-understood) traditional rankers, help understand the particular experimental setting of MS-Marco-based test collections.

Keywords: Neural IR · BERT · Sparse retrieval · BM25 · Analysis

1 Introduction

Neural information retrieval has recently experienced impressive performance gains over traditional term-based methods such as BM25 or Query-Likelihood [3, 4]. Nevertheless, its success comes with the caveat of extremely complex models that are hard to interpret and pinpoint their effectiveness.

With the arrival of large-scale ranking dataset MS MARCO [1] massive models such as BERT [5] found their successful application in text ranking. Due to the large capacity of BERT (110m+ parameters), it can deal with long-range dependencies and complex sentence structures. When applied to ranking BERT can build deep interactions between query and document that allow uncovering complex relevance patterns that go beyond the simple term matching. Up to this point, the large performance gains achieved by the BERT Cross-Encoder are not well understood. Little is known about underlying matching principles that BERT bases its estimate of relevance on, what features are encoded in the model, and how the ranking relates to traditional sparse rankers such as BM25 [12]. In this work, we focus on the Cross-Encoder (CE) BERT that captures relevance signals directly between query and document through term interactions between them and refer from now on to the BERT model as *CE*. First, we aim to gain a deeper understanding of how CE and BM25 rankings relate to each other, particularly for different levels of relevance by answering the following research questions:

RQ1: How do CE and BM25 rankings vary?
RQ1.2: Does CE better rank the same documents retrieved by BM25?
RQ1.3: Does CE better find documents missed by BM25?

Second, we isolate and quantify the contribution of exact- and soft-term matching to the overall performance. To examine those are particularly interesting as they pose the most direct contrast between the matching paradigms of sparse- and neural retrieval. More concretely, we investigate:

RQ2: Does CE incorporate "exact matching"?
RQ3: Can CE still find "impossible" relevant results?

2 Related Work

Even though little research has been done to understand the ranking mechanism of BERT previous work exists. [9,10,19], have undertaken initial efforts to open ranking with BERT as a black-box and empirically find evidence that exact term matching and term importance seem to play in an important role. Others have tested and defined well-known IR axioms [2,7,11] or tried to enforced those axioms through regularization [13]. Another interesting direction is to enforce sparse encoding and able to relate neural ranking to sparse retrieval [6,18]. Although related, the work in [16] differs in two important aspects. First, they examine dense BERT retrievers which encode queries and documents independently. Second, they focus rather on the interpolation between BERT and BM25, whereas we specifically aim to understand how the two rankings relate to each other.

3 Experimental Setup

The vanilla BERT Cross-Encoder (CE) encodes both queries and documents at the same time. Given input $x \in \{[CLS], q_1, \ldots, q_n [SEP], d_1, \ldots, d_m, [SEP]\}$, where q represents query tokens and d document tokens, the activations of the CLS token are fed to a binary classifier layer to classify a passage as relevant or non-relevant; the relevance probability is then used as a relevance score to re-rank the passages.

Table 1. Performance of BM25 and crossencoder rankers on the NIST judgements of the TREC Deep Learning Task 2020.

Ranker	NDCG@10	MAP	MRR
BM25	49.59	27.47	67.06
BERT Cross-Encoder (CE)	69.33	45.99	80.85

We conduct our experiments on the TREC 2020 Deep Learning Track's passage retrieval task on the MS MARCO dataset [1]. For our experiments, we use the pre-trained model released by [8]. To obtain the set of top-1000 documents we use anserini's [17] BM25 (default parameters) without stemming, following [4]. Table 1 shows the baseline performance of BM25 and a vanilla BERT based cross-ranker (CE), re-ranking the 1,000 passages.

4 Experiments

4.1 RQ1: How do CE and BM25 Rankings Vary?

CE outperforms BM25 by a large margin across all metrics (see Table 1). To understand the different nature of the CE we trace where documents were initially ranked in the BM25 ranking. For this we split the ranking in different in four rank-ranges: 1–*10*, 11–*100*, 101–*500*, 501–*1000* and will refer to them with ranges 10, 100, 500 and 1000 respectively from now on. We observe in which rank-range the documents were positioned with respect to the initial BM25 ranking. We show the results in form of heatmaps[1] in Fig. 1.

Our initial goal is to obtain general differences between the ranking of CE and BM25 by considering all documents of the test collection (see Fig. 1(a)). First, we note that CE and BM25 vary substantially on the top of the ranking (33% CE@10), whereas at low ranks (60% CE@1000) the opposite holds. Second, we note that CE is bringing many documents up to higher ranks. Third, we observe that documents ranked high by BM25 are rarely ranked low by CE, suggesting exact matching to be a main underlying ranking strategy.

4.2 RQ1.2: Does CE Better Rank the Same Documents Retrieved by BM25?

To answer RQ1.2 we consider documents that were judged highly relevant or relevant according to the NIST judgments 2020. The results can be found in Fig. 1(b), (c) respectively. Most strikingly, both rankers exhibit a low agreement (40%) on the documents in CE@10 for *highly relevant* documents hinting a substantial different notion of relevance for the top of the ranking of both methods.

---

[1] The code for reproducing the heat maps can be found under https://github.com/davidmrau/transformer-vs-bm25

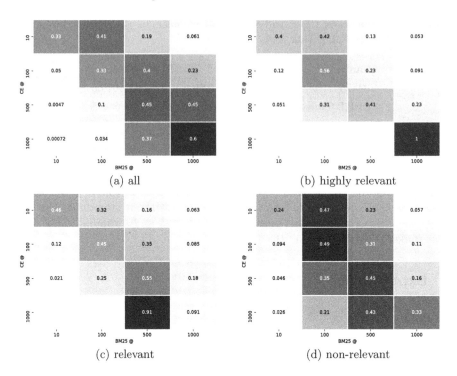

Fig. 1. Ranking differences between BERT Cross-Encoder (CE) and BM25: Origin of documents in CE ranking at different rank-ranges with respect to the initial BM25 ranking. More intuitively, each row indicates to what ratio documents stem from different rank-ranges. E.g., the top row can be read as the documents in rank 1–10 of the CE re-ranking originate 33% from rank 1–10, 41% from rank 11–100, 19% from rank 101–500 and 6.1% from rank 501–1000 in the initial BM25 ranking. The rank compositions are shown for (a) all, (b) highly relevant, (c) relevant, and (d) non-relevant documents according to the NIST 2020 relevant judgments.

For *relevant* documents we observe CE and BM25 overlap 46% at the top of the ranking and a large part (32%) comes from BM25@100, implying BM25 underestimated the relevance of many documents. The highest agreement between CE and BM25 here is in CE@500 (91%).

Interestingly, highly relevant documents that appear in lower ranks originate from high ranks in BM25 (CE@100: 12%, CE@500: 5%). This is an interesting finding as CE fails and underestimates the relevance of those documents, while BM25 - being a much simpler ranker - ranks them correctly. The same effect is also present for *relevant* documents. When considering documents that both methods ranked low we find a perfect agreement for @1000, showing that the two methods identify the same (highly-)relevant documents as irrelevant.

What about *non-relevant* documents that end up high in the ranking? CE brings up to CE@10 a large amount of non-relevant documents from low ranks (47% BM25@100, 23% BM25@500, and 5% BM@1000). Therewith overestimating the relevance of many documents that were correctly considered less relevant

Table 2. Performance of keeping only or removing the query terms from the input.

Model input	NDCG@10	MAP	MRR
Only Q	49.89	29.08	65.12
Drop Q	31.70	18.56	44.38

by BM25. We also note the little agreement of non-relevant documents @1000 (33%), hinting at a different notion of irrelevance.

4.3 RQ1.3: Does CE Better Find Documents Missed by BM25?

To answer RQ1.3 we again consider documents that were judged (b) highly relevant and (c) relevant and refer to Fig. 1, especially focusing on CE@10. The nature of CE, being too expensive for running it on the whole corpus, allows us to only study recall effects within the top-1000 documents. Hence, studying the top-10 results of CE will inform us best about the recall dynamics at high ranks. According to results in Fig. 1(b) almost half (42%) of the highly relevant documents that are missed by BM25 are brought up from BM25@100, 13% from range BM25@500, and 5% from range BM25@1000. The same effect can be observed for relevant documents. This demonstrates the superior ability of CE to pull up (highly)-relevant documents that are missed by BM25 even from very low ranks. This is the domain where the true potential of the neural models over exact matching techniques lies.

4.4 RQ2: Does CE Incorporate "Exact Matching"?

The presence of query words in the document is one of the strongest signals for relevance in ranking [14,15]. Our goal is to isolate the exact term matching effect, quantify its contribution to the performance, and relate it to sparse ranking. For this, we simply replace all non-query terms in the document with the [MASK] token leaving the model only with a skeleton of the original document and thus forcing it to rely on the exact term matches between query and document only. We do not fine-tune the model on this input. Note that there are no query document pairs within the underlying BM25 top-1000 run that have no term overlap. Results can be found in Table 2 under Only Q. CE with only the query words achieving comparable performance to BM25 finding clear support that CE is not only able to encode exact matches but also use them optimally.

As in view of finding potential ways to improve CE, our results suggest that exact term matching is already sufficiently encoded in CE.

4.5 RQ3: Can CE Still Find "Impossible" Relevant Results?

While CE can leverage both, exact term- as well as "soft" matches, the biggest advantage over traditional sparse retrievers holds the ability to overcome lexical matches and to take context into account. Through "soft" matches neural models can retrieve documents that are "impossible" to retrieve using traditional potentially resulting in high recall gains. To isolate and quantify the effect of "soft matches" we follow our previous experiment but this time mask the appearance of the query words in the document. The model has now to rely on the surrounding context only. We do not fine-tune the model on this input. Note that in this setting BM25 would score randomly. Results can be found in Table 2 under Drop Q.

We observe that CE can score documents sensibly with no overlapping query terms, although with a moderate drop in performance. The model scores 31.70 NDCG@10 points losing around 37 points with respect to non-manipulated input. CE might be able to guess the masked tokens from the context, as this makes up a main part of the Masked-Language modeling pre-training task. Therefore, to draw on its ability to create semantics through the contextualization of query and document and to leverage its associate memory.

5   Conclusions and Discussion

Our experiments find evidence that documents at the top of the ranking are generally ranked very differently while a stronger agreement at the bottom of the ranking seems to be present. By investigating the rankings for different relevance levels we gain further insight. Even though, for (highly-)relevant documents there exists a bigger consensus at the top of the ranking compared to the bottom we find a discrepancy in the notion of high relevance between them for some documents, highlighting core differences between the two rankers.

We discover that CE is dramatically underestimating some of the highly relevant documents that are correctly ranked by BM25. This sheds light on the sub-optimal ranking dynamics of CE, sparking clues to overcome current issues to improve ranking in the future. Our analysis finds further evidence that the main gain in precision stems from bringing (highly-)relevant documents up from lower ranks (early precision). On the other hand, CE overestimates the relevance of many non-relevant documents where BM25 scored them correctly lower.

Through masking all but the query words within the documents we show that CE is able to rank on the basis of only exact term matches scoring on par with BM25. By masking the query words in the document we demonstrate the ability of CE to score queries and documents without any lexical overlap with a moderate loss of performance, therefore demonstrating the true strength of neural models over traditional methods, that would completely fail in this scenario, in isolation.

We leave it to further research to qualitatively investigate the query-document pairs that BERT fails, but BM25 ranks correctly.

Acknowledgments. This research is funded in part by the Netherlands Organization for Scientific Research (NWO CI # CISC.CC.016), and the Innovation Exchange Amsterdam (POC grant).

References

1. Bajaj, P., et al.: MS MARCO: a human generated machine reading comprehension dataset (2016)
2. Câmara, A., Hauff, C.: Diagnosing bert with retrieval heuristics. Adv. Inf. Retrieval **12035**, 605 (2020)
3. Craswell, N., Mitra, B., Yilmaz, E., Campos, D.: Overview of the TREC 2020 deep learning track (2021)
4. Craswell, N., Mitra, B., Yilmaz, E., Campos, D., Voorhees, E.M.: Overview of the TREC 2019 deep learning track (2020)
5. Devlin, J., Chang, M., Lee, K., Toutanova, K.: BERT: pre-training of deep bidirectional transformers for language understanding (2018)
6. Formal, T., Lassance, C., Piwowarski, B., Clinchant, S.: Splade v2: sparse lexical and expansion model for information retrieval. arXiv preprint arXiv:2109.10086 (2021)
7. Formal, T., Piwowarski, B., Clinchant, S.: A white box analysis of ColBERT. In: Hiemstra, D., Moens, M.-F., Mothe, J., Perego, R., Potthast, M., Sebastiani, F. (eds.) ECIR 2021. LNCS, vol. 12657, pp. 257–263. Springer, Cham (2021). https://doi.org/10.1007/978-3-030-72240-1_23
8. Nogueira, R., Cho, K.: Passage re-ranking with bert (2019)
9. Padigela, H., Zamani, H., Croft, W.B.: Investigating the successes and failures of bert for passage re-ranking (2019)
10. Qiao, Y., Xiong, C., Liu, Z., Liu, Z.: Understanding the behaviors of bert in ranking. arXiv preprint arXiv:1904.07531 (2019)
11. Rennings, D., Moraes, F., Hauff, C.: An axiomatic approach to diagnosing neural IR models. In: Azzopardi, L., Stein, B., Fuhr, N., Mayr, P., Hauff, C., Hiemstra, D. (eds.) ECIR 2019. LNCS, vol. 11437, pp. 489–503. Springer, Cham (2019). https://doi.org/10.1007/978-3-030-15712-8_32
12. Robertson, S.E., Walker, S.: Some simple effective approximations to the 2-poisson model for probabilistic weighted retrieval. In: SIGIR 1994, pp. 232–241. Springer, Heidelberg (1994). https://doi.org/10.1007/978-1-4471-2099-5_24
13. Rosset, C., Mitra, B., Xiong, C., Craswell, N., Song, X., Tiwary, S.: An axiomatic approach to regularizing neural ranking models. In: Proceedings of the 42nd International ACM SIGIR Conference on Research and Development in Information Retrieval, pp. 981–984 (2019)
14. Salton, G., McGill, M.J.: Introduction to Modern Information Retrieval. McGraw-Hill Inc, USA (1986). https://sigir.org/resources/museum/
15. Saracevic, T.: Relevance: a review of and a framework for the thinking on the notion in information science. J. Am. Soc. Inf. Sci. **26**, 321–343 (1975). https://doi.org/10.1002/asi.4630260604
16. Wang, S., Zhuang, S., Zuccon, G.: Bert-based dense retrievers require interpolation with bm25 for effective passage retrieval. In: Proceedings of the 2021 ACM SIGIR International Conference on Theory of Information Retrieval, pp. 317–324 (2021)
17. Yang, P., Fang, H., Lin, J.: Anserini: Reproducible ranking baselines using lucene. J. Data Inf. Qual. **10**(4) (2018). https://doi.org/10.1145/3239571

18. Zamani, H., Dehghani, M., Croft, W.B., Learned-Miller, E., Kamps, J.: From neural re-ranking to neural ranking: Learning a sparse representation for inverted indexing. In: Proceedings of the 27th ACM International Conference on Information and Knowledge Management, pp. 497–506 (2018)
19. Zhan, J., Mao, J., Liu, Y., Zhang, M., Ma, S.: An analysis of bert in document ranking. In: Proceedings of the 43rd International ACM SIGIR Conference on Research and Development in Information Retrieval, pp. 1941–1944 (2020)

Comparing Intrinsic and Extrinsic Evaluation of Sensitivity Classification

Mahmoud F. Sayed$^{(\boxtimes)}$, Nishanth Mallekav, and Douglas W. Oard

University of Maryland, College Park, MD 20742, USA
{mfayoub,oard}@umd.edu, nishum@terpmail.umd.edu

Abstract. With accelerating generation of digital content, it is often impractical at the point of creation to manually segregate sensitive information from information which can be shared. As a result, a great deal of useful content becomes inaccessible simply because it is intermixed with sensitive content. This paper compares traditional and neural techniques for detection of sensitive content, finding that using the two techniques together can yield improved results. Experiments with two test collections, one in which sensitivity is modeled as a topic and a second in which sensitivity is annotated directly, yield consistent improvements with an intrinsic (classification effectiveness) measure. Extrinsic evaluation is conducted by using a recently proposed learning to rank framework for sensitivity-aware ranked retrieval and a measure that rewards finding relevant documents but penalizes revealing sensitive documents.

Keywords: Evaluation · Sensitivity · Classification

1 Introduction

The goal of information retrieval is to find things that a searcher wants to see. Present systems are fairly good, so content providers need to be careful to exclude things that should not be found from the content being searched. As content volumes increase, segregation of sensitive content becomes more expensive. One approach is to ask content producers to mark sensitive content, but that suffers from at least two problems. First, the producer's interests may differ from those of future searchers, so producers may not be incentivized to label sensitivity in ways that would facilitate future access to content that is not actually sensitive. As an example, some lawyers note at the bottom of every email message that the message may contain privileged content. Doing so serves the lawyer's general interest in protecting privileged content, but there is no incentive for the lawyer to decide in each case whether such a note should be added. Second, sensitivity can change over time, so something marked as sensitive today may not be sensitive a decade from now. For both reasons, post-hoc sensitivity classification is often required. This paper explores measurement of the utility of a post-hoc classifier, comparing intrinsic evaluation (asking whether the classifier decided correctly in each case) with extrinsic evaluation (measuring the

M. Hagen et al. (Eds.): ECIR 2022, LNCS 13186, pp. 215–222, 2022.
https://doi.org/10.1007/978-3-030-99739-7_25

effect of a sensitivity classifier on a search engine that seeks to protect sensitive content [1]).

2   Related Work

This problem of deciding what information can be shown in response to a request arises in many settings [2], including protection of attorney-client privilege [3], protection of government interests [4], and protection of personal privacy [5]. Three broad approaches have been tried. The first is detecting sensitive content at the point of creation, a type of pre-filtering. For example, social media posts can be checked before posting to detect inappropriate content [6]. One problem with pre-filtering is that the effort required to detect errors is spread equally over all content, including content nobody is ever likely to search for. A second approach is to review search results for sensitive content before their release. This post-filtering approach is used when searching for digital evidence in lawsuits or regulatory investigations [7–9] and for government transparency requests [10–13]. Post-filtering and pre-filtering yield similar results, but with different operational considerations. Some limitations of post-filtering are that the initial search must be performed by some intermediary on behalf of the person requesting the content, and that review of retrieved results for sensitivity may be undesirably slow. There has been some work on a third way, integrating sensitivity review more closely with the search process [1]. The basic idea in this approach is to train a search system to balance the imperatives to find relevant documents and to protect sensitive documents. In this paper we compare post-filtering with this approach of jointly modeling relevance and sensitivity.

Determining whether a document is sensitive is a special case of text classification [14]. Many such techniques are available; among them we use the sklearn implementation of logistic regression in this paper [15]. More recently, excellent results have been obtained using neural deep learning techniques, in particular using variants of Bidirectonal Encoder Representation from Transformer (BERT) models [16]. In this paper, we use the DistilBERT implementation [17]. In text classification, the most basic feature set is the text itself: the words in each document, and sometimes also word order. Additional features can also be useful in specific applications. For example, in email search, senders and recipients might be useful cues [18–21]. Similarly, in news the source of the story (e.g., the New York Times or the National Enquirer) and its date might be useful. For this paper we limit our attention to word presence and, for BERT, word order.

Research on jointly modeling relevance and sensitivity has been facilitated by test collections that model both factors. We are aware of four such collections. Two simulate sensitivity using topic annotations in large collections of public documents (news [8] or medical articles [1]). Although using topicality to simulate sensitivity may be a useful first-order approximation, higher fidelity models are also needed. Two email collections have been annotated for relevance and sensitivity (the Avocado collection [5,22] and the Enron collection [23,24]). However, the use of content that is actually sensitive requires policy protections.

Some sensitivities decline over time, so a third approach is to annotate content that was initially sensitive but is no longer so. We are aware of two collections annotated for former sensitivities (national security classification [25] and deliberative process privilege [12]), but neither case includes relevance annotations.

3   Test Collections

The test collections used to train and evaluate the models are the Avocado Research Email Collection, and the OHSUMED text classification test collection. The OHSUMED test collection is a set of 348,566 references from MEDLINE, an on-line medical information database, consisting of titles and abstracts from 270 medical journals for the period 1987 through 1991. Each document is categorized based on predefined Medical Subject Heading (MeSH) labels, from which Sayed and Oard [1] selected two categories to represent sensitivity: C12 (Male Urogenital Diseases) and C13 (Female Urogenital Diseases and Pregnancy Complications). The Avocado Research Email Collection consists of emails and attachments taken from 279 accounts of a defunct information technology company referred to as "Avocado". The collection includes messages, attachments, contacts, and tasks, of which we use only the messages and the attachments (concatenating the text in each message and all of its attachments). There are in total of 938,035 messages and 325,506 attachments. The collection is distributed by the Linguistic Data Consortium on a restricted research license that includes content nondisclosure provisions [26].

Sayed et al. [22] created a test collection based on the Avocado email collection. Each email that is judged for relevance to any topic is also judged for sensitivity according to one of two predefined personas [5]. The persona represents the sender if the email was sent from an Avocado employee, or the recipient if the email was sent from outside the company network. The sensitivity of an email was annotated based on the persona's expected decision whether to allow the email to appear in search results. The John Snibert persona was motivated to donate his email to an archive because it documents his career; he was careful in his use of email, but worried that he may have overlooked some kinds of information about which he was sensitive (e.g., romantic partners, peer reviews, and proprietary information). 3,045 messages are annotated for a total of 35 topics, 1,485 of which are sensitive. The Holly Palmer persona, by contrast, had originally been reluctant to donate her email because she knows how much sensitive information they contain (e.g., family matters, receipts that contain credit card numbers, and conversations that might be taken out of context). 2,869 messages were annotated for a total of 35 topics, 493 of which are sensitive.

4   Intrinsic Evaluation

In this section, we measure the effectiveness of three models for classifying sensitivity: logistic regression, DistilBERT, and a combination of the two. For the OHSUMED collection, all documents have sensitivity labels, but only a subset have relevance labels. So we use the subset that has both sensitivity and relevance labels as our test set, we use 85% of the documents that lack relevance labels for training sensitivity classifiers, and we use the remaining 15% of those documents that lack relevance labels as a validation set for sensitivity classifier parameter selection. Avocado is smaller, so in that case we evaluated classifiers using cross-validation. For each persona (John Snibert or Holly Palmer) in the Avocado collection, we first randomly split the annotated query-document pairs into 5 nearly equal partitions. We then iteratively chose one partition for evaluation and randomly selected 85% from the remaining four partitions as the corresponding training set, reserving the remaining 15% as a validation set.

Our logistic regression classifiers use sklearn's Logistic Regression library to estimate sensitivity probabilities [15]. The logistic regression model was trained on the union of the title and abstract for each document in the OHSUMED dataset, and on the union of the subject, body, and attachments for the Avocado collection. Our neural classifier estimates sensitivity probabilities using huggingface's DistilBERT, a pre-trained classification model trained on a large collection of English data in a self-supervised fashion [17]. DistilBERT is a distilled version of BERT large that runs 60% faster than BERT large while still retaining over 95% of its effectiveness. For the OHSUMED collection, fine-tuning DistilBERT for this classification task was performed using the training set. Many email messages have more text in the union of their subject, body and attachments than DistilBERT's 512-token limit, so for Avocado we divided the text of each item into 500-token passages with a 220-token stride. For fine-tuning each of the 5 Avocado classifiers for this task on the 5 training folds, we considered a passage sensitive if the document from which it had been extracted was marked as sensitive; for testing, we considered a document sensitive if any passage in that document was sensitive, and the probability of sensitivity for a document to be the maximum sensitivity probability for any passage in that document.

To assign a binary (yes/no) value for sensitivity to each test document, we learned one probability threshold for each classifier. We learned this threshold using the single validation partition on OHSUMED. For Avocado, we learned 5 thresholds, one for each validation fold. In each case, we used a grid search in the range $[0, 1]$ with step size 0.01 to find the threshold that optimized the F_1 measure.

Our third model, a disjunctive combination of our logistic regression and DistilBERT models, used the **or** function between the decisions of the DistilBERT and logistic regression models. For example, if Logistic Regression identified an Avocado email message as sensitive but DistilBERT classified it as not sensitive, the combined model would declare it as sensitive.

Table 1 reports four intrinsic measures of classification effectiveness: precision, recall, F_1, and F_2. Our experiment showed the DistilBERT classifier having the best F_1 score on the OHSUMED dataset, logistic regression having the best F_1 for John Snibert, and the combined model having the best F_1 for Holly Palmer. The reason for DistilBERT excelling on OHSUMED for F_1 is likely related to the number of training samples ($> 250k$). Neural methods

Table 1. Intrinsic sensitivity classification results (percent, ↑ indicates higher is better). Superscripts indicate statistically significant improvement in accuracy over that system by McNemar's test [27] at $p < 0.05$.

Classifier	OHSUMED				
	Precision↑	Recall↑	F_1↑	F_2↑	Accuracy↑
(a) LR	76.72	73.29	74.96	73.95	94.01
(b) DistilBERT	**82.75**	80.08	**81.39**	80.60	**95.52**^{a,c}
(c) Combined	74.61	**83.81**	78.94	**81.8**	94.53^a
Classifier	Avocado: Holly Palmer				
	Precision↑	Recall↑	F_1↑	F_2↑	Accuracy↑
(a) LR	**72.29**	69.98	71.12	70.43	**90.34**^{b,c}
(b) DistilBERT	66.20	67.85	67.02	67.52	88.65
(c) Combined	64.15	**80.11**	**71.25**	**76.31**	89.02
Classifier	Avocado: John Snibert				
	Precision↑	Recall↑	F_1↑	F_2↑	Accuracy↑
(a) LR	**80.53**	84.85	**82.63**	83.95	**83.06**^{b,c}
(b) DistilBERT	72.87	87.00	79.31	83.75	78.44
(c) Combined	70.86	**93.73**	80.71	**88.05**	78.72

tend to perform better with more data, and the Avocado collection only contains around 2,000 training samples. The combined model performed the best by F_2 for all three datasets. F_2 emphasizes recall, and as expected, the combined model yielded the best recall in every case.

5   Extrinsic Evaluation

In this section, we study the effect of sensitivity classification on search among sensitive content. The post-filter approach works by applying the sensitivity classifier on the ranking model's output as a filter, so that any result that is predicted to be sensitive is removed from the result list. We build ranking models using the Coordinate Ascent ranking algorithm [28], optimizing towards normalized Discounted Cumulative Gain (nDCG). The joint approach works by having the ranking model optimized towards a measure that balances between relevance and sensitivity. This can be achieved by leveraging listwise learning to rank (LtR) techniques. In our experiments, we used the Coordinate Ascent listwise LtR algorithm, which outperforms other alternatives on these collections [1]. We use the nCS-DCG@10 measure for both training and evaluating models in this approach [1].

For the Combined joint model, we calculate the sensitivity probability using an independence assumption as $P_{Combined} = 1 - (1 - P_{LR})(1 - P_{DistilBERT})$, where P_x is the sensitivity probability of classifier x. Logistic regression produces

Table 2. Extrinsic nCS-DCG@10 ($C_s = 12$) (percent, higher is better), 5-fold cross validation. Superscript: significant improvement over that system by 2-tailed paired t-test (p < 0.05) [30].

Collection: topics classifier	OHSUMED: 106		Holly Palmer: 35		John Snibert: 35	
	Post-filter	Joint	Post-filter	Joint	Post-filter	Joint
(a) LR	83.11	83.81	79.92	87.38	76.32	80.87
(b) DistilBERT	84.57[a]	**85.95**[a,c]	82.41	86.30	75.48	80.74
(c) Combined	**84.97**[a]	84.44	**84.40**[a]	**90.67**[a]	**79.65**	**83.46**[a]
Oracle	89.44	88.70	92.19	89.64	95.40	91.91

well calibrated probabilities, but DistilBERT probabilities can benefit from calibration. For this, we binned the DistilBERT sensitivity estimates on the validation set into 10 uniform partitions (0-10%, 10%-20%, ..., 90%-100%). The fraction of truly sensitive documents in each partition was then computed using validation set annotations. We then found an affine function to transform system estimates to ground truth values, minizing Mean Square Error (MSE) over the 10 points. At test time, this function was used to map DistilBERT sensitivity probability estimates to better estimates of the true sensitivity probability. This is similar to Platt scaling [29], but with a linear rather than sigmoid model.

Table 2 compares the effect of the three classifiers, and an oracle classifier that gives 100% probability to the ground truth annotation, on the two sensitivity-protecting search approaches. As expected, the oracle classifier with post-filtering consistently yields the best results because it never makes a mistake. However, with real classifiers, jointly modeling relevance and sensitivity consistently yields better results than post-filtering. We also see that using DistilBERT for sensitivity classification yields strong extrinsic evaluation results for our largest training data condition (OHSUMED). However, for both of our smaller training data conditions (Holly Palmer and John Snibert) the combined model outperforms DistilBERT numerically (although not statistically significantly). Looking back to Table 1, we see that it was F_2 that preferred the combined classifier on those two smaller test collections, suggesting that when training data is limited, F_2 might be a useful intrinsic measure with which to initially compare sensitivity classifiers when optimizing for measures such as nCS-DCG that penalize failures to detect sensitive content which is our ultimate goal.

6   Conclusions and Future Work

It is tempting to believe that better sensitivity classification will yield better results for search among sensitive content, but we have shown that the truth of that statement depends on how one measures "better." Of course, there is more to be done. Our current classifiers use only words and word sequences; additional features such as relationship graphs and temporal patterns might help to further improve classification accuracy [31]. For our email experiments, we have trained on sensitivity labels that are available only for documents that have been judged for relevance, but active learning might be used to extend the set of labeled documents in ways that could further improve classification accuracy. Finally, although we have tried a neural classification technique, we have combined this with a traditional approach to learning to rank for integrating search and protection. In future work, we plan to experiment with neural ranking as well.

Acknowledgments. This work has been supported in part by NSF grant 1618695.

References

1. Sayed, M.F., Oard, D.W.: Jointly modeling relevance and sensitivity for search among sensitive content. In: Proceedings of the 42nd International ACM SIGIR Conference on Research and Development in Information Retrieval, pp. 615–624. ACM (2019)
2. Thompson, E.D., Kaarst-Brown, M.L.: Sensitive information: a review and research agenda. J. Am. Soc. Inf. Sci. Technol. **56**(3), 245–257 (2005)
3. Gabriel, M., Paskach, C., Sharpe, D.: The challenge and promise of predictive coding for privilege. In: ICAIL 2013 DESI V Workshop (2013)
4. Mcdonald, G., Macdonald, C., Ounis, I.: How the accuracy and confidence of sensitivity classification affects digital sensitivity review. ACM Trans. Inf. Syst. (TOIS) **39**(1), 1–34 (2020)
5. Iqbal, M., Shilton, K., Sayed, M.F., Oard, D., Rivera, J.L., Cox, W.: Search with discretion: value sensitive design of training data for information retrieval. Proc. ACM Human Comput. Interact. 5, 1–20 (2021)
6. Biega, J.A., Gummadi, K.P., Mele, I., Milchevski, D., Tryfonopoulos, C., Weikum, G.: R-susceptibility: an IR-centric approach to assessing privacy risks for users in online communities. In: Proceedings of the 39th International ACM SIGIR conference on Research and Development in Information Retrieval, pp. 365–374 (2016)
7. Oard, D.W., Webber, W.: Information retrieval for e-discovery. Found. Trends Inf. Retrieval **7**(2–3), 99–237 (2013)
8. Oard, D.W., Sebastiani, F., Vinjumur, J.K.: Jointly minimizing the expected costs of review for responsiveness and privilege in e-discovery. ACM Trans. Inf. Syst. (TOIS) **37**(1), 11 (2018)
9. Vinjumur, J.K.: Predictive Coding Techniques with Manual Review to Identify Privileged Documents in E-Discovery. PhD thesis, University of Maryland (2018)
10. McDonald, G., Macdonald, C., Ounis, I.: Enhancing sensitivity classification with semantic features using word embeddings. In: Jose, J.M., Hauff, C., Altıngovde, I.S., Song, D., Albakour, D., Watt, S., Tait, J. (eds.) ECIR 2017. LNCS, vol. 10193, pp. 450–463. Springer, Cham (2017). https://doi.org/10.1007/978-3-319-56608-5_35
11. Abril, D., Navarro-Arribas, G., Torra, V.: On the declassification of confidential documents. In: Torra, V., Narakawa, Y., Yin, J., Long, J. (eds.) MDAI 2011. LNCS (LNAI), vol. 6820, pp. 235–246. Springer, Heidelberg (2011). https://doi.org/10.1007/978-3-642-22589-5_22
12. Baron, J.R., Sayed, M.F., Oard, D.W.: Providing more efficient access to government records: a use case involving application of machine learning to improve FOIA review for the deliberative process privilege. arXiv preprint arXiv:2011.07203, 2020

13. McDonald, G., Macdonald, C., Ounis, I., Gollins, T.: Towards a classifier for digital sensitivity review. In: de Rijke, M., Kenter, T., de Vries, A.P., Zhai, C.X., de Jong, F., Radinsky, K., Hofmann, K. (eds.) ECIR 2014. LNCS, vol. 8416, pp. 500–506. Springer, Cham (2014). https://doi.org/10.1007/978-3-319-06028-6_48

14. Sebastiani, F.: Machine learning in automated text categorization. ACM Comput. Surv. **34**(1), 1–47 (2002)

15. Feurer, M., Eggensperger, K., Falkner, S., Lindauer, M., Hutter, F.: Auto-sklearn 2.0: the next generation. arXiv preprint arXiv:2007.04074 (2020)

16. Adhikari, A., Ram, A., Tang, R., Lin, J.: DocBERT: BERT for document classification. arXiv preprint arXiv:1904.08398 (2019)

17. Sanh, V., Debut, L., Chaumond, J., Wolf, T.: DistilBERT, a distilled version of BERT: smaller, faster, cheaper and lighter. arXiv preprint arXiv:1910.01108 (2019)

18. Alkhereyf, S., Rambow, O.: Work hard, play hard: email classification on the Avocado and Enron corpora. In: Proceedings of TextGraphs-11: The Workshop on Graph-based Methods for Natural Language Processing, pp. 57–65 (2017)

19. Crawford, E., Kay, J., McCreath, E.: Automatic induction of rules for e-mail classification. In: Australian Document Computing Symposium (2001)

20. Sahami, M., Dumais, S., Heckerman, D., Horvitz, E.: A Bayesian approach to filtering junk e-mail. In: Learning for Text Categorization: Papers from the 1998 workshop, Madison, Wisconsin, vol. 62, pp. 98–105 (1998)

21. Wang, M., He, Y., Jiang, M.: Text categorization of Enron email corpus based on information bottleneck and maximal entropy. In: IEEE 10th International Conference on Signal Processing, pp. 2472–2475. IEEE (2010)

22. Sayed, M.F., et al.: A test collection for relevance and sensitivity. In: Proceedings of the 43rd International ACM SIGIR Conference on Research and Development in Information Retrieval, pp. 1605–1608 (2020)

23. Cormack, G.V., Grossman, M.R., Hedin, B., Oard, D.W.: Overview of the TREC 2010 legal track. In: TREC (2010)

24. Vinjumur, J.K., Oard, D.W., Paik, J.H.: Assessing the reliability and reusability of an e-discovery privilege test collection. In: Proceedings of the 37th International ACM SIGIR Conference on Research & Development in Information Retrieval, pp. 1047–1050 (2014)

25. Brennan, W.: The declassification engine: reading between the black bars. The New Yorker (2013). https://www.newyorker.com/tech/annals-of-technology/the-declassification-engine-reading-between-the-black-bars

26. Oard, D., Webber, W., Kirsch, D., Golitsynskiy, S.: Avocado research email collection. Linguistic Data Consortium, Philadelphia (2015)

27. McNemar, Q.: Note on the sampling error of the difference between correlated proportions or percentages. Psychometrika **12**(2), 153–157 (1947)

28. Metzler, D., Croft, W.B.: Linear feature-based models for information retrieval. Inf. Retrieval **10**(3), 257–274 (2007)

29. Platt, J.: Probabilistic outputs for support vector machines and comparisons to regularized likelihood methods. Adv. Large Margin Classifiers **10**(3), 61–74 (1999)

30. De Winter, J.C.F.: Using the Student's t-test with extremely small sample sizes. Pract. Assess. Res. Eval. **18**(1), 10 (2013)

31. Sayed, M.F.: Search Among Sensitive Content. PhD thesis, University of Maryland, College Park (2021)

Zero-Shot Recommendation as Language Modeling

Damien Sileo$^{(\boxtimes)}$ ⬩, Wout Vossen, and Robbe Raymaekers

KU Leuven, Leuven, Belgium
damien.sileo@kuleuven.be

Abstract. Recommendation is the task of ranking items (e.g. movies or products) according to individual user needs. Current systems rely on collaborative filtering and content-based techniques, which both require structured training data. We propose a framework for recommendation with off-the-shelf pretrained language models (LM) that only used unstructured text corpora as training data. If a user u liked *Matrix* and *Inception*, we construct a textual prompt, e.g. *"Movies like Matrix, Inception, $<m>$"* to estimate the affinity between u and m with LM likelihood. We motivate our idea with a corpus analysis, evaluate several prompt structures, and we compare LM-based recommendation with standard matrix factorization trained on different data regimes. The code for our experiments is publicly available (https://colab.research.google.com/drive/...?usp=sharing).

1 Introduction

Recommender systems predict an affinity score between users and items. Current recommender systems are based on content-based filtering (CB), collaborative filtering techniques (CF), or a combination of both. CF recommender systems rely on (USER, ITEM, INTERACTION) triplets. CB relies on (ITEM, FEATURES) pairs. Both system types require a costly structured data collection step. Meanwhile, web users express themselves about various items in an unstructured way. They share lists of their favorite items and ask for recommendations on web forums, as in (1)[1] which hints at a similarity between the enumerated movies.

(1) *Films like Beyond the Black Rainbow, Lost River, Suspiria, and The Neon Demon.*

The web also contains a lot of information about the items themselves, like synopsis or reviews for movies. Language models such as GPT-2 [14] are trained on large web corpora to generate plausible text. We hypothesize that they can make use of this unstructured knowledge to make recommendations by estimating the plausibility of items being grouped together in a prompt. LM can estimate the probability of a word sequence, $P(w_1, ...w_n)$. Neural language models are trained over a large corpus of

---

This work is part of the CALCULUS project, which is funded by the ERC Advanced Grant H2020-ERC-2017. ADG 788506 https://calculus-project.eu/

[1] https://www.reddit.com/r/MovieSuggestions/...lost_river/

M. Hagen et al. (Eds.): ECIR 2022, LNCS 13186, pp. 223–230, 2022.
https://doi.org/10.1007/978-3-030-99739-7_26

documents: to train a neural network, its parameters Θ are optimized for next word prediction likelihood maximization over k-length sequences sampled from a corpus. The loss writes as follows:

$$\mathcal{L}_{\mathrm{LM}} = -\log \sum_i P(w_i|w_{i-k}....w_{i-1}; \Theta) \qquad (1)$$

We rely on existing pretrained language models. To make a relevance prediction , we build a prompt for each user:

$$p_{u,i} = \textit{Movies like } <m_1>, ...<m_n>, <m_i> \qquad (2)$$

where $<m_i>$ is the name of the movie m_i and $<m_1...m_n>$ are those of randomly ordered movies liked by u. We then directly use $\widehat{R}_{u,i} = P_\Theta(p_{u,i})$ as a relevance score to sort items for user u.

Our contributions are as follow (i) we propose a model for recommendation with standard LM; (ii) we derive prompt structures from a corpus analysis and compare their impact on recommendation accuracy; (iii) we compare LM-based recommendation with next sentence prediction (NSP) [12] and a standard supervised matrix factorization method [9, 15].

2 Related Work

Language Models and Recommendation. Previous work leveraged language modeling techniques to perform recommendations. However, they do not rely on natural language: they use sequences of user/item interactions, and treat these sequences as sentences to leverage the architectures inspired by NLP, such as Word2Vec [1,4,7,11] or BERT [19].

Zero-Shot Prediction with Language Models. Neural language models have been used for zero-shot inference on many NLP tasks [2,14]. For example, they manually construct a prompt structure to translate text, e.g. *Translate english to french: "cheese"* =>, and use the language model completions to find the best translations. Petroni et al. [13] show that masked language models can act as a knowledge base when we use part of a triplet as input, e.g. *Paris in $<mask>$*. Here, we apply LM-based prompts to recommendation.

Hybrid and Zero-Shot Recommendation. The cold start problem [17], i.e. dealing with new users or items is a long-standing problem in recommender systems, usually addressed with hybridization of CF-based and CB-based systems. Previous work [5,6,10,20] introduced models for zero-shot recommendation, but they use zero-shot prediction with a different sense than ours. They train on a set of (USER, ITEM, INTERACTION) triplets, and perform zero-shot predictions on new users or items with known attributes. These methods still require (USER, ITEM, INTERACTION) or (ITEM, FEATURES) tuples for training. To our knowledge, the only attempt to perform recommendations without such data at all is from Penha et al. [12] who showed that BERT [3] next sentence prediction (NSP) can be used to predict the most plausible movie after a prompt. NSP is not available in all language models and requires a specific pretraining. Their work is designed as a probing of BERT knowledge about common items, and lacks comparison with a standard recommendation model, which we here address.

3   Experiments

3.1   Setup

Dataset. We use the standard the MovieLens 1M dataset [8] with 1M ratings from 0.5 to 5, 6040 users, and 3090 movies in our experiments. We address the relevance prediction task[2], so we consider a rating r as positive if $r \geq 4.0$, as negative if ≤ 2.5 and we discard the other ratings. We select users with at least 21 positive ratings and 4 negative ratings and thus obtain 2716 users. We randomly select 20% of them as test users[3]. 1 positive and 4 negative ratings are reserved for evaluation for each user, and the goal is to give the highest relevance score to the positively rated item. We use 5 positive ratings per user unless mentioned otherwise. We remove the years from the movie titles and reorder the articles (*a, the*) in the movie titles provided in the dataset (e.g. *Matrix, The (1999)* → *The Matrix*).

Evaluation Metric. We use the mean average precision at rank 1 (MAP@1) [18] which is the rate of correct first ranked prediction averaged over test users, because of its interpretability.

Pretrained Language Models. In our experiments we use the GPT-2 [14] language models, which are publicly available in several sizes. GPT-2 is trained with LM pre-training (Eq. 1) on the WebText corpus [14], which contains 8 million pages covering various domains. Unless mentioned otherwise, we use the GPT-base model, with 117M parameters.

3.2   Mining Prompts for Recommendation

Table 1. Occurrence counts of 3–6 grams that contain movie names in the Reddit corpus. $<m>$ denotes a movie name.

3–6 gram	#Count
$<m>$ and $<m>$	387
$<m>$, $<m>$, $<m>$	232
Movies like $<m>$	196
$<m>$, $<m>$, $<m>$, $<m>$	85
Movies similar to $<m>$	25

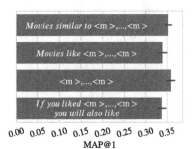

Fig. 1. Comparison of LM recommendations MAP@1 with different prompt structures.

---

[2] Item relevance could be mapped to ratings but we do not address rating prediction here.

[3] Training users are only used for the matrix factorization baseline.

We analyze the Reddit comments from May 2015[4] to find out how web users mention lists of movies in web text. This analysis will provide prompt candidates for LM-based recommendations. We select comments where a movie name of the MovieLens dataset is present and replace movies with a $<m>$ tag. This filtered dataset of comments has a size of $> 900k$ words. We then select the most frequent pattern with at least three words, as shown in Table 1. Movie names are frequently used in enumerations. The patterns *Movies like* $<m>$ and *Movies similar to* confirm that users focus on the similarity of movies.

Figure 1 shows that prompt design is important but not critical for high accuracy. Our corpus-derived prompts significantly outperform *if you like* $<m_1...m_n>$, *you will like* $<m_i>$ used in [12]. We will use $<m_1...m_n>$, $<m_i>$ in the remaining of the paper due to its superior results and its simplicity.

3.3   Effect of the Number of Ratings Per Test User

We investigate the effect of the number of mentioned movies in prompts. We expect the accuracy of the models in making recommendations to increase when they get more info about movies a user likes. We compare the recommendation accuracy on the same users 0,1,2,3,5,10,15 or 20 movies per prompt.

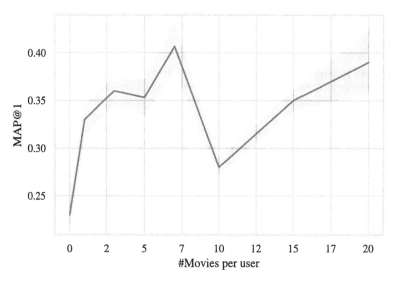

Fig. 2. MAP@1 of LM models with a varying number of movies per user sampled in the input prompt.

Figure 2 shows that increasing the number of ratings per user has diminishing returns and lead to increasing instability, so specifying $n \approx 5$ seems to lead to the best results with the least user input. After 5 items, adding more items might make the

---

[4] https://www.kaggle.com/reddit/reddit-comments-may-2015

prompt less natural, even though the LM seems to adapt when the number of items keeps increasing. It is also interesting to note that when we use an empty prompt, accuracy is above chance level because the LM captures some information about movie popularity.

3.4 Comparison with Matrix Factorization and NSP

We now use a matrix factorization as a baseline, with the Bayesian Personalised Ranking algorithm (BPR) [15]. Users and items are mapped to d randomly initialized latent factors, and their dot product is used as a relevance score trained with ranking loss. We use [16] implementation with default hyperparameters[5] $d = 10$ and a learning rate of 0.001.

We also compare GPT-2 LM to BERT next sentence prediction [12] which models affinity scores with $\widehat{R}_{u,i} = \text{BERT}_{\text{NSP}}(p_u, <m_i>)$, where p_u is a prompt containing movies liked by u. BERT was pretrained with contiguous sentence prediction task [3] and Penha et al. [12] proposed to use it as a way to probe BERT for recommendation capabilities.

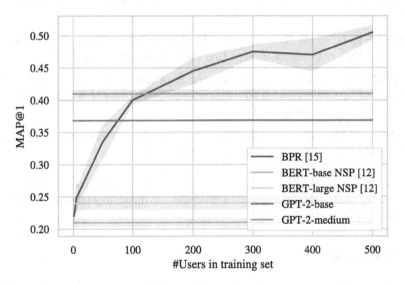

Fig. 3. MAP@1 for BPR models with increasing numbers of users compared the zero-shot language models (with 0 training user). BERT-base and BERT-large respectively have 110M and 340M parameters. GPT-2-base and GPT-2-medium have 117M and 345M parameters.

[5] https://cornac.readthedocs.io/en/latest/models.html#bayesian-personalized-ranking-bpr, we experimented with other hyperparameter configurations but did not observe significant changes.

Figure 3 shows that the proposed LM-based recommendation significantly outperforms BERT$_{NSP}$. We explain the difference by the fact that LM are generative and actually model text likelihood, while next sentence prediction is discriminative and can be based on simple discursive coherence features. It is also interesting to note that LM-based models outperform matrix factorization when there are few users, i.e. $<$50 and $<$100 for BASE and MEDIUM GPT-2, which demonstrates that LM-based recommendation is viable for cold start regimes. Using models larger than the BASE versions lead to better results, however when we evaluated with larger versions (we did not perform the full experiments due to memory limitations), we did not see additional improvement, which could be explained by overfitting.

3.5   Qualitative Analysis of Generations

Up until there, we have used LM to score the likelihood of sequences. LM can also be used directly for text generation, unlike BERT. We here show LM-generated prompt completions randomly sampled in our dataset, using greedy decoding.

Prompt (P1): *Forrest Gump, Blade Runner, Modern Times, Amelie, Lord of the Rings The Return of the King, Shaun of the Dead, Alexander, Pan's Labyrinth, Cashback, Avatar:*

Completion (C1): *3, The Hunger Games: Mockingjay Part 2, King Arthur, A Feast for Crows, The Hunger Games: Catching Fire, Jackass, Jackass 2, King Arthur*

Prompt (P2): *Independence Day, Winnie the Pooh and the Blustery Day, Raiders of the Lost Ark, Star Wars Episode VI - Return of the Jedi, Quiet Man, Game, Labyrinth, Return to Oz, Song of the South, Matrix:*

Completion (C2): *and many more. The list can be read by clicking on the relevant section at the left of the image. To access the list of releases*

Some prompts, i.e. (P1) generate valid movie names, but others, like (P2), do not. LM-based recommender do need a post-processing to match movie names in the possible sampled generations.

4   Conclusion

We showed that standard language models can be used to perform item recommendations without any adaptation and that they are competitive with supervised matrix factorization when the number of users is very low (less than 100 users). LM can therefore be used to kickstart recommender systems if items are frequently discussed in the training corpora. Further research could explore ways to adjust LM for recommendation purposes or to combine LM with matrix factorization into hybrid systems. Another way to use of our findings would be to generate movie recommendation datasets by mining web data which could feed standard supervised recommendation techniques.

References

1. Barkan, O., Koenigstein, N.: Item2vec: neural item embedding for collaborative filtering. In: 2016 IEEE 26th International Workshop on Machine Learning for Signal Processing (MLSP), pp. 1–6 (2016). https://doi.org/10.1109/MLSP.2016.7738886
2. Brown, T.B., et al.: Language models are few-shot learners (2020)
3. Devlin, J., Chang, M.W., Lee, K., Toutanova, K.: BERT: pre-training of deep bidirectional transformers for language understanding. In: Proceedings of the 2019 Conference of the North American Chapter of the Association for Computational Linguistics: Human Language Technologies, vol. 1 (Long and Short Papers), pp. 4171–4186. Association for Computational Linguistics, Minneapolis (2019). https://doi.org/10.18653/v1/N19-1423, https://aclanthology.org/N19-1423
4. Devooght, R., Bersini, H.: Long and short-term recommendations with recurrent neural networks. In: UMAP 2017, pp. 13–21. Association for Computing Machinery, New York (2017). https://doi.org/10.1145/3079628.3079670
5. Ding, H., Ma, Y., Deoras, A., Wang, Y., Wang, H.: Zero-shot recommender systems (2021)
6. Feng, P.J., Pan, P., Zhou, T., Chen, H., Luo, C.: Zero shot on the cold-start problem: Model-agnostic interest learning for recommender systems. In: Proceedings of the 30th ACM International Conference on Information & Knowledge Management, CIKM 2021, pp. 474–483. Association for Computing Machinery, New York (2021). https://doi.org/10.1145/3459637.3482312
7. Guàrdia-Sebaoun, E., Guigue, V., Gallinari, P.: Latent trajectory modeling: a light and efficient way to introduce time in recommender systems. In: Proceedings of the 9th ACM Conference on Recommender Systems, pp. 281–284 (2015)
8. Harper, F.M., Konstan, J.A.: The movielens datasets: history and context. ACM Trans. Interact. Intell. Syst. **5**(4) (2015). https://doi.org/10.1145/2827872
9. Koren, Y., Bell, R., Volinsky, C.: Matrix factorization techniques for recommender systems. Computer **42**(8), 30–37 (2009)
10. Li, J., Jing, M., Lu, K., Zhu, L., Yang, Y., Huang, Z.: From zero-shot learning to cold-start recommendation. In: Proceedings of the AAAI Conference on Artificial Intelligence, vol. 33, no. 01, pp. 4189–4196 (2019). https://doi.org/10.1609/aaai.v33i01.33014189, https://ojs.aaai.org/index.php/AAAI/article/view/4324
11. Li, Z., Zhao, H., Liu, Q., Huang, Z., Mei, T., Chen, E.: Learning from history and present: next-item recommendation via discriminatively exploiting user behaviors. In: Proceedings of the 24th ACM SIGKDD International Conference on Knowledge Discovery & Data Mining, pp. 1734–1743 (2018)
12. Penha, G., Hauff, C.: What does bert know about books, movies and music? probing bert for conversational recommendation. In: Fourteenth ACM Conference on Recommender Systems, RecSys 2020, pp. 388–397. Association for Computing Machinery, New York (2020). https://doi.org/10.1145/3383313.3412249
13. Petroni, F., et al.: Language models as knowledge bases? In: Proceedings of the 2019 Conference on Empirical Methods in Natural Language Processing and the 9th International Joint Conference on Natural Language Processing (EMNLP-IJCNLP), pp. 2463–2473. Association for Computational Linguistics, Hong Kong (2019). https://doi.org/10.18653/v1/D19-1250, https://aclanthology.org/D19-1250
14. Radford, A., Wu, J., Child, R., Luan, D., Amodei, D., Sutskever, I.: Language models are unsupervised multitask learners (2019). https://openai.com/blog/better-language-models/
15. Rendle, S., Freudenthaler, C., Gantner, Z., Schmidt-Thieme, L.: BPR: bayesian personalized ranking from implicit feedback. In: UAI 2009, pp. 452–461. AUAI Press, Arlington (2009)

16. Salah, A., Truong, Q.T., Lauw, H.W.: Cornac: a comparative framework for multimodal recommender systems. J. Mach. Learn. Res. **21**(95), 1–5 (2020)

17. Schein, A.I., Popescul, A., Ungar, L.H., Pennock, D.M.: Methods and metrics for cold-start recommendations. In: Proceedings of the 25th Annual International ACM SIGIR Conference on Research and Development in Information Retrieval, SIGIR 2002, pp. 253–260. Association for Computing Machinery, New York (2002). https://doi.org/10.1145/564376.564421

18. Schröder, G., Thiele, M., Lehner, W.: Setting goals and choosing metrics for recommender system evaluations. In: UCERSTI2 Workshop at the 5th ACM Conference on Recommender Systems, Chicago, USA, vol. 23, p. 53 (2011)

19. Sun, F., Let al.: Bert4rec: sequential recommendation with bidirectional encoder representations from transformer. In: Proceedings of the 28th ACM International Conference on Information and Knowledge Management, CIKM 2019, pp. 1441–1450. Association for Computing Machinery, New York (2019). https://doi.org/10.1145/3357384.3357895

20. Volkovs, M., Yu, G., Poutanen, T.: Dropoutnet: addressing cold start in recommender systems. In: Guyon, I., Luxburg, U.V., Bengio, S., Wallach, H., Fergus, R., Vishwanathan, S., Garnett, R. (eds.) Advances in Neural Information Processing Systems, vol. 30. Curran Associates, Inc. (2017). https://proceedings.neurips.cc/paper/2017/file/dbd22ba3bd0df8f385bdac3e9f8be207-Paper.pdf

What Matters for Shoppers: Investigating Key Attributes for Online Product Comparison

Nikhita Vedula[✉], Marcus Collins, Eugene Agichtein, and Oleg Rokhlenko

Amazon, Seattle, USA
{veduln,collmr,eugeneag,olegro}@amazon.com

Abstract. Before making high-consideration purchase decisions, shoppers generally need to identify and evaluate products' key differentiating features or attributes. Many customers, however, lack the knowledge required to do so for all product domains. In this work, we investigate and analyze alternatives for identifying important product attributes, which customers can then use to compare candidate products. We propose an unsupervised attribute-ranking approach ReBARC, that combines both objective data from structured product catalogs, and subjective information from unstructured customer reviews, to suggest to the shopper the most important attributes to consider. Our detailed analysis of product attribute importance across various domains on a shopping website shows that ReBARC significantly outperforms prior efforts judged by both automated and human evaluation metrics. We also analyze the correlation and overlap between key product attributes detected by ReBARC, and those visible to customers during online product search.

1 Introduction and Background

E-commerce web sites contain a wealth of information describing the products they sell in the form of product features or attributes, which is largely factual and objective, and which is organized in a structured catalog. For instance, commonly available attributes for laptops include *brand*, *screen dimensions* and *memory*. These catalog attributes describe the product characteristics and help customers find and evaluate products for purchase. However, each product may have a large number of attributes, not all of which are equally useful. Searchers may not know in advance which attributes or product features are *important* to evaluate a given product. An important resource for customers to learn what attributes or features are most important is the opinion of other customers, in the form of *reviews* [6,8,14,31]. Often, these reviews are quite detailed, and cover multiple product characteristics, attributes, and features useful to the review author. It is not feasible for customers to read a large set of reviews and aggregate multiple opinions to identify key product attributes.

A method to detect and suggest to the searcher the most important attributes can significantly improve their search and shopping experience in several ways. It

M. Hagen et al. (Eds.): ECIR 2022, LNCS 13186, pp. 231–239, 2022.
https://doi.org/10.1007/978-3-030-99739-7_27

would guide manufacturers to better help customers by choosing which attributes to highlight in product titles and descriptions. Important attributes could be used as hints to help customers navigate retail websites or refine searches; or to offer appropriate product recommendations. For many customers, identifying these key attributes will educate them about the key considerations for the product category, and guide their comparison of multiple similar products, in product categories they are not yet familiar with (*e.g.*, Electronics). Such a method requires a high-quality, complete set of attributes, a way to rank them, and a source of data from which to compute the ranking. Several attempts have been made to extract product attribute names and their values from web pages using rule-based techniques [19,27], naive Bayes and EM-based algorithms [11,28], co-training [36], external dictionaries [29], feature engineering [17,22,24], active learning [39] and aspect extraction [7,10,25,26,37]. These methods do not generalize well across domains, and the expense of procuring manually labeled data makes them infeasible to be used at the scale of e-commerce. Distant supervision using general-purpose, open-source knowledge bases have been proposed to alleviate this cost [12,38]. But they are limited by the accuracy and completeness of the external sources, to tackle which additional efforts would have to be made [15,33,35]. Unsupervised extraction of popular attributes or aspects from review text has been studied before [1,12,13,21,34]. But directly using keywords mentioned in customer reviews as aspects or attributes also often leads to a lower domain coverage and noisy, incoherent and redundant aspects which need further manual clean-up to avoid downstream errors.

We propose an approach, ReBARC (Review Based Attribute Ranker for Product Comparison) that ranks objective product catalog attributes and subjective product aspects, based on their presence in customer reviews and the sentiment of review authors towards these attributes. We use catalog data provided by product manufacturers, which is likely to be accurate and complete. ReBARC is *domain-agnostic* and *unsupervised*, requiring no manually labeled training data. ReBARC also avoids direct use of noisy review data, by mapping attribute mentions in reviews back to the more reliable structured catalog data.

2   ReBARC: Review-Based Attribute Ranking for Product Comparison

Data Collection: We utilize the Amazon Product Reviews (APR) [23] data as our primary source to develop and evaluate ReBARC. APR consists of a set of products, their associated customer reviews, and some metadata for each product, (i.e. product categories, similar products, and catalog attributes). We consider all reviews that other users have marked helpful at least once. Available catalog attributes include both generic attributes common across product categories such as *price, item weight, etc.*, and product- or category-specific attributes such as *screen size*. We also performed a web crawl on product pages from www.amazon.com to obtain the names of specific aspects or features separately rated by customers who bought a given product, and added these to our set of potential product attributes. We manually removed any attributes that are unlikely to influence users' buying choices (e.g. *date first available*). We experiment on more than 10,000 unique products with an average of 64 attributes and 116,700 reviews per product category, as shown in Table 1. We now present our proposed method ReBARC, which involves ranking attributes based on their presence in reviews, as well as customer sentiment towards the attributes.

Popularity Based Attribute Ranking: We use a light-weight unsupervised key-phrase extraction method based on text statistical features, YAKE [4], to extract a set of useful terms from the reviews linked to each product. We then segment each review into sentences, and extract only those sentences that either contain a useful term or a product attribute. This yields R useful review sentences per product. Since the product title is likely to contain useful attribute information identified by its sellers or manufacturers, we also append the product title to a sample of the R review sentences. We then use the pre-trained Sentence-BERT [30] model to compute embeddings for these sentences. We also compute Sentence-BERT embeddings for each attribute associated with these products. For each review sentence t in R, we select the top 3 catalog attributes for the product, using Maximal Marginal Relevance (MMR) [5] to rank the attributes for each product, based on their cosine similarity with the transformed review sentences:

$$MMR = \arg\max_{a_i \in A-S}[\lambda(sim(a_i, t) - (1 - \lambda) \max_{a_j \in S} sim(a_i, a_j)]$$

where $a_i \in A$ and $a_j \in A$ denote attributes being ranked. S denotes the subset of attributes already selected for ranking. λ trades off between the similarity of the ranked attributes to the transformed review sentences and to each other. This ensures that highly ranked attributes are similar to both the review sentences and product titles, and are also diverse from each other to avoid redundancy. Of the resulting $3R$ attribute-sentence pairs, we pick the k most frequent attributes as the highest ranked attributes based on review popularity, for the given product.

Opinion Based Attribute Re-ranking: From Sect. 2, we obtain a list of k highly ranked attributes associated with each product, and also the review sentences that are relevant to each attribute. This approach considers both the seller-identified attributes with respect to the product title and catalog, as well as specific aspect features rated by customers, but does not yet consider customer *sentiment* with respect to the attributes mentioned in reviews. To perform a secondary ranking using sentiment, we utilize a RoBERTa [20] model fine-tuned on the SST-2 sentiment detection benchmark corpus [32]. This model outputs a sentiment score for each review sentence relevant to an attribute, i.e. how positive or negative the sentence is. We assume that the sentiment score of a sentence relevant to attribute a represents the sentiment towards a itself. We then average the absolute sentiment scores for each attribute a over all sentences linked with a. We use the absolute value because we want to find attributes customers feel strongly about, whether they feel negatively or positively. Finally, we re-rank the list of attributes obtained earlier using the aggregated sentiment scores, yielding the final ranked list of *top-k* attributes for each product. Therefore, ReBARC ensures that attribute importance is evaluated based on direct and indirect mentions of product attributes by buyers in their reviews; as well as the (positive or negative) opinions of customers towards these attributes. Highly-ranked, key attribute values for similar products can then be compared by users to make purchase decisions. We can improve our technique of finding the sentiment of a review sentence by extracting sub-sentence fragments, and aggregating the sentiment score of each fragment as the sentiment score of the sentence. There are also unusual ways of mentioning certain attributes that might be missed by our sentence embedding technique, which might need to be solved by manual intervention or supervision. We leave these directions for future work.

3   Experiments and Results

Experimental Setup: Our first baseline (*SRA*) consists of those attributes customers used to refine or filter product searches on an online shopping website which most frequently lead to customers purchases or adding products to their shopping cart. Our second baseline (*QAC*) consists of the top attributes identified from query auto-completion [3] logs of the same online shopping website, that assist in automatically completing customers' search queries for a product. The above two approaches are state-of-the-art, optimal indicators of attribute importance based on real-world search and purchase behavior of millions of customers using this shopping website. All data was aggregated, anonymized and limited to targeted and relevant information (product names, attribute names and values), to protect customer privacy. We also compare ReBARC with a recent high-performing unsupervised aspect extraction technique, CAt [34]. We evaluated other existing techniques [1,12], using TF-IDF weighting of attributes extracted from reviews, and unsupervised methods using CRFs [18] to extract attributes from review text. These methods did not outperform any of our other baselines, so to save space we omit their results in Table 2. Since ReBARC is completely unsupervised, we do not compare it with any supervised methods.

We performed multiple crowd-sourced user studies to assess the performance of ReBARC, and followed recommended practices [2] to ensure good quality output from crowd workers. For a given product, we presented to annotators one randomly chosen attribute from the top 5 important attributes identified by our model, and asked the annotators if they would consider that attribute important if purchasing that product (Table 1, inter-annotator agreement Cohen's $\kappa = 0.77$). We also combined and shuffled the top 5 attributes each from ReBARC and the baselines for specific products, and presented a list of about 15 unique attributes to crowd workers. We asked them to pick the top 3 and top 5 attributes that they thought would help them the most in buying that particular product, or in comparing other similar product options of the same type (Table 2, Cohen's $\kappa = 0.65$, which indicates a substantial inter-annotator agreement [9]).

Table 1. Key attributes detected by ReBARC and chosen as important by annotators. 'Num.' and 'cat.' denote numerical and categorical valued attributes.

Product category (#Products, #Attributes, #Reviews)	Human imp. attrs.	Human imp. num. attrs.	Human imp. cat. attrs.	Sample key attributes frequently detected by ReBARC per category
Home (2319, 61, 150K)	0.66	0.78	0.55	*Color, assembly, easy-ToClean*
Electronics (3267, 84, 339K)	0.71	0.82	0.56	*Price, display, color, resolution*
Tools (1218,76,291K)	0.73	0.82	0.55	*Durability, easy to install rating*
Beauty (546, 48, 66K)	0.77	0.82	0.71	*Brand, skinType, valueForMoney*
Appliances (1104, 78, 91K)	0.81	0.84	0.72	*Batteries, price, brand, rating*
Avg (all 10 categories)	0.71	0.77	0.61	N/A

Table 2. Evaluating the top k ranked important attributes using human evaluation and the metrics MAP@k and NDCG@k, for $k = 3$ and 5. Best performances are in bold. R, S, Q and C denote ReBARC, SRA, QAC, and CAt respectively.

Product category	MAP@5				MAP@3				NDCG@5				NDCG@3			
	R	S	Q	C	R	S	Q	C	R	S	Q	C	R	S	Q	C
Home	**0.51**	0.38	0.34	0.42	**0.32**	0.24	0.19	0.26	**0.57**	0.36	0.12	0.45	**0.43**	0.25	0.08	0.36
Electronics	**0.5**	0.35	0.36	0.4	**0.4**	0.2	0.24	0.26	**0.48**	0.27	0.14	0.39	**0.45**	0.13	0.05	0.34
Tools	**0.5**	0.36	0.35	0.39	**0.3**	0.21	0.18	0.21	**0.55**	0.32	0.18	0.44	**0.44**	0.19	0.1	0.34
Pets	**0.52**	0.37	0.34	0.41	**0.42**	0.33	0.25	0.31	**0.6**	0.36	0.17	0.5	**0.48**	0.34	0.1	0.37
Beauty	**0.6**	0.32	0.32	0.43	**0.35**	0.14	0.15	0.22	**0.65**	0.13	0.12	0.5	**0.52**	0.1	0.05	0.41
Grocery	**0.6**	0.33	0.35	0.46	**0.5**	0.22	0.23	0.37	**0.68**	0.11	0.18	0.51	**0.6**	0.1	0.13	0.47
Appliances	**0.57**	0.37	0.35	0.41	**0.48**	0.28	0.21	0.33	**0.7**	0.28	0.19	0.57	**0.64**	0.17	0.1	0.5

We manually inspected and cleaned the task to ensure that crowd workers were not asked to judge attributes that required any specialized domain expertise. All parameters of ReBARC were tuned based on performance on a validation set, which we created based on the above ground truth human annotations.

Experimental Results: About 54% of the important attributes ranked highly by ReBARC were numerically-valued. Table 1 shows that annotators chose 71% of our key product attributes as useful for making purchase decisions. We observe that more than 60% of attributes available as search refinement filters or recommended during query auto-completion are categorical-valued. On the contrary, ReBARC detects a good mix of categorical and numerical valued key attributes, across different product groups. Overall, annotators preferred numerically- over categorically-valued attributes. Customers are thus likely to benefit from access to more numerically-valued attributes during their product search and comparison process. Table 2 evaluates ReBARC and three baselines in ranking important product attributes. CAt [34] outperforms SRA and QAC for most product categories. ReBARC significantly outperforms all baselines by 10–20% across product categories, as per Mean Average Precision [40] and Normalized Discounted Cumulative Gain [16]. Inspecting a random sample of products also showed that key attributes detected by ReBARC are diverse with less repetition.

Discussion: Our results show that more than 70% of the review sentences we analyzed either explicitly mention the names of attributes (58% of the time), or have a high cosine similarity >0.7 to a catalog attribute (42% of the time). A wide range of numerical and categorical attributes identified by ReBARC were found useful by our human annotators. Most prior work extracts attribute names directly from review text, which is further used to identify key product attributes. This can cause ambiguity and redundancy in the important attributes detected, since the same catalog attribute can be referred to by different names in reviews (e.g. for a laptop, both *performance* and *speed* refer to a single *processor* attribute). In contrast, ReBARC links customer opinions taken from user reviews to existing product attributes identified by product retailers or manufacturers. Thus, it maintains consistency in the detected key attributes and avoids ambiguity and redundancy despite being completely unsupervised. Interestingly, sentiment-based attribute re-ranking improves performance for specific product categories only. For instance, in *Electronics*, nearly 60% of the attributes are discussed in a neutral, descriptive, rather than opinionated, way. Some attributes are also more frequently referred to than others in reviews (e.g. *network speed* vs *frequency band* for routers). In these cases, signals from reviews could be combined with search-based popularity for additional improvements.

Our evaluation reveals that the overlap between important attributes detected by ReBARC, and those sourced by search refinements or query auto-completions, is lower than 50% across various product categories. The search logs of the shopping website under consideration show that a large fraction of the search filters and query auto-completions suggest generic attributes (e.g. *price, brand, delivery speed*). In contrast, our model identifies both generic and product-specific attributes. Annotators perceived a product-specific attribute as more useful than a generic attribute for product comparison in more than 65% cases. For instance, ReBARC identified *wireless network speed* as a popular and important attribute based on reviews for routers. However, the e-commerce engine does not suggest anything related to "network speed" as a filter or auto-completion suggestion when searching for any of the diverse queries *'router', 'router wifi', 'router speed', 'router internet', 'router wireless', or 'router network'*. Incorporating attributes identified from reviews into the search interface could improve the search and shopping experience, especially for more technical product categories such as *Electronics* or *Tools and Home Improvement*. Understanding the meaning or values of catalog attributes for certain product categories may require the searcher to possess domain knowledge. Such attributes could be referred to by more common, easier to understand terminology from reviews, captured by ReBARC. For example, the term *image quality* from reviews can refer to more technical attributes such as *refresh rate or resolution*. Thus, our insights imply that automatic product comparison and customer education would benefit from a diverse set of both generic and product-specific attributes.

4 Conclusions

We presented an unsupervised approach, ReBARC, that uses data from structured product catalogs and customer opinions from reviews to automatically identify key product features useful for online shopping and product comparison. ReBARC significantly outperforms strong baselines on diverse metrics and product domains. We also studied the correlation between product attributes of interest to customers based on reviews, and those available to them for search on shopping websites. In future, we plan to actively use customer behavior and shopping history for detecting key attributes, and personalizing attribute ranking for customers.

References

1. Bing, L., Wong, T., Lam, W.: Unsupervised extraction of popular product attributes from e-commerce web sites by considering customer reviews. ACM TOIT (2016)
2. Buhrmester, M., Kwang, T., Gosling, S.: Amazon's mechanical Turk: a new source of inexpensive, yet high-quality data? (2016)
3. Cai, F., de Rijke, M.: A survey of query auto completion in information retrieval. Found. Trends Inf. Retrieval (2016)
4. Campos, R., Mangaravite, V., Pasquali, A., Jorge, A., Nunes, C., Jatowt, A.: Yake! keyword extraction from single documents using multiple local features. Inf. Sci. (2020)
5. Carbonell, J., Goldstein, J.: The use of MMR, diversity-based reranking for reordering documents and producing summaries. In: ACM SIGIR (1998)
6. Carmel, D., Lewin-Eytan, L., Maarek, Y.: Product question answering using customer generated content-research challenges. In: ACM SIGIR (2018)
7. Chen, G., Tian, Y., Song, Y.: Joint aspect extraction and sentiment analysis with directional graph convolutional networks. In: COLING (2020)
8. Chen, S., Li, C., Ji, F., Zhou, W., Chen, H.: Driven answer generation for product-related questions in e-commerce. In: ACM WSDM (2019)
9. Cohen, J.: A coefficient of agreement for nominal scales. Educ. Psychol. Meas. **20**(1), 37–46 (1960)
10. Da'u, A., Salim, N.: Aspect extraction on user textual reviews using multi-channel convolutional neural network. PeerJ Comput. Sci. **5** (2019)
11. Ghani, R., Probst, K., Liu, Y., Krema, M., Fano, A.: Text mining for product attribute extraction. ACM SIGKDD Explor. Newsl. (2006)
12. Giannakopoulos, A., Musat, C., Hossmann, A., Baeriswyl, M.: Unsupervised aspect term extraction with B-LSTM & CRF using automatically labelled datasets. arXiv preprint arXiv:1709.05094 (2017)
13. He, R., Lee, W.S., Ng, H.T., Dahlmeier, D.: An unsupervised neural attention model for aspect extraction. In: ACL (2017)

14. Hirschmeier, S., Egger, M.: Social product search-enhancing product search with mined (sparse) product features (2018)
15. Huynh, V.P., Papotti, P.: A benchmark for fact checking algorithms built on knowledge bases. In: Proceedings of the 28th ACM International Conference on Information and Knowledge Management, pp. 689–698 (2019)
16. Järvelin, K., Kekäläinen, J.: Cumulated gain-based evaluation of IR techniques. TOIS (2002)
17. Kozareva, Z., Li, Q., Zhai, K., Guo, W.: Recognizing salient entities in shopping queries. In: ACL (2016)
18. Lafferty, J., McCallum, A., Pereira, F.C.: Conditional random fields: probabilistic models for segmenting and labeling sequence data (2001)
19. Liu, B., Hu, M., Cheng, J.: Opinion observer: analyzing and comparing opinions on the web. In: WWW (2005)
20. Liu, Y., et al.: Roberta: a robustly optimized Bert pretraining approach. arXiv preprint arXiv:1907.11692 (2019)
21. Luo, L., et al.: Unsupervised neural aspect extraction with sememes. In: IJCAI (2019)
22. More, A.: Attribute extraction from product titles in ecommerce. arXiv preprint arXiv:1608.04670 (2016)
23. Ni, J., Li, J., McAuley, J.: Justifying recommendations using distantly-labeled reviews and fine-grained aspects. In: EMNLP-IJCNLP (2019)
24. Petrovski, P., Bizer, C.: Extracting attribute-value pairs from product specifications on the web. In: International Conference on Web Intelligence (2017)
25. Pontiki, M., et al.: SemEval-2016 task 5: aspect based sentiment analysis. In: Proceedings of the 10th International Workshop on Semantic Evaluation (2016)
26. Pontiki, M., Galanis, D., Pavlopoulos, J., Papageorgiou, H., Androutsopoulos, I., Manandhar, S.: SemEval-2014 task 4: aspect based sentiment analysis. In: Proceedings of the 8th International Workshop on Semantic Evaluation (2014)
27. Popescu, A., Etzioni, O.: Extracting product features and opinions from reviews. In: Kao, A., Poteet, S.R. (eds.) Natural Language Processing and Text Mining. Springer, London (2007). https://doi.org/10.1007/978-1-84628-754-1_2
28. Probst, K., Ghani, R., Krema, M., Fano, A., Liu, Y.: Semi-supervised learning of attribute-value pairs from product descriptions. In: IJCAI (2007)
29. Putthividhya, D., Hu, J.: Bootstrapped named entity recognition for product attribute extraction. In: EMNLP (2011)
30. Reimers, N., Gurevych, I.: Sentence-Bert: sentence embeddings using Siamese Bert-networks. In: EMNLP-IJCNLP (2019)
31. Retail, T.: They say they want a revolution - price water house (2016). https://www.pwc.es/es/publicaciones/retail-y-consumo/assets/total-retail-2016.pdf
32. Socher, R., et al.: Recursive deep models for semantic compositionality over a sentiment treebank. In: EMNLP (2013)
33. Thorne, J., Vlachos, A.: Evidence-based factual error correction. arXiv preprint arXiv:2106.01072 (2021)
34. Tulkens, S., van Cranenburgh, A.: Embarrassingly simple unsupervised aspect extraction. ArXiv abs/2004.13580 (2020)
35. Vedula, N., Parthasarathy, S.: Face-keg: fact checking explained using knowledge graphs. In: Proceedings of the 14th ACM International Conference on Web Search and Data Mining, pp. 526–534 (2021)
36. Wu, B., Cheng, X., Wang, Y., Guo, Y., Song, L.: Simultaneous product attribute name and value extraction from web pages. In: ACM WI-IAT (2009)

37. Xu, H., Liu, B., Shu, L., Yu, P.S.: Double embeddings and CNN-based sequence labeling for aspect extraction. ArXiv abs/1805.04601 (2018)
38. Yang, Y., Chen, W., Li, Z., He, Z., Zhang, M.: Distantly supervised NER with partial annotation learning and reinforcement learning. In: International Conference on Computational Linguistics (2018)
39. Zheng, G., Mukherjee, S., Dong, X., Li, F.: OpenTag: open attribute value extraction from product profiles. In: ACM SIGKDD (2018)
40. Zhu, M.: Recall, precision and average precision. Department of Statistics and Actuarial Science, University of Waterloo, Waterloo (2004)

Evaluating Simulated User Interaction and Search Behaviour

Saber Zerhoudi[1(✉)] ⓘ, Michael Granitzer[1] ⓘ, Christin Seifert[2] ⓘ, and Joerg Schloetterer[2] ⓘ

[1] University of Passau, Passau, Germany
{saber.zerhoudi,michael.granitzer}@uni-passau.de
[2] University of Duisburg-Essen, Duisburg, Germany
{christin.seifert,joerg.schloetterer}@uni-due.de

Abstract. Simulating user sessions in a way that comes closer to the original user interactions is key to generating user data at any desired volume and variety such that A/B-testing in domain-specific search engines becomes scalable. In recent years, research on evaluating Information Retrieval (IR) systems has mainly focused on simulation as means to improve users models and evaluation metrics about the performance of search engines using test collections and user studies. However, test collections contain no user interaction data and user studies are expensive to conduct. Thus there is a need in developing a methodology for evaluating simulated user sessions. In this paper, we propose evaluation metrics to assess the realism of simulated sessions and describe a pilot study to assess the capability of generating simulated search sequences representing an approximation of real behaviour. Our findings highlight the importance of investigating and utilising classification-based metrics besides the distribution-based ones in the evaluation process.

Keywords: Evaluation metrics · Simulating user session · Simulation evaluation · User search behaviour · User modelling

1 Introduction

Developing evaluation methods to help improve the performance of Information Retrieval (IR) systems has been the focal point of researchers in IR community for many years [8,14,21]. Originally, the Cranfield evaluation methodology [12], which is so far the leading methodology for evaluating an IR system, is designed to evaluate the performance of a system using a test collection comprising a sample of queries, documents and a set of relevance judgments (indicating which documents are relevant/non-relevant to which queries) but lacks user interaction data (indicating the interaction sequences generated by users while expressing their information needs). Furthermore, users are represented in a highly abstracted form without considering the complexities of their interactions. As users' information needs become more complex in sophisticated IR systems, assessing the performance of a system needs to be assessed over an entire interactive session. Such sophisticated IR systems have been so far evaluated mainly using controlled user studies [14] or using the history log of user interactions. However, the experimental results obtained in such a way would be expensive and hard to reproduce using the former method, or would require search logs that are mostly inaccessible due to users' privacy using the latter.

TREC Session Track [5] and Dynamic Test Collections [4] are two important attempts to evaluate IR system performance over an entire search session. The evaluation metrics they adopted (e.g. P@k and nDCG@k) are cheap and reusable, but they cannot cope with users' dynamic information need (i.e. query reformulation behaviors). Jiang et al. [11] explored the correlation between user models and metrics. They examined several evaluation metrics and showcased that session Rank-Biased Precision (sRBP) [15] and session-based DCG (sDCG) [10] have stronger correlations with user satisfaction compared with existing session-based metrics. Following their finding, Zhang et al. introduced Recency-aware Session-based Metrics (RSMs) [24] which characterise users' cognitive process in search sessions by incorporating the recency effect.

There has been growing interest in the generation of simulated interaction data, and in particular how to develop more realistic models of search [2,18], for multiple reasons: First, simulation offers a way to overcome the lack of experimental real-world data, especially when the acquisition of such data is costly or challenging. Second, simulation helps reducing the amount of collected user data while preserving the profiling efficiency and protecting the privacy and the confidentiality of users' personal information.

While most related studies focus on the browsing model (i.e., user browsing behavior when consulting a page of search results), querying model and document relevance model for search evaluation, few studies have investigated the utility of evaluating simulated user interactions. Carterette et al. [4] suggested a meta-evaluation methodology using session histories and evaluated the simulation model based on its effectiveness at predicting actual user interactions, using standard classification evaluation metrics (i.e., precision, recall, accuracy and AUC). However, as it will be discussed in this paper, the used classification-based metrics are difficult to be justified and lack an evaluation of distributional properties of the data. Inspired by Carterette et al. [4], we propose a method to evaluate simulated user sessions' realism in the context of a search session. Realism represents the level of authenticity that simulated sessions present compared to the real log data. We model users' browsing patterns using Markov models. Markov models have been widely used for discovering meaningful patterns in browsing data due to their good interpretability [16,23]. In particular, they capture sequences in search patterns using transitional probabilities between states and translate user sessions into Markov processes.

To summarise, the main contributions of our work are twofold: (1) We model users' browsing patterns using a first-order Markov approach and a contextual Markov model that utilises user's browsing context based on common sense assumptions. We then conduct experiments on a real-world dataset and simulate user search behaviour by the two approaches. (2) We propose a method to evaluate the realism of simulated user interactions in the context of a search session. We first utilise the Kolmogorov-Smirnov statistical test as an empirical validation to compare the similarity between data log and simulated sessions distribution, and then employ a classification-based evaluation technique to assess the quality of simulated search session.

2   Evaluation Methods

Our goal is to develop an evaluation methodology to evaluate to which extend simulated user models can replace or complement sample-based ones. The quality of simulated user search sessions is usually evaluated by comparing real log and simulated data. In fact, simulated data are expected to be similar to real data as we do not want them to be distinguishable. Our evaluation method assumes the following user models:

First-Order Markov Model: We propose investigating the use of Markov Chains to model the search dynamics. The theoretical model is based on first-order Markov models [22]. Let X_k be the random variable that models actions in a user search session. The transition probability is modelled using maximum likelihood estimation: $P(X_k = A_j | X_{k-1} = A_i) = \frac{N_{A_i, A_j}}{N_{A_i}}$, where N_{A_i} is the total amount of how many times the action A_i occurred in the training data and N_{A_i, A_j} is the amount of how many times the transition from action i to action j has been observed.

Contextual Markov Model: During a search session, a user performs different search actions to find documents that fulfil their information needs. The technique that we propose here aims to categorise users into different groups based on their search behaviour. Search tasks are commonly divided into two major types of user's behaviour [1,17]: i) *Exploratory:* where users are more likely to formulate more queries as they learn about the topic and explore the search result list exhaustively, ii) *Lookup:* where users only investigate the first few results and rephrase their queries quickly.

Kumaripaba et al. [1] extended the work of Marchionini [17] and provided a few simple indicators of information search behaviours (e.g. query length, maximum scroll depth, completion time) to categorise users into exploratory and lookup searchers. We utilise these indicators to split the training data into smaller portions. We build a first-order Markov model for each type (i.e. exploratory and lookup) and we compare them to the Markov model built from the whole data (i.e. first-order Markov model). This would allow us to evaluate the impact of context on the accuracy of simulated sessions.

2.1   Kolmogorov-Smirnov-Based Evaluation

The two-sample Kolmogorov-Smirnov (KS-2) goodness-of-fit test [13] is one of
the most useful and non-parametric methods for comparing two datasets. It is
a convenient method for investigating whether two probability distributions can
be regarded as indistinguishable. Essentially, we test the null hypothesis that the
two independent samples are drawn from the same distribution and proceed with
calculating the absolute value of the distance between two data samples which
we refer to as the test statistic d to compare their distribution for similarities.

We derive two separate simulation models for context-aware approaches (i.e.,
dividing the dataset into lookup and exploratory subsets and then for each subset
we construct a Markov model to simulate user-type specific search sessions) and
one global simulation model that is trained on whole dataset.

2.2   Classification-Based Evaluation

Additionally, we define a classification-based evaluation to evaluate the simulation realism of our models. We first develop a set of features that represent
the sequential nature of a user search session in the form of a feature vector.
Then we train a classifier to distinguish simulated sessions from real log data
sessions and report the results. Building upon previous work [9] about what
kinds of engineered features are best suited to various machine learning model
types, we developed a set of features that represent the sequentiality of the
search session (i.e., typing a query; reformulating the query; clicking, viewing
and exporting actions) and discarded those that only describe the user's overall
search behaviour (e.g., tally of search actions, queries formulation and clicks).
We used a binary vector to indicate the presence of a feature (i.e., (0) if present
and (1) if not) and ordered features in the sequence (i.e., i_feature where i refer
to the sequence order of the query in a session, e.g., 1_search, 2_view_record).

Each user session is converted to a feature vector, labelled and fed to a classifier. This process was repeated separately for each of the Markov approaches,
i.e., first-order and contextual. We created an equal amount of simulated sessions
as real log sessions for a balanced classification and evaluated three classifiers
with tenfold cross-validation. As per the classifier, we used the most popular
algorithms in binary classification, namely, Support Vector Machine [6], Decision Trees [20] (XGBoost), Random Forests [3] and reported the average score.
We also used automated machine learning (Auto-sklearn [7]) as it employs an
ensemble of top performing models discovered during the optimisation process.
Since we are interested in finding a classifier that is close to 100% Recall on
the real log sessions (i.e., successful in detecting all real log sessions) and a high
recall on the simulated sessions (i.e., good at detecting most of simulated sessions), we incorporate a bias in the classifier by weighting the class of real data
($w_{real} = 10^4$, $w_{simulated} = 1$) to penalise bad real log sessions predictions.

To evaluate the realism of our models, we use metrics *Precision*, *Recall*, *F-
score* and *Accuracy* common for objectively measuring the classifier's performance. In our case, we consider *True Positive* (TP) to be the scenario where

the model classifies simulated sessions as simulated. A score of 0 means that the classifier cannot distinguish between simulated and real log sessions and therefore the simulated sessions are similar to real log data sessions, whereas with a score of 1, simulated sessions and log data are completely different. Since we can distinguish between real log and simulated sessions, reporting the accuracy alone can obfuscate some of the performance that F-score would highlight. F-score tells how precise the classifier is (i.e., how many instances it classifies correctly), as well as how robust it is (i.e., does not miss a significant number of instances). In fact, if F-score showed low precision/recall along with a low accuracy, we can have better confidence in the results. Therefore, we utilise all four metrics to demonstrate relative performance and consistency of the results.

3   DataSet

We use Sowiport *User Search Session Data Set (SUSS)*[1] [19] for our experiments, which includes 484,437 individual search sessio0ns, 179,796 queries and around 8 million log entries that was collected over a period of one year (from April 2014 to April 2015). SUSS describes users' search actions using a list of 58 different actions that covers all user's activities while interacting with the interface of the search engine (e.g., formulating a query, clicking on a document, viewing the full document's content, selecting a facet, using search filters). For each user interaction, a session id, date stamp, length of the action and other additional information are stored to describe user's path during the search process. From the 484,437 individual search sessions in the dataset, we filter sessions that do not contain a query (i.e., users having searched nothing) or have invalid query annotations and we sample 100,000 sessions which we refer to as SUSS$^-$.

4   Results

For this evaluation test, we derived two separate simulation models (i.e., exploratory and lookup) and one global simulation model (i.e., first-order) that is trained on whole SUSS$^-$ dataset. For each model, we utilise the transition probabilities between states which are drawn from the log sessions and the simulated sessions separately to generate two independent samples. By feeding these data points to KS-2 we obtain the test statistic value (i.e., 0.00417 first-order, 0.00381 and 0.00302 for exploratory and lookup respectively) and compare it to the critical value for the two samples (i.e., 0.00421 first-order, 0.00389 and 0.00356 for exploratory and lookup respectively).

Results show that the statistical value is smaller than the critical value across all models, hence we retain the null hypothesis. Therefore, we conclude that the simulated and the real log sessions belong to the same distribution.

Since the KS-2 critical values are all significant, it means that query change as context factor does not improve the simulation or at least it is hard to quantify the improvement using a KS-2 test. Therefore, we need to adopt a second

---

[1] The dataset is publicly available at http://dx.doi.org/10.7802/1380.

evaluation method: we investigate whether we can train a classifier and try to distinguish between real log and simulated sessions through controlled scenarios. For each scenario, we simulate an equal amount of sessions as present in the log data to balance class distribution.

Table 1. Classification of real log sessions vs simulated sessions using first-order and contextual Markov model (CMM) approaches. We report the accuracy, recall, precision and F-score across 10-CV folds (while (1) averaging over three classifiers defined in Subsect. 2.2 and (2) using Auto-sklearn (AS.). Bold indicates the best result in terms of the corresponding metric. Lowest results are the best as we aim to reduce the classifier's capability to distinguish between real log and simulated sessions.

Approach		Size	Accuracy		Recall		Precision		F-score	
			Avg.	AS.	Avg.	AS.	Avg.	AS.	Avg.	AS.
First-order Markov model		1	0.661	0.660	0.814	0.796	0.543	0.558	0.651	0.656
CMM	Exploratory	0.39	0.611	0.625	0.628	0.673	0.506	0.502	0.560	0.575
	Lookup	0.61	**0.572**	**0.577**	**0.612**	**0.624**	**0.452**	**0.463**	**0.519**	**0.531**

Table 1 shows that when using contextual Markov with the exploratory-lookup approach, the model did better while simulating sessions for *"Lookup"* with an F-score of 0.519 in comparison to *"Exploratory"* with a score of 0.560. One possible explanation for this is that lookup sessions are probably easier to simulate since there is less variation. The exploratory group of users generate longer sessions, thus higher total of state transitions which results a diverse number of simulated sessions.

In summary, we report that grouping user search sessions depending on their behavioural characteristics helps improving the simulation quality (i.e., reducing the accuracy of the classifier which is translated by lower F-score, recall and precision values).

5   Conclusion

In this paper, we propose a method to evaluate simulated user interactions in the context of a search session, which can be used as economic alternatives of user studies. We performed experiments using a real-world academic dataset with contextual Markov models and provided empirical results showing that the context-aware models allow to account for finer context granularity, i.e., more specific models. The proposed evaluation methods represents a theoretical foundation for experimental studies of sophisticated IR systems and opens up many new research directions. For example, we can use the classification-based methods to derive potentially better metrics than the existing ones that we proposed. The evaluation methods also opens up many interesting opportunities to leverage search log data to generate various realistic user simulators for evaluating complicated search systems.

Acknowledgments. This work has been partially carried out within the project "SINIR: Simulating INteractive Information Retrieval" funded by the Deutsche Forschungsgemeinschaft (DFG, German Research Foundation) - 408022022.

References

1. Athukorala, K., Głowacka, D., Jacucci, G., Oulasvirta, A., Vreeken, J.: Is exploratory search different? A comparison of information search behavior for exploratory and lookup tasks. J. Assoc. Inf. Sci. Technol. **67**(11), 2635–2651 (2016)
2. Baskaya, F., Keskustalo, H., Järvelin, K.: Time drives interaction: simulating sessions in diverse searching environments. In: Proceedings of the 35th International ACM SIGIR Conference on Research and Development in Information Retrieval, pp. 105–114 (2012)
3. Belgiu, M., Drăguţ, L.: Random forest in remote sensing: a review of applications and future directions. ISPRS J. Photogrammetry Remote Sens. **114**, 24–31 (2016)
4. Carterette, B., Bah, A., Zengin, M.: Dynamic test collections for retrieval evaluation. In: Proceedings of the 2015 International Conference on the Theory of Information Retrieval, pp. 91–100 (2015)
5. Carterette, B., Kanoulas, E., Hall, M., Clough, P.: Overview of the TREC 2014 session track. Technical report, Delaware Univ Newark, Department of Computer and Information Sciences (2014)
6. Cortes, C., Vapnik, V.: Support-vector networks. Mach. Learn. **20**(3), 273–297 (1995)
7. Feurer, M., Klein, A., Eggensperger, K., Springenberg, J.T., Blum, M., Hutter, F.: Auto-sklearn: efficient and robust automated machine learning. In: Hutter, F., Kotthoff, L., Vanschoren, J. (eds.) Automated Machine Learning. TSSCML, pp. 113–134. Springer, Cham (2019). https://doi.org/10.1007/978-3-030-05318-5_6
8. Harman, D.: Information retrieval evaluation. Synth. Lect. Inf. Concepts Retrieval Serv. **3**(2), 1–119 (2011)
9. Heaton, J.: An empirical analysis of feature engineering for predictive modeling. In: SoutheastCon 2016, March 2016
10. Järvelin, K., Price, S.L., Delcambre, L.M.L., Nielsen, M.L.: Discounted cumulated gain based evaluation of multiple-query IR sessions. In: Macdonald, C., Ounis, I., Plachouras, V., Ruthven, I., White, R.W. (eds.) ECIR 2008. LNCS, vol. 4956, pp. 4–15. Springer, Heidelberg (2008). https://doi.org/10.1007/978-3-540-78646-7_4
11. Jiang, J., Allan, J.: Correlation between system and user metrics in a session. In: Proceedings of the 2016 ACM on Conference on Human Information Interaction and Retrieval, pp. 285–288 (2016)
12. Jones, K.S., Willett, P.: Readings in Information Retrieval. Morgan Kaufmann, Burlington (1997)
13. Massey Jr., F.J.: The Kolmogorov-Smirnov test for goodness of fit. J. Am. Stat. Associ. **46**(253), 68–78 (1951)
14. Kelly, D.: Methods for Evaluating Interactive Information Retrieval Systems with Users. Now Publishers Inc., Delft (2009)
15. Lipani, A., Carterette, B., Yilmaz, E.: From a user model for query sessions to session rank biased precision (sRBP). In: Proceedings of the 2019 ACM SIGIR International Conference on Theory of Information Retrieval, pp. 109–116 (2019)
16. Manavoglu, E., Pavlov, D., Giles, C.L.: Probabilistic user behavior models. In: Third IEEE International Conference on Data Mining, pp. 203–210. IEEE (2003)

17. Marchionini, G.: Exploratory search: from finding to understanding. Commun. ACM **49**(4), 41–46 (2006)
18. Maxwell, D., Azzopardi, L.: Agents, simulated users and humans: an analysis of performance and behaviour. In: Proceedings of the 25th ACM International Conference on information and Knowledge Management, pp. 731–740 (2016)
19. Mayr, P.: Sowiport User Search Sessions Data Set (SUSS) (Version: 1.0.0) (2016)
20. Quinlan, J.R.: Simplifying decision trees. Int. J. Man Mach. Stud. **27**(3), 221–234 (1987)
21. Sanderson, M.: Test Collection Based Evaluation of Information Retrieval Systems. Now Publishers Inc., Delft (2010)
22. Shamshad, A., Bawadi, M.A., Hussin, W.W., Majid, T.A., Sanusi, S.A.M.: First and second order Markov chain models for synthetic generation of wind speed time series. Energy **30**, 693–708 (2005)
23. Tran, V., Maxwell, D., Fuhr, N., Azzopardi, L.: Personalised search time prediction using Markov chains. In: Proceedings of the ACM SIGIR International Conference on Theory of Information Retrieval, pp. 237–240 (2017)
24. Zhang, F., Mao, J., Liu, Y., Ma, W., Zhang, M., Ma, S.: Cascade or recency: constructing better evaluation metrics for session search. In: Proceedings of the 43rd International ACM SIGIR Conference on Research and Development in Information Retrieval, pp. 389–398 (2020)

Multilingual Topic Labelling of News Topics Using Ontological Mapping

Elaine Zosa(✉) , Lidia Pivovarova , Michele Boggia ,
and Sardana Ivanova

University of Helsinki, Helsinki, Finland
{elaine.zosa,lidia.pivovarova,michele.boggia,sardana.ivanova}@helsinki.fi

Abstract. The large volume of news produced daily makes topic modelling useful for analysing topical trends. A topic is usually represented by a ranked list of words but this can be difficult and time-consuming for humans to interpret. Therefore, various methods have been proposed to generate labels that capture the semantic content of a topic. However, there has been no work so far on coming up with multilingual labels which can be useful for exploring multilingual news collections. We propose an ontological mapping method that maps topics to concepts in a language-agnostic news ontology. We test our method on Finnish and English topics and show that it performs on par with state-of-the-art label generation methods, is able to produce multilingual labels, and can be applied to topics from languages that have not been seen during training without any modifications.

Keywords: Topic labelling · Ontology linking · Cross-lingual embeddings

1 Introduction

Topic models uncover the latent themes in a document collection through the co-occurrences of words in documents[4]. The large volume of news produced daily makes topic models especially useful for tracking and analysing news trends [12,14,17]. A topic is usually represented by a ranked list of words but these words can be difficult and time-consuming to interpret for humans [10]. Therefore various methods have been proposed to assign concise labels to topics to improve interpretability [1,3,16,18]. However, there has been no work so far on coming up with multilingual topic labels. Generating labels in multiple languages allows users to compare topical trends across linguistic boundaries without having to align topics and to explore news collections by users who might not have the necessary linguistic skills to do otherwise.

In this work we are interested in assigning concise multilingual labels to news topics. We propose an ontological mapping method that maps topics to concepts in a language-agnostic news ontology. These concepts have labels in multiple languages that we use as topic labels. We approach ontology mapping

© The Author(s), under exclusive license to Springer Nature Switzerland AG 2022
M. Hagen et al. (Eds.): ECIR 2022, LNCS 13186, pp. 248–256, 2022.
https://doi.org/10.1007/978-3-030-99739-7_29

as a multilabel classification task where a topic can be classified as belonging to multiple concepts.

We train our classifier on a dataset of Finnish news and test it on Finnish and English topics, using the distant supervision approach proposed in Ref. [1], where articles are used as training data. Our method produces results that are on par with state-of-the-art label generation methods, produces multilingual labels and can be used for topics in languages that have not been used during training without any modification. The contributions in this paper are: (1) an ontological mapping approach that can produce topic labels in multiple languages; (2) a method based on contextualised cross-lingual embeddings that works in a zero-shot setting, assigning labels to topics in languages not seen during training; and (3) a novel dataset of Finnish news topics with gold standard labels.[1]

2   Related Work

Several existing methods for automatic topic labelling generate candidate labels either by extracting short phrases from topic-related documents [2,9,16] or from external sources such as Wikipedia [1,9] and then ranking the candidates according to their relevance to the topic using distance metrics such as cosine distance [3] or the Kullback-Leibler divergence [8,16].

Wikipedia is a popular external corpora for topic labelling, using article titles as candidate labels [3,9]. However, Ref. [9] argues that the broad domain covered by Wikipedia make it unsuitable for labelling topics from a domain-specific corpus, such as biomedical research papers. Moreover, Wikipedia sizes vary widely across different languages. Some previous work have also used ontologies [5,7] but their methods rely on network analysis techniques to extract labels from the ontologies.

A more recent development is using deep learning to directly generate labels. Ref. [1] proposes a sequence-to-sequence model (seq2seq) trained on a synthetic dataset of Wikipedia articles and titles while Ref. [18] finetune BART, a pretrained transformer-based language model [11], with topic keywords and candidate labels from weak labellers to generate labels.

3   Experimental Setup

3.1   Models

Ontology Mapping. We propose an ontological mapping method that maps topics to concepts in a language-agnostic news ontology and use the corresponding labels for these concepts—available in multiple languages—as topic labels. We treat the ontology mapping problem as a multilabel classification task where a topic can be classified as belonging to one or more concepts in the ontology.

The classifier takes as an input a sequence $X = (x_1, \ldots, x_n)$ of the n top terms of a topic, and predicts $P(c_i|X)$, the probabilities for each ontology concept $c_i \in$

---

[1] Our code and dataset are available: https://github.com/ezosa/topic-labelling.

C. The topic labels are obtained from the distribution $P(c_i|X)$ as follows: First, a list of label candidates is obtained by considering all c_i such that $P(c_i|X) > t$, where t is the classification threshold. Then, we propagate the predicted concepts to the top of the ontology. For instance, if a topic is classified as belonging to concept 01005000:CINEMA, it also belongs to concept 01000000:ARTS, CULTURE AND ENTERTAINMENT, the parent of 01005000:CINEMA. Lastly, we obtain the top topic labels by taking the most frequent concepts among the candidates and taking the labels of these concepts in the preferred language.

To compute the probabilities $P(c_i|X)$, we encode the top terms $(x_1, \ldots, x_n)$ using SBERT [19][2] and pass this representation to a classifier composed of two fully-connected layers with a ReLU non-linearity and a softmax activation. We set the classification threshold t to 0.03 as determined by the validation set. We refer to this as the **ontology** model. We illustrate this model in Fig. 1.

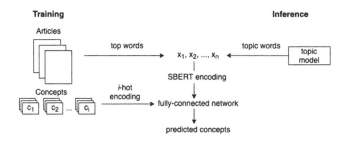

Fig. 1. News concepts prediction pipeline.

Comparisons to State-of-the-Art. We also investigate how our ontology mapping method compares to methods that directly generate topic labels. Ref. [1] uses an RNN-based encoder-decoder architecture with attention as a seq2seq model while Ref. [18] finetunes a pretrained BART model. Both methods have reported state-of-the-art results on English topics from multiple domains.

We implement a RNN seq2seq model using the same hyperparameters as [1]: 300-dim for the embedding layer and a hidden dimension of 200. We refer to this as the **rnn** model. We also implement a slightly modified model where we replace RNN with transformers, which has yielded state-of-the-art results in many NLP tasks. We use the hyperparameters from the original transformers model [22]: 6 layers for the encoder and decoder with 8 attention heads and an embedding dimension of 512. We refer to this as the **transformer** model.

Instead of BART which is trained only on English, we finetune a multilingual version, mBART [13], and set the source and target languages to Finnish. We finetuned mBART-25 from HuggingFace[3] for 5 epochs. We use the AdamW optimizer with weight decay set to 0.01. We refer to this as the **mbart** model [4].

---

[2] We use the multilingual model *distiluse-base-multilingual-cased*.

[3] https://huggingface.co/facebook/mbart-large-cc25.

[4] While the mBART encoder is in a multilingual space, it cannot be used directly for cross-lingual language generation [15].

For consistency, all the models except mbart are trained using Adam optimizer for 30 epochs with early stopping based on the validation loss.

3.2 Datasets

News Ontology. We use the IPTC Subject Codes as our news ontology.[5] This is a language-agnostic ontology designed to organise news content. Labels for concepts are available in multiple languages—in this work we focus specifically on Finnish and English. This ontology has three levels with 17 high-level concepts, 166 mid-level concepts and 1,221 fine-grained concepts. Mid-level concepts have exactly one parent and multiple children.

Training Data. We use news articles from 2017 of the Finnish News Agency (STT) dataset [20,21] which have been tagged with IPTC concepts and lemmatized with the Turku neural parser [6]. Following the distant-supervision approach in [1], we construct a dataset where the top n words of an article are treated as input $X = (x_1, \ldots, x_n)$ and the tagged concepts are the target C; an article can be mapped to multiple concepts. Top words can either be the top 30 scoring words by tf-idf (**tfidf** dataset) or the first 30 unique content words in the article (**sent** dataset). All models are trained on both datasets. For each dataset, we have 385,803 article-concept pairs which we split 80/10/10 into train, validation and test sets.

Test Data. For Finnish topics, we train an LDA model for 100 topics on the articles from 2018 of the Finnish news dataset and select 30 topics with high topic coherence for evaluation. We also check that the topics are diverse enough such that they cover a broad range of subjects.

To obtain gold standard labels for these topics, we recruited three fluent Finnish speakers to provide labels for each of the selected topics. For each topic, the annotators received the top 20 words and three articles closely associated with the topic. We provided the following instructions to the annotators:

Given the words associated with a topic, provide labels (in Finnish) for that topic. There are 30 topics in all. You can propose as many labels as you want, around 1 to 3 labels is a good number. We encourage concise labels (maybe 1–3 words) but the specificity of the labels is up to you. If you want to know more about a topic, we also provide some articles that are closely related to the topic. These articles are from 2018.

We reviewed the given labels to make sure the annotators understood the task and the labels are relevant to the topic. We use all unique labels as our gold standard, which resulted in seven labels for each topic on average. While previous studies on topic labelling mainly relied on having humans evaluate the labels outputted by their methods, we opted to have annotators *provide* labels instead because this will give us an insight into how someone would interpret a topic[6]. During inference, the input X are the top 30 words for each topic.

---

[5] https://cv.iptc.org/newscodes/subjectcode/.

[6] Volunteers are compensated for their efforts. We limited our test data to 30 topics due to budget constraints.

Table 1. Averaged BERTScores between labels generated by the models and the gold standard labels for Finnish and English news topics.

	PREC	REC	F-SCORE
Finnish news			
baseline: top 5 terms	*89.47*	*88.08*	*88.49*
ontology-tfidf	94.54	95.42	94.95
ontology-sent	95.18	**95.96**	95.54
mbart-tfidf	93.99	94.56	94.19
mbart-sent	94.02	95.04	94.51
rnn-tfidf	**96.15**	95.61	**95.75**
rnn-sent	95.1	94.63	94.71
transformer-tfidf	94.26	94.42	94.30
transformer-sent	95.45	94.73	94.98
English news			
baseline: top 5 terms	*98.17*	*96.58*	*97.32*
ontology-tfidf	97.00	95.25	96.04
ontlogy-sent	97.18	95.43	96.21

To test our model in a cross-lingual zero-shot setting, we use the English news topics and gold standard labels from the NETL dataset [3]. These gold labels were obtained by generating candidate labels from Wikipedia titles and asking humans to evaluate the labels on a scale of 0–3. This dataset has 59 news topics with 19 associated labels but we only take as gold labels those that have a mean rating of at least 2.0, giving us 330 topic-label pairs. We use default topic labels—top five terms of each topic—as the baselines.

4   Results and Discussion

We use BERTScore [23] to evaluate the labels generated by the models with regards to the gold standard labels. BERTScore finds optimal correspondences between gold standard tokens and generated tokens and from these correspondences, recall, precision, and F-score are computed. For each topic, we compute the pairwise BERTScores between the gold labels and the labels generated by the models and take the maximum score. We then average the scores for all topics and report this as the model score.

We show the BERTScores for the Finnish news topics at the top of Table 1. All models outperform the baseline by a large margin which shows that labels to ontology concepts are more aligned with human-preferred labels than the top topic words. The rnn-tfidf model obtained the best scores followed by ontology-sent. The transformer-sent and mbart-sent models also obtain comparable results. We do not see a significant difference in performance between training on the tfidf or sent datasets. In Table 2 (top), we show an example of the labels generated by the models and the gold standard labels. All models give sufficiently suitable labels, focusing on motor sports. However only the ontology-sent model was able to output 'formula 1' as one of its labels.

Table 2. Generated labels for selected topics. Finnish labels are manually translated except for ontology-sent. For ontology-sent, we provide the concept ID and the corresponding Finnish and English labels.

	Finnish topic
Topic	räikkönen, bottas, ajaa *(to drive)*, hamilto, mercedes
Gold	formula, formulat, formula 1, f1, formula-auto, aika-ajot *(time trial)*, moottoriurheilu *(motor sport)*
rnn-tfidf	autourheilu *(auto sport)*, urheilutapahtumat *(sports event)*, mm-kisat *(world championship)*, urheilu *(sport)*, urheilijat *(athletes)*
transformer-sent	urheilutapahtumat *(sports event)*, mm-kisat *(world championship)*, urheilu *(sport)*, autourheilu *(auto sport)*, kansainväliset *(international)*
mbart-sent	autourheilu moottoriurheilu, urheilutapahtumat, mm-kisat, urheilijat pelaajat, urheilu
ontology-sent	ID: 15000000, fi: urheilu, en: sport; ID: 15039000, fi: autourheilu moottoriurheilu, en: motor racing; ID: 15073000, fi: urheilutapahtumat, en: sports event; ID: 15039001, fi: formula 1, en: formula one; ID: 15073026, fi: mm-kisat, en: world championship
	English topic
Topic	film, movie star, director, hollywood, actor, minute, direct, story, witch
Gold	fantasy film, film adaptation, quentin tarantino, a movie, martin scorsese, film director, film
ontology-sent	ID: 01005001, en: film festival, fi: elokuvajuhlat; ID: 04010003, en: cinema industry, fi: elokuvateollisuus; ID: 08000000, en: human interest, fi: human interest; ID: 01022000, en: culture (general), fi: kulttuuri yleistä; ID: 04010000, en: media, fi: mediatalous

We also demonstrate the ability of the ontology models to label topics in a language it has not seen during training by testing it on English news topics from the NETL dataset [3]. This dataset was also used in Ref. [1] for testing but our results are not comparable since they present the scores for topics from all domains while we only use the news topics. The results are shown at the bottom of Table 1. Although the ontology models do not outperform the baseline, they are still able to generate English labels that are very close to the gold labels considering that the models have been trained only on Finnish data. From the example in Table 2 (bottom), we also observe that the gold labels are overly specific, suggesting names of directors as labels when the topic is about the film industry in general. We believe this is due to the procedure used to obtain the gold labels, where the annotators were asked to *rate* labels rather than propose their own.

5   Conclusion

We propose a straightforward ontology mapping method for producing multi-lingual labels for news topics. We cast ontology mapping as a multilabel classification task, represent topics as contextualised cross-lingual embeddings with SBERT and classify them into concepts from a language-agnostic news ontology where concepts have labels in multiple languages. Our method performs on par with state-of-the-art topic label generation methods, produces multilingual labels, and works on multiple languages without additional training. We also show that labels of ontology concepts correlate highly with labels preferred by humans. In future, we plan to adapt this model for historical news articles and also test it on more languages.

Acknowledgements. We would like to thank our annotators: Valter Uotila, Sai Li, and Emma Vesakoivu. This work has been supported by the European Union's Horizon 2020 research and innovation programme under grant 770299 (NewsEye) and 825153 (EMBEDDIA).

References

1. Alokaili, A., Aletras, N., Stevenson, M.: Automatic generation of topic labels. In: Proceedings of the 43rd International ACM SIGIR Conference on Research and Development in Information Retrieval, pp. 1965–1968 (2020)
2. Basave, A.E.C., He, Y., Xu, R.: Automatic labelling of topic models learned from twitter by summarisation. In: Proceedings of the 52nd Annual Meeting of the Association for Computational Linguistics (Volume 2: Short Papers), pp. 618–624 (2014)
3. Bhatia, S., Lau, J.H., Baldwin, T.: Automatic labelling of topics with neural embeddings. In: Proceedings of COLING 2016, the 26th International Conference on Computational Linguistics: Technical Papers, pp. 953–963 (2016)
4. Blei, D.M., Ng, A.Y., Jordan, M.I.: Latent Dirichlet allocation. J. Mach. Learn. Res. **3**, 993–1022 (2003)
5. Hulpus, I., Hayes, C., Karnstedt, M., Greene, D.: Unsupervised graph-based topic labelling using dbpedia. In: Proceedings of the Sixth ACM International Conference on Web Search and Data Mining, pp. 465–474 (2013)
6. Kanerva, J., Ginter, F., Miekka, N., Leino, A., Salakoski, T.: Turku neural parser pipeline: an end-to-end system for the conll 2018 shared task. In: Proceedings of the CoNLL 2018 Shared Task: Multilingual Parsing from Raw Text to Universal Dependencies. Association for Computational Linguistics (2018)

7. Kim, H.H., Rhee, H.Y.: An ontology-based labeling of influential topics using topic network analysis. J. Inf. Process. Syst. **15**(5), 1096–1107 (2019)
8. Kou, W., Li, F., Baldwin, T.: Automatic labelling of topic models using word vectors and letter trigram vectors. In: Zuccon, G., Geva, S., Joho, H., Scholer, F., Sun, A., Zhang, P. (eds.) AIRS 2015. LNCS, vol. 9460, pp. 253–264. Springer, Cham (2015). https://doi.org/10.1007/978-3-319-28940-3_20
9. Lau, J.H., Grieser, K., Newman, D., Baldwin, T.: Automatic labelling of topic models. In: Proceedings of the 49th Annual Meeting of the Association for Computational Linguistics: Human Language Technologies, pp. 1536–1545 (2011)
10. Lau, J.H., Newman, D., Baldwin, T.: Machine reading tea leaves: automatically evaluating topic coherence and topic model quality. In: Proceedings of the 14th Conference of the European Chapter of the Association for Computational Linguistics, pp. 530–539 (2014)
11. Lewis, M., et al.: BART: denoising sequence-to-sequence pre-training for natural language generation, translation, and comprehension. In: Proceedings of the 58th Annual Meeting of the Association for Computational Linguistics, pp. 7871–7880 (2020)
12. Li, Y., et al.: Global surveillance of covid-19 by mining news media using a multi-source dynamic embedded topic model. In: Proceedings of the 11th ACM International Conference on Bioinformatics, Computational Biology and Health Informatics, pp. 1–14 (2020)
13. Liu, Y., et al.: Multilingual denoising pre-training for neural machine translation. Trans. Assoc. Comput. Linguist. **8**, 726–742 (2020)
14. Marjanen, J., Zosa, E., Hengchen, S., Pivovarova, L., Tolonen, M.: Topic modelling discourse dynamics in historical newspapers. arXiv preprint arXiv:2011.10428 (2020)
15. Maurya, K.K., Desarkar, M.S., Kano, Y., Deepshikha, K.: ZmBART: an unsupervised cross-lingual transfer framework for language generation. In: Findings of the Association for Computational Linguistics: ACL-IJCNLP 2021, pp. 2804–2818. Association for Computational Linguistics, Online, August 2021. https://doi.org/10.18653/v1/2021.findings-acl.248, https://aclanthology.org/2021.findings-acl.248
16. Mei, Q., Shen, X., Zhai, C.: Automatic labeling of multinomial topic models. In: Proceedings of the 13th ACM SIGKDD International Conference on Knowledge Discovery and Data Mining, pp. 490–499 (2007)
17. Mueller, H., Rauh, C.: Reading between the lines: prediction of political violence using newspaper text. Am. Polit. Sci. Rev. **112**(2), 358–375 (2018)
18. Popa, C., Rebedea, T.: BART-TL: weakly-supervised topic label generation. In: Proceedings of the 16th Conference of the European Chapter of the Association for Computational Linguistics: Main Volume, pp. 1418–1425 (2021)
19. Reimers, N., Gurevych, I.: Sentence-BERT: sentence embeddings using siamese BERT-networks. In: Proceedings of the 2019 Conference on Empirical Methods in Natural Language Processing and the 9th International Joint Conference on Natural Language Processing (EMNLP-IJCNLP), pp. 3982–3992 (2019)
20. STT: Finnish news agency archive 1992–2018 (2019). source (http://urn.fi/urn:nbn:fi:lb-2019041501)
21. STT, Helsingin yliopisto, Alnajjar, K.: Finnish News Agency Archive 1992-2018, CoNLL-U (2020). source http://urn.fi/urn:nbn:fi:lb-2020031201

22. Vaswani, A., et al.: Attention is all you need. In: Advances in Neural Information Processing Systems, pp. 5998–6008 (2017)
23. Zhang, T., Kishore, V., Wu, F., Weinberger, K.Q., Artzi, Y.: BERTScore: evaluating text generation with BERT. In: International Conference on Learning Representations (2019)

Demonstration Papers

ranx: A Blazing-Fast Python Library for Ranking Evaluation and Comparison

Elias Bassani[1,2(✉)] [ID]

[1] Consorzio per il Trasferimento Tecnologico - C2T, Milan, Italy
[2] University of Milano-Bicocca, Milan, Italy
e.bassani3@campus.unimib.it

Abstract. This paper presents `ranx`, a Python evaluation library for Information Retrieval built on top of `Numba`. `ranx` provides a user-friendly interface to the most common ranking evaluation metrics, such as MAP, MRR, and NDCG. Moreover, it offers a convenient way of managing the evaluation results, comparing different runs, performing statistical tests between them, and exporting LATEX tables ready to be used in scientific publications, all in a few lines of code. The efficiency brought by `Numba`, a *just-in-time* compiler for Python code, makes the adoption `ranx` convenient even for industrial applications.

Keywords: Information Retrieval · Evaluation · Comparison

1 Introduction

Offline evaluation and comparison of different Information Retrieval (IR) systems is a fundamental step in developing new solutions [5,18]. The introduction of `trec_eval`[1] by the Text Retrieval Conference (TREC) [21] allowed standardizing evaluation measures in IR. This handy tool comes as a standalone C executable that researchers and practitioners must compile and run through a command-line interface. Unfortunately, it does not provide additional functionalities, such as comparing results from different IR systems or exporting the evaluation response to specific formats (e.g., LATEX).

Nowadays, the large majority of IR researchers rely on Python as their primary coding language. Because of that, many recent tools [2,4,9–13,17] provide experimentation and evaluation utilities in Python, such as IR evaluation measures. Nevertheless, we think there is still the need for a user-friendly Python library following a truly *Plug & Play* paradigm, which can also be helpful for young researchers with different backgrounds. For this reason, here we present `ranx`[2]. `ranx` lets the user calculate multiple evaluation measures, run statistical tests, and visualize comparison summaries, all in a few lines of code. Furthermore, it offers a convenient way of managing the evaluation results, allowing

---

[1] https://github.com/usnistgov/trec_eval.
[2] https://github.com/AmenRa/ranx.

© The Author(s), under exclusive license to Springer Nature Switzerland AG 2022
M. Hagen et al. (Eds.): ECIR 2022, LNCS 13186, pp. 259–264, 2022.
https://doi.org/10.1007/978-3-030-99739-7_30

the user to export them in LATEX format for scientific publications. We built **ranx** on top of **Numba** [8], a *just-in-time* compiler [1] for Python [16,20] and NumPy [6,15,22] code that allows high-speed vector operations and automatic parallelization. To the best of our knowledge, none of the available tools support multi-threading, which can vastly improve efficiency and grants **ranx** the ability to scale on large industrial datasets.

2   System Overview

In this section, we present the main functionalities provided by **ranx**: the **Qrels** and **Run** classes, the **evaluate** method, the **compare** method, and the **Report** class. More details and examples are available in the official repository.

2.1   Qrels and Run Classes

The first step in the offline evaluation of the effectiveness of an IR system is the definition of a list of *query relevance judgments* and the ranked lists of documents retrieved for those queries by the system. To ease the managing of these data, **ranx** implements two dedicated Python classes: 1) **Qrels** for the query relevance judgments and 2) **Run** for the computed ranked lists. As shown in Listing 1, the users can manually define the evaluation data and add new queries and the associated ranked lists dynamically. They can also import them from TREC-style files, Python dictionaries, and Pandas DataFrames [14]. **ranx** takes care of sorting and checking the data so that the user does not need to. Finally, **Qrels** and **Run** can be saved as TREC-style files for sharing. Every time the user evaluates a metric over a **Qrels-Run** pair, **ranx** stores the metrics' scores for each query and their averages in the Run instance so that they can be accessed later on, as shown in Listing 2.

```
1   # Instantiate an empty Qrels object
2   qrels = Qrels()
3
4   # Add query to qrels
5   qrels.add(q_id="q_1", doc_ids=["doc_12", "doc_25"], scores=[5, 3])
6
7   # Add multiple queries to qrels
8   qrels.add_multi(
9       q_ids=["q_1", "q_2"],
10      doc_ids=[["doc_12", "doc_25"], ["doc_11", "doc_2"]],  # Relevant document ids
11      scores=[[5, 3], [6, 1]],                              # Relevance judgements
12  )
13
14  # Import qrels from TREC-Style file
15  qrels = Qrels.from_file(path_to_qrels)
16  # Import qrels from Python Dictionary
17  qrels = Qrels.from_dict(qrels_dict)
18  # Import qrels from Pandas DataFrame
19  qrels = Qrels.from_df(qrels_df, q_id_col="q_id", doc_id_col="doc_id", score_col="score")
20
21  # Save as TREC-Style file
22  qrels.save("qrels.txt")
```

Listing 1: **Qrels**' methods. *The same methods are also available for* **Run**.

2.2 Metrics

ranx provides a series of *very* fast ranking evaluation metrics implemented in Numba. It currently supports *Hits, Precision, Recall, r-Precision, Reciprocal Rank* (and *Mean Reciprocal Rank*), *Average Precision* (and *Mean Average Precision*), and *Normalized Discounted Cumulative Gain* (both the original formulation from Järvelin et al. [7] and the variant introduced by Burges et al. [3]). For each metric but *r-Precision*, the "at *k*" variants are also available. We tested all the metrics against trec_eval for correctness. ranx provides access to all the implemented metrics by a single interface, the method evaluate, as shown in Listing 2. As mentioned above, Numba allows ranx to compute the scores for the evaluation metrics very efficiently, as shown in Table 1.

```
1   # Compute score for a single metric
2   evaluate(qrels, run, "ndcg@5")
3   >>> 0.7861
4
5   # Compute scores for multiple metrics at once
6   evaluate(qrels, run, ["map@5", "mrr"])
7   >>> {"map@5": 0.6416, "mrr": 0.75}
8
9   # Computed metric scores are saved in the Run object
10  run.mean_scores
11  >>> {"ndcg@5": 0.7861, "map@5": 0.6416, "mrr": 0.75}
12
13  # Access scores for each query
14  dict(run.scores)
15  >>> {"ndcg@5": {"q_1": 0.9430, "q_2": 0.6292},
16        "map@5": {"q_1": 0.8333, "q_2": 0.4500},
17        "mrr": {"q_1": 1.0000, "q_2": 0.5000}}
```

Listing 2: Usage example of the evaluate method.

Table 1. Efficiency comparison between ranx (using different number of threads) and pytrec_eval (pytrec), a Python interface to trec_eval. The comparison was conducted with synthetic data. Queries have 1-to-10 relevant documents. Retrieved lists contain 100 documents. NDCG, MAP, and MRR were computed on the entire lists. Results are reported in milliseconds. Speed-ups were computed w.r.t. pytrec_eval.

Metric	Queries	Pytrec	ranx t = 1		ranx t = 2		ranx t = 4		ranx t = 8	
NDCG	1 000	28	4	7.0×	3	9.3×	2	14.0×	2	14.0×
	10 000	291	35	8.3×	24	12.1×	18	16.2×	15	19.4×
	100 000	2 991	347	8.6×	230	13.0×	178	16.8×	152	19.7×
MAP	1 000	27	2	13.5×	2	13.5×	1	27.0×	1	27.0×
	10 000	286	21	13.6×	13	22.0×	9	31.8×	7	40.9×
	100 000	2 950	210	14.0×	126	23.4×	84	35.1×	69	42.8×
MRR	1 000	28	1	28.0×	1	28.0×	1	28.0×	1	28.0×
	10 000	283	7	40.4×	6	47.2×	4	70.8×	4	70.8×
	100 000	2 935	74	39.7×	57	51.5×	44	66.7×	38	77.2×

2.3   Comparison and Statistical Testing

As comparison is one of the fundamental steps in retrieval systems' evaluation, ranx implements a functionality—compare—for comparing multiple Runs. It computes the scores for a list of metrics provided by the user and performs statistical testing on those scores through Fisher's Randomization Test [19], as shown in Listing 3. The compare method outputs an object of the Report class.

2.4   The Report Class

The Report class store all the data produced by performing a *comparison* as described previously. The user can access this information by simply printing a Report in a Python shell, as shown in Listing 3. It also allows exporting a LaTeX table presenting the average scores for each computed metric for each of the compared models, enriched with superscripts denoting the statistical significance of the improvements (if any), as well as a pre-defined caption. Table 2 was generated by report.to_latex().

```
1   # Compare different runs and perform statistical tests
2   report = compare(
3       qrels=qrels,
4       runs=[run_1, run_2, run_3, run_4, run_5],
5       metrics=["map@100", "mrr@100", "ndcg@10"],
6       max_p=0.01  # P-value threshold
7   )
8
9   print(report)
10  >>>
11  #    Model      MAP@100      MRR@100      NDCG@10
12  ---  -------    ---------    ---------    ---------
13  a    model_1    0.3202ᵇ      0.3207ᵇ      0.3684ᵇᶜ
14  b    model_2    0.2332       0.2339       0.2390
15  c    model_3    0.3082ᵇ      0.3089ᵇ      0.3295ᵇ
16  d    model_4    0.3664ᵃᵇᶜ    0.3668ᵃᵇᶜ    0.4078ᵃᵇᶜ
17  e    model_5    0.4053ᵃᵇᶜᵈ   0.4061ᵃᵇᶜᵈ   0.4512ᵃᵇᶜᵈ
```

Listing 3: Usage example of the compare method.

Table 2. Overall effectiveness of the models. Best results are highlighted in boldface. Superscripts denote statistically significant differences in Fisher's Randomization Test with $p \leq 0.01$.

#	Model	MAP@100	MRR@100	NDCG@10
a	model_1	0.3202^{b}	0.3207^{b}	0.3684^{bc}
b	model_2	0.2332	0.2339	0.239
c	model_3	0.3082^{b}	0.3089^{b}	0.3295^{b}
d	model_4	0.3664^{abc}	0.3668^{abc}	0.4078^{abc}
e	model_5	$\mathbf{0.4053}^{abcd}$	$\mathbf{0.4061}^{abcd}$	$\mathbf{0.4512}^{abcd}$

3   Conclusion and Future Works

In this paper, we presented `ranx`, a Python library for the evaluation and comparison of retrieval results powered by `Numba`. It provides a user-friendly interface to the most commonly used ranking evaluation metrics and a procedure for comparing the results of multiple models and export them as a LaTeX table. We plan to add many features to the current functionalities offered by `ranx`, such as new evaluation metrics and statistical tests, new LaTeX templates for tables formatting, the possibility to fuse different `Runs` for the same query set, and a command-line interface. We are open to feature requests and suggestions from the community.

References

1. Aycock, J.: A brief history of just-in-time. ACM Comput. Surv. **35**(2), 97–113 (2003)
2. Breuer, T., Ferro, N., Maistro, M., Schaer, P.: `repro_eval`: a python interface to reproducibility measures of system-oriented IR experiments. In: Hiemstra, D., Moens, M.-F., Mothe, J., Perego, R., Potthast, M., Sebastiani, F. (eds.) ECIR 2021. LNCS, vol. 12657, pp. 481–486. Springer, Cham (2021). https://doi.org/10.1007/978-3-030-72240-1_51
3. Burges, C.J.C., et al.: Learning to rank using gradient descent. In: ICML, ACM International Conference Proceeding Series, vol. 119, pp. 89–96. ACM (2005)
4. Gysel, C.V., de Rijke, M.: Pytrec_eval: an extremely fast python interface to trec_eval. In: SIGIR, pp. 873–876. ACM (2018)
5. Harman, D.: Information Retrieval Evaluation. Synthesis Lectures on Information Concepts, Retrieval, and Services. Morgan & Claypool Publishers, San Rafael (2011)
6. Harris, C.R.: Array programming with numpy. Nature **585**, 357–362 (2020)
7. Järvelin, K., Kekäläinen, J.: Cumulated gain-based evaluation of IR techniques. ACM Trans. Inf. Syst. **20**(4), 422–446 (2002)
8. Lam, S.K., Pitrou, A., Seibert, S.: Numba: a llvm-based python JIT compiler. In: LLVM@SC, pp. 7:1–7:6. ACM (2015)
9. Lucchese, C., Muntean, C.I., Nardini, F.M., Perego, R., Trani, S.: Rankeval: an evaluation and analysis framework for learning-to-rank solutions. In: SIGIR, pp. 1281–1284. ACM (2017)
10. Lucchese, C., Muntean, C.I., Nardini, F.M., Perego, R., Trani, S.: Rankeval: evaluation and investigation of ranking models. SoftwareX **12**, 100614 (2020)
11. MacAvaney, S., Yates, A., Feldman, S., Downey, D., Cohan, A., Goharian, N.: Simplified data wrangling with ir_datasets. In: SIGIR, pp. 2429–2436. ACM (2021)
12. Macdonald, C., Tonellotto, N.: Declarative experimentation in information retrieval using pyterrier. In: ICTIR, pp. 161–168. ACM (2020)
13. Macdonald, C., Tonellotto, N., MacAvaney, S., Ounis, I.: Pyterrier: declarative experimentation in python from BM25 to dense retrieval. In: CIKM, pp. 4526–4533. ACM (2021)
14. McKinney, W., et al.: Pandas: a foundational python library for data analysis and statistics. Python High Perf. Sci. Comput. **14**(9), 1–9 (2011)
15. Oliphant, T.E.: A guide to NumPy, vol. 1. Trelgol Publishing USA (2006)

16. Oliphant, T.E.: Python for scientific computing. Comput. Sci. Eng. **9**(3), 10–20 (2007)
17. Palotti, J.R.M., Scells, H., Zuccon, G.: Trectools: an open-source python library for information retrieval practitioners involved in trec-like campaigns. In: SIGIR, pp. 1325–1328. ACM (2019)
18. Sanderson, M.: Test collection based evaluation of information retrieval systems. Found. Trends Inf. Retr. **4**(4), 247–375 (2010)
19. Smucker, M.D., Allan, J., Carterette, B.: A comparison of statistical significance tests for information retrieval evaluation. In: CIKM, pp. 623–632. ACM (2007)
20. Van Rossum, G., Drake Jr, F.L.: Python reference manual. Centrum voor Wiskunde en Informatica Amsterdam (1995)
21. Voorhees, E., Harman, D.: Experiment and evaluation in information retrieval (2005)
22. van der Walt, S., Colbert, S.C., Varoquaux, G.: The numpy array: a structure for efficient numerical computation. Comput. Sci. Eng. **13**(2), 22–30 (2011)

DuoSearch: A Novel Search Engine for Bulgarian Historical Documents

Angel Beshirov[(✉)] [iD], Suzan Hadzhieva[iD], Ivan Koychev[iD],
and Milena Dobreva[iD]

Faculty of Mathematics and Informatics, Sofia University "St. Kliment Ohridski",
Sofia, Bulgaria
angel.beshirov@abv.bg

Abstract. Search in collections of digitised historical documents is hindered by a two-prong problem, orthographic variety and optical character recognition (OCR) mistakes. We present a new search engine for historical documents, DuoSearch, which uses ElasticSearch and machine learning methods based on deep neural networks to offer a solution to this problem. It was tested on a collection of historical newspapers in Bulgarian from the mid-19th to the mid-20th century. The system provides an interactive and intuitive interface for the end-users allowing them to enter search terms in modern Bulgarian and search across historical spellings. This is the first solution facilitating the use of digitised historical documents in Bulgarian.

Keywords: Historical newspapers search engine · Orthographic variety · Post-OCR text correction · BERT

1 Introduction and Related Work

Many libraries and archives digitised sizeable collections and made an additional step towards datafication applying Optical Character Recognition (OCR). This is a process that identifies characters within a document and converts them into computer codes, which can then be processed by other programs and applications. However, digitised information is not easily accessible by end-users due to two main hindering blocks. First, the OCR process introduces errors in recognition because of challenging layouts and orthographic variety due to the nine language reforms applied to the Bulgarian language. Second, because the collections of historical documents include texts in a mixture of orthographic conventions the users should be able to use the historical forms of the search keywords. We are applying a novel approach that builds upon the automated techniques for post-OCR text correction in combination with spelling conversion and extended searching to tackle both issues (OCR errors and orthographic variety). Our search engine was used for a case study with a historical newspaper collection from The National Library "Ivan Vazov" (NLIV) in Plovdiv.

M. Hagen et al. (Eds.): ECIR 2022, LNCS 13186, pp. 265–269, 2022.
https://doi.org/10.1007/978-3-030-99739-7_31

The purpose of our research was to build a prototype search engine which addresses the two issues mentioned above. DuoSearch can be used for all kinds of digitised historical Bulgarian documents within the same period where OCR was not followed by quality control. The tool uses dictionary for Bulgarian but can be modified and adapted for other languages.

The paper would be useful for anyone who is developing search tools for historical collections of texts with errors and/or linguistic variance. The system can perform a search in a collection of documents that are written in different spellings and have a relatively high number of erroneous tokens. It was implemented using ElasticSearch and uses machine learning techniques to improve the quality of the indexed data. It also provides an intuitive and interactive web interface, export of the results and easy navigation across the returned documents.

The interest in developing digital resources and tools which answer historians' needs for searching across collections of newspapers developed over the last decades [4]. Some researchers focused on newspaper collections metadata as a solution for improved search [5]. Many challenges arise when working with cultural heritage, like historical newspapers, some of which are described in [6]. Common interests around these challenges in different disciplines have led to projects such as the European project NewsEye [1]. It uses data in a few languages and provides enhanced access to historical newspapers for a wide range of users. In Bulgaria, there is already a collection of digitised historical newspapers, but access to it is cumbersome, due to OCR errors and the multiple language reforms. There was a competition [2] supported by NewsEye whose purpose was to compare and evaluate automatic approaches for correcting OCR-ed texts from historical documents. The competition provides data in 10 European languages including Bulgarian.

2   System Design

The design of the system is shown on Fig. 1. The system uses a three-tier architecture, where we have the user interface, the business logic and the data as independent modules. The presentation layer is responsible for the direct interaction between the user and the system, sending requests to the system and visualizing results. The search API component transforms and proxies the request to and from ElasticSearch in order to retrieve the relevant documents for the search query. To handle the linguistic variance, the Search API component contains a converter, which transforms the text entered by the user into the historical spelling and sends two requests.

The processor component does the data preprocessing before sending it to ElasticSearch for indexing. The preprocessing includes correcting mistakes from the post-OCRed text and removing redundant metadata from the pages. The post-OCR text correction process can be divided into two phases: error detection and error correction. For error detection we have used pretrained multilingual model BERT [3] which is context-aware and helps in identifying both syntactical

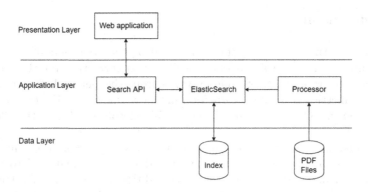

Fig. 1. System architecture

and grammatical mistakes. The data is first tokenized into words, which are then transformed into BERT sub-tokens. Afterwards, we feed these sub-tokens to the BERT model and the output for each sub-token is passed to 5 convolutional layers with different kernel sizes: 2, 3, 4, 5 and 6, each one with 32 filters. After the convolutional layers, we have maxpool layers with a stride of 1, which intuitively represents the information exchange between the n-grams. The outputs from these layers are then concatenated and fed to the last layer, which is linear and does the final classification of the sub-token. The BERT sub-token predictions are then used to classify each token. More specifically, if at least one sub-token is predicted to be erroneous, then the whole token is erroneous.

For error correction we use character-level sequence to sequence model which takes the classified as erroneous tokens from the error detection model and corrects them. For further improvement over the text correction we have used the CLaDA-BG dictionary [7] which contains over 1.1 million unique words in all of their correct forms. To address different orthographic varieties during post-OCR text correction we need to know the version of the language in which the document is written and adapt the dictionary in our model to that orthographic convention.

DuoSearch supports two search types: regular search and extended search. The regular search is a full-text search using the standard match query from ElasticSearch. The extended search supports additional options, like proximity searching, boolean operators support, regex matching and others. It is implemented using the query DSL of ElasticSearch.

For testing the current search engine prototype we have used a set of Bulgarian newspapers provided by NLIV, which is around 4 GB from the period 1882–1930. The source code is available on GitHub[1] and live demo version on AWS.[2]

---

[1] https://github.com/angelbeshirov/DuoSearch.
[2] http://ec2-3-250-238-254.eu-west-1.compute.amazonaws.com:8080.

3   Evaluation

We have used the Bulgarian dataset provided by the organizers of the ICDAR 2019 competition [2] to evaluate our post-OCR text correction model. It contains 200 files with around 32,000 unique words from the period before 1945. For evaluation metrics of our text correction model we have used F-measure and % of improvement. Our results are shown in Table 1 compared with the Clova AI model.

Our model is similar to the one of the Clova AI team from the competition with some improvements for the Bulgarian dataset. We managed to achieve an improvement of 9% over it, using the additional dictionary [7]. The % of improvement is measured by comparing the sum of the Levenshtein distances for each token in the raw OCR text and the corrected one with the gold standard. We used the trained model afterwards to correct the mistakes from the PDF documents before indexing by the search engine. In future, additional evaluation will be done with the users of the system.

Table 1. Evaluation results

Model	F-score	% of improvement
Clova AI	0.77	9%
DuoSearch	0.79	18.7%

4   Conclusion and Contributions

In this paper, we have described a new search engine that combines various technologies to allow for fast searching across a collection of historical newspapers. This contributes to the overall problem in Bulgaria with improving the access to historical documents. It has been acknowledged by Europeana as an example of successful partnership between universities and libraries.

In future, we will work on the text correction by developing a model which predicts the language revision version in which the text is written and apply different models for each orthography, as some of the words are being flagged as incorrect when they are written in different language version from the one the model is trained on. The system also has to be installed on the servers in the library and index the whole collection of newspapers, which is around 200 GB.

Acknowledgements. This research is partially supported by Project UNITe BG05M2OP001-1.001-0004 funded by the OP "Shcience and Education for Smart Growth" and co-funded by the EU through the ESI Funds. The contribution of M. Dobreva is supported by the project KP-06-DB/6 DISTILL funded by the NSF of Bulgaria.

References

1. Doucet, A., et al.: NewsEye: a digital investigator for historical newspapers. In: Digital Humanities 2020, DH 2020, Conference Abstracts, Ottawa, Canada, 22–24 July 2020. Alliance of Digital Humanities Organizations (ADHO). https://zenodo.org/record/3895269 (2020)
2. Rigaud, C., Doucet, A., Coustaty, M., Moreux, J.P.: ICDAR 2019 competition on post-OCR text correction. In: 2019 International Conference on Document Analysis and Recognition (ICDAR) (2019). https://doi.org/10.1109/icdar.2019.00255
3. Devlin, J., Chang, M.-W., Lee, K., Toutanova, K.: BERT: pre-training of deep bidirectional transformers for language understanding arxiv:1810.04805 (2018)
4. Allen, R.B.: Toward a platform for working with sets of digitized historical newspapers. In: IFLA International Newspaper Conference: Digital Preservation and Access to News and Views, New Delhi, pp. 54–59 (2010)
5. Cole, B.: The National Digital Newspaper Program. Organization of American Historians Newsletter 32 (2004)
6. Oberbichler, S., et al.: Integrated interdisciplinary workflows for research on historical newspapers: perspectives from humanities scholars, computer scientists, and librarians. J. Assoc. Inf. Sci. Technol. 1–15 (2021). https://doi.org/10.1002/asi.24565
7. CLaDA-BG dictionary, Grant number DO01-164/28.08.2018. https://clada-bg.eu

Tweet2Story: A Web App to Extract Narratives from Twitter

Vasco Campos[1,2(✉)], Ricardo Campos[1,3], Pedro Mota[1,2], and Alípio Jorge[1,2]

[1] INESC TEC, Porto, Portugal
vasco.m.campos@inesctec.pt
[2] FCUP, University of Porto, Porto, Portugal
amjorge@fc.up.pt
[3] Polytechnic Institute of Tomar, Ci2 - Smart Cities Research Center,
Tomar, Portugal
ricardo.campos@ipt.pt

Abstract. Social media platforms are used to discuss current events with very complex narratives that become difficult to understand. In this work, we introduce Tweet2Story, a web app to automatically extract narratives from small texts such as tweets and describe them through annotations. By doing this, we aim to mitigate the difficulties existing on creating narratives and give a step towards deeply understanding the actors and their corresponding relations found in a text. We build the web app to be modular and easy-to-use, which allows it to easily incorporate new techniques as they keep getting developed.

Keywords: Narrative extraction · Twitter · Information retrieval

1 Introduction

Modern social media platforms, such as *Twitter*, are used to discuss current events in real time. *Twitter*, in particular, is a valuable platform for common people, but even more so for journalists [1]. Given the nature of tweets, they can individually contain relevant information, but stacked together they can be cumbersome and redundant. In addition to this, they are often written in colloquial language, which makes it hard to follow up on the different dimensions of events and opinions revolving around it. To make sense of this data, researchers often make use of automatic summarization processes [2], yet the generated summaries are still far away from the human-generated ones [3]. Recent years have also shown an interest towards a better understanding of the text, that goes beyond simply summarizing its contents, leading to a new research area known as narrative extraction [4,5]. Some of the works proposed so far, provide a solution to partially extract a narrative [6]. Others, extract narratives from long texts and focus on understanding its structure [7,8]. However, as far as we know, none of these works provide a framework or a live demo that can be used to reproduce and test the advances stated in their paper. To answer these

M. Hagen et al. (Eds.): ECIR 2022, LNCS 13186, pp. 270–275, 2022.
https://doi.org/10.1007/978-3-030-99739-7_32

problems, we propose a methodology to automatically extract narratives from tweet posts. We call it Tweet2Story. Alongside this methodology, we take a step towards the understanding of the narratives behind tweet posts about an event, by proposing a webapp, which, given a text, produces an annotation file that can be used for subsequent processes such as visualization.

2   Tweet2Story Framework

The Tweet2Story framework has a modular architecture which is divided in three components (Annotators, Core and Webapp) as detailed bellow:

1. **Annotators:** The *Annotators* contain the logic of the extraction of narratives consisting of 5 tasks: (1) the Named Entity Recognition (NER) to retrieve named entities from the text[1]; the (2) Temporal Entity Extraction to perform temporal normalization[2]; the (3) Co-reference Resolution to find co-occurrences about actors in the entire document and to group them into clusters[3]; the (4) Event extraction to detect events in the text typically through verbs and modifiers[4]; and the (5) Entity Relation Extraction which uses the semantic role classification of each word/expression on a sentence to extract relations (triples) between entities (an actor and an event)[5]. Each one can be performed by one or more tools. For instance the NER module can be operated individually through Spacy [9], NLTK [10] or SparkNLP [11], or together, by combining the results of the 3 different tools. For the temporal module, we resort to Heideltime [12,13] temporal tagger. Finally, the co-reference resolution [14], the event extraction [15] and the entity relation extraction [15] are operated by AllenNLP [16].

2. **Core:** In the *Core* component we take the results from the *Annotators* component and join them to create the narrative in the form of annotations. For example, the extracted events and the entity relations are joined together in an ".ann" annotation file to form a semantic relation annotation (triple).

3. **Webapp:** The *Webapp* is a **user interface** that receives texts as input from a user on a *Web Browser* and communicates with the *Core* to show the user an ".ann" annotation file (partially shown in Fig. 1 (e)) that describes the narrative. While the system does not put any restriction on the type, length and domain of text, it is particularly suited to small texts such as tweets. A formal evaluation on top of this kind of texts has been conducted, but it is out of the scope of this work.

---

[1] in the sentence "Steve Jobs was the CEO of Apple", the entity "Steve Jobs" fits the category of "person".

[2] the expression "last year" would be parsed as "2020" as of this writing.

[3] in the sentence "Sally lives in Paris. She lives in France", both "Sally" and "She" refer to the same entity and, therefore, belong to the same cluster.

[4] in the sentence "Sally lives in Paris", the event is expressed through the verb "lives".

[5] in the sentence "Sally lives in Paris" the triple "Sally - lives - in Paris" is categorized as a location triple.

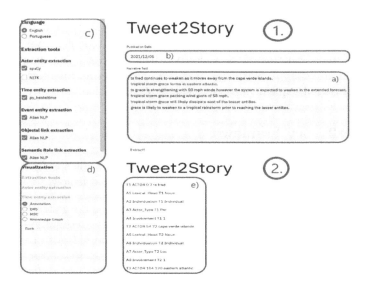

Fig. 1. Webapp showcasing the extraction of a narrative using Tweet2Story

3   Demonstration

Figure 1 shows the user interface of the Tweet2Story web app. Its source code[6] is publicly available and users can run a local version of the demo by following the provided instructions. A live demo[7] is also available for anyone that wants to test the Tweet2Story framework with any set of tweets or texts.

The remainder of this section will describe the features of the page by following their alphabetical order displayed in Fig. 1. When the user opens the app, the system displays the first panel (Fig. 1 - (1.)) where he/she can input a set of tweet posts (a) and provide its publication date (b). In order to extract the narrative, the system needs to know the language of the text and what tools to use for each task of the pipeline. Figure 1 (c) shows how the user can choose between the different tools available. In case of a multiple selection, the framework uses both to extract the part of the narrative in question. Finally, the user can click the *Extract!* button to extract the narrative. Once the narrative extraction process is concluded, the user is taken to a second panel (Fig. 1 - (2.)). Here, he/she can choose how to visualize the narrative (d). Figure 1 (e) shows the visualization of the narrative through the annotation file. The annotation follows the brat format[8]. For example, in the first line of Fig. 1 (e), "T1" is the unique identifier of the annotation, "ACTOR" indicates that the entity is an actor, "0 7" is the span of characters where the entity can be found in the text and "ts fred" is the actual entity in the text.

---

[6] https://github.com/LIAAD/Tweet2Story-demo.
[7] http://tweet2story.inesctec.pt/.
[8] https://brat.nlplab.org/standoff.html.

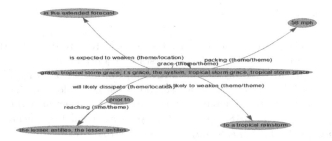

Fig. 2. Knowledge graph made from the annotations of tweets about the grace storm

Furthermore, our framework uses the annotations it extracts from the texts to generate visualizations of the narratives. First, it generates a DRS (Discourse Representation Schema) [17], which is used as a base to build the MSC (Message Sequence Chart) [18] and the Knowledge Graph (Fig. 2), where the nodes represent a text entity and the edges represent the events they take part in. The MSC, for example, is a UML diagram that represents the narrative, its events and how the actors interact with each other. The knowledge graph, instead, aims to provide a simple and intuitive way of visualizing the narrative. To achieve this, we highlight the relations between entities using triples (e.g. grace - likely to weaken - to a tropical rainstorm). Finally, we display the type of each relation, for example "the lesser antilles" displays a "location" relation with "grace storm".

4   Summary and Future Work

In this demo, we have presented Tweet2Story, a modular framework to extract narratives from sets of tweets. Its modularity is a powerful asset that lets it grow upon future developments. Yet, we are aware of its limitations. For example, users are only able to provide one publication date for the tweets, which restricts them to using tweets from the same day. Extracting the narrative of a very large number of tweets is also a problem, in terms of computation time. As future work, we plan to give users the chance to feed the system with real-time tweet posts, by enabling hashtag querying and publication date detection. However, moving from an offline collection to a real-time one, raises other challenges, such as tweet salience and volume. To this extent, one future direction would be to integrate a new summarization component in the pipeline.

Acknowledgements. Vasco Campos and Pedro Mota were financed by National Funds through the Portuguese funding agency, FCT - Fundação para a Ciência e a Tecnologia, within project UIDB/50014/2020 and LA/P/0063/2020. Ricardo Campos and Alípio Jorge were financed by the ERDF - European Regional Development Fund through the North Portugal Regional Operational Programme (NORTE 2020), under the PORTUGAL 2020 and by National Funds through the Portuguese funding agency, FCT - Fundação para a Ciência e a Tecnologia within project PTDC/CCI-COM/31857/2017 (NORTE-01-0145-FEDER-03185). This funding fits under the research line of the Text2Story project.

References

1. MuckRack: The state of journalism 2021. MUCK RACK Blog, 15 March 2021. https://muckrack.com/blog/2021/03/15/state-of-journalism-2021
2. Rudra, K., Goyal, P., Ganguly, N., Mitra, P., Imran, M.: Identifying sub-events and summarizing disaster-related information from microblogs. In: The 41st International ACM SIGIR Conference on Research and Development in Information Retrieval, ser. SIGIR 2018, pp. 265–274. Association for Computing Machinery, New York (2018). https://doi.org/10.1145/3209978.3210030
3. El-Kassas, W.S., Salama, C.R., Rafea, A.A., Mohamed, H.K.: Automatic text summarization: a comprehensive survey. Expert Syst. Appl. **165**, 113679 (2021). https://www.sciencedirect.com/science/article/pii/S0957417420305030
4. Jorge, A., Campos, R., Jatowt, A., Nunes, S.: Information processing & management journal special issue on narrative extraction from texts (Text2Story). Inf. Process. Manag. **56**, 1771–1774 (2019)
5. Campos, R., Jorge, A., Jatowt, A., Bhatia, S., Finlayson, M.: The 4th international workshop on narrative extraction from texts: Text2Story 2021. In: Hiemstra, D., Moens, M.-F., Mothe, J., Perego, R., Potthast, M., Sebastiani, F. (eds.) ECIR 2021. LNCS, vol. 12657, pp. 701–704. Springer, Cham (2021). https://doi.org/10.1007/978-3-030-72240-1_84
6. Metilli, D., Bartalesi, V., Meghini, C.: Steps towards a system to extract formal narratives from text. In: Text2Story 2019 - Second Workshop on Narrative Extraction From Texts, Cologne, Germany, 14 April 2019, pp. 53–61. CEUR-WS.org, Aachen, DEU (2019)
7. Vargas, J.V.: Narrative information extraction with non-linear natural language processing pipelines. Ph.D. dissertation, Drexel University (2017)
8. Eisenberg, J.D.: Automatic extraction of narrative structure from long form text. Ph.D. dissertation, Florida International University (2018)
9. Honnibal, M., Montani, I.: spaCy 2: Natural language understanding with bloom embeddings, convolutional neural networks and incremental parsing (2017). To appear
10. Bird, S., Loper, E.: NLTK: the natural language toolkit. In: Proceedings of the ACL Interactive Poster and Demonstration Sessions, pp. 214–217. Association for Computational Linguistics, Barcelona, July 2004. https://aclanthology.org/P04-3031
11. Kocaman, V., Talby, D.: Spark NLP: natural language understanding at scale. CoRR, abs/2101.10848 (2021). arXiv:2101.10848
12. Strotgen, J., Gertz, M.: Heideltime: high quality rule-based extraction and normalization of temporal expressions. In: Proceedings of the 5th International Workshop on Semantic Evaluation, ser. SemEval 2010, pp. 321–324. Association for Computational Linguistics, USA (2010)
13. Strotgen, J., Gertz, M.: Multilingual and cross-domain temporal tagging. Lang. Resour. Eval. **47**(2), 269–298 (2013)
14. Lee, K., He, L., Zettlemoyer, L.: Higher-order coreference resolution with coarse-to-fine inference. In: Proceedings of the 2018 Conference of the North American Chapter of the Association for Computational Linguistics: Human Language Technologies, Volume 2 (Short Papers), pp. 687–692. Association for Computational Linguistics, New Orleans, June 2018. https://aclanthology.org/N18-2108
15. Shi, P., Lin, J.: Simple Bert models for relation extraction and semantic role labeling. ArXiv, abs/1904.05255 (2019)

16. Gardner, M., et al.: Allennlp: a deep semantic natural language processing platform. CoRR, abs/1803.07640 (2018). arXiv:1803.07640
17. Kamp, H., Genabith, J., Reyle, U.: Discourse representation theory. **11**, 125–394 (2010). Springer
18. Letier, E., Kramer, J., Magee, J., Uchitel, S.: Monitoring and control in scenario-based requirements analysis. In: Proceedings of the 27th International Conference on Software Engineering, pp. 382–391. Association for Computing Machinery, June 2005

Patapasco: A Python Framework for Cross-Language Information Retrieval Experiments

Cash Costello[(✉)][iD], Eugene Yang[iD], Dawn Lawrie[iD], and James Mayfield[iD]

Human Language Technology Center of Excellence, Johns Hopkins University,
Baltimore, MD 21211, USA
{ccostel2,eugene.yang,lawrie,mayfield}@jhu.edu
https://hltcoe.jhu.edu/

Abstract. While there are high-quality software frameworks for information retrieval experimentation, they do not explicitly support cross-language information retrieval (CLIR). To fill this gap, we have created Patapsco, a Python CLIR framework. This framework specifically addresses the complexity that comes with running experiments in multiple languages. Patapsco is designed to be extensible to many language pairs, to be scalable to large document collections, and to support reproducible experiments driven by a configuration file. We include Patapsco results on standard CLIR collections using multiple settings.

Keywords: Cross-language information retrieval · CLIR · Experimental framework · Reproducible experiments

1 Introduction

The introduction of neural ranking methods to information retrieval (IR) research has led to the adoption of two-stage pipelines. In the first stage, documents are retrieved from an inverted index and ranked according to a scoring function such as BM25. Those documents are then re-ranked in a second stage using a slower neural model trained for the task. Several software frameworks have been created around multi-stage ranking, including Pyserini [3], PyTerrier [5], and OpenNIR [4]. These frameworks standardize the first stage, allowing researchers to focus on developing better-performing neural ranking models. They also support reproducibility and comparison of performance across different models with standardized data sets and settings. These frameworks are designed with the assumption that the queries and documents are written in the same language.

Cross-language information retrieval (CLIR) is the retrieval of documents in one language based on a search query in another language. In this setting, the system needs to be aware of the language of the queries and the documents, what types of processing are supported in those languages, and how the language

M. Hagen et al. (Eds.): ECIR 2022, LNCS 13186, pp. 276–280, 2022.
https://doi.org/10.1007/978-3-030-99739-7_33

barrier is to be crossed. Existing frameworks were not designed to handle these complexities. However, with the advent of high-quality machine translation and the success of neural ranking methods for IR, neural CLIR experimentation needs to be supported. The Patapsco framework implements the modern retrieve-and-rerank approach for the CLIR use case by using the existing Pyserini framework and extending it to support CLIR. Patapsco was successfully used in the summer of 2021 at a CLIR workshop involving more than fifty participants. Such a large scale workshop of which many participants were new to CLIR research demonstrated how Patapsco enables sophisticated CLIR experiments and lowers barriers to entry for newcomers to the field

2   System Overview

Patapsco[1] is built on top of Pyserini and maintains Pyserini's design goal of reproducible experiments. Patapsco adds extensive language-specific preprocessing, and scalability through parallel processing. An experiment is described in a configuration file. This file is used to generate the pipeline, which begins with a standard information retrieval setup of a document collection, a topic file, and, if available, relevance judgments. The pipeline is reproducible from that configuration file. For CLIR support, Patapsco maintains the language metadata for documents and queries throughout the pipeline to enable language-specific processing. This is a key feature needed to support CLIR experimentation that is not present in existing frameworks.

Patapsco handles conversion of topics into queries, ingest and normalization of documents and queries, inverted index construction, retrieval of an initial results set, reranking, and scoring. A pipeline can be run to completion to compute scores through pytrec_eval [8], or it can be stopped early to create artifacts that support development and training. A pipeline can also start where a previous pipeline stopped. For example, a set of experiments may use the same preprocessing and indexing, and differ only in retrieval and reranking.

2.1   Text Preprocessing

As is standard in IR systems, Patapsco normalizes the documents and queries used in an experiment. Character-level normalization, such as correcting Unicode encoding issues and standardizing diacritics, is applied first. This is followed by token-level normalization, which includes stop word removal and stemming. Preprocessing is language-dependent and specified in the configuration file:

```
documents:
  input:
    format: json
    lang: rus
```

---

[1] Patapsco is available at https://github.com/hltcoe/patapsco. A video demonstration is at https://www.youtube.com/watch?v=jYj1GAbABBc.

```
  path: /data/clef02/rus_docs.jsonl
process:
  normalize:
    lowercase: true
  tokenize: spacy
  stopwords: lucene
  stem: false
```

Patapsco supports both the rule-based tokenization and stemming used by most IR systems, and neural models from toolkits like spaCy [2] and Stanza [7].

2.2 Retrieval

Patapsco uses Pyserini to retrieve initial document results sets from the inverted index. Users can select between BM25 and a query likelihood model using Dirichlet smoothing. Query expansion using RM3 is also available.

Two approaches to CLIR are document translation and query translation (both of which need to be produced externally). To support the latter use case, Patapsco's default JSONL topic format can hold multiple translations per query. A third CLIR approach is to project the query into the language of the documents using a probabilistic translation table. This method, called Probabilistic Structured Query (PSQ) [1], projects each term in the query to multiple target language terms with associated weights. Patapsco implements PSQ as a extension on Lucene.

2.3 Reranking

Researchers working on neural reranking have the option of registering their reranker with Patapsco and having it run directly in the pipeline, or implementing a command-line interface that is executed from the pipeline. A primary advantage of the command line interface is the avoidance of any dependency conflicts with Patapsco, which is a common issue with machine learning frameworks in Python.

The reranker is passed the query, the list of document identifiers, and access to a document database. This database contains the original documents, since the tokenization and normalization required by the word embedding is likely different than that used for building the inverted index.

2.4 Parallel Processing

To support large document collections with the added computation required by neural text processing models, Patapsco includes two ways to use parallel processing: multiprocessing and grid computing. Each divides the documents into chunks that are processed in separate processes and then assembled in a map-reduce job. Both slurm and qsub are supported for grid computing; Patapsco manages the entire job submission process; the user only has to select the queue and the number of jobs.

Table 1. Baseline runs using Patapsco. QT/DT/HT stand for machine query translation, machine document translation, and human query translation.

Model	CLEF Persian			CLEF Russian			NTCIR Chinese		
	MAP	nDCG	R@1k	MAP	nDCG	R@1k	MAP	nDCG	R@1k
PSQ	0.1370	0.3651	0.4832	0.2879	0.4393	0.7441	0.1964	0.3752	0.5867
QT	0.2511	0.5340	0.6945	0.3857	0.5527	0.9268	0.2186	0.3953	0.6201
DT	0.3420	0.6808	0.8501	0.3408	0.5151	0.8881	0.3413	0.5627	0.8110
HT	0.4201	0.7476	0.9175	0.3975	0.5623	0.9401	0.4810	0.6840	0.9125

3   Evaluation

A set of baseline experiments conducted with Patapsco is presented in Table 1, which reports mean average precision (MAP), nDCG at rank 1000, and recall at rank 1000, where we observe the classic results of translating the documents yields better effectiveness than translating the queries. We evaluate the probabilistic structured query [1] and translation approaches on three widely-used CLIR collections: CLEF Russian and Persian [6], and NTCIR Chinese collections[2]. Each query is formed by concatenating the topic title and description. The experiments are executed on a research cluster with 20 parallel jobs during indexing and one job for retrieval. The effectiveness is on par with the implementations from other studies. The running time ranges from several minutes to an hour, depending on the size of the collection and the tokenization used in the experiment. This fast running time enables large ablation studies. The index is supported by Pyserini [3], which is a framework designed for large-scale IR experiments. The memory footprint is consequently minimal, ranging from 2 to 3 GB in our experiments.

4   Conclusion

Patapsco brings recent advances in IR software frameworks and reproducibility to the CLIR research community. It provides configuration-driven experiments, parallel processing for scalability, a flexible pipeline, and a solid baseline for performance evaluations. Patapsco's configuration file fully documents each experiment, making them simple to reproduce. Patapsco enables sophisticated CLIR experiments and lowers barriers to entry for newcomers to the field. During a CLIR workshop in the summer of 2021, researchers from outside of information retrieval successfully ran CLIR experiments because of the ease of experimentation provided by Patapsco.

---

[2] http://research.nii.ac.jp/ntcir/permission/ntcir-8/perm-en-ACLIA.html.

References

1. Darwish, K., Oard, D.W.: Probabilistic structured query methods. In: Proceedings of the 26th Annual International ACM SIGIR Conference on Research and Development in Information Retrieval, pp. 338–344 (2003)
2. Honnibal, M., Montani, I., Landeghem, S.V., Boyd, A.: SPACY: industrial-strength natural language processing in python (2020). https://doi.org/10.5281/zenodo.1212303
3. Lin, J., Ma, X., Lin, S.C., Yang, J.H., Pradeep, R., Nogueira, R.: Pyserini: a python toolkit for reproducible information retrieval research with sparse and dense representations. In: Proceedings of the 44th Annual International ACM SIGIR Conference on Research and Development in Information Retrieval (SIGIR 2021) (2021)
4. MacAvaney, S.: Opennir: a complete neural ad-hoc ranking pipeline. In: Proceedings of the 13th International Conference on Web Search and Data Mining, pp. 845–848 (2020)
5. Macdonald, C., Tonellotto, N.: Declarative experimentation in information retrieval using pyterrier. In: Proceedings of the 2020 ACM SIGIR on International Conference on Theory of Information Retrieval, pp. 161–168 (2020)
6. Peters, C., Braschler, M.: European research letter: cross-language system evaluation: the CLEF campaigns. J. Am. Soc. Inf. Sci. Technol. **52**(12), 1067–1072 (2001)
7. Qi, P., Zhang, Y., Zhang, Y., Bolton, J., Manning, C.D.: Stanza: a python natural language processing toolkit for many human languages. In: Proceedings of the 58th Annual Meeting of the Association for Computational Linguistics: System Demonstrations (2020)
8. Van Gysel, C., de Rijke, M.: Pytrec_eval: an extremely fast python interface to trec_eval. In: SIGIR. ACM (2018)

City of Disguise:
A Query Obfuscation Game
on the ClueWeb

Maik Fröbe[1(✉)], Nicola Lea Libera[2], and Matthias Hagen[1]

[1] Martin-Luther-Universität Halle-Wittenberg, Halle, Germany
maik.froebe@informatik.uni-halle.de
[2] Bauhaus-Universität Weimar, Weimar, Germany

Abstract. We present City of Disguise, a retrieval game that tests how well searchers are able to reformulate some sensitive query in a 'Taboo'-style setup but still retrieve good results. Given one of 200 sensitive information needs and a relevant example document, the players use a special ClueWeb12 search interface that also hints at potentially useful search terms. For an obfuscated query, the system assigns points depending on the result quality and the formulated query. In a pilot study with 72 players, we observed that they find obfuscations to retrieve relevant documents but often only when they relied on the suggested terms.

Keywords: Query obfuscation · Private information retrieval · Gamification

1 Introduction

Retrieving relevant results without revealing private or confidential information is a current challenge in information retrieval [9]. Search engines can use innovative techniques to collect data while ensuring privacy [12,14,23]. Still, those privacy techniques are applied on the side of the search engines, requiring searchers to trust them. This trust might be unacceptable for searchers with a very sensitive information need, especially given the recent news that the police can access query logs.[1]

An option for someone not trusting a search engine but still wanting to retrieve results for some sensitive information need is to try to submit less sensitive but similar queries [3,11]. In a way, this resembles the popular game Taboo where players try to explain a private word without using the word or some related ones. In this spirit, we present City of Disguise,[2] a game inspired by Taboo and PageHunt [15]. Players have to obfuscate a sensitive query and only

---

[1] cnet.com/news/google-is-giving-data-to-police-based-on-search-keywords-court-docs-show/
[2] Demo: https://demo.webis.de/city-of-disguise
Screencast: https://demo.webis.de/city-of-disguise/screencast
Code and Data: https://github.com/webis-de/ecir22-query-obfuscation-game.

M. Hagen et al. (Eds.): ECIR 2022, LNCS 13186, pp. 281–287, 2022.
https://doi.org/10.1007/978-3-030-99739-7_34

(a) The search interface in City of Disguise for the sensitive query `bph treatment`.

(b) Categories in City of Disguise.   (c) Scoring for a successful obfuscation.

Fig. 1. The main elements in City of Disguise: (a) the search interface, (b) the city map where players select a category, and (c) the scoring scheme shown to players after query submission.

submit "harmless" queries that still retrieve relevant results. The idea is that the sensitive query itself never appears in the search engine's log, which would happen for other privacy techniques like hiding the actual query in a stream of fake queries [1,10,17–19,22]—an attacker would then still know that the sensitive query exists. In a gamification sense, the game's point system will particularly reward less sensitive alternative queries that still return results relevant to the original query.

Playing City of Disguise is pretty simple. In the city map (cf. Fig. 1 (b)), a player chooses a category (e.g., health-related topics) and the search interface (cf. Fig. 1 (a)) is opened for a random "unsolved" sensitive information need from that category. In the search interface, the to-be-obfuscated sensitive query, a relevant target document for the underlying information need, and a list of suggested terms are shown (terms from the target document with highest TF · IDF scores). When a player submits a query, a score is derived (cf. Fig. 1 (c)) with which they may compete in the public leaderboard or they may choose to try to further improve their score.

All interactions in the game are logged in an anonymized way. Besides showing scores in the leaderboards, the logging also enables analyses of human obfuscation strategies. Successful players' strategies can then be compared to automatic query obfuscation approaches [2–5,11] which even might inspire improvements.

2   Search System and Game Design

Our new query obfuscation game comes with 200 sensitive queries in six categories that we have manually selected from a pool of 700 candidate queries. Each sensitive query has one relevant document assigned that we show as a target document to players to simplify the obfuscation process. To test their obfuscations with fast response times, players submit their queries to a search engine with 0.6 million ClueWeb12 documents.

Document Collection and Rendering Target Documents. We use the ClueWeb as a resource because it is widely used in research in information retrieval [7,8], and since we can reuse relevance judgments. We show a relevant document for each sensitive query to the players to simplify the query obfuscation process. Since the ClueWeb does not include all resources to nicely display a page, we replace links in the documents to stylesheets and images with links to the Wayback machine to the corresponding snapshots (if available). We render the final HTML into a PDF document (using the Wayback Machine) so that the target pages still render nicely inside our preview image and allow zooming, while the players can still copy terms from the relevant document.

Selection of Sensitive Queries. We select the 200 pairs of sensitive queries with a respective relevant document for the obfuscation game from 750 candidate queries that we extract from the 96 sensitive queries published by Arampatzis et al. [3], 65 sensitive TREC Web track queries on the ClueWebs, and 589 sensitive queries from the AOL query log [16]. Especially from the 96 sensitive queries by Arampatzis et al. [3], we remove those with pornographic or hateful intent because we show a document relevant to each query in our game. For each of the remaining sensitive queries, we retrieve the top-3 ClueWeb12 documents with ChatNoir [6] and render the documents (including CSS and image resources from the Wayback Machine). For each sensitive query, we review the top-3 documents and omit documents looking odd (due to missing CSS or images) or documents that are irrelevant to the sensitive query, retaining only the most relevant document for each query. We assign the remaining 204 valid query-document pairs into six categories, selecting 200 final queries that provide the best balance of all six categories (Fig. 1 (b) gives an overview of the available categories).

Scoring Query Obfuscations. To motivate players to improve their obfuscated query multiple times, we show a score composed of four subscores to suggest potential ways for improvement. We calculate the score by submitting

Table 1. Overview of the effectiveness of obfuscated queries in ChatNoir and the games' document sample ('Sample'). We report the MRR, the number of documents retrieved for the original query ('Ori.'), and the number of retrieved relevant documents ('Rel.'). We show results for automatically obfuscated queries and four different types of queries submitted by players.

				Our Sensitive Queries				Sensitive Web Track Queries			
		Obfuscated Queries			ChatNoir		Sample		ChatNoir		Sample	
		Count	Length	Time	MRR	Ori.	MRR	Ori.	MRR	Rel.	MRR	Rel.
Players	Only Suggestions	130/21	2.42	40.50 s	0.093	5.223	0.325	67.592	0.010	3.094	0.152	3.691
	Some Suggestions	556/125	4.53	42.45 s	0.046	4.667	0.258	85.829	0.013	3.632	0.038	3.568
	No Suggestions	576/157	2.88	44.39 s	0.029	2.935	0.082	38.932	0.015	1.783	0.024	3.316
	New Word	559/158	3.57	46.27 s	0.002	1.517	0.051	49.992	0.002	1.235	0.005	2.790
	Automatic	1025/327	2.91	—	0.088	9.229	0.420	84.264	0.014	2.872	0.042	3.743

the obfuscated query (not allowing any queries that reuse terms from the sensitive query) against a test search engine indexing a 0.6 million document sample. The score combines the position of the relevant document, the query length, the recall, and the mean average precision for the 100 documents retrieved for the sensitive query. We show all four subscores to the players to indicate whether a query could be improved (cf. Fig. 1 (c)).

To allow fast feedback cycles for players, we use a setup similar to Arampatzis et al. [3] and submit obfuscated queries against a search engine with a small sample of 0.6 million ClueWeb12 documents. To ensure that each of our 200 sensitive queries has enough relevant documents, we include the top-1000 ChatNoir [6] results for each sensitive query into the sample. We complement those 0.2 million documents by sampling 0.4 million documents from the ClueWeb12 with the sampling strategy of Arampatzis et al. [3]. We index this document sample with the BM25 implementation of Anserini [20] using the default settings (stemming with the Porter stemmer and removing stopwords using Lucene's default stopword list for English).

3   Evaluation

We test our query obfuscation game and the ability of players to obfuscate sensitive information needs in a pilot study with 72 participants. We recruited players from two information retrieval courses and mailing lists at our universities. We logged 1,462 obfuscated queries, with an average of 43 s to formulate an obfuscated query.

Table 1 compares the effectiveness of the obfuscated queries that the players formulated in our pilot study to queries automatically obfuscated by formulating keyqueries for the target document [11] using the suggested terms as vocabulary. We report the MRR for finding the given relevant document, the number of documents retrieved from the top-100 ranking when submitting the sensitive query (Ori.), and the number of retrieved relevant documents (Rel.) for queries with relevance judgments from the ClueWeb tracks. We split the human obfuscations into four categories: (1) queries where all terms come from the game's suggestions (Only Suggestions), (2) queries with at least one term from the game's suggestions and one new term (Some Suggestions), (3) queries without suggestions (No Suggestions), and (4) queries with a term outside the shown relevant document (New Word). Overall, we find that the obfuscation effectiveness decreases the more the players deviate from the game's term suggestions. While players who use only suggestions slightly improve upon automatic obfuscation (MRR of 0.093 vs. 0.088), creative obfuscations that include new words are rather ineffective (MRR of 0.002). Those observations are also confirmed by the evaluations using the sensitive Web Track queries with real relevance judgments.

4 Conclusion and Future Work

Our new query obfuscation game tests a player's ability to hide sensitive information needs from a search engine while still retrieving relevant results. We plan to maintain the game as part of ChatNoir [6] and to add more topics and different search engines in the future (e.g., Transformer-based re-rankers [21] or dense retrieval models [13]).

From the game's logs, we want to learn how searchers manually obfuscate sensitive information needs. This knowledge could help to improve automatic query obfuscation approaches that one could apply when querying an untrusted search engine for a sensitive information need that should not appear "unencrypted" in the engine's log files.

References

1. Ahmad, W.U., Rahman, M., Wang, H.: Topic model based privacy protection in personalized web search. In: Perego, R., Sebastiani, F., Aslam, J.A., Ruthven, I., Zobel, J. (eds.) Proceedings of the 39th International ACM SIGIR conference on Research and Development in Information Retrieval, SIGIR 2016, Pisa, Italy, pp. 1025–1028. ACM (2016)
2. Arampatzis, A., Drosatos, G., Efraimidis, P.S.: A versatile tool for privacy-enhanced web search. In: Serdyukov, P., et al. (eds.) ECIR 2013. LNCS, vol. 7814, pp. 368–379. Springer, Heidelberg (2013). https://doi.org/10.1007/978-3-642-36973-5_31
3. Arampatzis, A., Drosatos, G., Efraimidis, P.S.: Versatile query scrambling for private web search. Inf. Retr. J. 18(4), 331–358 (2015)
4. Arampatzis, A., Efraimidis, P., Drosatos, G.: Enhancing deniability against query-logs. In: Clough, P., et al. (eds.) ECIR 2011. LNCS, vol. 6611, pp. 117–128. Springer, Heidelberg (2011). https://doi.org/10.1007/978-3-642-20161-5_13
5. Arampatzis, A., Efraimidis, P.S., Drosatos, G.: A query scrambler for search privacy on the internet. Inf. Retr. 16(6), 657–679 (2013)
6. Bevendorff, J., Stein, B., Hagen, M., Potthast, M.: Elastic ChatNoir: search engine for the ClueWeb and the common crawl. In: Pasi, G., Piwowarski, B., Azzopardi, L., Hanbury, A. (eds.) ECIR 2018. LNCS, vol. 10772, pp. 820–824. Springer, Cham (2018). https://doi.org/10.1007/978-3-319-76941-7_83

7. Collins-Thompson, K., Bennett, P.N., Diaz, F., Clarke, C., Voorhees, E.M.: TREC 2013 Web track overview. In: Voorhees, E.M. (ed.) Proceedings of The Twenty-Second Text REtrieval Conference, TREC 2013, Gaithersburg, Maryland, USA, November 19–22, 2013, NIST Special Publication, vol. 500–302, National Institute of Standards and Technology (NIST) (2013)
8. Collins-Thompson, K., Macdonald, C., Bennett, P.N., Diaz, F., Voorhees, E.M.: TREC 2014 Web track overview. In: Voorhees, E.M., Ellis, A. (eds.) Proceedings of The Twenty-Third Text REtrieval Conference, TREC 2014, Gaithersburg, Maryland, USA, November 19–21, 2014, NIST Special Publication, vol. 500–308, National Institute of Standards and Technology (NIST) (2014)
9. Culpepper, J.S., Diaz, F., Smucker, M.D.: Research frontiers in information retrieval: Report from the third strategic workshop on information retrieval in Lorne (SWIRL 2018). SIGIR Forum 52(1), 34–90 (2018)
10. Domingo-Ferrer, J., Solanas, A., Castellà-Roca, J.: H(k)-private information retrieval from privacy-uncooperative queryable databases. Online Inf. Rev. 33(4), 720–744 (2009)
11. Fröbe, M., Schmidt, E.O., Hagen, M.: Efficient query obfuscation with keyqueries. In: 20th International IEEE/WIC/ACM Conference on Web Intelligence (WI-IAT 2021). ACM, December 2021. https://doi.org/10.1145/3486622.3493950, https://dl.acm.org/doi/10.1145/3486622.3493950
12. Hong, Y., He, X., Vaidya, J., Adam, N.R., Atluri, V.: Effective anonymization of query logs. In: Cheung, D.W., Song, I., Chu, W.W., Hu, X., Lin, J.J. (eds.) Proceedings of the 18th ACM Conference on Information and Knowledge Management, CIKM 2009, Hong Kong, China, pp. 1465–1468. ACM (2009)
13. Karpukhin, V., et al.: Dense passage retrieval for open-domain question answering. In: Webber, B., Cohn, T., He, Y., Liu, Y. (eds.) Proceedings of the 2020 Conference on Empirical Methods in Natural Language Processing, EMNLP 2020, Online, November 16–20, 2020, pp. 6769–6781. Association for Computational Linguistics (2020)
14. Kumar, R., Novak, J., Pang, B., Tomkins, A.: On anonymizing query logs via token-based hashing. In: Proceedings of the 16th International Conference on World Wide Web, WWW 2007, Banff, Alberta, Canada, pp. 629–638 (2007)
15. Ma, H., Chandrasekar, R., Quirk, C., Gupta, A.: Page hunt: improving search engines using human computation games. In: Allan, J., Aslam, J.A., Sanderson, M., Zhai, C., Zobel, J. (eds.) Proceedings of the 32nd Annual International ACM SIGIR Conference on Research and Development in Information Retrieval, SIGIR 2009, Boston, MA, USA, July 19–23, 2009, pp. 746–747. ACM (2009)
16. Pass, G., Chowdhury, A., Torgeson, C.: A picture of search. In: Jia, X. (ed.) Proceedings of the 1st International Conference on Scalable Information Systems, Infoscale 2006, Hong Kong, 30 May–1 June, 2006. ACM International Conference Proceeding Series, vol. 152, p. 1. ACM (2006)
17. Peddinti, S.T., Saxena, N.: On the privacy of web search based on query obfuscation: a case study of TrackMeNot. In: Atallah, M.J., Hopper, N.J. (eds.) PETS 2010. LNCS, vol. 6205, pp. 19–37. Springer, Heidelberg (2010). https://doi.org/10.1007/978-3-642-14527-8_2
18. Peddinti, S.T., Saxena, N.: Web search query privacy: evaluating query obfuscation and anonymizing networks. J. Comput. Secur. 22(1), 155–199 (2014)
19. Toubiana, V., Subramanian, L., Nissenbaum, H.: TrackMeNot: enhancing the privacy of web search. CoRR arXiv:1109.4677 (2011)

20. Yang, P., Fang, H., Lin, J.: Anserini: Enabling the use of Lucene for information retrieval research. In: Kando, N., Sakai, T., Joho, H., Li, H., de Vries, A.P., White, R.W. (eds.) Proceedings of the 40th International ACM SIGIR Conference on Research and Development in Information Retrieval, Shinjuku, Tokyo, Japan, pp. 1253–1256. ACM (2017)

21. Yates, A., Nogueira, R., Lin, J.: Pretrained transformers for text ranking: BERT and beyond. In: Diaz, F., Shah, C., Suel, T., Castells, P., Jones, R., Sakai, T. (eds.) SIGIR '21: The 44th International ACM SIGIR Conference on Research and Development in Information Retrieval, Virtual Event, Canada, July 11–15, 2021, pp. 2666–2668. ACM (2021)

22. Yu, P., Ahmad, W.U., Wang, H.: Hide-n-Seek: an intent-aware privacy protection plugin for personalized web search. In: Collins-Thompson, K., Mei, Q., Davison, B.D., Liu, Y., Yilmaz, E. (eds.) The 41st International ACM SIGIR Conference on Research & Development in Information Retrieval, SIGIR 2018, Ann Arbor, MI, USA, pp. 1333–1336. ACM (2018)

23. Zhang, S., Yang, G.H., Singh, L.: Anonymizing query logs by differential privacy. In: Perego, R., Sebastiani, F., Aslam, J.A., Ruthven, I., Zobel, J. (eds.) Proceedings of the 39th International ACM SIGIR Conference on Research and Development in Information Retrieval, SIGIR 2016, Pisa, Italy, pp. 753–756. ACM (2016)

DocTAG: A Customizable Annotation Tool for Ground Truth Creation

Fabio Giachelle[(✉)], Ornella Irrera, and Gianmaria Silvello

Department of Information Engineering, University of Padua, Padua, Italy
{fabio.giachelle,ornella.irrera,gianmaria.silvello}@unipd.it

Abstract. *Information Retrieval* (IR) is a discipline deeply rooted on evaluation that in many cases relies on annotated data as ground truth. Manual annotation is a demanding and time-consuming task, involving human intervention for topic-document assessment. To ease and possibly speed up the work of the assessors, it is desirable to have easy-to-use, collaborative and flexible annotation tools. Despite their importance, in the IR domain no open-source fully customizable annotation tool has been proposed for topic-document annotation and assessment, so far. In this demo paper, we present DocTAG, a portable and customizable annotation tool for ground-truth creation in a web-based collaborative setting.

Keywords: Annotation tool · Passage annotation · Evaluation · Ground-truth creation

1 Motivation and Background

Ground-truth creation is an expensive and time-consuming task, involving human experts to produce richly-annotated datasets that are fundamental for training and evaluation purposes. In IR, gold standard relevance judgments are essential for the evaluation of retrieval models. The creation of experimental collections in the context of large scale evaluation campaigns (e.g., *Text Retrieval Conference* (TREC)[1] and *Cross Lingual Evaluation Forum* (CLEF)[2]) requires a huge deal of human effort to manually create high quality annotations. To this aim, evaluation campaigns usually adopt custom made annotation and assessment tools to support human assessors and ease their workload [1,8–10,14,16]. Since, the relevance assessment process is usually carried out in a short time, an effective annotation tool can be of great help to speed up the overall process or at least to reduce the annotation bargain. However, in the typical IR scenario, it is common to develop a custom annotation tool for a specific evaluation task or campaign; available annotation tools are tailored for specific tasks, thus making them difficult to reuse for others without a significant overhaul.

---

[1] https://trec.nist.gov.
[2] http://www.clef-initiative.eu.

© The Author(s), under exclusive license to Springer Nature Switzerland AG 2022
M. Hagen et al. (Eds.): ECIR 2022, LNCS 13186, pp. 288–293, 2022.
https://doi.org/10.1007/978-3-030-99739-7_35

The annotation software currently available [11] can be divided into general-purpose and domain-specific tools. General-purpose ones provide a set of common features that cover most of the typical annotation scenarios and use-cases [5,15,18] but require a great deal of customization to fit in a domain-specific setting – e.g., the typical topic-document pair is not handled by these systems. In contrast, domain-specific tools provide ad-hoc functionalities that meet the needs of very specific fields, focusing especially in the biomedical domain [2–4,6,7,12,13].

A recent exhaustive comparison of the major annotation tools [11] points out that choosing the best suitable tool is a demanding task, since each one presents specific advantages and disadvantages in terms of the functionalities provided. In addition, even the most comprehensive tool may present drawbacks such as a tricky installation procedure, no support for online use or a complex user interface. In addition, adapting existing tools not designed for a specific domain is a burdensome process requiring not naive programming skills.

For these reasons, we propose DocTAG, an annotation tool designed specifically for the typical IR annotation tasks. DocTAG provides a streamlined user interface in a collaborative web-based setting. DocTAG provides several features to support human annotators, including: (i) topic-document annotation with customized labels (binary or graded relevance judgements or other custom labels for instance for sentiment/emotion classification) or based on custom defined ontological concepts; (ii) passage-level annotation; (iii) inter-annotation agreement via majority vote; (iv) collaborative facilities (e.g., annotation sharing between assessors); (v) annotation statistics; (vi) responsive interface for long document visualization; (vii) download of ground-truths in CSV and JSON formats; (viii) customizable parsing and ingestion of document corpus, runs and topic files in several formats; (ix) annotation highlighting; (x) topic-document matching words emphasized (i.e. TF-IDF weighted highlight of the words present in the topic-document pair) and (xi) multi-lingual support – i.e. users can annotate the same topic-document pair in different languages (if provided). In case of multiple languages, the documents are grouped by language, so that users can search and filter them accordingly.

DocTAG is portable since it is provided as a Docker container, that ensures code isolation and dependencies packaging. Thus, it can be either installed as a local Webapp or deployed in a cloud container orchestration service.

The rest of the paper is organized as follows: Sect. 2 describes the annotation tool and the main aspect of the demo we present, and Sect. 3 draws some final remarks.

2   DocTAG

DocTAG is a web-based annotation tool specifically designed to support human annotators in the IR domain. The DocTAG source code is publicly available at https://github.com/DocTAG/doctag-core. In addition, to present the main

DocTAG features, we provide a demonstration video[3] and a step-by-step "tutorial" section, included in the DocTAG web interface. DocTAG allows the users to customize several annotation aspects including the set of labels or ontological concepts used for both document-level and passage-level annotation, and the document fields to be visualized and/or annotated. The users can specify all the setting parameters via a wizard configuration procedure[4]. There is no limit to the number of labels and concepts that can be used for the annotation. Since the concepts are custom, the users can specify also concepts defined in external ontologies and terminological resources.

In addition, the configuration interface allows the users to specify (i) the document corpus to be annotated in CSV or JSON format; (ii) the topic files in CSV or JSON format and (iii) the runs (to build the pool to be annotated) in CSV, JSON or plain text.

Architecture and Implementation. DocTAG architecture consists of (i) a web-based front-end interface built with *React.js*; (ii) a back-end for REST API and services built with the Python web framework Django; (iii) a PostgreSQL database to guarantee the persistence of the annotated data.

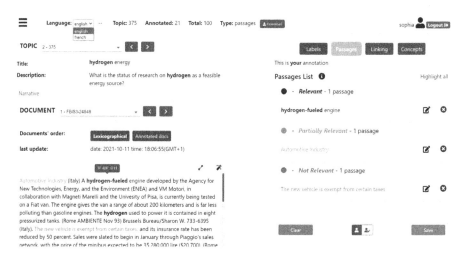

Fig. 1. DocTAG interface, with the *Passages*-level annotation annotation mode (yellow button) active. (Color figure online)

User Interface and Interaction. Figure 1 shows the main DocTAG annotation interface. In the upper part, the header shows the current annotation statistics (i.e., the number of annotated documents for the selected topic out of the total number of documents) for each annotation type (i.e. labels, passages, concepts and linking). The header includes also the button to download

---

[3] https://bit.ly/3pqwHtF.
[4] https://github.com/DocTAG/doctag-core#customize-doctag.

the ground-truth in CSV and JSON formats. On the left side of the header, the menu button allows us to access the DocTAG options and settings (e.g. configuration, inter-annotator agreement and annotation statistics). On the right side instead, the user section shows the username of the current user and the log out button. The interface body is divided in two sections: the document and the annotation sections. The first one (left-side), presents the information concerning both the current topic (e.g. title and description) and the document (e.g. document identifier and text). To switch between documents, users can use either the *next-previous* buttons or the keyboard arrows. The annotation section (right-side), shows the annotations (e.g., labels and concepts) made for the selected document. The users can visualize their own annotations and also the ones made by other annotators, by clicking on the user icons in the lower part of the right-side of the interface. In addition, assessors can import and edit (in their own profiles) the annotations made by other assessors, by clicking on the *Upload and transfer* menu option.

DocTAG users can use four annotation modes: (i) *Labels* where each topic-document pair can be associated with a label (only a single label is allowed since a document cannot be marked, for instance, as *relevant* and *not relevant* at the same time); (ii) *Passages* where document passages can be marked with labels (one label per topic-passage pair) highlighted with different colors; (iii) *Linking* where each passage can be linked to user-defined or ontological concepts (one or many) and (iv) *Concepts* where each document can be associated with several user-defined or ontological concepts. Figure 1 shows the *Passages* annotation mode with several passages annotated. For instance, *hydrogen-fueled engine* (highlighted in green) is labelled with *Relevant* for the considered topic. All the passages marked with the same label are highlighted with a label-specific color to facilitate their recognition in the text. To quickly annotate long passages, users can click on the first passage word and on the last word, DocTAG automatically identifies the words in-between as a unique passage. By default, DocTAG provides automatic saving; nevertheless, manual saving is allowed as well, via *Save* button. Finally, to remove all the annotations made for the current annotation mode, users can click on the *Clear* button.

3   Final Remarks

In this paper, we present DocTAG, a web-based annotation tool specifically designed to ease the ground-truth creation process and support human annotators, with regards to the IR domain. DocTAG is an open-source, portable and customizable annotation tool that aims to be a reusable solution, for instance, in the context of IR evaluation campaigns. For the demo, we plan to showcase the annotation tool instantiated with the TIPSTER document collection along with the TREC 7 topics [17], since it is a very well-known collection in the IR domain. As future work, we plan to conduct a user study to improve the annotation interface, in terms of accessibility and inclusive design.

Acknowledgements. This work was partially supported by the ExaMode Project, as a part of the European Union Horizon 2020 Program under Grant 825292.

References

1. Biega, A.J., Diaz, F., Ekstrand, M.D., Kohlmeier, S.: Overview of the TREC 2019 fair ranking track. CoRR abs/2003.11650 (2020)
2. Cejuela, J.M., et al.: tagtog: interactive and text-mining-assisted annotation of gene mentions in PLOS full-text articles. Database J. Biol. Databases Curation **2014** (2014)
3. Dogan, R.I., Kwon, D., Kim, S., Lu, Z.: TeamTat: a collaborative text annotation tool. Nucleic Acids Res. **48**(Webserver-Issue), W5–W11 (2020)
4. Giachelle, F., Irrera, O., Silvello, G.: MedTAG: a portable and customizable annotation tool for biomedical documents. BMC Med. Inform. Decis. Making **21**, 352 (2021)
5. Klie, J.C., Bugert, M., Boullosa, B., de Castilho, R.E., Gurevych, I.: The inception platform: machine-assisted and knowledge-oriented interactive annotation. In: Proceedings of the 27th International Conference on Computational Linguistics: System Demonstrations, pp. 5–9. Association for Computational Linguistics, June 2018
6. Kwon, D., Kim, S., Shin, S., Chatr-aryamontri, A., Wilbur, W.J.: Assisting manual literature curation for protein-protein interactions using BioQRator. Database J. Biol. Databases Curation **2014**, bau067 (2014)
7. Kwon, D., Kim, S., Wei, C., Leaman, R., Lu, Z.: ezTag: tagging biomedical concepts via interactive learning. Nucleic Acids Res. **46**(Webserver-Issue), W523–W529 (2018)
8. Lin, J., et al.: Overview of the TREC 2017 real-time summarization track. In: Voorhees, E.M., Ellis, A. (eds.) Proceedings of The Twenty-Sixth Text REtrieval Conference, TREC 2017, Gaithersburg, Maryland, USA, 15–17 November 2017. NIST Special Publication, vol. 500–324. National Institute of Standards and Technology (NIST) (2017)
9. Lin, J., Roegiest, A., Tan, L., McCreadie, R., Voorhees, E.M., Diaz, F.: Overview of the TREC 2016 real-time summarization track. In: Voorhees, E.M., Ellis, A. (eds.) Proceedings of the Twenty-Fifth Text REtrieval Conference, TREC 2016, Gaithersburg, Maryland, USA, 15–18 November 2016. NIST Special Publication, vol. 500–321. National Institute of Standards and Technology (NIST) (2016)
10. Lin, J., Wang, Y., Efron, M., Sherman, G.: Overview of the TREC-2014 microblog track. In: Voorhees, E.M., Ellis, A. (eds.) Proceedings of the Twenty-Third Text REtrieval Conference, TREC 2014, Gaithersburg, Maryland, USA, 19–21 November 2014. NIST Special Publication, vol. 500–308. National Institute of Standards and Technology (NIST) (2014)
11. Neves, M., Ševa, J.: An extensive review of tools for manual annotation of documents. Brief. Bioinform. **22**(1), 146–163 (2021)
12. Neves, M.L., Leser, U.: A survey on annotation tools for the biomedical literature. Briefings Bioinform. **15**(2), 327–340 (2014)
13. Salgado, D., et al.: MyMiner: a web application for computer-assisted biocuration and text annotation. Bioinform. **28**(17), 2285–2287 (2012)

14. Sequiera, R., Tan, L., Lin, J.: Overview of the TREC 2018 real-time summarization track. In: Voorhees, E.M., Ellis, A. (eds.) Proceedings of the Twenty-Seventh Text REtrieval Conference, TREC 2018, Gaithersburg, Maryland, USA, 14–16 November 2018. NIST Special Publication, vol. 500–331. National Institute of Standards and Technology (NIST) (2018)
15. Stenetorp, P., Pyysalo, S., Topic, G., Ohta, T., Ananiadou, S., Tsujii, J.: BRAT: a web-based tool for NLP-assisted text annotation. In: Daelemans, W., Lapata, M., Màrquez, L. (eds.) EACL 2012, 13th Conference of the European Chapter of the Association for Computational Linguistics, Avignon, France, 23–27 April 2012, pp. 102–107. The Association for Computer Linguistics (2012)
16. Voorhees, E.M., et al.: TREC-COVID: constructing a pandemic information retrieval test collection. SIGIR Forum **54**(1), 1:1–1:12 (2020)
17. Voorhees, E.M., Harman, D.K.: Overview of the seventh text retrieval conference (TREC-7) (1999)
18. Yimam, S.M., Gurevych, I., de Castilho, R.E., Biemann, C.: WebAnno: a flexible, web-based and visually supported system for distributed annotations. In: 51st Annual Meeting of the Association for Computational Linguistics, ACL 2013, Proceedings of the Conference System Demonstrations, Sofia, Bulgaria, 4–9 August 2013, pp. 1–6. The Association for Computer Linguistics (2013)

ALWars: Combat-Based Evaluation of Active Learning Strategies

Julius Gonsior[1]([⊠])(iD), Jakob Krude[1], Janik Schönfelder[1], Maik Thiele[2](iD), and Wolgang Lehner[1](iD)

[1] Technische Universität Dresden, Dresden, Germany
{julius.gonsior,jakob.krude,janik.schonfelder,
wolgang.lehner}@tu-dresden.de
[2] Hochschule für Technik und Wirtschaft Dresden, Dresden, Germany
maik.thiele@htw-dresden.de

Abstract. The demand for annotated datasets for supervised *machine learning* (ML) projects is growing rapidly. Annotating a dataset often requires domain experts and is a timely and costly process. A premier method to reduce this overhead drastically is *Active Learning* (AL). Despite a tremendous potential for annotation cost savings, AL is still not used universally in ML projects. The large number of available AL strategies has significantly risen during the past years leading to an increased demand for thorough evaluations of AL strategies. Existing evaluations show in many cases contradicting results, without clear superior strategies. To help researchers in taming the AL zoo we present ALWARS: an interactive system with a rich set of features to compare AL strategies in a novel replay view mode of all AL episodes with many available visualization and metrics. Under the hood we support a rich variety of AL strategies by supporting the API of the powerful AL framework ALiPy [21], amounting to over 25 AL strategies out-of-the-box.

Keywords: Active learning · Python · GUI · Machine learning · Demo

1 Introduction

Machine learning (ML) is a popular and powerful approach to deal with the rapidly increasing availability of otherwise unusable datasets. Usually, an annotated set of data is required for an initial training phase before being applicable. In order to gain high quality data, these annotation tasks need to be performed by domain experts, who unfortunately dispose of a limited amount of working time and who are costly. The standard approach to reduce human labor cost massively is *Active Learning* (AL). During recent years the amount of proposed AL strategies has increased significantly [3,6,7,9,11,12,20,24]. Evaluations often show contradicting and mixed results, without any clearly superior strategies [13,16]. Very often, the strategies struggle even in beating the most naïve baselines e.g. [4,5,9,11,12,22]. Also, most evaluations are based on simple learning curves and only give a glimpse of the possibilities to compare AL strategies. A very important, and often left out, aspect of AL is the time dependency of metrics and visualizations during the iterations of the AL loop. Often, strategies undergo a change during the AL cycles and should therefore not be judged in the light of the final result. We present therefore ALWARS, an interactive demo application with a feature-rich *battle mode* to put AL strategies to the test in a novel and time-sensitive simulation replay mode. We included different metrics and visualization methods like the newly proposed *data maps* [19], classification boundaries, and manifold metrics in the comparison. We based our battle mode on top of the annotation web application ETIKEDI[1] which uses itself the popular AL framework ALiPy [21]. Thereby ALWARS can compare over 25 AL strategies[2] out-of-the-box and can be easily extended by additional strategies.

2 Active Learning 101

AL is the process of iteratively selecting those documents to be labeled first that improve the quality of the classification model the most. The basic AL cycle starts with a small labeled dataset $\mathcal{L}$ and a large unlabeled dataset $\mathcal{U}$. In a first step, a learner model θ is trained on the labeled set $\mathcal{L}$. Subsequently, a query strategy selects a set of unlabeled samples $\mathcal{U}_q$ to be annotated by the domain experts. This cycle repeats until the annotation budget is exhausted. Thus, by using a clever AL strategy, many samples that are not adding significant value to the classification model can be left unlabeled, while still achieving the same classification quality. AL strategies often use the confidence of the learner model to select those samples, the model is most uncertain about [10,14,18], a query-by-committee approach combining the uncertainty of many learner models [17], or the diversity of the vector space [15]. There are also many more complex strategies that apply for example *Reinforcement Learning* or *Imitation Learning* and use deep neural networks at the core of AL strategies [1–3,8,9,11,12,23].

3 Battle Mode

The battle mode enables researchers to compare two AL strategies side-by-side by showing different plots and metrics for each AL cycle separately in a replay simulation. In the following, the possible metrics and visualization tools as well as their relevance to AL research are described, referring to the components shown in the exemplary battle in Fig. 1:

Metrics: ALWARS displays metrics calculated per each AL strategy (Ⓒ) as well as metrics computed for both of them (Ⓓ). The latter ones include the

---

[1] https://github.com/etikedi/etikedi.

[2] Note that BatchBALD [6] and LAL-RL [9] are, as of now, submitted by us as a Pull-Request to ALiPy, and are not yet part of the upstream AL framework.

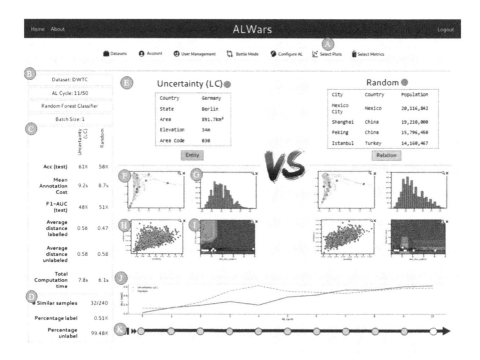

Fig. 1. Screenshot of ALWARS between uncertainty and random

percentage of similar samples annotated by both AL strategies or the percentage of the labeled and unlabeled samples. Metrics calculated for both separately are standard ML metrics such as precision, recall, accuracy, or F1-Score, available for the training and the test dataset. All these metrics are also available in an AUC-variant, defined as the proportion of the area under the AL learning curve with respect to the optimal learning curve, used as a summary representation of the learning curve. Interestingly, for the displayed example in Fig. 1, the Uncertainty strategy is better than Random according to the final test accuracy, but worse according to the AUC-value, as Random performed much better for the early AL cycles. More advanced metrics are the mean annotation cost, the average distance in the vector space across all labeled or all unlabeled samples, the average uncertainty or confidence of the learner model for the training or test samples, or the total computation time of the AL strategies.

Learning Curves: The most common evaluation visualization found in AL papers are learning curves (Ⓙ). The x-axis, often referred to as time, displays the AL cycles. The y-axis shows ML metrics such as accuracy or F1-score. Optimally, the learning curve goes straight up in the beginning and stays on top, maximizing the area under the curve.

Data Maps: A newly proposed visualization tool for datasets are so-called *Data Maps* [19] (Ⓗ). In a data map are the mean and the standard deviation of the confidence of the learner model over all AL cycles so far, defined as *confidence* and *variability*, displayed in a scatter plot for all training samples. The percentage of correctness of the predictions of the learner model during all AL cycles is used as color encoding. Data maps can be used to locate three distinct sample regions: *easy-to-learn*, *ambiguous*, and *hard-to-learn* samples. For the displayed battle in Fig. 1, it is apparent that the plot for Uncertainty contains more samples to the top left und less to the bottom left than Random, indicating a focus on labeling more easy-to-learn and less hard-to-learn samples.

Vector Space: The often high-dimensional feature vector space can be plotted using either manual selection of two important features, or automatic vector space transformation tools such as *PCA* or *t-SNE* as a 2D-plot (Ⓕ). Color coded are the labeled and unlabeled samples, as well as the samples, which have been selected in the current AL cycle as $\mathcal{U}_q$. This visualization is useful to understand, if AL strategies focus more on specific regions in the vector space, or evenly distributed, as is the case for both strategies in the example screenshot.

Classification Boundaries: In addition to the 2D representation of the vector space the classification boundaries of the learner model can be included as a surface plot overlay in an additional plot (Ⓘ). This plot is useful to analyze in depth how the learner model behaves regarding specific features.

Uncertainty Histogram: Similar to the classification boundaries plot, the uncertainty or confidence of the learner model can be displayed as a histogram for the training or test set (Ⓖ). For the displayed example in Fig. 1 the Uncertainty strategy leads to an overall slightly more confident learner model indicated by the flatter histogram in contrast to the Random strategy.

4 Demo Walkthrough

At the beginning, the visitors of ALWARS are requested to select two AL strategies, to upload the evaluation dataset (if not already present on the server), to set common AL configuration options like the AL batch size, the learner model, the amount of AL cycles to simulate, the train-test split ratio, or to configure the desired plots and metrics (Ⓐ). After the simulation is finished, the visitors of the demo are presented with the screenshot displayed in Fig. 1. At its core the researchers can see the samples of $\mathcal{U}_q$ (Ⓔ). Next to them are different plots and metrics about the current state of the AL strategies to be found. Using the timeline slider at the bottom (Ⓚ) the users can navigate through the AL cycles of the simulation. The plots can be maximized to get a more detailed look at them, or they can be reconfigured to display f. e. different features. The used dataset and the current AL cycle are displayed to the left (Ⓑ).

5   Conclusion

ALWARS enables fellow AL researchers to gain a deep and novel understanding on how AL strategies behave differently over the course of all AL cycles by displaying metrics and visualizations separately for each AL cycle. This leads to unique and more detailed time-sensitive evaluations of AL strategies, helping researchers in deciding which AL strategies to use for their ML projects, and opening the door for further improved AL strategies.

Acknowledgements. This research and development project is funded by the German Federal Ministry of Education and Research (BMBF) and the European Social Funds (ESF) within the "Innovations for Tomorrow's Production, Services, and Work" Program (funding number 02L18B561) and implemented by the Project Management Agency Karlsruhe (PTKA). The author is responsible for the content of this publication.

References

1. Bachman, P., Sordoni, A., Trischler, A.: Learning algorithms for active learning. In: Precup, D., Teh, Y.W. (eds.) Proceedings of the 34th International Conference on Machine Learning. Proceedings of Machine Learning Research, vol. 70, pp. 301–310. PMLR, International Convention Centre, Sydney, Australia, 06–11 August 2017
2. Fang, M., Li, Y., Cohn, T.: Learning how to active learn: a deep reinforcement learning approach. In: Proceedings of the 2017 Conference on Empirical Methods in Natural Language Processing, Copenhagen, Denmark, pp. 595–605. Association for Computational Linguistics, September 2017. https://doi.org/10.18653/v1/D17-1063
3. Gonsior, J., Thiele, M., Lehner, W.: ImitAL: learning active learning strategies from synthetic data (2021)
4. Hsu, W.N., Lin, H.T.: Active learning by learning. In: Proceedings of the Twenty-Ninth AAAI Conference on Artificial Intelligence, AAAI 2015, pp. 2659–2665. AAAI Press (2015)
5. Huang, S.j., Jin, R., Zhou, Z.H.: Active learning by querying informative and representative examples. In: Lafferty, J., Williams, C., Shawe-Taylor, J., Zemel, R., Culotta, A. (eds.) Advances in Neural Information Processing Systems, vol. 23, pp. 892–900. Curran Associates, Inc. (2010)
6. Kirsch, A., van Amersfoort, J., Gal, Y.: BatchBALD: efficient and diverse batch acquisition for deep Bayesian active learning. In: NIPS, vol. 32, pp. 7026–7037. Curran Associates, Inc. (2019)
7. Kirsch, A., Rainforth, T., Gal, Y.: Active learning under pool set distribution shift and noisy data. arXiv preprint arXiv:2106.11719 (2021)
8. Konyushkova, K., Sznitman, R., Fua, P.: Learning active learning from data. In: Guyon, I., Luxburg, U.V., Bengio, S., Wallach, H., Fergus, R., Vishwanathan, S., Garnett, R. (eds.) Advances in Neural Information Processing Systems, vol. 30, pp. 4225–4235. Curran Associates, Inc. (2017)
9. Konyushkova, K., Sznitman, R., Fua, P.: Discovering general-purpose active learning strategies. arXiv preprint arXiv:1810.04114 (2018)

10. Lewis, D.D., Gale, W.A.: A sequential algorithm for training text classifiers. In: Croft, B.W., van Rijsbergen, C.J. (eds.) SIGIR 1994, pp. 3–12. Springer, London (1994). https://doi.org/10.1007/978-1-4471-2099-5_1

11. Liu, M., Buntine, W., Haffari, G.: Learning how to actively learn: a deep imitation learning approach. In: Proceedings of the 56th Annual Meeting of the Association for Computational Linguistics (Volume 1: Long Papers), Melbourne, Australia, pp. 1874–1883. Association for Computational Linguistics, July 2018. https://doi.org/10.18653/v1/P18-1174

12. Pang, K., Dong, M., Wu, Y., Hospedales, T.: Meta-learning transferable active learning policies by deep reinforcement learning. arXiv preprint arXiv:1806.04798 (2018)

13. Ren, P., et al.: A survey of deep active learning. arXiv preprint arXiv:2009.00236 (2020)

14. Scheffer, T., Decomain, C., Wrobel, S.: Active hidden Markov models for information extraction. In: Hoffmann, F., Hand, D.J., Adams, N., Fisher, D., Guimaraes, G. (eds.) IDA 2001. LNCS, vol. 2189, pp. 309–318. Springer, Heidelberg (2001). https://doi.org/10.1007/3-540-44816-0_31

15. Sener, O., Savarese, S.: Active learning for convolutional neural networks: a core-set approach (2018)

16. Settles, B.: Active learning literature survey. Computer Sciences Technical report 1648 (2010)

17. Seung, H.S., Opper, M., Sompolinsky, H.: Query by committee. In: Proceedings of the Fifth Annual Workshop on Computational Learning Theory, COLT 1992, pp. 287–294, Association for Computing Machinery, New York (1992). https://doi.org/10.1145/130385.130417

18. Shannon, C.E.: A mathematical theory of communication. Bell Syst. Tech. J. **27**(3), 379–423 (1948). https://doi.org/10.1002/j.1538-7305.1948.tb01338.x

19. Swayamdipta, S., Schwartz, R., Lourie, N., Wang, Y., Hajishirzi, H., Smith, N.A., Choi, Y.: Dataset cartography: mapping and diagnosing datasets with training dynamics. In: Proceedings of the 2020 Conference on Empirical Methods in Natural Language Processing (EMNLP), pp. 9275–9293. Association for Computational Linguistics, November 2020. https://doi.org/10.18653/v1/2020.emnlp-main.746

20. Tang, Y.P., Huang, S.J.: Self-paced active learning: query the right thing at the right time. In: Proceedings of the AAAI Conference on Artificial Intelligence, vol. 33, no. 01, pp. 5117–5124 (2019). https://doi.org/10.1609/aaai.v33i01.33015117

21. Tang, Y.P., Li, G.X., Huang, S.J.: ALiPy: active learning in Python. arXiv preprint arXiv:1901.03802 (2019)

22. Wang, Z., Ye, J.: Querying discriminative and representative samples for batch mode active learning. ACM Trans. Knowl. Discov. Data **9**(3) (2015). https://doi.org/10.1145/2700408

23. Woodward, M., Finn, C.: Active one-shot learning. arXiv preprint arXiv:1702.06559 (2017)

24. Zhang, M., Plank, B.: Cartography active learning. arXiv preprint arXiv:2109.04282 (2021)

INForex: Interactive News Digest for Forex Investors

Chih-Hen Lee[(✉)], Yi-Shyuan Chiang, and Chuan-Ju Wang

Research Center for Information Technology Innovation, Academia Sinica, Taipei, Taiwan
{ch.lee,cjwang}@citi.sinica.edu.tw, yschiang@gapp.nthu.edu.tw

Abstract. As foreign exchange (Forex) markets reflect real-world events, locally or globally, financial news is often leveraged to predict Forex trends. In this demonstration, we propose INForex, an interactive web-based system that displays a Forex plot alongside related financial news. To our best knowledge, this is the first system to successfully align the presentation of two types of time-series data—Forex data and textual news data—in a unified and time-aware manner and as well as the first Forex-related online system leveraging deep learning techniques. The system can be of great help in revealing valuable insights and relations between the two types of data and is thus valuable for decision making not only for professional financial analysts or traders but also for common investors. The system is available online at http://cfda.csie.org/forex/, and the introduction video is at https://youtu.be/ZhFqQamTFY0.

Keywords: FOREX · Attention mechanism · Web system

1 Introduction

Forex markets are influenced by numerous factors, including gross domestic product (GDP), interest rates, and politics. In order to consider as many factors as possible to gain a more comprehensive view for Forex investment, most investors stay informed by relying on news.

Considering the critical role of news for Forex investment, in this demonstration, we propose INForex, an interactive web-based system that displays a Forex plot alongside related financial news. With the proposed system we seek to align the timing of Forex and news data to highlight the connection between these two types of time-series data. In INForex the release time of news stories is displayed as vertical lines on the Forex plot to emphasize the order of breaking news and exchange rates. Also, when users hover on a vertical line, the background color of the corresponding news will change.

Moreover, we propose using two event detection methods based on the standard deviation (SD) and directional change (DC) [1] of the Forex data to locate specific periods during which the prices or trends in Forex markets change dramatically. Via this design, we assist users to make trading strategies by putting the focus on changes of trend or periods of large price volatility. In addition, with the great advancement in natural language processing (NLP) and deep learning (DL) in recent years, in the proposed system, we utilize attention mechanisms [2] to further exploit the semantics

M. Hagen et al. (Eds.): ECIR 2022, LNCS 13186, pp. 300–304, 2022.
https://doi.org/10.1007/978-3-030-99739-7_37

in financial news. Specifically, by leveraging the financial news as training corpus and the aforementioned DC events as labels, our proposed system train an attention-based classification model that can be used to highlight worth-noting news and the potential keywords to users. Last but not least, INForex also comes with a keyword filtering function that allows users to filter news with self-defined keywords, where they can focus on searching related news of their interests.

To our best knowledge, this is the first system that effectively visualizes these two types of time-series data—Forex data and textual news data—in a unified and time-aware manner and as well as the first Forex system leveraging deep learning techniques. The proposed INForex can be of great help in revealing valuable insight and relations between the two types of data and is thus valuable for decision-making not only for professional financial analysts or traders but also for common investors.

2   System Description

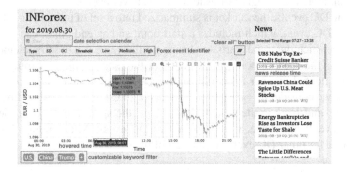

Fig. 1. The interfaces of INForex. Note that news is highlighted when the cursor hovers over the corresponding timeline on the chart. (Color figure online)

2.1   Data Collection

The Forex data was collected from philipperemy/FX-1-Minute-Data.[1] The training period is from January 1, 2010 to December 31, 2018, and the testing period is from August 4, 2019 to September 13, 2019 (which is the time span shown in the system).

As for news, we crawled articles from The Wall Street Journal.[2] Following the above setting, we separate these news into training, validation, and testing sets, the number of news corresponds to which are 80,000, 2,766, and 1,021, respectively Note that the validation set is the last 2,766 news in the training period, and only the testing batch of news is displayed on our system.

---

[1] https://github.com/philipperemy/FX-1-Minute-Data.
[2] https://www.wsj.com.

2.2   Event Detection Algorithms

The proposed system comes with two event detection algorithms: one based on the standard deviation of Forex price differences (referred as SD events hereafter) and one based on the directional change of Forex prices (referred as DC events hereafter) [1]. Specifically, SD events indicate time points during which the Forex prices dramatically change, whereas DC events show time points where the trends of Forex prices change.

We here briefly describe the two algorithms. Let $P = [p_1, p_2, ..., p_n]$ denote a series of Forex prices, where p_t stands for the CLOSE Bid Quote at minute t and n equals the total numbers of daily Forex data; also, the series of difference rates between two Forex prices is defined as $\mathcal{D} = [d_1, d_2, ..., d_m]$, where $d_t = (p_{t+w} - p_t)/p_t$ and $w \in \mathbb{Z}^+$ denotes the range in minutes. Note that we set $w = 20$ in our prototype system. We first calculate the standard deviation of differences in list $\mathcal{D}$ and denote this as $\sigma_{\mathcal{D}}$; we then use the z-score of each difference to determine the events (i.e., $z_{d_i} = (|d_i - m_{\mathcal{D}}|)/\sigma_{\mathcal{D}} > \tau$), where $m_{\mathcal{D}}$ denotes the mean of the list $\mathcal{D}$.) In this demonstration, we consider $\tau = 1, 2, 3$ as our thresholds corresponding to labels *Low*, *Medium*, and *High* in the system. Note that a higher threshold locates fewer events.

As for the DC events, the market is summarized into a set of uptrend and downtrend events [1]. A DC event, which includes a start point and an end point, can be seen as a period during which the trend starts to change. Initially, we use the first element in the price list p_1 as the starting point; then, the algorithm starts looking through the price list P in a sequential manner until it locates a price p_c for which $|(p_c - p_1)/p_1| > \theta$. In this demo, we consider $\theta = \sigma_{\mathcal{D}}, 2\sigma_{\mathcal{D}}, 3\sigma_{\mathcal{D}}$ as our thresholds, corresponding to labels *Low*, *Medium*, and *High* in the system. If $p_c > p_1$, the span from p_1 to p_c denotes an upward DC event; otherwise, it is a downward DC event. After finding the first DC event, we look for other DC events in the rest of the price sequence.

2.3   Attention Model

In this subsection, we introduce the pipeline of our model in detail. First, we represent each word in news titles and content with the pre-trained embedding from GloVe.[3] We later construct an attention layer [2] to aggregate the words embeddings in each news into one document embedding for each news. According to [2], the attention mechanism is formulated as $\text{Attention}(Q, K, V) = \text{softmax}(QK^T/\sqrt{d_k})V$, where d_k is the dimension of keys and queries. In our settings, the queries(Q) are initially randomized, while the keys(K) and values(V) are the GloVe embeddings for words. Here, the main concept is that every word contributes to the document embeddings by different levels; therefore, if a word accounts for a much greater weight than other words in the document do, it might be a potential keyword. After getting the document embeddings, we first apply a ReLU activation function, and then pass the embeddings to a linear classification layer to predict the final label. In terms of labeling training data, a piece of news is labeled as positive when there is a start of any high-threshold DC event in the interval $(t, t + 30)$, where t is the news release time in minutes. Otherwise, it will be labeled as negative.

---

[3] https://nlp.stanford.edu/projects/glove/.

Other experimental details are listed as follows. We implement the model under the PyTorch[4] framework. To reduce the training time, we only use the first 200 words in the concatenated news title and content as the textual input. For the training process, we select Adam [3] as the optimizer with learning rate 0.0001, and train data with 120 epochs and batch size 20.

2.4 Interfaces and Features

The INForex interface can be divided into two main parts: 1) the Forex chart (in the left panel of the system) and 2) the news section (the right panel).

Interactive Forex Chart. Forex prices are displayed by a standard candlestick chart, powered by Plotly graphing libraries.[5] Users can specify the date they are interested in with the calendar icon above the chart and then explore the chart freely by zooming in or dragging out the time frame of interest. Except for the Forex chart, the blue vertical lines correspond to the news story release time points. As shown in Fig. 1, the corresponding news in the right panel is highlighted if users hover over the corresponding blue line on the chart. Moreover, users can easily spot specific time ranges of interest by single-clicking the start and end points of a period on the Forex chart (resulting in the yellow span in the figure); the news section in the right panel changes correspondingly according to the selected time span.

Forex Event Identifier. Users specify event types (i.e., SD or DC) and the corresponding thresholds (i.e., Low, Medium, or High) with the Forex event identifier (see the buttoms below the date selection calendar in Fig. 1), after which the algorithm output is shown on the Forex chart as gray dots for the SD events and span lines for the DC events. Note that once done with a specific setting, users clear the setting with the "clear all" button on the top-right of the left panel; a higher threshold locates fewer events.

Fig. 2. Attention weights. The word "clashes" is a dark red word, "pessimism" and "tariff" are red words, and "Beating" and "statisticians" are pink words.

---

[4] https://pytorch.org/.

[5] https://plotly.com/graphing-libraries/.

Attention Weight Visualization. If a news is predicted as positive (i.e., potentially highly correlated with the DC events), the news title is highlighted in red. After clicking the positive news title, a window showcasing the news title and content with highlighted keywords will pop up. To help visualize the importance of each word, three classes of keywords are defined. Let ℓ be the word count of a concatenated news title and content, and the maximum of ℓ is set to 200 as mentioned above. The dark red words are the most worth-noting words whose attention weights are greater than $40/\ell$, while the attention weights of red words and pink words are greater than $20/\ell$ and $10/\ell$, respectively. As shown in Fig. 2, our model is able to capture keywords related to the Forex market in this case.

3 Conclusion and Future Work

In this paper, we propose INForex, an interactive system that aligns the Forex data and textual news data in a unified and time-aware manner. By placing the chart and related news side by side and displaying the potential keywords, users, financial professionals, and amateur investors can all understand Forex trends quicker and easier.

With limited financial resources, we can only propose a static prototype system with data made available online. However, since the real-time feature is crucial for investors, we will thrive to make INForex become a real-time system in the future, either with real-time industrial APIs. We hope that this demonstration can facilitate more research on predicting Forex markets through deep learning models in the future.

References

1. Bakhach, A., Tsang, E.P.K., Jalalian, H.: Forecasting directional changes in the FX markets. In: Proceedings of the 2016 IEEE Symposium Series on Computational Intelligence (SSCI), pp. 1–8 (2010)
2. Vaswani, A., et al.: Attention is all you need. In: Proceedings of the 31st International Conference on Neural Information Processing Systems (NeurIPS), pp. 6000–6010 (2017)
3. Kingma, D.P., Ba, J.: Adam: a method for stochastic optimization. In: Proceedings of the 3rd International Conference on Learning Representations (ICLR) (2015)

Streamlining Evaluation with `ir-measures`

Sean MacAvaney[(✉)], Craig Macdonald, and Iadh Ounis

University of Glasgow, Glasgow, UK
{sean.macavaney,craig.macdonald,iadh.ounis}@glasgow.ac.uk

Abstract. We present `ir-measures`, a new tool that makes it convenient to calculate a diverse set of evaluation measures used in information retrieval. Rather than implementing its own measure calculations, `ir-measures` provides a common interface to a handful of evaluation tools. The necessary tools are automatically invoked (potentially multiple times) to calculate all the desired metrics, simplifying the evaluation process for the user. The tool also makes it easier for researchers to use recently-proposed measures (such as those from the C/W/L framework) alongside traditional measures, potentially encouraging their adoption.

1 Introduction

The field of Information Retrieval (IR) is fortunate to have a vibrant and diverse ecosystem of tools and resources. This is particularly true for evaluation tools; there exists a variety of fully-fledged evaluation suites capable of calculating a wide array of measures (e.g., `trec_eval` [24], `cwl_eval` [2], `trectools` [25], and RankEval [17]) as well as single-purpose scripts that are usually designed for the evaluation of specific tasks or datasets.[1] However, none of these tools themselves provide comprehensive coverage of evaluation metrics, so researchers often need to run multiple tools to get all the desired results. Even when a single tool can provide the desired measures, it can sometimes require multiple invocations with different settings to get all desired results (e.g., the TREC Deep Learning passage ranking task [10] requires multiple invocations of `trec_eval` with different relevance cutoff thresholds).

In this demonstration, we present a new evaluation tool: `ir-measures`.[2] Unlike prior tools, which provide their own measure implementations, `ir-measures` operates as an abstraction over multiple evaluation tools. Researchers are able to simply describe *what* evaluation measures they want in natural syntax (e.g., `nDCG@20` for nDCG [13] with a rank cutoff of 20, or `AP(rel=2)` for Average Precision [12] with a binary relevance cutoff of 2), without necessarily needing to concern themselves with which specific tools provide the functionality they are looking for or what settings would give the desired results. By providing both a Python and command line interface and accepting multiple input and output formats, the tool is convenient to use in a variety

---

[1] For instance, the MSMARCO MRR evaluation script: https://git.io/JKG1S.

[2] Docs: https://ir-measur.es/, Source: https://github.com/terrierteam/ir_measures.

M. Hagen et al. (Eds.): ECIR 2022, LNCS 13186, pp. 305–310, 2022.
https://doi.org/10.1007/978-3-030-99739-7_38

of environments (e.g., both as a component of larger IR toolkits, or for simply doing *ad hoc* evaluation). By interfacing with existing evaluation toolkits, ir-measures is more sustainable than efforts that re-implement evaluation measures, especially given the ongoing debate over the suitability of some measures (e.g., [11,27]) and the proliferation of new measures (e.g., [1,9]). An interactive demonstration of the software is available at: https://git.io/JMt6G.

2   Background

Recently, there have been several efforts to resolve incompatibilities for other IR tools. trectools [25] provides Python implementations of numerous IR-related functions including pooling, fusion, and evaluation techniques (including a handful of evaluation measures). PyTerrier [20] provides a Python interface to a myriad of retrieval, rewriting, learning-to-rank, and neural re-ranking techniques as well as an infrastructure for conducting IR experiments. CIFF [16] defines a common index interchange format, for compatibility between search engines. ir_datasets [19] provides a common interface to access and work with document corpora and test collections. ir-measures is complementary to efforts like these. In Sect. 4, we show that ir-measures can easily be integrated into other tools, bolstering their evaluation capacity.

3   ir-measures

ir-measures provides access to over 30 evaluation measures. Table 1 provides a summary of the supported measures, which span a variety of categories and applications (e.g., intent-aware measures, set measures, etc.) Measures are referenced by name and a measure-dependent set of parameters (e.g., AP(rel=2) specifies a minimum relevance level and nDCG@10 specifies a rank cutoff). We refer the reader to the measure documentation[3] for further details.

Table 1. Measures provided by ir-measures, along with their providers.

Measure	Provided by...	Measure	Provided by...
alpha_nDCG [7]	ndeval	NERR [1]	cwl_eval
(M)AP(@k) [12]	cwl_eval, trec_eval, trectools	NRBP [8]	ndeval
(M)AP_IA	ndeval	NumQ, NumRel, NumRet	trec_eval
BPM [32]	cwl_eval	P(recision)@k [29]	trec_eval, cwl_eval, trectools
Bpref [4]	trec_eval, trectools	P_IA@k	ndeval
Compat [9]	Compatibility script	R(ecall)@k	trec_eval
ERR@k [6]	gdeval	RBP [8]	cwl_eval, trectools
ERR_IA [6]	ndeval	Rprec [5]	trec_eval, trectools
infAP [31]	trec_eval	(M)RR [15]	trec_eval, cwl_eval, trectools, MSMARCO
INSQ [21], INST [23]	cwl_eval	SDCG@k	cwl_eval
IPrec@i	trec_eval	SetAP, SetF, SetP, SetR	trec_eval
Judged@k	OpenNIR script	STREC	ndeval
nDCG(@k) [13]	trec_eval, gdeval, trectools	Success@k	trec_eval

---

[3] https://ir-measur.es/en/latest/measures.html.

Providers. The calculation of measure values themselves are implemented by *providers*. Not all providers are able to calculate all measures (or all parameters of a measure). The current version of `ir-measures` includes eight providers:

`trec_eval` [24] is a well-known IR evaluation tool that is used for calculating a variety of measures for TREC tasks. We use Python bindings adapted from the `pytrec_eval` [28] package.

`cwl_eval` [2] provides an implementation of a variety of measures that adhere to the C/W/L framework [22], such as `BPM` and `RBP`.

`ndeval`[4] enables the calculation of measures that consider multiple possible query intents (i.e., diversity measures), such as `alpha_nDCG` and `ERR_IA`.

The `trectools` [25] tookit includes Python implementations of a variety of evaluation measures, including `AP`, `Bpref`, and others. `gdeval`[5] includes an implementation of `ERR@k` and an alternative formulation of `nDCG@k` that places additional weight on high relevance.

The **OpenNIR** `Judged@k` script[6] was adapted from the OpenNIR toolkit [18] to calculate `Judged@k`, a measure of the proportion of top-ranked documents that have relevance assessments.

The **MSMARCO** `RR@k` script[7] is an interface to the official evaluation script for the MSMARCO dataset [3], with minor adjustments to allow for the configuration of the measure parameters and handling of edge cases.

The **Compatibility script**[8] provides `Compat` [9], a recently-proposed measure that calculates the Rank Biased Overlap of a result set compared to the closest ideal ranking of qrels, which can consider preference judgments.

Interfaces. `ir-measures` can be installed using `pip install ir-measures`, which provides both a command line interface and a Python package. The command line interface is similar to that of `trec_eval`, accepting a TREC-formatted relevance judgments (qrels) file, a run file, and the desired measures:

```
$ ir_measures path/to/qrels path/to/run 'nDCG@10 P(rel=2)@5 Judged@10'
nDCG@10    0.6251
P(rel=2)@5    0.6000
Judged@10    0.9486
```

Command line arguments allow the user to get results by query, use a particular provider, and control the output format. If the `ir-datasets` [19] package is installed, a dataset identifier can be used in place of the qrels path.

The Python API makes it simple to calculate measures from a larger toolkit. A variety of input formats are accepted, including TREC-formatted files, dictionaries, Pandas dataframes, and `ir-datasets` iterators. The Python API also can

---

[4] https://git.io/JKG94, https://git.io/JKCTo.
[5] https://git.io/JKCT1.
[6] https://git.io/JKG9O.
[7] https://git.io/JKG1S.
[8] https://git.io/JKCT5.

provide results by query and can reuse evaluation objects for improved efficiency over multiple runs. Here is a simple example that calculates four measures:

```
> import ir_measures
> from ir_measures import * # import natural measure names
> qrels = ir_measures.read_trec_qrels('path/to/qrels')
> run = ir_measures.read_trec_run('path/to/run')
> ir_measures.calc_aggregate([nDCG@10, P(rel=2)@5, Judged@10], qrels, run)
{nDCG@10: 0.6251, P(rel=2)@5: 0.6000, Judged@10: 0.9486}
```

4   Adoption of `ir-measures`

ir-measures is already in use by several tools, demonstrating its utility. It recently replaced `pytrec_eval` in PyTerrier [20], allowing retrieval pipelines to easily be evaluated on a variety of measures. For instance, the following example show an experiment on the TREC COVID [30] dataset. ir-measures allows the evaluation measures to be expressed clearly and concisely, and automatically invokes the necessary tools to compute the desired metrics:

```
import pyterrier as pt
dataset = pt.get_dataset('trec-covid')
pt.Experiment(
  [pt.TerrierRetrieve.from_dataset(dataset, "terrier_stemmed")],
  dataset.get_topics("round5"),
  dataset.get_qrels("round5"),
  eval_metrics=[nDCG@10, P(rel=2)@5, Judged@10])
```

It is also used by OpenNIR [18], Experimaestro [26], and DiffIR [14]. The ir-datasets [19] package uses ir-measures notation to provide documentation of the official evaluation measures for test collections.

A core design decision of ir-measures is to limit the required dependencies to the Python Standard Library and the packages for the measure providers (which can be omitted, but will degrade functionality). This should encourage the adoption of the tool by reducing the chance of package incompatibilities.

5   Conclusion

We demonstrated the new ir-measures package, which simplifies the computation of a variety of evaluation measures for IR researchers. We believe that by leveraging a variety of established tools (rather than providing its own implementations), ir-measures can be a salable and appealing choice for evaluation. We expect that our tool will also encourage the adoption of new evaluation measures, since they can be easily computed alongside long-established measures.

Acknowledgements.. We thank the contributors to the `ir-measures` repository. We acknowledge EPSRC grant EP/R018634/1: Closed-Loop Data Science for Complex, Computationally- & Data-Intensive Analytics.

References

1. Azzopardi, L., Mackenzie, J., Moffat, A.: ERR is not C/W/L: exploring the relationship between expected reciprocal rank and other metrics. In: ICTIR (2021)
2. Azzopardi, L., Thomas, P., Moffat, A.: Cwl_eval: an evaluation tool for information retrieval. In: SIGIR (2019)
3. Bajaj, P., et al.: MS MARCO: a human generated machine reading comprehension dataset. In: CoCo@NIPS (2016)
4. Buckley, C., Voorhees, E.M.: Retrieval evaluation with incomplete information. In: SIGIR (2004)
5. Buckley, C., Voorhees, E.M.: Retrieval System Evaluation. MIT Press, Cambridge (2005)
6. Chapelle, O., Metlzer, D., Zhang, Y., Grinspan, P.: Expected reciprocal rank for graded relevance. In: CIKM (2009)
7. Clarke, C.L.A., et al.: Novelty and diversity in information retrieval evaluation. In: SIGIR (2008)
8. Clarke, C.L.A., Kolla, M., Vechtomova, O.: An effectiveness measure for ambiguous and underspecified queries. In: ICTIR (2009)
9. Clarke, C.L.A., Vtyurina, A., Smucker, M.D.: Assessing top-k preferences. TOIS **39**(3), 1–21 (2021)
10. Craswell, N., Mitra, B., Yilmaz, E., Campos, D., Voorhees, E.: Overview of the TREC 2019 deep learning track. In: TREC (2019)
11. Fuhr, N.: Some common mistakes in ir evaluation, and how they can be avoided. SIGIR Forum **51**, 32–41 (2018)
12. Harman, D.: Evaluation issues in information retrieval. IPM **28**(4), 439–440 (1992)
13. Järvelin, K., Kekäläinen, J.: Cumulated gain-based evaluation of ir techniques. TOIS **20**(4), 422–446 (2002)
14. Jose, K.M., Nguyen, T., MacAvaney, S., Dalton, J., Yates, A.: Diffir: exploring differences in ranking models' behavior. In: SIGIR (2021)
15. Kantor, P., Voorhees, E.: The TREC-5 confusion track. Inf. Retr. **2**(2–3), 165–176 (2000)
16. Lin, J., et al.: Supporting interoperability between open-source search engines with the common index file format. In: SIGIR (2020)
17. Lucchese, C., Muntean, C.I., Nardini, F.M., Perego, R., Trani, S.: Rankeval: an evaluation and analysis framework for learning-to-rank solutions. In: SIGIR (2017)
18. MacAvaney, S.: OpenNIR: a complete neural ad-hoc ranking pipeline. In: WSDM (2020)
19. MacAvaney, S., Yates, A., Feldman, S., Downey, D., Cohan, A., Goharian, N.: Simplified data wrangling with ir_datasets. In: SIGIR (2021)
20. Macdonald, C., Tonellotto, N.: Declarative experimentation ininformation retrieval using PyTerrier. In: Proceedings of ICTIR 2020 (2020)
21. Moffat, A., Bailey, P., Scholer, F., Thomas, P.: Inst: an adaptive metric for information retrieval evaluation. In: Australasian Document Computing Symposium (2015)
22. Moffat, A., Bailey, P., Scholer, F., Thomas, P.: Incorporating user expectations and behavior into the measurement of search effectiveness. TOIS **35**(3), 1–38 (2017)

23. Moffat, A., Scholer, F., Thomas, P.: Models and metrics: IR evaluation as a user process. In: Australasian Document Computing Symposium (2012)
24. National Institute of Standards and Technology: trec_eval. https://github.com/usnistgov/trec_eval (1993–2021)
25. Palotti, J., Scells, H., Zuccon, G.: TrecTools: an open-source python library for information retrieval practitioners involved in TREC-like campaigns. In: SIGIR (2019)
26. Piwowarski, B.: Experimaestro and datamaestro: experiment and dataset managers (for IR). In: SIGIR (2020)
27. Sakai, T.: On Fuhr's guideline for IR evaluation. SIGIR Forum **54**, 1–8 (2020)
28. Van Gysel, C., de Rijke, M.: Pytrec_eval: an extremely fast python interface to trec_eval. In: SIGIR (2018)
29. Van Rijsbergen, C.J.: Information retrieval (1979)
30. Voorhees, E., et al.: Trec-covid: constructing a pandemic information retrieval test collection. ArXiv abs/2005.04474 (2020)
31. Yilmaz, E., Aslam, J.A.: Estimating average precision with incomplete and imperfect judgments. In: CIKM (2006)
32. Zhang, F., Liu, Y., Li, X., Zhang, M., Xu, Y., Ma, S.: Evaluating web search with a bejeweled player model. In: SIGIR (2017)

Turning News Texts into Business Sentiment

Kazuhiro Seki$^{(\boxtimes)}$

Konan University, Hyogo 658-8501, Japan
seki@konan-u.ac.jp

Abstract. This paper describes a demonstration system for our project on news-based business sentiment nowcast. Compared to traditional business sentiment indices which rely on a time-consuming survey and are announced only monthly or quarterly, our system takes advantage of news articles continually published on the Web and updates the estimate of business sentiment as the latest news come in. Additionally, it provides functionality to search any keyword and temporally visualize how much it influenced business sentiment, which can be a useful analytical tool for policymakers and economists. The codes and demo system are available at https://github.com/kazuhiro-seki/sapir-web.

Keywords: Sentiment analysis · Economic index · Neural network

1 Introduction

Business sentiment is an overall impression of the economical situation, typically measured by periodical surveys on business conditions and expectations of, for example, business managers. The resulting index, called *business sentiment index* (BSI for short), is an important indicator in making actions and plans for governmental and monetary policies and industrial production.

In the case of Japan, there exist BSIs publicly announced by financial authorities, including the Economy Watchers Survey and Short-term Economic Survey of Principal Enterprise. As with other BSIs, they rely on surveys, hence time-consuming as well as costly to calculate. In fact, they are only reported monthly or quarterly, which is not ideal when the economic situation rapidly changes due to unexpected events, such as the COVID-19 pandemic.

To remedy the problem, there have been research efforts to estimate BSIs by analyzing already available textual data (e.g., financial reports) [1,5,10–12, 15]. In our previous work [11], we proposed an approach to nowcasting a BSI using an outlier detection model and a prediction model based on an attention-based language representation model [14]. We demonstrated that the news-based BSI, named S-APIR, has a strong positive correlation with an existing survey-based BSI (Economy Watchers DI or EWDI for short) as high as $r = 0.937$. Additionally, we also proposed a simple approach to temporally analyzing the influence of any given word(s) on the estimated business sentiment.

M. Hagen et al. (Eds.): ECIR 2022, LNCS 13186, pp. 311–315, 2022.
https://doi.org/10.1007/978-3-030-99739-7_39

This paper introduces a demonstration system built upon these approaches with additional dynamicity; it crawls web news articles and updates the estimate of a BSI as the latest news are retrieved, providing the current snapshot of business sentiment much earlier than the existing monthly or quarterly BSIs.

It is important to note that business sentiment is different from market sentiment [2,4,6,8,9,13], although the same technologies can be applied for analysis. Market sentiment is an overall attitude of investors to a particular company or the market as a whole, which is valuable to automate stock trade. On the other hand, analyzing business sentiment could offer a more timely and less costly alternative to conventional BSIs, which would help policymakers and economists assess the current/future economic condition so as to take prompt actions and make appropriate plans accordingly as exemplified in our previous work [11].

2   S-APIR Demo System

2.1   Business Sentiment Analysis

The system estimates business sentiment, i.e., the S-APIR index, from news texts based on the approach proposed in our work. The approach and its validity are detailed in our recent paper [11]. Thus, we present only a brief summary of the approach below.

We first predict the economic status or a *business sentiment score* of a given news text. For this purpose, we use the Bidirectional Encoder Representation from Transformers (BERT) [3], specifically, a pre-trained Japanese BERT model[1] with an additional output layer to output a sentiment score. The entire model was fine-tuned on the Economy Watchers Survey as training data, which contain the current **economic status** each respondent observed on a five-point scale from -2 to 2 as well as the **reason of his/her judgment**. For example, a respondent may judge the economic status as very good ("2") because "Automobile exports to the United States are increasing". The former is used as a label and the latter as an input to learn the model parameters.

The learned model can predict a business sentiment score for *any* input text. However, news articles are in various genres (e.g., sports and entertainment), some of which may be irrelevant or even harmful in estimating proper business sentiment. Therefore, we filter out such texts as outliers by a one-class support vector machine (SVM) [7], which is also trained on Economy Watchers Survey. News texts that passed the filter are fed to the BERT model and the resulting business sentiment scores are averaged monthly to form monthly S-APIR index.

In addition to presenting the S-APIR index as an alternative to survey-based BSIs, the demo system provides a function for word-level temporal analysis, which is also described in our previous work [11]. Simply put, it assumes that the business sentiment of a text is the sum of the sentiments of the words composing the text and computes the contributions of individual words based on the estimated business sentiment of the text.

---

[1] https://github.com/cl-tohoku/bert-japanese.

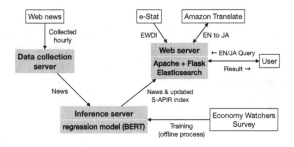

Fig. 1. An overview of the system.

2.2   Implementation

We developed a demonstration system composed of three servers as illustrated in Fig. 1. The data collection server fetches web news periodically. The newly collected news texts are sent to the inference server to predict the business sentiment score for each news text. The predicted scores are then averaged monthly and used as the updated S-APIR index of the current month. The updated index along with the news texts and their predicted business sentiment scores are sent to the webserver running an Apache HTTP server and also an Elasticsearch server, where Python Flask is used as a web application framework.

The web server obtains the EWDI from e-Stat[2] and presents the plots of EWDI and the predicted S-APIR. A user can send a query of his/her interest to the system to temporally analyze the economic impact of the query word(s). While the system is based on Japanese news articles and targets Japanese users, a search query can be also given in English (EN), which is translated to Japanese (JA) by Amazon Translate[3] on the fly for demonstration purposes.

The Elasticsearch index contains news texts with their release dates and predicted business sentiment scores. When a search query is issued, it is searched against the index and its sentiment in every time unit (currently a month) is computed. The computed sentiments are stored in a relational database (SQLAlchemy) for faster access for the same query in future. Also, the query word(s) is added to the user's watch list, which is stored in the browser's cookies, so that returning users would not need to send the same queries next time.

2.3   System Walkthrough

Figure 2 shows a screenshot of the demo system after a search query "ホテル" (hotels) is issued, where the top plot is a result of temporal analysis of the query, and the bottom plot shows S-APIR and EWDI. The S-APIR index was computed on news texts collected from the Nikkei websites[4] from March 2020 onward. While EWDI is updated monthly, S-APIR (of the current month) is

---

[2] e-Stat is a portal site for official Japanese government statistics.
[3] https://aws.amazon.com/translate.
[4] The Nikkei is the world's largest financial newspaper.

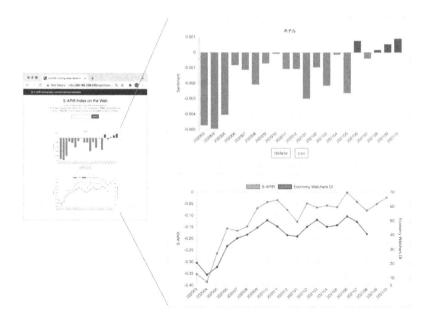

Fig. 2. A screenshot after a query "ホテル" (hotels) was issued on Oct. 21, 2021.

updated daily reflecting newly collected articles. Note that S-APIR has more recent data than EWDI as the latter is survey-based and has a certain delay. For the last two months (where EWDI is not available), S-APIR goes up steadily, showing important indication that economic conditions are improving.

Now, let us illustrate how the word-level analysis is performed. Suppose that the policymakers would like to examine the economic situation of the hotel business under the COVID-19 pandemic. Then, they may issue a query "ホテル" (hotels) and a bar plot appears as in Fig. 2. From the chart, one can observe that it had been damaged severely from 2020 but the situation is gradually recovering in the last three months. On the other hand, the restaurant business ("飲食店"), for example, is still struggling (not shown due to the limited space). The policymakers could allocate their limited budget for supporting such businesses. For a more in-depth analysis, the raw data of the bar chart can be downloaded as a CSV file by clicking the "csv" button.

Economists and policymakers could benefit from this system to overview the general trend of business sentiment in near real-time and to examine any factors which possibly affect business sentiment.

Acknowledgments. This work was done partly as a research project at APIR and was partially supported by JCER, JSPS KAKENHI #JP18K11558, #20H05633, and MEXT, Japan.

References

1. Aiba, Y., Yamamoto, H.: Data science and new financial engineering. Bus. Obs. **81**(2), 30–41 (2018). (in Japanese)
2. Derakhshan, A., Beigy, H.: Sentiment analysis on stock social media for stock price movement prediction. Eng. Appl. Artif. Intell. **85**, 569–578 (2019)
3. Devlin, J., Chang, M.W., Lee, K., Toutanova, K.: BERT: pre-training of deep bidirectional transformers for language understanding. In: Proceedings of the 2019 Conference of the North American Chapter of the Association for Computational Linguistics: Human Language Technologies, pp. 4171–4186, June 2019. https://doi.org/10.18653/v1/N19-1423,https://www.aclweb.org/anthology/N19-1423
4. Ge, Y., Qiu, J., Liu, Z., Gu, W., Xu, L.: Beyond negative and positive: exploring the effects of emotions in social media during the stock market crash. Inf. Process. Manage. **57**(4), 102218 (2020)
5. Goshima, K., Takahashi, D., Yamada, T.: Construction of business news index by natural language processing and its application to volatility prediction. Fin. Res. **38**(3) (2019). (in Japanese)
6. Li, X., Wu, P., Wang, W.: Incorporating stock prices and news sentiments for stock market prediction: a case of Hong Kong. Inf. Process. Manage. **57**(5), 102212 (2020)
7. Manevitz, L.M., Yousef, M.: One-class SVMs for document classification. J. Mach. Learn. Res. **2**, 139–154 (2002)
8. Picasso, A., Merello, S., Ma, Y., Oneto, L., Cambria, E.: Technical analysis and sentiment embeddings for market trend prediction. Expert Syst. Appl. **135**, 60–70 (2019)
9. Ren, J., Dong, H., Padmanabhan, B., Nickerson, J.V.: How does social media sentiment impact mass media sentiment? A study of news in the financial markets. J. Assoc. Inf. Sci. Technol. (2021). https://doi.org/10.1002/asi.24477. first published online
10. Seki, K., Ikuta, Y.: S-APIR: news-based business sentiment index. In: Darmont, J., Novikov, B., Wrembel, R. (eds.) ADBIS 2020. CCIS, vol. 1259, pp. 189–198. Springer, Cham (2020). https://doi.org/10.1007/978-3-030-54623-6_17
11. Seki, K., Ikuta, Y., Matsubayashi, Y.: News-based business sentiment and its properties as an economic index. Inf. Process. Manage. **59**(2) (to appear)
12. Shapiro, A.H., Sudhof, M., Wilson, D.J.: Measuring news sentiment. J. Econ. (2020). https://doi.org/10.1016/j.jeconom.2020.07.053. published first online
13. Tu, W., Yang, M., Cheung, D.W., Mamoulis, N.: Investment recommendation by discovering high-quality opinions in investor based social networks. Inf. Syst. **78**, 189–198 (2018)
14. Vaswani, A., et al.: Attention is all you need. In: Proceedings of the 31st International Conference on Neural Information Processing Systems, pp. 6000–6010 (2017). http://dl.acm.org/citation.cfm?id=3295222.3295349
15. Yamamoto, Y., Matsuo, Y.: Sentiment summarization of financial reports by LSTM RNN model with the Japan Economic Watcher Survey Data. In: Proceedings of the 30th JSAI (2016). (in Japanese)

SolutionTailor: Scientific Paper Recommendation Based on Fine-Grained Abstract Analysis

Tetsuya Takahashi$^{(\boxtimes)}$ and Marie Katsurai$^{(\boxtimes)}$ (ID)

Doshisha University, 1-3 Tatara Miyakodani, Kyotanabe-shi, Kyoto, Japan
{takahashi,katsurai}@mm.doshisha.ac.jp

Abstract. Locating specific scientific content from a large corpora is crucial to researchers. This paper presents SolutionTailor (The demo video is available at: https://mm.doshisha.ac.jp/sci2/SolutionTailor. html), a novel system that recommends papers that provide diverse solutions for a specific research objective. The proposed system does not require any prior information from a user; it only requires the user to specify the target research field and enter a research abstract representing the user's interests. Our approach uses a neural language model to divide abstract sentences into "Background/Objective" and "Methodologies" and defines a new similarity measure between papers. Our current experiments indicate that the proposed system can recommend literature in a specific objective beyond a query paper's citations compared with a baseline system.

Keywords: Paper recommendation system · Sentence classification · Sentence embedding

1 Introduction

Researchers usually spend a great deal of time searching for useful scientific papers for their continued research from a constantly growing number of publications in various academic fields. To assist academic search, various methods have been presented to recommend papers that approximate the user's interests and expertise. For example, content-based approaches often calculate sentence similarity between papers using natural language processing techniques, such as TF-IDF [9] and BERT [6]. Another line of research focuses on the user's past research history alongside co-authorship and citations and uses collaborative filtering or graph-based algorithms [3]. However, these methods usually focus on the semantic similarity of the overall content. Hence, the recommendation results often list papers that the user can easily access using a combination of research term queries or citation information in search engines. When considering the practicality of the usual literature survey, the search system must explain "why" and "how similar" the recommendated papers are to the user's interests.

M. Hagen et al. (Eds.): ECIR 2022, LNCS 13186, pp. 316–320, 2022.
https://doi.org/10.1007/978-3-030-99739-7_40

To clarify the recommendation intention, this paper presents SolutionTailor, a novel system that recommends literature on diverse research methodologies in a specific research objective. Given a research abstract as a query representing the user's interests, the proposed system searches for papers whose research problems are similar but whose solution strategies are significantly different. To achieve this, we divide abstract sentences into background and methodology parts and propose a novel scoring function to answer the reason for the similarity. Users can specify a target research field for the search, so the system provides insights beyond the users' expertise.

The main contributions of this paper are twofold. First, we apply fine-grained analysis to abstracts to clarify the recommendation intentions. Second, we present evaluation measures that characterize the proposed system compared with a baseline system that uses full abstract sentences.

2   SolutionTailor Framework

2.1   Dataset Construction

First, we prepared a list of research fields and their typical publication venues by referring to the rankings of the h5-index in Google Scholar Metrics[2]. Our current study uses the field "Engineering and Computer Science," comprising 56 subcategories, such as artificial intelligence, robotics, and sustainable energy. Then, from the Semantic Scholar corpus [2], we extracted papers whose publication venues were listed in the top-20 journals in each subcategory. The resulting dataset comprised 805,063 papers, all of which had English abstracts.

2.2   Sentence Labeling and Score Calculation

Owing to the recent advances in neural language models, there are several studies on fine-grained scientific text analyses, such as classifying sentences into problems and solution parts [8] and classifying citation intentions [7]. Following this line of research, our study uses a BERT-based pretrained model [5] to classify abstract sentences into categories of "Background," "Objective," and "Method." If no sentence is clearly assigned to these categories, we choose the sentence having the highest probability of the corresponding labels. Then, we concatenate the sentences labeled with Background and Objective into a single sentence, from which we extract the embedded context vector using SciBERT, a BERT model pretrained using scientific text [4]. Using the resulting vectors, we compute the cosine similarity, cos_{BO}, between the Background/Objective sentences of the query abstract and the abstract of each paper in the database. SolutionTailor extracts the top-100 abstracts having the highest cos_{BO} from the target research field. These abstracts are recommendation candidates whose backgrounds are

---

[2] https://scholar.google.co.jp/citations?view_op=top_venues.

Fig. 1. Interface of the proposed system.

similar to the focus of the query abstract. Finally, for each recommendation candidate, we compute the final similarity score with the query abstract as follows:

$$score = cos_{BO} - cos_M, \tag{1}$$

where cos_M denotes the cosine similarity between the method sentence vectors. A high score implies that the two abstracts have similar backgrounds, but presented different methodologies. Our system recommends the top-10 papers having the highest scores. This two-step search filters papers with irrelevant backgrounds and re-ranks the remaining ones in terms of methodological differences.

2.3   Recommendation Interface

Figure 1 shows the interface of the proposed system. When a user inputs a research abstract as a query to the text box and selects a target category from the pull-down menu, SolutionTailor displays a list of top-ranked papers in terms of the similarity measure. By clicking on the title, the user can jump to the paper page in Semantic Scholar. The results of abstract sentence labels are highlighted in yellow and green, corresponding to Background/Objective and Method, respectively, to facilitate the interpretation of the recommendation results. The system shows detailed bibliographic information of each recommended paper as well as its abstract labeling result by clicking the toggle to the right of the score.

3   Evaluation

SolutionTailor uses the original similarity measure, whereas we can construct a baseline system that uses the embedded context vectors extracted from the full abstracts and their cosine similarities. We quantified the characteristics of the proposed system compared with the baseline system using two types of quantitative measures.

i) Overlap with citations: The first experiment used an abstract of an existing paper as a query and investigated whether the query paper itself cited the papers in the recommendation results. If the overlap between the query's citations and the recommended papers is low, it implies that the system provides new insights for the user into a specific research background. We selected "artificial intelligence" from the categories and calculated MAP@10 by querying 100 papers that contained at least five citations in the category. The total number of papers to be searched was 36,355. The MAP@10 scores of the proposed and baseline systems were 0.003 and 0.076, respectively. We also applied BM25 [10], which used the title text as the query, as a comparative method; its score was 0.049. We should emphasize that the MAP may not be necessarily be high because we do not aim to predict the citations; this performance measure just characterizes the recommendation results. The lower MAP score implied that our system provides literature beyond the query paper's knowledge by using fine-grained sentence analysis.

ii) Similarity of objectives: The second experiment evaluated whether the similarity measure, cos_{BO}, used in the proposed system could actually find the same objective papers. Focusing on the fact that papers reporting results at a conference competition generally target the same research objective, we used the proceedings of SemEval-2021 [1], a workshop for the evaluation of computational semantic analysis systems. Each of the 11 tasks at SemEval-2021 had a single task description paper, and the papers submitted to a task were called "system description" papers. We used 11 task description papers as queries and evaluated whether the similarity measure, cos_{BO}, could appropriately recommend their system description papers.[3] The testing dataset for each query comprised $36,455$ $(36,355 + 100)$ papers used in the first experiment in addition to the system description papers of the target task. Each task had 15.91 system description papers on average, and a system should rank these higher than other papers. We removed the target task name and the competition name SemEval from all abstracts for fair experimental settings. As a result, the MAP scores obtained by our similarity measure, cos_{BO}, and the baseline system were 0.141 and 0.115, respectively, which demonstrated that our fine-grained sentence analysis could find the similarity of the objective more effectively than embedding the full abstracts.

4   Conclusion

This paper presented SolutionTailor, a novel system that recommends papers that provide diverse solutions to the same research background/objective. The system only requires text that summarizes the user's interests as a query, providing simple utility. The two types of the evaluation showed that our similarity

---

[3] We evaluated not the final similarity score but only cos_{BO} because the competition papers do not always have significantly different solutions.

measure provides insight beyond the user's cited papers and identifies the same-objective papers more effectively compared with the whole-text embedding. In future work, we will extend the dataset from engineering and computer science by adding more journals to be covered and conduct a subjective evaluation.

Acknowledgment. This research was partly supported by JSPS KAKENHI Grant Number 20H04484 and JST ACT-X grant number JPMJAX1909.

References

1. Semeval-2021 tasks. https://semeval.github.io/SemEval2021/tasks.html (2020). Accessed 20 Oct 2021
2. Ammar, W., et al.: Construction of the literature graph in semantic scholar. In: Proceedings of the 2018 Conference of the North American Chapter of the Association for Computational Linguistics: Human Language Technologies, Volume 3 (Industry Papers), pp. 84–91 (2018)
3. Asabere, N.Y., Xia, F., Meng, Q., Li, F., Liu, H.: Scholarly paper recommendation based on social awareness and folksonomy. Int. J. Parallel Emergent Distrib. Syst. **30**(3), 211–232 (2015)
4. Beltagy, I., Lo, K., Cohan, A.: SciBERT: a pretrained language model for scientific text. In: Proceedings of the 2019 Conference on Empirical Methods in Natural Language Processing and the 9th International Joint Conference on Natural Language Processing (EMNLP-IJCNLP), pp. 3615–3620 (2019)
5. Cohan, A., Beltagy, I., King, D., Dalvi, B., Weld, D.S.: Pretrained language models for sequential sentence classification. In: Proceedings of the 2019 Conference on Empirical Methods in Natural Language Processing and the 9th International Joint Conference on Natural Language Processing (EMNLP-IJCNLP) (2019)
6. Devlin, J., Chang, M.W., Lee, K., Toutanova, K.: BERT: Pre-training of deep bidirectional transformers for language understanding. In: Proceedings of the 2019 Conference of the North American Chapter of the Association for Computational Linguistics: Human Language Technologies, Volume 1 (Long and Short Papers), pp. 4171–4186. Association for Computational Linguistics, Minneapolis, Minnesota (2019)
7. Ferrod, R., Di Caro, L., Schifanella, C.: Structured semantic modeling of scientific citation intents. In: Verborgh, R., Hose, K., Paulheim, H., Champin, P.-A., Maleshkova, M., Corcho, O., Ristoski, P., Alam, M. (eds.) ESWC 2021. LNCS, vol. 12731, pp. 461–476. Springer, Cham (2021). https://doi.org/10.1007/978-3-030-77385-4_27
8. Heffernan, K., Teufel, S.: Identifying problems and solutions in scientific text. Scientometrics **116**(2), 1367–1382 (2018)
9. Jomsri, P., Sanguansintukul, S., Choochaiwattana, W.: A framework for tag-based research paper recommender system: an IR approach. In: 2010 IEEE 24th International Conference on Advanced Information Networking and Applications Workshops, pp. 103–108 (2010)
10. Robertson, S., Zaragoza, H.: The probabilistic relevance framework: Bm25 and beyond. Found. Trends Inf. Retr. **3**(4), 333–389 (2009). https://doi.org/10.1561/1500000019

Leaf: Multiple-Choice Question Generation

Kristiyan Vachev[1]([✉]), Momchil Hardalov[1], Georgi Karadzhov[2], Georgi Georgiev[3], Ivan Koychev[1], and Preslav Nakov[4]

[1] Faculty of Mathematics and Informatics, Sofia University "St. Kliment Ohridski", Sofia, Bulgaria
kdvachev@uni-sofia.bg
[2] Department of Computer Science and Technology, University of Cambridge, Cambridge, UK
[3] Releva.ai, Sofia, Bulgaria
[4] Qatar Computing Research Institute, HBKU, Doha, Qatar

Abstract. Testing with quiz questions has proven to be an effective way to assess and improve the educational process. However, manually creating quizzes is tedious and time-consuming. To address this challenge, we present Leaf, a system for generating multiple-choice questions from factual text. In addition to being very well suited for the classroom, Leaf could also be used in an industrial setting, e.g., to facilitate onboarding and knowledge sharing, or as a component of chatbots, question answering systems, or Massive Open Online Courses (MOOCs). The code and the demo are available on GitHub (https://github.com/KristiyanVachev/Leaf-Question-Generation).

Keywords: Multiple-choice questions · Education · Self-assessment · MOOCs

1 Introduction

Massive Open Online Courses (MOOCs) have revolutionized education by offering a wide range of educational and professional training. However, an important issue in such a MOOC setup is to ensure an efficient student examination setup. Testing with quiz questions has proven to be an effective tool, which can help both learning and student retention [38]. Yet, preparing such questions is a tedious and time-consuming task, which can take up to 50% of an instructor's time [41], especially when a large number of questions are needed in order to prevent students from memorizing and/or leaking the answers.

To address this issue, we present an automated multiple-choice question generation system with focus on educational text. Taking the course text as an input, the system creates question–answer pairs together with additional incorrect options (distractors). It is very well suited for a classroom setting, and the generated questions could also be used for self-assessment and for knowledge gap

M. Hagen et al. (Eds.): ECIR 2022, LNCS 13186, pp. 321–328, 2022.
https://doi.org/10.1007/978-3-030-99739-7_41

detection, thus allowing instructors to adapt their course material accordingly. It can also be applied in industry, e.g., to produce questions to enhance the process of onboarding, to enrich the contents of massive open online courses (MOOCs), or to generate data to train question–answering systems [10] or chatbots [22].

2 Related Work

While Question Generation is not as popular as the related task of Question Answering, there has been a steady increase in the number of publications in this area in recent years [1,18]. Traditionally, rules and templates have been used to generate questions [29]; however, with the rise in popularity of deep neural networks, there was a shift towards using recurrent encored–decoder architectures [2,8,9,33,40,46,47] and large-scale Transformers [7,20,23,27,36].

The task is often formulated as one of generating a question given a target answer and a document as an input. Datasets such as SQuAD1.1 [37] and NewsQA [44] are most commonly used for training, and the results are typically evaluated using measures such as BLEU [32], ROUGE [25], and METEOR [21]. Note that this task formulation requires the target answer to be provided beforehand, which may not be practical for real-world situations. To get over this limitation, some systems extract all nouns and named entities from the input text as target answers, while other systems train a classifier to label all word n-grams from the text and to pick the ones with the highest probability to be answers [45]. To create context-related wrong options (i.e., distractors), typically the RACE dataset [19] has been used along with beam search [3,11,31]. Note that MOOCs pose additional challenges as they often cover specialized content that goes beyond knowledge found in Wikipedia, and can be offered in many languages; there are some open datasets that offer such kinds of questions in English [5,6,19,28,42] and in other languages [4,12,13,15,16,24,26,30].

Various practical systems have been developed for question generation. Web-Experimenter [14] generates Cloze-style questions for English proficiency testing. AnswerQuest [39] generates questions for better use in Question Answering systems, and SQUASH [17] decomposes larger articles into paragraphs and generates a text comprehension question for each one; however, both systems lack the ability to generate distractors. There are also online services tailored to teachers. For example, Quillionz [35] takes longer educational texts and generates questions according to a user-selected domain, while Questgen [34] can work with texts up to 500 words long. While these systems offer useful question recommendations, they also require paid licenses. Our Leaf system offers a similar functionality, but is free and open-source, and can generate high-quality distractors. It is trained on publicly available data, and we are releasing our training scripts, thus allowing anybody to adapt the system to their own data.

3 System

System architecture: Leaf has three main modules as shown in Fig. 1. Using the *Client*, an instructor inputs a required number of questions and her educational text. The text is then passed through a REST API to the *Multiple-Choice Question (MCQ) Generator Module*, which performs pre-processing and then generates and returns the required number of question–answer pairs with distractors. To achieve higher flexibility and abstraction, the models implement an interface that allows them to be easily replaced.

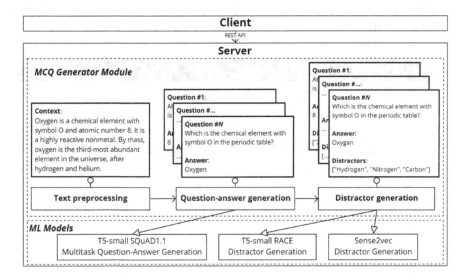

Fig. 1. The general architecture of Leaf.

Question and Answer Generation: To create the question–answer pairs, we combined the two tasks into a single multi-task model. We fine-tuned the small version of the T5 Transformer, which has 220M parameters, and we used the SQuAD1.1 dataset [37], which includes 100,000 question–answer pairs. We trained the model to output the question and the answer and to accept the passage and the answer with a 30% probability for the answer to be replaced by the [MASK] token. This allows us to generate an answer for the input question by providing the [MASK] token instead of the target answer. We trained the model for five epochs, and we achieved the best validation cross-entropy loss of 1.17 in the fourth epoch. We used a learning rate of 0.0001, a batch size of 16, and a source and a target maximum token lengths of 300 and 80, respectively. For question generation, we used the same data split and evaluation scripts as in [9]. For answer generation, we trained on the modified SQuAD1.1 Question Answering dataset as proposed in our previous work [45], achieving an Exact Match of 41.51 and an F1 score of 53.26 on the development set.

Distractor Generation: To create contextual distractors for the question–answer pairs, we used the RACE dataset [19] and the small pre-trained T5 model. We provided the question, the answer, and the context as an input, and obtained three distractors separated by a [SEP] token as an output. We trained the model for five epochs, achieving a validation cross-entropy loss of 2.19. We used a learning rate of 0.0001, a batch size of 16, and a source and a target maximum token lengths of 512 and 64, respectively. The first, the second, and the third distractor had BLEU1 scores of 46.37, 32.19, and 34.47, respectively. We further extended the variety of distractors with context-independent proposals, using sense2vec [43] to generate words or multi-word phrases that are semantically similar to the answer.

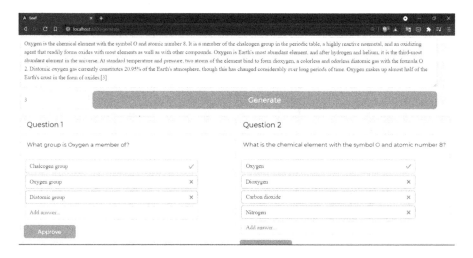

Fig. 2. Screenshot of Leaf showing the generated questions for a passage from the Wikipedia article on Oxygen. All distractors in *Question 1* are generated by the T5 model, and the last two distractors in *Question 2* are generated by the sense2vec model.

User Interface: Using the user interface shown on Fig. 2, the instructor can input her educational text, together with the desired number of questions to generate. Then, she can choose some of them, and potentially edit them, before using them as part of her course.

4   Conclusion and Future Work

We presented Leaf, a system to generate multiple-choice questions from text. The system can be used both in the classroom and in an industrial setting to detect knowledge gaps or as a self-assessment tool; it could also be integrated as part of other systems. With the aim to enable a better educational process, especially in the context of MOOCs, we open-source the project, including all training scripts and documentation.

In future work, we plan to experiment with a variety of larger pre-trained Transformers as the underlying model. We further plan to train on additional data. Given the lack of datasets created specifically for the task of Question Generation, we plan to produce a new dataset by using Leaf in real university courses and then collecting and manually curating the question–answer pairs Leaf generates over time.

Acknowledgements. This research is partially supported by Project UNITe BG05M 2OP001-1.001-0004 funded by the Bulgarian OP "Science and Education for Smart Growth."

References

1. Amidei, J., Piwek, P., Willis, A.: Evaluation methodologies in automatic question generation 2013–2018. In: Proceedings of the 11th International Conference on Natural Language Generation, INLG 2020, Tilburg University, The Netherlands , pp. 307–317. Association for Computational Linguistics (2018)
2. Bao, H., et al.: UniLMv2: pseudo-masked language models for unified language model pre-training. In: Proceedings of the 37th International Conference on Machine Learning, vol. 119 of ICML 2020, pp. 642–652. PMLR (2020)
3. Chung, H.L., Chan, Y.H., Fan, Y.C.: A BERT-based distractor generation scheme with multi-tasking and negative answer training strategies. In: Findings of the Association for Computational Linguistics: EMNLP 2020, pp. 4390–4400. Association for Computational Linguistics (2020)
4. Clark, J.H., et al.: TyDi QA: a benchmark for information-seeking question answering in typologically diverse languages. Trans. Assoc. Comput. Linguist. **8**, 454–470 (2020)
5. Clark, P., et al.: Think you have solved question answering? try ARC, the AI2 Reasoning Challenge. arXiv:1803.05457 (2018)
6. Clark, P., et al.: From 'F' to 'A' on the N.Y. regents science exams: an overview of the aristo project. AI Mag. **41**(4), 39–53 (2020)
7. Devlin, J., Chang, M.W., Lee, K., Toutanova, K.: BERT: pre-training of deep bidirectional transformers for language understanding. In: Proceedings of the 2019 Conference of the North American Chapter of the Association for Computational Linguistics: Human Language Technologies, NAACL-HLT 2019, Minneapolis, Minnesota, USA, pp. 4171–4186. Association for Computational Linguistics (2019)
8. Li, D., et al.: Unified language model pre-training for natural language understanding and generation. In: Advances in Neural Information Processing Systems 32: Annual Conference on Neural Information Processing Systems 2019, NeurIPS 2019, Vancouver, British Columbia, Canada, pp. 13042–13054 (2019)
9. Du, X., Shao, J., Cardie, C.: Learning to ask: neural question generation for reading comprehension. In: Proceedings of the 55th Annual Meeting of the Association for Computational Linguistics, ACL 2017, Vancouver, Canada, pp. 1342–1352. Association for Computational Linguistics (2017)
10. Duan, N., Tang, D., Chen, P., Zhou, M.: Question generation for question answering. In: Proceedings of the 2017 Conference on Empirical Methods in Natural Language Processing, EMNLP 2017, Copenhagen, Denmark, pp. 866–874. Association for Computational Linguistics (2017)

11. Gao, Y., Bing, L., Li, P., King, I., Lyu, M.R.: Generating distractors for reading comprehension questions from real examinations. In: Proceedings of the AAAI Conference on Artificial Intelligence, vol. 33 of AAAI 2019, pp. 6423–6430 (2019)

12. Hardalov, M., Koychev, I., Nakov, P.: Beyond English-only reading comprehension: experiments in zero-shot multilingual transfer for Bulgarian. In: Proceedings of the International Conference on Recent Advances in Natural Language Processing, RANLP 2019, Varna, Bulgaria, pp. 447–459. INCOMA Ltd. (2019)

13. Hardalov, M., Mihaylov, T., Zlatkova, D., Dinkov, Y., Koychev, I., Nakov, P.: EXAMS: a multi-subject high school examinations dataset for cross-lingual and multilingual question answering. In: Proceedings of the 2020 Conference on Empirical Methods in Natural Language Processing, EMNLP 2020, pp. 5427–5444. Association for Computational Linguistics (2020)

14. Hoshino, A., Nakagawa, H.: WebExperimenter for multiple-choice question generation. In: Proceedings of HLT/EMNLP 2005 Interactive Demonstrations, HLT/EMNLP 2005, Vancouver, British Columbia, Canada, pp. 18–19. Association for Computational Linguistics (2005)

15. Hu, J., Ruder, S., Siddhant, A., Neubig, G., Firat, O., Johnson, M.: XTREME: a massively multilingual multi-task benchmark for evaluating cross-lingual generalisation. In: Proceedings of the 37th International Conference on Machine Learning, vol. 119 of ICML 2020, pp. 4411–4421. PMLR (2020)

16. Jing, Y., Xiong, D., Yan, Z.: BiPaR: a bilingual parallel dataset for multilingual and cross-lingual reading comprehension on novels. In: Proceedings of the 2019 Conference on Empirical Methods in Natural Language Processing and the 9th International Joint Conference on Natural Language Processing, EMNLP-IJCNLP 2019, Hong Kong, China, pp. 2452–2462. Association for Computational Linguistics (2019)

17. Krishna, K., Iyyer, M.: Generating question-answer hierarchies. In: Proceedings of the 57th Annual Meeting of the Association for Computational Linguistics, Florence, Italy, ACL 2019, pp. 2321–2334. Association for Computational Linguistics (2019)

18. Kurdi, G., Leo, J., Parsia, B., Sattler, U., Al-Emari, S.: A systematic review of automatic question generation for educational purposes. Int. J. Artif. Intell. Educ. **30**, 121–204 (2019)

19. Lai, G., Xie, G., Liu, H., Yang, Y., Hovy, E.: RACE: large-scale reading comprehension dataset from examinations. In: Proceedings of the 2017 Conference on Empirical Methods in Natural Language Processing, EMNLP 2017, Copenhagen, Denmark, pp. 785–794. Association for Computational Linguistics (2017)

20. Lan, Z., Chen, M., Goodman, S., Gimpel, K., Sharma, P., Soricut, R.: ALBERT: a lite BERT for self-supervised learning of language representations. In: Proceedings of the 8th International Conference on Learning Representations, ICLR 2020, Addis Ababa, Ethiopia. OpenReview.net (2020)

21. Lavie, A., Agarwal, A.: METEOR: an automatic metric for MT evaluation with high levels of correlation with human judgments. In: Proceedings of the Second Workshop on Statistical Machine Translation, WMT 2007, Prague, Czech Republic, pp. 228–231. Association for Computational Linguistics (2007)

22. Lee, J., Liang, B., Fong, H.: Restatement and question generation for counsellor chatbot. In: Proceedings of the 1st Workshop on NLP for Positive Impact, pp. 1–7. Association for Computational Linguistics (2021)

23. Lewis, M., et al.: BART: denoising sequence-to-sequence pre-training for natural language generation, translation, and comprehension. In: Proceedings of the 58th Annual Meeting of the Association for Computational Linguistics, ACL 2020, pp. 7871–7880. Association for Computational Linguistics (2020)
24. Lewis, P., Oguz, B., Rinott, R., Riedel, S., Schwenk, H.: MLQA: evaluating cross-lingual extractive question answering. In: Proceedings of the 58th Annual Meeting of the Association for Computational Linguistics, ACL 2020, pp. 7315–7330. Association for Computational Linguistics (2020)
25. Lin, C.Y.: ROUGE: a package for automatic evaluation of summaries. In: Proceedings of the Workshop on Text Summarization Branches Out, Barcelona, Spain, pp. 74–81. Association for Computational Linguistics (2004)
26. Lin, X.V., et al.: Few-shot learning with multilingual language models. arXiv:2112.10668 (2021)
27. Liu, Y., et al.: RoBERTa: a robustly optimized BERT pretraining approach. arXiv:1907.11692 (2019)
28. Mihaylov, T., Clark, P., Khot, T., Sabharwal, A.: Can a suit of armor conduct electricity? a new dataset for open book question answering. In: Proceedings of the 2018 Conference on Empirical Methods in Natural Language Processing, EMNLP 2018, Brussels, Belgium, pp. 2381–2391. Association for Computational Linguistics (2018)
29. Mitkov, R., Ha, L.A.: Computer-aided generation of multiple-choice tests. In: Proceedings of the HLT-NAACL 03 Workshop on Building Educational Applications Using Natural Language Processing, BEA 2003, Edmonton, Alberta, Canada, pp. 17–22 (2003)
30. Van Nguyen, K., Tran, K.V., Luu, S.T., Nguyen, A.G.T., Nguyen, N.L.T.: Enhancing lexical-based approach with external knowledge for Vietnamese multiple-choice machine reading comprehension. IEEE Access 8, 201404–201417 (2020)
31. Offerijns, J., Verberne, S., Verhoef, T.: Better distractions: transformer-based distractor generation and multiple choice question filtering. arXiv:2010.09598 (2020)
32. Papineni, K., Roukos, S., Ward, T., Zhu, W.J.: Bleu: a method for automatic evaluation of machine translation. In: Proceedings of the 40th Annual Meeting of the Association for Computational Linguistics, ACL 2002, Philadelphia, Pennsylvania, USA, pp. 311–318. Association for Computational Linguistics (2002)
33. Qi, W., et al.: ProphetNet: predicting future n-gram for sequence-to-sequence pre-training. In: Findings of the Association for Computational Linguistics: EMNLP 2020, pp. 2401–2410. Association for Computational Linguistics (2020)
34. Questgen. Questgen: AI powered question generator. http://questgen.ai/, Accessed 05 Jan 2022
35. Quillionz. Quillionz - world's first AI-powered question generator. https://www.quillionz.com/, Accessed 05 Jan 2022
36. Raffel, C., et al.: Exploring the limits of transfer learning with a unified text-to-text transformer. J. Mach. Learn. Res. 21(140), 1–67 (2020)
37. Rajpurkar, P., Zhang, J., Lopyrev, K., Liang, P.: SQuAD: 100,000+ questions for machine comprehension of text. In: Proceedings of the 2016 Conference on Empirical Methods in Natural Language Processing, EMNLP 2016, Austin, Texas, USA, pp. 2383–2392. Association for Computational Linguistics (2016)
38. Roediger, H.L., III., Putnam, A.L., Smith, M.A.: Chapter one - ten benefits of testing and their applications to educational practice. Psychol. Learn. Motiv. 55, 1–36 (2011)

39. Roemmele, M., Sidhpura, D., DeNeefe, S., Tsou, L.: AnswerQuest: a system for generating question-answer items from multi-paragraph documents. In: Proceedings of the 16th Conference of the European Chapter of the Association for Computational Linguistics: System Demonstrations, EACL 2021, pp. 40–52. Association for Computational Linguistics (2021)

40. Song, L., Wang, Z., Hamza, W., Zhang, Y., Gildea, D.: Leveraging context information for natural question generation. In: Proceedings of the 2018 Conference of the North American Chapter of the Association for Computational Linguistics: Human Language Technologies, NAACL-HLT 2018, New Orleans, Louisiana, USA, pp. 569–574. Association for Computational Linguistics (2018)

41. Susanti, Y., Tokunaga, T., Nishikawa, H., Obari, H.: Evaluation of automatically generated English vocabulary questions. Res. Pract. Technol. Enhan. Learn. **12**(1), 1–21 (2017)

42. Tafjord, O., Clark, P., Gardner, M., Yih, W.T., Sabharwal, A.: Quarel: a dataset and models for answering questions about qualitative relationships. In: Proceedings of the AAAI Conference on Artificial Intelligence, vol. 33 of AAAI '19, pp. 7063–7071 (2019)

43. Trask, A., Michalak, P., Liu, J.: sense2vec - a fast and accurate method for word sense disambiguation in neural word embeddings. arXiv:1511.06388 (2015)

44. Trischler, A., et al.: NewsQA: a machine comprehension dataset. In: Proceedings of the 2nd Workshop on Representation Learning for NLP, RepL4NLP 2017, Vancouver, Canada, pp. 191–200. Association for Computational Linguistics (2017)

45. Vachev, K., Hardalov, M., Karadzhov, G., Georgiev, G., Koychev, I., Nakov, P.: Generating answer candidates for quizzes and answer-aware question generators. In: Proceedings of the Student Research Workshop Associated with RANLP 2021, RANLP 2021, pp. 203–209. INCOMA Ltd. (2021)

46. Xiao, D., et al.: ERNIE-GEN: an enhanced multi-flow pre-training and fine-tuning framework for natural language generation. In: Proceedings of the Twenty-Ninth International Joint Conference on Artificial Intelligence, IJCAI 2020, pp. 3997–4003. ijcai.org (2020)

47. Zhou, Q., Yang, N., Wei, F., Tan, C., Bao, H., Zhou, M.: Neural question generation from text: a preliminary study. In: Huang, X., Jiang, J., Zhao, D., Feng, Y., Hong, Yu. (eds.) NLPCC 2017. LNCS (LNAI), vol. 10619, pp. 662–671. Springer, Cham (2018). https://doi.org/10.1007/978-3-319-73618-1_56

CLEF 2022 Lab Descriptions

Overview of PAN 2022: Authorship Verification, Profiling Irony and Stereotype Spreaders, Style Change Detection, and Trigger Detection
Extended Abstract

Janek Bevendorff[1], Berta Chulvi[2], Elisabetta Fersini[3], Annina Heini[4],
Mike Kestemont[5], Krzysztof Kredens[4], Maximilian Mayerl[6],
Reyner Ortega-Bueno[2], Piotr Pęzik[4], Martin Potthast[7], Francisco Rangel[8],
Paolo Rosso[2], Efstathios Stamatatos[9], Benno Stein[1], Matti Wiegmann[1],
Magdalena Wolska[1(✉)], and Eva Zangerle[6]

[1] Bauhaus-Universität Weimar, Weimar, Germany
pan@webis.de, magdalena.wolska@uni-weimar.de
[2] Universitat Politècnica de València, Valencia, Spain
[3] Universitty Milano-Bicocca, Milan, Italy
[4] Aston University, Birmingham, UK
[5] University of Antwerp, Antwerp, Belgium
[6] University of Innsbruck, Innsbruck, Austria
[7] Leipzig University, Leipzig, Germany
[8] Symanto Research, Nuremberg, Germany
[9] University of the Aegean, Mytilene, Greece
http://pan.webis.de

Abstract. The paper gives a brief overview of the four shared tasks to be organized at the PAN 2022 lab on digital text forensics and stylometry hosted at the CLEF 2022 conference. The tasks include authorship verification across discourse types, multi-author writing style analysis, author profiling, and content profiling. Some of the tasks continue and advance past editions (authorship verification and multi-author analysis) and some are new (profiling irony and stereotypes spreaders and trigger detection). The general goal of the PAN shared tasks is to advance the state of the art in text forensics and stylometry while ensuring objective evaluation on newly developed benchmark datasets.

1 Introduction

PAN is a workshop series and a networking initiative for stylometry and digital text forensics. The workshop's goal is to bring together scientists and practitioners studying technologies which analyze texts with regard to originality, authorship, trust, and ethicality. Since its inception 15 years back PAN has included shared tasks on specific computational challenges related to authorship analysis, computational ethics, and determining the originality of a piece of writing.

© The Author(s), under exclusive license to Springer Nature Switzerland AG 2022
M. Hagen et al. (Eds.): ECIR 2022, LNCS 13186, pp. 331–338, 2022.
https://doi.org/10.1007/978-3-030-99739-7_42

Over the years, the respective organizing committees of the 54 shared tasks have assembled evaluation resources for the aforementioned research disciplines that amount to 51 datasets plus nine datasets contributed by the community.[1] Each new dataset introduced new variants of author verification, profiling, or author obfuscation tasks as well as multi-author analysis and determining the morality, quality, or originality of a text. The 2022 edition of PAN continues in the same vein, introducing new resources as well as previously unconsidered problems to the community. As in earlier editions, PAN is committed to reproducible research in IR and NLP therefore all shared tasks will ask for software submissions on our TIRA platform [7]. We briefly outline the upcoming tasks in the sections that follow.

2 Authorship Verification

Authorship verification is a fundamental task in author identification and all questioned authorship cases, be it closed-set or open-set scenarios, can be decomposed into a series of verification instances [6]. Previous editions of PAN included across-domain authorship verification tasks where texts of known and unknown authorship come from different domains [2,3,21]. In most of the examined cases, domains corresponded to topics (or thematic areas) and fandoms (non-professional fiction that is nowadays published online in significant quantities by fans of high-popularity authors or works, so-called fanfiction). The obtained results of the latest editions have demonstrated that it is feasible to handle such cases with relatively high performance [2,3]. In addition, at PAN'15, cross-genre authorship verification was partially studied using datasets in Dutch and Spanish covering essays and reviews [21]. However, these are relatively similar genres with respect to communication purpose, intended audience, or level of formality. On the other hand, it is not clear yet how to handle more difficult authorship verification cases where texts of known and unknown authorship belong to different discourse types (DTs), especially when these DTs have few similarities (e.g., argumentative essays vs. text messages to family members). In such cases, it is very challenging to distinguish the authorial characteristics that remain intact along DTs.

Cross-DT Author Verification at PAN'22

For the 2022 edition, we will focus on the cross-DT authorship verification scenario. In more detail, we will use a new corpus in English comprising writing samples from around 100 individuals composing texts in the following DTs: essays, emails, text messages, and business memos. All individuals have similar age (18-22) and are native English speakers. The topic of text samples is not restricted while the level of formality can vary within a certain DT (e.g., text messages may be addressed to family members or non-familial acquaintances). The new edition of author verification task at PAN'22 will allow us to

---

[1] https://pan.webis.de/data.html.

study the ability of stylometric approaches to capture elements of authorial style that remain stable across DTs even when very different forms of expression are imposed by the DT norms. The task will also focus on the ability of the submitted approaches to compare long texts of known authorship with short texts of unknown authorship. As concerns the experimental setup, it will be similar to the last edition of PAN and the same evaluation measures (AUROC, $c@1$, F_1, $F_{0.5u}$, and Brier score) will be used [2].

3   Author Profiling

Author profiling is the problem of distinguishing between classes of authors by studying how language is shared by people. This helps in identifying authors' individual characteristics, such as age, gender, or language variety, among others. During the years 2013-2021 we addressed several of these aspects in the shared tasks organised at PAN.[2] In 2013 the aim was to identify gender and age in social media texts for English and Spanish [14]. In 2014 we addressed age identification from a continuous perspective (without gaps between age classes) in the context of several genres, such as blogs, Twitter, and reviews (in Trip Advisor), both in English and Spanish [12]. In 2015, apart from age and gender identification, we addressed also personality recognition on Twitter in English, Spanish, Dutch, and Italian [16]. In 2016, we addressed the problem of cross-genre gender and age identification (training on Twitter data and testing on blogs and social media data) in English, Spanish, and Dutch [17]. In 2017, we addressed gender and language variety identification in Twitter in English, Spanish, Portuguese, and Arabic [15]. In 2018, we investigated gender identification in Twitter from a multimodal perspective, considering also the images linked within tweets; the dataset was composed of English, Spanish, and Arabic tweets [13]. In 2019 the focus was on profiling bots and discriminating bots from humans on the basis of textual data only [11]. We used Twitter data both in English and Spanish. Bots play a key role in spreading inflammatory content and also fake news. Advanced bots that generated human-like language, also with metaphors, were the most difficult to profile. It is interesting to note that when bots were profiled as humans, they were mostly confused with males. In 2020 we focused on profiling fake news spreaders [9]. The easiness of publishing content in social media has led to an increase in the amount of disinformation that is published and shared. The goal was to profile those authors who have shared some fake news in the past. Early identification of possible fake news spreaders on Twitter should be the first step towards preventing fake news from further dissemination. In 2021 the focus was on profiling hate speech spreaders in social media [8]. The goal was to identify Twitter users who can be considered haters, depending on the number of tweets with hateful content that they had spread. The task was set in English and Spanish.

---

[2] To generate the datasets, we have followed a methodology that complies with the EU General Data Protection Regulation [10].

Profiling Irony and Stereotype Spreaders on Twitter (IROSTEREO)

With irony, language is employed in a figurative and subtle way to mean the opposite to what is literally stated [18]. In case of sarcasm, a more aggressive type of irony, the intent is to mock or scorn a victim without excluding the possibility to hurt [4]. Stereotypes are often used, especially in discussions about controversial issues such as immigration [22] or sexism [19] and misogyny [1]. At PAN'22 we will focus on profiling ironic authors in Twitter. Special emphasis will be given to those authors that employ irony to spread stereotypes, for instance, towards women or the LGTB community. The goal will be to classify authors as ironic or not depending on their number of tweets with ironic content. Among those authors we will consider a subset that employs irony to convey stereotypes in order to investigate if state-of-the-art models are able to distinguish also these cases. Therefore, given authors together with their tweets, the goal will be to profile those authors that can be considered as ironic, and among them those that employ irony to convey stereotypical messages. As an evaluation setup, we will create a collection that contains tweets posted by users in Twitter. One document will consist of a feed of tweets written by the same user.

4   Multi-author Writing Style Analysis

The goal of the style change detection task is to identify—based on an intrinsic style analysis—the text positions within a given multi-author document at which the author switches. Detecting these positions is a crucial part of the authorship identification process and multi-author document analysis; multi-author documents have been largely understudied in general. This task has been part of PAN since 2016, with varying task definitions, data sets, and evaluation procedures. In 2016, participants were asked to identify and group fragments of a given document that correspond to individual authors [20]. In 2017, we asked participants to detect whether a given document is multi-authored and, if this is indeed the case, to determine the positions at which authorship changes [23]. However, since this task was deemed as highly complex, in 2018 its complexity was reduced to asking participants to predict whether a given document is single- or multi-authored [5]. Following the promising results achieved, in 2019 participants were asked first to detect whether a document was single- or multi-authored and if it was indeed written by multiple authors, to then predict the number of authors [26]. Based on the advances made over the previous years, in 2020 we decided to go back towards the original definition of the task, i.e., finding the positions in a text where authorship changes. Participants first had to determine whether a document was written by one or by multiple authors and, if it was written by multiple authors, they had to detect between which paragraphs the authors change [25]. In the 2021 edition, we asked the participants first to detect whether a document was authored by one or multiple authors. For two-author documents, the task was to find the position of the authorship change and for multi-author documents, the task was to find all positions of authorship change [24].

Multi-author Writing Style Analysis at PAN'22

The analysis of author writing styles is the foundation for author identification. As previous research shows, it also allows distinguishing between authors in multi-authored documents. In this sense, methods for multi-author writing style analysis can pave the way for authorship attribution at the sub-document level and thus, intrinsic plagiarism detection (i.e., detecting plagiarism without the use of a reference corpus). Given the importance of these tasks, we foster research in this direction through our continued development of benchmarks: the ultimate goal is to identify the exact positions within a document at which authorship changes based on an intrinsic style analysis. Based on the progress made towards this goal in previous years and to entice novices and experts, we extend the set of challenges: (i) Style Change Basic: given a text written by two authors and that contains a single style change only, find the position of this change, i.e., cut the text into the two authors' texts on the paragraph-level, (ii) Style Change Advanced: given a text written by two or more authors, find all positions of writing style change, i.e., assign all paragraphs of the text uniquely to some author out of the number of authors assumed for the multi-author document, (iii) Style Change Real-World: given a text written by two or more authors, find all positions of writing style change, where style changes now not only occur between paragraphs but at the sentence level. For this year's edition, we will additionally introduce a new corpus that is based on publicly available social media data to show the performance of the approaches based on different data sources.

5   Trigger Detection

A trigger in psychology is a stimulus that elicits negative emotions or feelings of distress. In general, triggers include a broad range of stimuli—such as smells, tastes, sounds, textures, or sights—which may relate to possibly distressing acts or events of whatever type, for instance, violence, trauma, death, eating disorders, or obscenity. In order to proactively apprise audience that a piece of media (writing, audio, video, etc.) contains potentially distressing material, the use of "trigger warnings"—labels indicating the type of triggering content present—have become common not only in online communities, but also in institutionalized education, making it possible for sensitive audience to prepare for the content and better manage their reactions. In the planned series of shared tasks on triggers, we propose a computational problem of identifying whether or not a given document contains triggering content, and if so, of what type.

Identifying Violent Content at PAN'22

In the first pilot edition of the task, we will focus on a single trigger type: violence. As data we will use a corpus of fanfiction (millions of stories crawled from fanfiction.net and archiveofourown.org (Ao3)) in which trigger warnings have been assigned by the authors, that is, we do not define "violence" as a construct ourselves here, but rather rely on user-generated labels. We unify the set of label names where necessary and create a balanced corpus of positive and negative examples. The problem is formulated as binary classification at document level as follows: Given a piece of fanfiction discourse, classify it as triggering or not triggering. Standard measures of classifier quality will be used for evaluation.

Acknowledgments. The contributions from Bauhaus-Universität Weimar and Leipzig University have been partially funded by the German Ministry for Science and Education (BMBF) project "Shared Tasks as an innovative approach to implement AI and Big Data-based applications within universities (SharKI)" (grant FKZ 16DHB4021). The Cross-DT corpus was developed at the Aston Institute for Forensic Linguistics with funding from Research England's Expanding Excellence in England (E3) Fund. The work of the researchers from the Universitat Politècnica de València was partially funded by the Spanish MICINN under the project MISMIS-FAKEnHATE on MISinformation and MIScommunication in social media: FAKE news and HATE speech (PGC2018-096212-B-C31), and by the Generalitat Valenciana under the project DeepPattern (PROMETEO/2019/121). The work of Francisco Rangel has been partially funded by the Centre for the Development of Industrial Technology (CDTI) of the Spanish Ministry of Science and Innovation under the research project IDI-20210776 on Proactive Profiling of Hate Speech Spreaders - PROHATER (Perfilador Proactivo de Difusores de Mensajes de Odio).

References

1. Anzovino, M., Fersini, E., Rosso, P.: Automatic identification and classification of misogynistic language on twitter. In: Silberztein, M., Atigui, F., Kornyshova, E., Métais, E., Meziane, F. (eds.) NLDB 2018. LNCS, vol. 10859, pp. 57–64. Springer, Cham (2018). https://doi.org/10.1007/978-3-319-91947-8_6
2. Bevendorff, J., et al.: Overview of PAN 2021: authorship verification, profiling hate speech spreaders on twitter, and style change detection. In: Candan, K.S., et al. (eds.) CLEF 2021. LNCS, vol. 12880, pp. 419–431. Springer, Cham (2021). https://doi.org/10.1007/978-3-030-85251-1_26
3. Bevendorff, J., et al.: Overview of PAN 2020: authorship verification, celebrity profiling, profiling fake news spreaders on twitter, and style change detection. In: Arampatzis, A., et al. (eds.) CLEF 2020. LNCS, vol. 12260, pp. 372–383. Springer, Cham (2020). https://doi.org/10.1007/978-3-030-58219-7_25
4. Frenda, S., Cignarella, A., Basile, V., Bosco, C., Patti, V., Rosso, P.: The unbearable hurtfulness of sarcasm. Expert Syst. Appl. (2022). https://doi.org/10.1016/j.eswa.2021.116398
5. Kestemont, M., et al.: Overview of the author identification task at PAN 2018: cross-domain authorship attribution and style change detection. In: CLEF 2018 Labs and Workshops, Notebook Papers (2018)
6. Koppel, M., Winter, Y.: Determining if two documents are written by the same author. J. Assoc. Inf. Sci. Technol. **65**(1), 178–187 (2014)
7. Potthast, M., Gollub, T., Wiegmann, M., Stein, B.: TIRA integrated research architecture. In: Ferro, N., Peters, C. (eds.) Information Retrieval Evaluation in a

Changing World. TIRS, vol. 41, pp. 123–160. Springer, Cham (2019). https://doi.org/10.1007/978-3-030-22948-1_5

8. Rangel, F., De-La-Peña-Sarracén, G.L., Chulvi, B., Fersini, E., Rosso, P.: Profiling hate speech spreaders on twitter task at PAN 2021. In: Faggioli, G., Ferro, N., Joly, A., Maistro, M., Piroi, F. (eds.) CLEF 2021 Labs and Workshops, Notebook Papers, CEUR-WS.org (2021)

9. Rangel, F., Giachanou, A., Ghanem, B., Rosso, P.: Overview of the 8th author profiling task at PAN 2019: profiling fake news spreaders on twitter. In: CLEF 2020 Labs and Workshops, Notebook Papers. CEUR Workshop Proceedings (2020)

10. Rangel, F., Rosso, P.: On the implications of the general data protection regulation on the organisation of evaluation tasks. Lang. Law/Linguagem e Direito 5(2), 95–117 (2019)

11. Rangel, F., Rosso, P.: Overview of the 7th author profiling task at pan 2019: bots and gender profiling. In: CLEF 2019 Labs and Workshops, Notebook Papers (2019)

12. Rangel, F., et al.: Overview of the 2nd author profiling task at PAN 2014. In: CLEF 2014 Labs and Workshops, Notebook Papers (2014)

13. Rangel, F., Rosso, P., Montes-y-Gómez, M., Potthast, M., Stein, B.: Overview of the 6th author profiling task at PAN 2018: multimodal gender identification in twitter. In: CLEF 2019 Labs and Workshops, Notebook Papers (2018)

14. Rangel, F., Rosso, P., Moshe Koppel, M., Stamatatos, E., Inches, G.: Overview of the author profiling task at PAN 2013. In: CLEF 2013 Labs and Workshops, Notebook Papers (2013)

15. Rangel, F., Rosso, P., Potthast, M., Stein, B.: Overview of the 5th author profiling task at PAN 2017: gender and language variety identification in twitter. In: Working Notes Papers of the CLEF (2017)

16. Rangel, F., Rosso, P., Potthast, M., Stein, B., Daelemans, W.: Overview of the 3rd author profiling task at PAN 2015. In: CLEF 2015 Labs and Workshops, Notebook Papers (2015)

17. Rangel, F., Rosso, P., Verhoeven, B., Daelemans, W., Potthast, M., Stein, B.: Overview of the 4th author profiling task at PAN 2016: cross-genre evaluations. In: CLEF 2016 Labs and Workshops, Notebook Papers (2016). ISSN 1613–0073

18. Reyes, A., Rosso, P.: On the difficulty of automatically detecting irony: beyond a simple case of negation. Knowl. Inf. Syst. 40(3), 595–614 (2014)

19. Rodríguez-Sánchez, F., et al.: Overview of exist 2021: sexism identification in social networks. In: Procesamiento del Lenguaje Natural (SEPLN), no. 67, pp. 195–207 (2021)

20. Rosso, P., Rangel, F., Potthast, M., Stamatatos, E., Tschuggnall, M., Stein, B.: Overview of PAN 2016–new challenges for authorship analysis: cross-genre profiling, clustering, diarization, and obfuscation. In: 7th International Conference of the CLEF Initiative on Experimental IR Meets Multilinguality, Multimodality, and Interaction (CLEF 2016) (2016)

21. Stamatatos, E., Potthast, M., Rangel, F., Rosso, P., Stein, B.: Overview of the PAN/CLEF 2015 evaluation lab. In: Mothe, J., et al. (eds.) CLEF 2015. LNCS, vol. 9283, pp. 518–538. Springer, Cham (2015). https://doi.org/10.1007/978-3-319-24027-5_49

22. Sánchez-Junquera, J., Chulvi, B., Rosso, P., Ponzetto, S.: How do you speak about immigrants? taxonomy and stereoimmigrants dataset for identifying stereotypes about immigrants. Appl. Sci. 11(8), 3610 (2021)

23. Tschuggnall, M., et al.: Overview of the author identification task at PAN 2017: style breach detection and author clustering. In: CLEF 2017 Labs and Workshops, Notebook Papers (2017)

24. Zangerle, E., Mayerl, M., Potthast, M., Stein, B.: Overview of the style change detection task at PAN 2021. In: Faggioli, G., Ferro, N., Joly, A., Maistro, M., Piroi, F. (eds.) CLEF 2021 Labs and Workshops, Notebook Papers, CEUR-WS.org (2021)
25. Zangerle, E., Mayerl, M., Specht, G., Potthast, M., Stein, B.: Overview of the style change detection task at PAN 2020. In: CLEF 2020 Labs and Workshops, Notebook Papers (2020)
26. Zangerle, E., Tschuggnall, M., Specht, G., Stein, B., Potthast, M.: Overview of the style change detection task at PAN 2019. In: CLEF 2019 Labs and Workshops, Notebook Papers (2019)

Overview of Touché 2022:
Argument Retrieval
Extended Abstract

Alexander Bondarenko[1]([⊠]), Maik Fröbe[1], Johannes Kiesel[2], Shahbaz Syed[3],
Timon Gurcke[4], Meriem Beloucif[5], Alexander Panchenko[6], Chris Biemann[5],
Benno Stein[2], Henning Wachsmuth[4], Martin Potthast[3], and Matthias Hagen[1]

[1] Martin-Luther-Universität Halle-Wittenberg, Halle, Germany
`touche@webis.de`
[2] Bauhaus-Universität Weimar, Weimar, Germany
[3] Leipzig University, Leipzig, Germany
[4] Paderborn University, Paderborn, Germany
[5] Universität Hamburg, Hamburg, Germany
[6] Skolkovo Institute of Science and Technology, Moscow, Russia

Abstract. The goal of the Touché lab on argument retrieval is to foster and support the development of technologies for argument mining and argument analysis. In the third edition of Touché, we organize three shared tasks: (a) argument retrieval for controversial topics, where participants retrieve a gist of arguments from a collection of online debates, (b) argument retrieval for comparative questions, where participants retrieve argumentative passages from a generic web crawl, and (c) image retrieval for arguments, where participants retrieve images from a focused web crawl that show support or opposition to some stance. In this paper, we briefly summarize the results of two years of organizing Touché and describe the planned setup for the third edition at CLEF 2022.

1 Introduction

Decision making and opinion formation are routine human tasks that often involve weighing pro and con arguments. Since the Web is full of argumentative texts on almost any topic, in principle, everybody has the chance to acquire knowledge to come to informed decisions or opinions by simply using a search engine. However, large amounts of the easily accessible arguments may be of low quality. For example, they may be irrelevant, contain incoherent logic, provide insufficient support, or use foul language. Such arguments should rather remain "invisible" in search results which implies several retrieval challenges—regardless of whether a query is about socially important topics or "only" about personal decisions. The challenges include assessing an argument's relevance to a query, deciding what is an argument's main "gist" in terms of the take-away, and estimating how well an implied stance is justified but also range to finding images that help to illustrate some stance. Still, today's popular web search engines do

© The Author(s), under exclusive license to Springer Nature Switzerland AG 2022
M. Hagen et al. (Eds.): ECIR 2022, LNCS 13186, pp. 339–346, 2022.
https://doi.org/10.1007/978-3-030-99739-7_43

not really address these challenges and lack a sophisticated support for searchers in argument retrieval scenarios—a gap we aim to close with the Touché lab.[1]

In the spirit of the two successful Touché labs on argument retrieval at CLEF 2020 and 2021 [6,7], we propose a third lab edition to again bring together researchers from the fields of information retrieval and natural language processing who work on argumentation. At Touché 2022, we organize the following three shared tasks, the last of which being fully new to this edition:

1. Argumentative sentence retrieval from a focused collection (crawled from debate portals) to support argumentative conversations on controversial topics.
2. Argument retrieval from a large collection of text passages to support answering comparative questions in the scenario of personal decision making.
3. Image retrieval to corroborate and strengthen textual arguments and to provide a quick overview of public opinions on controversial topics.

As part of the previous Touché labs, we evaluated about 130 submissions from 44 teams; the majority submitted their software using the tira.io platform. Many of the submissions improved over the "official" argumentation-agnostic DirichletLM- and BM25-based baselines. In total, we manually assessed more than 11,000 argumentative texts and web documents for 200 search topics. All topics and judgments are publicly available at https://touche.webis.de.

While the first two Touché editions focused on retrieving complete arguments and documents, the third edition focuses on more refined problems. Three shared tasks explore whether argument retrieval can support decision making and opinion formation more directly by extracting the argumentative gist from documents, by classifying their stance as pro or con towards the issue in question, and by retrieving images that show support or opposition to some stance.

2   Task Definition

In the Touché lab, we follow the classic TREC-style[2] methodology: documents and topics are provided to the participants who then submit their ranked results (up to five runs) for every topic to be judged by human assessors. The third lab edition includes the three complementary tasks already sketched above and further detailed in the following: (1) argument retrieval for controversial questions, (2) argument retrieval for comparative questions, and (3) image retrieval for arguments. The unit of retrieval of our previous tasks were always entire documents, whereas now we focus on the retrieval of relevant argumentative sentences, passages, and images as well as their stance detection.

---

[1] 'touché' is commonly "used to acknowledge a hit in fencing or the success or appropriateness of an argument" [https://merriam-webster.com/dictionary/touche].
[2] https://trec.nist.gov/tracks.html

2.1   Task Description

Task 1: Argument Retrieval for Controversial Questions. Given a controversial topic and a collection of arguments, the task is to retrieve sentence pairs that represent one argument's gist (e.g., a claim in one sentence and a premise in the other), and to rank these pairs according to their relevance to the topic. The argument collection for Task 1 is the args.me corpus [1]. A pre-processed version of the args.me corpus with each argument split into its constituent sentences is provided and can be indexed easily by the participants.

The pairs retrieved by the participants will be evaluated by human assessors with respect to topical relevance and argument quality. As for quality, there are three key properties: (1) each sentence in the pair must be argumentative (e.g., a claim, a premise, or a conclusion), (2) the sentence pair must form a coherent text (e.g., sentences in a pair must not contradict each other), and (3) the sentence pair constitutes a short summary of a single argument (i.e., the major claim of an argument and the best premise supporting this claim are good candidates).

The participants may use a number of previously compiled resources to lower the entry barrier of this task. These include the document-level relevance and quality judgments from the previous Touché editions, and a sample of sentence pairs from the snippet generation framework of Alshomary et al. [3], enabling a basic understanding of the task and the evaluation during development. For the identification of claims and premises, the participants can use any existing argument tagging tool, such as the API[3] of TARGER [9] hosted on our own servers, or develop an own method if necessary.

Task 2: Argument Retrieval for Comparative Questions. Given a comparison search topic with two comparison objects and a collection of text passages, the task is to retrieve relevant argumentative passages for one or both objects, and to detect the passages' stances with respect to the two objects. The collection for Task 2 is a focused collection of 868,655 passages extracted from the ClueWeb12 for the 50 search topics of the task (cf. Sect. 2.2). Near-duplicates are already removed with CopyCat [12] to mitigate negative impacts [13, 14].

The relevance of the top-k ranked passages of a system ($k \geq 5$ determined based on assessor load) will be assessed by human annotators ('not relevant', 'relevant', or 'highly relevant') along with the rhetorical quality [22] ('no arguments or low quality', 'average quality', or 'high quality'). Stance detection effectiveness will be evaluated in terms of the accuracy of distinguishing 'pro first compared object', 'pro second compared object', 'neutral', and 'no stance'.

The participants may use a number of previously compiled resources to lower the entry barrier of this task. These include the document-level relevance and argument quality judgments from the previous Touché editions as well as, for passage-level relevance judgments, a subset of MS MARCO [19] with comparative questions identified by our ALBERT-based [17] classifier (about 40,000 questions are comparative) [5]. Each comparative question in MS MARCO contains

---

[3] Also available as a Python library: https://pypi.org/project/targer-api/

10 text passages with relevance labels. For stance detection, a dataset comprising 950 comparative questions and answers extracted from Stack Exchange is provided [5]. For the identification of arguments in texts (e.g., claims and premises), the participants can use any existing argument tagging tool, such as the TARGER API hosted on our own servers, or develop their own tools.

Task 3: Image Retrieval for Arguments (New Task). Given a controversial topic and a collection of web documents with images, the task is to retrieve images that show support for each stance (pro/con the topic). The collection for Task 3 is a focused crawl of 10,000 images with the documents that contain them; for the retrieval, also the textual content of the web documents can be used.

A system's results should provide a searcher with a visual overview of public opinions on a controversial topic; we envision systems that juxtapose images for each stance. The approaches will be evaluated in terms of precision, namely by the ratio of relevant images among 20 retrieved images, 10 per stance.

Participants may use our available image-level relevance judgments [16]; The format is aligned with the format of the task's collection. Similar to the other Touché tasks, participants are free to use any additional existing tools and datasets or develop their own. Moreover, our goal is to collect a software suite for extracting various features—both for the images and web documents. Participants are encouraged to contribute Docker containers to this suite.

2.2 Search Topics

For the tasks on controversial questions (Task 1) and image retrieval (Task 3), we provide 50 search topics that represent a variety of debated societal matters. Each of these topics has a *title* in terms of a question on a controversial issue, a *description* specifying the particular search scenario, and a *narrative* that serves as a guideline for the human assessors:

```
<title> Should teachers get tenure? </title>
<description> A user has heard that some countries do give teachers
tenure and others don't. Interested in the reasoning for or against
tenure, the user searches for positive and negative arguments. [...]
</description>
<narrative> Highly relevant statements clearly focus on tenure for
teachers in schools or universities. Relevant statements consider tenure
more generally, not specifically for teachers, or [...] </narrative>
```

For the task on comparative questions (Task 2), we provide 50 search topics that describe scenarios of personal decision making. Each of these topics has a *title* in terms of a comparative question, *comparison objects* for the stance detection of the retrieved passages, a *description* specifying the particular search scenario, and a *narrative* that serves as a guideline for the assessors:

```
<title> Should I major in philosophy or psychology? </title>
<objects> major in philosophy, psychology </objects>
<description> A soon-to-be high-school graduate finds themselves at a
```
crossroad in their live. Based on their interests, majoring in philosophy
or in psychology are the potential options and the graduate is searching
for information about the differences and [...] </description>

    `<narrative>` Relevant passages will overview one of the two majors in
terms of career prospects or developed new skills, or they will provide
reasons [...] `</narrative>`

3 Touché at CLEF 2021: Results and Findings

At Touché 2021, we received 36 registrations (compared to 28 registrations in
the first year); aligned with the lab's fencing-related title, the participants were
asked to select a real or fictional fencer or swordsman character (e.g., Zorro) as
their team name upon registration. We received result submissions from 27 of
the 36 registered teams (after 17 active teams in the first year) that resulted in
88 valid runs (after 41 in 2020; participants were allowed to submit up to 5 result
rankings in both years). Touché aims to foster the reproducibility of submissions
by asking participants to submit their approaches via the TIRA platform [20],
which allows easy software submission and automatic evaluation.

Task 1: Argument Retrieval for Controversial Questions. In the first two Touché
editions, Task 1 was stated as follows: given a question on a controversial topic,
retrieve relevant and high-quality arguments from a focused crawl of online
debate portals—the args.me corpus [1]. The submissions in 2021 [7] partly continued the trend of Touché 2020 [6] by deploying "traditional" retrieval models,
however, with an increased focus on machine learning models (especially for
query expansion and for argument quality assessment). Overall, there were two
main trends in the participants' retrieval pipelines: (1) reproducing and fine-
tuning approaches from the previous year by increasing their robustness, and
(2) developing new, mostly neural approaches for argument retrieval by fine-
tuning pre-trained models for the domain-specific search task at hand.

    Like in the first year, combining "traditional" retrieval models with various
query expansion methods and domain-specific re-ranking features remained a
frequent choice for Task 1. Not really surprising—given its top effectiveness as
the 2020 baseline—, DirichletLM was employed most often as the initial retrieval
model, followed by BM25. For query expansion (e.g., with synonyms), most
participating teams continued to use WordNet [11], however, Transformer-based
approaches received increased attention [2]. Moreover, many approaches tried to
use some form of argument quality estimation in the (re-)ranking.

    The approaches in 2021 benefited from the relevance judgments collected at
Touché in 2020. Many teams used them for general parameter optimization but
also to evaluate intermediate results of their approaches, to select preprocessing
methods, and to fine-tune or select the best configurations.

Task 2: Argument Retrieval for Comparative Questions. In the first two Touché editions, Task 2 was stated as follows: given a comparative question, retrieve documents from the ClueWeb12 that help to answer the comparative question. The participants' approaches submitted in 2021 all used the ChatNoir search engine [4] for an initial document retrieval, either by submitting the original topic titles as queries, or by applying query preprocessing (e.g., lemmatization and POS-tagging) and query expansion techniques (e.g., synonyms from WordNet [11], or generation based on word2vec [18] or sense2vec embeddings [21]). Most teams then applied a document "preprocessing" (e.g., removing HTML markup) before re-ranking the ChatNoir results with feature-based or neural classifiers trained on the Touché 2020 judgments (e.g., using argumentativeness, credibility, or comparativeness scores as features). The teams predicted document relevance labels by using a random forest classifier, XGBoost [8], LightGBM [15], or a fine-tuned BERT [10].

Overall, in both tasks, many more approaches submitted in 2021 could improve upon the argumentation-agnostic baselines (DirichletLM for Task 1 and BM25 for Task 2) than in the first year, indicating that progress was achieved.

4   Conclusion

At Touché, we continue our activities to establish a collaborative platform for researchers in the area of argument retrieval, and organize respective shared tasks for the third time. By providing submission and evaluation tools as well as by organizing collaborative events such as workshops, Touché aims to foster the accumulation of knowledge and development of new approaches in the field. All evaluation resources developed at Touché are shared freely, including search queries (topics), the assembled manual relevance and argument quality judgments (qrels), and the ranked result lists submitted by the participants (runs).

Acknowledgments. This work was partially supported by the Deutsche Forschungsgemeinschaft (DFG) through the projects "ACQuA" and "ACQuA 2.0" (Answering Comparative Questions with Arguments; grants HA 5851/2-1, HA 5851/2-2, BI 1544/7-1, BI 1544/7-2) and "OASiS: Objective Argument Summarization in Search" (grant WA 4591/3-1), all part of the priority program "RATIO: Robust Argumentation Machines" (SPP 1999), and the German Ministry for Science and Education (BMBF) through the project "Shared Tasks as an Innovative Approach to Implement AI and Big Data-based Applications within Universities (SharKI)" (grant FKZ 16DHB4021). We are also grateful to Jan Heinrich Reimer for developing the TARGER Python library.

References

1. Ajjour, Y., Wachsmuth, H., Kiesel, J., Potthast, M., Hagen, M., Stein, B.: Data acquisition for argument search: the args.me corpus. In: Benzmüller, C., Stuckenschmidt, H. (eds.) KI 2019. LNCS (LNAI), vol. 11793, pp. 48–59. Springer, Cham (2019). https://doi.org/10.1007/978-3-030-30179-8_4

2. Akiki, C., Potthast, M.: Exploring argument retrieval with transformers. In: Working Notes Papers of the CLEF 2020 Evaluation Labs, vol. 2696 (2020), ISSN 1613–0073. http://ceur-ws.org/Vol-2696/

3. Alshomary, M., Düsterhus, N., Wachsmuth, H.: Extractive snippet generation for arguments. In: Proceedings of the 43nd International ACM Conference on Research and Development in Information Retrieval, SIGIR 2020, pp. 1969–1972, ACM (2020). https://doi.org/10.1145/3397271.3401186

4. Bevendorff, J., Stein, B., Hagen, M., Potthast, M.: Elastic ChatNoir: search engine for the ClueWeb and the common crawl. In: Pasi, G., Piwowarski, B., Azzopardi, L., Hanbury, A. (eds.) ECIR 2018. LNCS, vol. 10772, pp. 820–824. Springer, Cham (2018). https://doi.org/10.1007/978-3-319-76941-7_83

5. Bondarenko, A., Ajjour, Y., Dittmar, V., Homann, N., Braslavski, P., Hagen, M.: Towards understanding and answering comparative questions. In: Proceedings of the 15th ACM International Conference on Web Search and Data Mining, WSDM 2022. ACM (2022). https://doi.org/10.1145/3488560.3498534

6. Bondarenko, A., et al.: Overview of touché 2020: argument retrieval. In: Working Notes Papers of the CLEF 2020 Evaluation Labs, CEUR Workshop Proceedings, vol. 2696 (2020). https://doi.org/10.1007/978-3-030-58219-7_26

7. Bondarenko, A., et al.: Overview of touché 2021: argument retrieval. In: Candan, K.S., Ionescu, B., Goeuriot, L., Larsen, B., Müller, H., Joly, A., Maistro, M., Piroi, F., Faggioli, G., Ferro, N. (eds.) CLEF 2021. LNCS, vol. 12880, pp. 450–467. Springer, Cham (2021). https://doi.org/10.1007/978-3-030-85251-1_28

8. Chen, T., Guestrin, C.: XGBoost: a scalable tree boosting system. In: Proceedings of the 22nd ACM SIGKDD International Conference on Knowledge Discovery and Data Mining, KDD 2016, pp. 785–794, ACM (2016). https://doi.org/10.1145/2939672.2939785

9. Chernodub, A., et al.: TARGER: neural argument mining at your fingertips. In: Proceedings of the 57th Annual Meeting of the Association for Computational Linguistics, ACL 2019, pp. 195–200. ACL (2019). https://doi.org/10.18653/v1/p19-3031

10. Devlin, J., Chang, M., Lee, K., Toutanova, K.: BERT: pre-training of deep bidirectional transformers for language understanding. In: Proceedings of the 2019 Conference of the North American Chapter of the Association for Computational Linguistics: Human Language Technologies, NAACL-HLT 2019, pp. 4171–4186. ACL (2019). https://doi.org/10.18653/v1/n19-1423

11. Fellbaum, C.: WordNet: An Electronic Lexical Database. Bradford Books (1998)

12. Fröbe, M., Bevendorff, J., Gienapp, L., Völske, M., Stein, B., Potthast, M., Hagen, M.: CopyCat: near-duplicates within and between the ClueWeb and the common crawl. In: Proceedings of the 44th International ACM Conference on Research and Development in Information Retrieval, SIGIR 2021, pp. 2398–2404. ACM (2021). https://dl.acm.org/doi/10.1145/3404835.3463246

13. Fröbe, M., Bevendorff, J., Reimer, J., Potthast, M., Hagen, M.: Sampling Bias due to near-duplicates in learning to rank. In: Proceedings of the 43rd International ACM Conference on Research and Development in Information Retrieval, SIGIR 2020. ACM (2020). https://dl.acm.org/doi/10.1145/3397271.3401212

14. Fröbe, M., Bittner, J.P., Potthast, M., Hagen, M.: The effect of content-equivalent near-duplicates on the evaluation of search engines. In: Jose, J.M., Yilmaz, E., Magalhães, J., Castells, P., Ferro, N., Silva, M.J., Martins, F. (eds.) ECIR 2020. LNCS, vol. 12036, pp. 12–19. Springer, Cham (2020). https://doi.org/10.1007/978-3-030-45442-5_2

15. Ke, G., et al.: LightGBM: a highly efficient gradient boosting decision tree. In: Proceedings of the Annual Conference on Neural Information Processing Systems, NeurIPS 2017, pp. 3146–3154 (2017). https://proceedings.neurips.cc/paper/2017/hash/6449f44a102fde848669bdd9eb6b76fa-Abstract.html

16. Kiesel, J., Reichenbach, N., Stein, B., Potthast, M.: Image retrieval for arguments using stance-aware query expansion. In: Proceedings of the 8th Workshop on Argument Mining, ArgMining 2021 at EMNLP, pp. 36–45. ACL (2021)

17. Lan, Z., Chen, M., Goodman, S., Gimpel, K., Sharma, P., Soricut, R.: ALBERT: a lite BERT for self-supervised learning of language representations. In: Proceedings of the 8th International Conference on Learning Representations, ICLR 2020, OpenReview.net (2020). https://openreview.net/forum?id=H1eA7AEtvS

18. Mikolov, T., Chen, K., Corrado, G., Dean, J.: Efficient estimation of word representations in vector space. In: Proceedings of the 1st International Conference on Learning Representations, ICLR 2013 (2013). http://arxiv.org/abs/1301.3781

19. Nguyen, T., Rosenberg, M., Song, X., Gao, J., Tiwary, S., Majumder, R., Deng, L.: MS MARCO: a human generated machine reading comprehension dataset. In: Proceedings of the Workshop on Cognitive Computation: Integrating Neural and Symbolic Approaches 2016 at NIPS, CEUR Workshop Proceedings, vol. 1773, CEUR-WS.org (2016). http://ceur-ws.org/Vol-1773/CoCoNIPS_2016_paper9.pdf

20. Potthast, M., Gollub, T., Wiegmann, M., Stein, B.: TIRA integrated research architecture. In: Information Retrieval Evaluation in a Changing World. TIRS, vol. 41, pp. 123–160. Springer, Cham (2019). https://doi.org/10.1007/978-3-030-22948-1_5

21. Trask, A., Michalak, P., Liu, J.: sense2vec - a fast and accurate method for word sense disambiguation in neural word embeddings. CoRR abs/1511.06388 (2015). http://arxiv.org/abs/1511.06388

22. Wachsmuth, H., et al.: Computational argumentation quality assessment in natural language. In: Proceedings of the 15th Conference of the European Chapter of the Association for Computational Linguistics, EACL 2017, pp. 176–187 (2017). http://aclweb.org/anthology/E17-1017

Introducing the HIPE 2022 Shared Task: Named Entity Recognition and Linking in Multilingual Historical Documents

Maud Ehrmann[1]([✉]) [iD], Matteo Romanello[2] [iD], Antoine Doucet[3] [iD], and Simon Clematide[4] [iD]

[1] Digital Humanities Laboratory, EPFL, Vaud, Switzerland
maud.ehrmann@epfl.ch
[2] University of Lausanne, Lausanne, Switzerland
matteo.romanello@unil.ch
[3] University of La Rochelle, La Rochelle, France
antoine.doucet@univ-lr.fr
[4] Department of Computational Linguistics, University of Zurich, Zurich, Switzerland
simon.clematide@uzh.ch

Abstract. We present the HIPE-2022 shared task on named entity processing in multilingual historical documents. Following the success of the first CLEF-HIPE-2020 evaluation lab, this edition confronts systems with the challenges of dealing with more languages, learning domain-specific entities, and adapting to diverse annotation tag sets. HIPE-2022 is part of the ongoing efforts of the natural language processing and digital humanities communities to adapt and develop appropriate technologies to efficiently retrieve and explore information from historical texts. On such material, however, named entity processing techniques face the challenges of domain heterogeneity, input noisiness, dynamics of language, and lack of resources. In this context, the main objective of the evaluation lab is to gain new insights into the *transferability* of named entity processing approaches across languages, time periods, document types, and annotation tag sets.

Keywords: Named entity processing · Information extraction · Text understanding · Historical documents · Digital humanities

1 Introduction

Through decades of massive digitisation, an unprecedented amount of historical documents became available in digital format, along with their machine-readable texts. While this represents a major step forward in terms of preservation and accessibility, it also bears the potential for new ways to engage with historical documents' contents. The application of machine reading to historical documents is potentially transformative and the next fundamental challenge is to adapt and

M. Hagen et al. (Eds.): ECIR 2022, LNCS 13186, pp. 347–354, 2022.
https://doi.org/10.1007/978-3-030-99739-7_44

develop appropriate technologies to efficiently search, retrieve and explore information from this 'big data of the past' [9]. Semantic indexing of historical documents is in great demand among humanities scholars, and the interdisciplinary efforts of the digital humanities (DH), natural language processing (NLP), computer vision and cultural heritage communities are progressively pushing forward the processing of facsimiles, as well as the extraction, linking and representation of the complex information enclosed in transcriptions of digitised collections [14]. In this regard, information extraction techniques, and particularly named entity (NE) processing, can be considered among the first and most crucial processing steps.

Yet, the recognition, classification and disambiguation of NEs in historical texts are not straightforward, and performances are not on par with what is usually observed on contemporary well-edited English news material [3]. In particular, NE processing on historical documents faces the challenges of domain heterogeneity, input noisiness, dynamics of language, and lack of resources [6]. Although some of these issues have already been tackled in isolation in other contexts (with e.g., user-generated text), what makes the task particularly difficult is their simultaneous combination and their magnitude: texts are severely noisy, and domains and time periods are far apart.

In this regard, the first CLEF-HIPE-2020 edition[1] [5] proposed the tasks of NE recognition and classification (NER) and entity linking (EL) in ca. 200 years of historical newspapers written in English, French and German and successfully showed that the progress in neural NLP – specifically driven by Transformer-based approaches – also translates into improved performances on historical material, especially for NER. In the meantime, several European cultural heritage projects have prepared additional annotated text material, thereby opening a unique window of opportunity for organising a second edition of the HIPE evaluation lab in 2022.

2   Motivation and Objectives

As the first evaluation campaign of its kind on multilingual historical newspaper material, HIPE-2020 brought together 13 enthusiastic teams who submitted a total of 75 runs for 5 different task bundles. The main conclusion of this edition was that neural-based approaches can achieve good performances on historical NERC when provided with enough training data, but that progress is still needed to further improve performances, adequately handle OCR noise and small-data settings, and better address entity linking. HIPE-2022 will attempt to drive further progress on these points, and also confront systems with new challenges.

HIPE-2022[2] will focus on named entity processing in historical documents covering the period from the 18th to the 20th century and featuring several languages. Compared to the first edition, HIPE-2022 introduces several novelties, with:

---

[1] https://impresso.github.io/CLEF-HIPE-2020.
[2] https://hipe-eval.github.io/HIPE-2022/.

- the addition of a new type of document alongside historical newspapers, namely classical commentaries[3];
- the consideration of a broader language spectrum, with 5 languages for historical newspapers (3 for the previous edition), and 3 for classical commentaries;
- the confrontation with the issue of the heterogeneity of annotation tag sets and guidelines.

Overall, HIPE-2022 will confront participants with the challenges of dealing with more languages, learning domain-specific entities, and adapting to diverse annotation schemas. The evaluation lab will therefore contribute to gain new insights on how best to ensure the transferability of NE processing approaches across languages, time periods, document and annotation types, and to answer the question whether one architecture/model can be optimised to perform well across settings and annotation targets in a cultural heritage context. In particular, the following research questions will be addressed:

1. How well can general prior knowledge transfer to historical texts?
2. Are in-domain language representations (i.e. language models learned on the historical document collections) beneficial, and under which conditions?
3. How can systems adapt and integrate training material with different annotations?
4. How can systems, with limited additional in-domain training material, (re)-target models to produce a certain type of annotation?

Recent work on NERC showed encouraging progress on several of these topics: Beryozkin et al. [1] proposed a method to deal with related, but heterogeneous tag sets. Several researchers successfully applied meta-learning strategies to NERC in order to improve transfer learning: Li et al. [10] improved results for extreme low-resource few-shot settings where only a handful of annotated examples for each entity class are used for training; Wu et al. [17] presented techniques to improve cross-lingual transfer; and Li et al. [11] tackled the problem of domain shifts and heterogeneous label sets using meta-learning, proposing a highly data-efficient domain adaptation approach.

2.1 Significance of the Evaluation Lab

HIPE-2022 will benefit the NLP and DH communities, as well as cultural heritage professionals.

Benefits for the NLP community - NLP and information extraction practitioners will have the possibility to test the robustness of existing approaches and to experiment with transfer learning and domain adaptation methods, whose performances could be systematically evaluated and compared on broad historical and multilingual data sets. Beside gaining new insights with respect to

---

[3] Classical commentaries are scholarly publications dedicated to the in-depth analysis and explanation of ancient literary works. As such, they aim to facilitate the reading and understanding of a given literary text. More information on the HIPE-2022 classical commentaries corpus in Sect. 3.2.

domain and language adaptation and advancing the state of the art in semantic indexing of historical material, the lab will also contribute a set of multilingual NE-annotated datasets that could be used for further training and benchmarking.

Benefits for the DH community - DH researchers are in need of support to explore the large quantities of text they currently have at hand, and NE processing is high on their wish list. Such processing can support research questions in various domains (e.g. history, political science, literature and historical linguistics) and knowing about performances is a must in order to do an informed usage of the enriched data. This lab's outcome (datasets and systems) will be beneficial to DH practitioners insofar as it will help identify state-of-the-art solutions for NE processing of historical texts.

Benefits for cultural heritage professionals - Libraries, archives and museums (LAM) increasingly focus on advancing the usage of artificial intelligence methods on cultural heritage text collections, in particular NE processing [7,13]. This community is eager to collaborate and provide data (when copyright allows) for high-quality semantic enrichment.

3 Overview of the Evaluation Lab

3.1 Task Description

HIPE-2022 focuses on the same tasks as CLEF-HIPE-2020, namely:

Task 1: Named Entity Recognition and Classification (NERC)

Subtask 1.1 - NERC-Coarse: this task includes the recognition and classification of high-level entity types (Person, Organisation, Location, Product and domain-specific entities, e.g. mythological characters or literary works in classical commentaries).

Subtask 1.2 - NERC-Fine: includes 'NERC-Coarse', plus the detection and classification at sub-type level and the detection of NE components (e.g. function, title, name). This subtask will be proposed for English, French and German only.

Task 2: Named Entity Linking (EL). This task corresponds to the linking of named entity mentions to a unique item ID in Wikidata, our knowledge base of choice, or to a NIL node if the mention does not have a corresponding item in the KB. We will allow submissions of both end-to-end systems (NERC and EL) and of systems performing exclusively EL on gold entity mentions provided by the organizers (EL-only).

3.2 Data Sets

Corpora. The lab's corpora will be composed of historical newspapers and classic commentaries covering ca. 200 years. We benefit from published and to-date unpublished NE-annotated data from organisers' previous research project, from the previous HIPE-2020 campaign, as well as from several ongoing research projects which agreed to postpone the publication of 10% to 20% of their annotated material in order to support HIPE-2022.

Historical Newspapers. The historical newspaper data is composed of several datasets in English, Finnish, French, German and Swedish which originate from various projects and national libraries in Europe:

- *Le Temps* data: an unpublished, annotated diachronic dataset composed of historical newspaper articles from two Swiss newspapers in French (19C-20C) [3]. This dataset contains 10,580 entity mentions and will be part of the training, dev and test sets.
- *HIPE-2020* data: the datasets used during the first HIPE-2020 campaign, composed of newspaper articles from Swiss, Luxembourgish and American newspapers in French, German and English (19C-20C). These datasets contain 19,848 linked entities and will be part of the training sets.
- *HIPE-2020* unpublished data: a set of unpublished diachronic annotated data composed of newspaper articles from Swiss newspapers in French and German (19C-20C). These data will be part of the test sets.
- *NewsEye* data⁴: a partially published annotated dataset composed of newspaper articles from newspapers in French, German, Finnish and Swedish (19C-20C) [8]. The already published part contains 30,580 entities and will be part of the training and dev sets. The unpublished one (roughly 20% of the total) will be part of the test set.
- *SoNAR* data: an annotated dataset composed of newspaper articles from the Berlin State library newspaper collections in German (19C-20C), produced in the context of the SoNAR project⁵. The (soon to be) published part of this dataset will be part of the training and dev sets, while the unpublished lot will integrate the HIPE-2022 test set.
- *Living With Machines* data⁶: an annotated dataset composed of newspaper articles from the British Library newspapers in English (18C-19C), and annotated exclusively with geographical locations following ad-hoc annotation guidelines. The already published portion of the data [2] contains 3,355 annotated toponyms and will be included in the training and dev sets. The unpublished portion will be part of the test set.

Historical Commentaries. The classical commentaries data originates from the *Ajax Multi-Commentary* project and is composed of OCRed 19C commentaries published in French, German and English [15], annotated with both universal NEs (person, location, organisation) and domain-specific NEs (bibliographic references to primary and secondary literature). In the field of classical studies, commentaries constitute one of the most important and enduring forms of scholarship, together with critical editions and translations. They are information-rich texts, characterised by a high density of NEs.

Annotation. In terms of annotation, the common requirement for most of these datasets is to have person, location and organisation entity types, and

---

⁴ https://www.newseye.eu/.
⁵ https://sonar.fh-potsdam.de/.
⁶ https://livingwithmachines.ac.uk/.

entity links towards Wikidata. The guidelines used to annotate many datasets derive from related directives (Quaero and *impresso*-HIPE-2020 guidelines)[7], yet they do differ in some respects such as granularity of annotations (coarse vs fine-grained), treatment of metonymy and inclusion of entity components.

3.3 Evaluation

To accommodate the different dimensions that characterise our datasets (languages, document types, domains, entity tag sets) and foster research on transferability, the evaluation lab will be organised around two main 'challenges', namely a *multilingual challenge* and an *adaptation challenge*, each featuring several task 'tracks'. They will ensure that participants will have to work across settings, e.g. with documents in at least two different languages or annotated according to two different tag sets or guidelines, while keeping a clear and defined evaluation frame.

Evaluation will be performed with the open source HIPE scorer[8], which was developed for the first edition of the shared task. Evaluation metrics implemented in the scorer include (macro and micro) Precision, Recall, and F-measure and evaluation settings will include strict (exact matching) and relaxed (fuzzy matching) evaluation scenarios.

4   Conclusion

Following a first and successful shared task on NE processing on historical newspapers, the HIPE-2022 evaluation lab proposes to confront systems with the new challenges of dealing with more languages, learning domain-specific entities, and adapting to diverse annotation tag sets. The overall objective is to assess and advance the development of robust, adaptable and transferable named entity processing systems in order to support information extraction and text understanding of cultural heritage data.

Acknowledgements. We are grateful to the research project consortia and teams who kindly accepted to retain the publication of part of their NE-annotated datasets to support HIPE-2022: the NewsEye project (The NewsEye project has received funding from the European Union's Horizon 2020 research and innovation programme under grant agreement No 770299); the Living with Machine project, in particular Mariona Coll'Ardanuy; and the SoNAR project, in particular Clemens Neudecker. We also thank Sally Chambers, Clemens Neudecker and Frédéric Kaplan for their support and guidance as part of the lab's advisory board.

---

[7] *Impresso* [4] and SoNAR guidelines [12] were derived from Quaero guidelines [16], while NewsEye guidelines correspond to a subset of the *impresso* guidelines.

[8] https://github.com/impresso/CLEF-HIPE-2020-scorer.

References

1. Beryozkin, G., Drori, Y., Gilon, O., Hartman, T., Szpektor, I.: A joint named-entity recognizer for heterogeneous tag-sets using a tag hierarchy. In: Proceedings of the 57th Annual Meeting of the Association for Computational Linguistics, pp. 140–150, Florence, Italy, July 2019. https://aclanthology.org/P19-1014

2. Coll Ardanuy, M., Beavan, D., Beelen, K., Hosseini, K., Lawrence, J.: Dataset for Toponym Resolution in Nineteenth-Century English Newspapers (2021). https://doi.org/10.23636/b1c4-py78

3. Ehrmann, M., Colavizza, G., Rochat, Y., Kaplan, F.: Diachronic evaluation of NER systems on old newspapers. In: Proceedings of the 13th Conference on Natural Language Processing (KONVENS 2016), pp. 97–107, Bochum (2016). Bochumer Linguistische Arbeitsberichte. https://infoscience.epfl.ch/record/221391

4. Ehrmann, M., Romanello, M., Flückiger, A., Clematide, S.: Impresso Named Entity Annotation Guidelines. Annotation guidelines, Ecole Polytechnique Fédérale de Lausanne (EPFL) and Zurich University (UZH), January 2020. https://zenodo.org/record/3585750

5. Ehrmann, M., Romanello, M., Flückiger, A., Clematide, S.: Extended Overview of CLEF HIPE 2020: named entity processing on historical newspapers. In: Cappellato, L., Eickhoff, C., Ferro, N., Névéol, A., (eds.), Working Notes of CLEF 2020 - Conference and Labs of the Evaluation Forum, vol. 2696, p. 38, Thessaloniki, Greece (2020). CEUR-WS. https://doi.org/10.5281/zenodo.4117566, https://infoscience.epfl.ch/record/281054

6. Ehrmann, M., Hamdi, A., Pontes, E.L., Romanello, M., Doucet, A.: Named Entity Recognition and Classification on Historical Documents: A Survey. arXiv:2109.11406 [cs], September 2021

7. Markus, G., Neudecker, C., Isaac, A., Bergel, G., et al.: AI in relation to GLAMs task FOrce - Report and Recommendations. Technical report, Europeana Network ASsociation (2021). https://pro.europeana.eu/project/ai-in-relation-to-glams

8. Hamdi, A., et al.: A multilingual dataset for named entity recognition, entity linking and stance detection in historical newspapers. In: Proceedings of the 44th International ACM SIGIR Conference on Research and Development in Information Retrieval, SIGIR 2021, pp. 2328–2334, New York, NY, USA, July 2021. Association for Computing Machinery. ISBN 978-1-4503-8037-9. https://doi.org/10.1145/3404835.3463255

9. Kaplan, F., di Lenardo, I.: Big data of the past. Front. Digit. Hum. 4:1–21 (2017). ISSN 2297–2668. https://doi.org/10.3389/fdigh.2017.00012. Publisher: Frontiers

10. Li, J., Chiu, B., Feng, S., Wang, H.: Few-shot named entity recognition via meta-learning. IEEE Trans. Knowl. Data Eng. 1 (2020)

11. Li, J., Shang, S., Shao, L.: Metaner: Named entity recognition with meta-learning. In: Proceedings of The Web Conference 2020, WWW 2020, pp. 429–440, New York, NY, USA (2020). Association for Computing Machinery. ISBN 9781450370233. https://doi.org/10.1145/3366423.3380127

12. Menzel, S., Zinck, J., Schnaitter, H., Petras, V.: Guidelines for Full Text Annotations in the SoNAR (IDH) Corpus. Technical report, Zenodo, July 2021. https://zenodo.org/record/5115933

13. Padilla, T.: Responsible Operations: Data Science, Machine Learning, and AI in Libraries. Technical report, OCLC Research, USA, May 2020. https://www.oclc.org/content/research/publications/2019/oclcresearch-responsible-operations-data-science-machine-learning-ai.html

14. Ridge, M., Colavizza, G., Brake, L., Ehrmann, M., Moreux, J.P., Prescott, A.: The past, present and future of digital scholarship with newspaper collections. In: DH 2019 Book of Abstracts, pp. 1–9, Utrecht, The Netherlands (2019). http://infoscience.epfl.ch/record/271329

15. Matteo, R., Sven, N.-M., Bruce, R.: Optical character recognition of 19th century classical commentaries: the current state of affairs. In: The 6th International Workshop on Historical Document Imaging and Processing (HIP 2021), Lausanne, September 2021. Association for Computing Machinery. https://doi.org/10.1145/3476887.3476911

16. Rosset, S., Grouin, C., Zweigenbaum, P.: Entités nommées structurées : Guide d'annotation Quaero. Technical Report 2011–04, LIMSI-CNRS, Orsay, France (2011)

17. Wu, Q.: Enhanced meta-learning for cross-lingual named entity recognition with minimal resources. CoRR, abs/1911.06161 (2019). http://arxiv.org/abs/1911.06161

CLEF Workshop JOKER: Automatic Wordplay and Humour Translation

Liana Ermakova[1]([envelope]) [ID], Tristan Miller[2] [ID], Orlane Puchalski[1],
Fabio Regattin[3] [ID], Élise Mathurin[1], Sílvia Araújo[4] [ID] , Anne-Gwenn Bosser[5],
Claudine Borg[6] [ID], Monika Bokiniec[7], Gaelle Le Corre[8], Benoît Jeanjean[1],
Radia Hannachi[9], Ġorġ Mallia[6], Gordan Matas[10], and Mohamed Saki[1]

[1] Université de Bretagne Occidentale, HCTI EA-4249/MSHB, 29200 Brest, France
`liana.ermakova@univ-brest.fr`
[2] Austrian Research Institute for Artificial Intelligence, Vienna, Austria
[3] Dipartimento DILL, Università degli Studi di Udine, 33100 Udine, Italy
[4] Universidade do Minho, CEHUM, Rua da Universidade, 4710-057 Braga, Portugal
[5] École Nationale d'Ingénieurs de Brest,Lab-STICC CNRS UMR 6285, Plouzané,
France
[6] University of Malta, Msida MSD 2020, Malta
[7] University of Gdansk, Gdańsk, Poland
[8] Université de Bretagne Occidentale, CRBC, 29200 Brest, France
[9] Université de Bretagne Sud, HCTI EA-4249, 56321 Lorient, France
[10] University of Split, Split, Croatia

Abstract. Humour remains one of the most difficult aspects of intercultural communication: understanding humour often requires understanding implicit cultural references and/or double meanings, and this raises the question of its (un)translatability. Wordplay is a common source of humour in due to its attention-getting and subversive character. The translation of humour and wordplay is therefore in high demand. Modern translation depends heavily on technological aids, yet few works have treated the automation of humour and wordplay translation, or the creation of humour corpora. The goal of the JOKER workshop is to bring together translators and computer scientists to work on an evaluation framework for wordplay, including data and metric development, and to foster work on automatic methods for wordplay translation. We propose three pilot tasks: (1) classify and explain instances of wordplay, (2) translate single words containing wordplay, and (3) translate entire phrases containing wordplay.

Keywords: Machine translation · Humour · Wordplay · Puns · Parallel corpora · Evaluation metrics · Creative language analysis

1 Introduction

Intercultural communication relies heavily on translation. Humour remains by far one of its most difficult aspects; to understand humour, one often has to

M. Hagen et al. (Eds.): ECIR 2022, LNCS 13186, pp. 355–363, 2022.
https://doi.org/10.1007/978-3-030-99739-7_45

grasp implicit cultural references and/or capture double meanings, which of course raises the question of the (un)translatability of humour. One of the most common sources of humour is wordplay, which involves the creative application or bending of rules governing word formation, choice, or usage. Wordplay is used by novelists, poets, playwrights, scriptwriters, and copywriters, and is often employed in titles, headlines, proper nouns, and slogans for its ability to grab attention and for its mnemonic, playful, or subversive character. The translation of wordplay is therefore in high demand. But while modern translation is heavily aided by technological tools, virtually none has any specific support for humour and wordplay, and there has been very little research on the automation of humour and wordplay translation. Furthermore, most AI-based translation tools require a quality and quantity of training data (e.g., parallel corpora) that has historically been lacking of humour and wordplay.

Preserving wordplay can be crucial for maintaining the pragmatic force of discourse. Consider the following pun from *Alice's Adventures in Wonderland* by Lewis Carroll, which exploits the homophony of *lessons* and *lessens* for humorous effect: " 'That's the reason they're called lessons,' the Gryphon remarked: 'because they lessen from day to day.' " Henri Parisot's French translation manages to preserve both the sound and meaning correspondence by using the pair *cours/courts*: *"C'est pour cette raison qu'on les appelle des cours: parce qu'ils deviennent chaque jour un peu plus courts."* By contrast, Google Translate uses the pair *leçons/diminuent* and the sentence becomes nonsensical: *" 'C'est la raison pour laquelle on les appelle leçons, remarqua le Griffon: parce qu'elles diminuent de jour en jour.' "*

The goal of the JOKER workshop is to bring together translators, linguists and computer scientists to work on an evaluation framework for creative language. All types of contributions will be welcomed: this includes research, survey, position, discussion, and demo papers, as well as extended abstracts of published papers. We will also oversee pilot tasks making use of a new, multilingual parallel corpus of wordplay and humour that we have produced: **Pilot Task 1** is to classify single words containing wordplay according to a given typology, and provide lexical-semantic interpretations; **Pilot Task 2** is to translate single words containing wordplay; and **Pilot Task 3** is to translate entire phrases that subsume or contain wordplay. The two translation tasks will initially target English and French but may be expanded to further languages as data becomes available. As discussed in Sect. 3 below, the consideration of appropriate evaluation metrics for these tasks is one of the goals of the workshop.

To encourage participants to use our data in creative ways and to collect ideas for future editions of the workshop, we also propose an **Unshared Task**. We particularly welcome ideas from researchers in the humanities on how we can promote deeper linguistic and social-scientific analysis of our data.

2   Background

Automatic Humour Analysis and Related Campaigns. To date, there have been only a handful of studies on the machine translation (MT) of wordplay. Farwell and Helmreich [7] proposed a pragmatics-based approach to MT that accounts for the author's locutionary, illocutionary, and perlocutionary intents (that is, the "how", "what", and "why" of the text), and discuss how it might be applied to puns. However, no working system appears to have been implemented. Miller [19] proposed an interactive method for the computer-assisted translation of puns, an implementation and evaluation of which was described by Kolb and Miller [16]. Their study was limited to a single language pair (English to German) and translation strategy (namely, the PUN→PUN strategy described below).

Numerous studies have been conducted for the related tasks of humour generation and detection. Pun generation systems have often been based on template approaches. Valitutti and al. [26] used lexical constraints to generate adult humour by substituting one word in a pre-existing text. Hong and Ong [11] trained a system to automatically extract humorous templates which were then used for pun generation. Some current efforts to tackle this difficult problem more generally using neural approaches have been hindered by the lack of a sizable pun corpus [29]. Meanwhile, the recent rise of conversational agents and the need to process large volumes of social media content point to the necessity of automatic humour recognition [21]. Humour and irony studies are now crucial when it comes to social listening [9,13,14,25], dialogue systems (chatbots), recommender systems, reputation monitoring, and the detection of fake news [10] and hate speech [8].

There do exist a few monolingual humour corpora exist: for example, the datasets created for shared tasks of the International Workshop on Semantic Evaluation (SemEval): #HashtagWars: Learning a Sense of Humor [23], Detection and Interpretation of English Puns [20], Assessing Humor in Edited News Headlines [12], and HaHackathon: Detecting and Rating Humor and Offense [17]. Mihalcea and Strapparava [18] collected 16,000 humorous sentences and an equal number of negative samples from news titles, proverbs, the British National Corpus, and the Open Mind Common Sense dataset, while another dataset contains 2,400 puns and non-puns from news sources, Yahoo! Answers, and proverbs [3,28]. Most datasets are in English, with some notable exceptions for Italian [24], Russian [1,6], and Spanish [2].

Strategies for Wordplay Translation. Humorous wordplay often exploits the confrontation of similar forms but different meanings, evoking incongruity between expected and presented stimuli, and this makes it particularly important in NLP to study the strategies that human translators use for dealing with it [4,27]. On the one hand, this is because MT is generally ignorant of pragmatics and assumes that words in the source text are formed and used in a conventional manner. MT systems fail to recognize the deliberate ambiguity of puns or the unorthodox morphology of neologisms, leaving such terms untranslated or else translating them in ways that lose the humorous aspect [19]. Apart from these implementation issues, human translation strategies could also inform the evaluation of machine-translated wordplay, since existing metrics based on lexical overlap [15,22] are not applicable.

Perhaps the most commonly cited typology of wordplay translation strategies is that of Dirk Delabastita [4,5]. This typology was developed on the basis of parallel corpus analysis and therefore reflects the techniques used by working translators. And while the typology was developed specifically for puns (a type of wordplay that exploits multiple meanings of a term or of similar-sounding words), many of the strategies are applicable to other forms not based on ambiguity. Delabastita's basic options are the following:

PUN→PUN: The source text pun is translated by a target language pun.

PUN→NON-PUN: The pun is translated by a non-punning phrase, which may reproduce all senses of the wordplay or just one of them.

PUN→RELATED RHETORICAL DEVICE: The pun is replaced by some other rhetorically charged phrase (involving repetition, alliteration, rhyme, etc.)

PUN→ZERO: The part of text containing the pun is omitted altogether.

PUN ST=PUN TT: The pun is reproduced verbatim, without attempting a target-language rendering.

NON-PUN→PUN: A pun is introduced in the target text where no wordplay was present in the source text.

ZERO→PUN: New textual material involving wordplay is added in the target text, which bears no correspondence whatsoever in the source text.

EDITORIAL TECHNIQUES: Use of some paratextual strategy for explaining the pun of the source text (footnote, preface, etc.).

Delabastita insists on one further point: the techniques are by no means exclusive. A translator could, for instance, suppress a pun somewhere in their target text (PUN→NON-PUN), explain it in a footnote (EDITORIAL TECHNIQUES), then try to compensate for the loss by adding another pun somewhere else in the text (NON-PUN→PUN or ZERO→PUN). The very typology of translation strategies drawn in [5] directly points to the main reason for the difficulty of conceiving a working MT of puns. Translating wordplay does not involve recourse to what we may commonly think of as translation, but to (almost) autonomous creative writing activities starting from a situation determined by the source text. Therefore, a more realistic goal for NLP might probably be machine detection, followed by human or computer-assisted translation of wordplay.

3   Task Setup

Data. Wordplay includes a wide variety of phenomena that exploit or subvert the phonological, orthographical, morphological, and semantic conventions of a language. We have collected over two thousand translated examples of wordplay, in English and French, from video games, literature, and other sources. Each example has been manually classified according to a well-defined, multi-label inventory of wordplay types and structures, and annotated according to its lexical-semantic or morphosemantic components.

The type inventory covers phenomena such as puns; alliteration, assonance, and consonance (repetition of sounds across nearby words); portmanteaux (combining parts of multiple words into a new word); malapropisms (the erroneous use of a word in place of a similar-sounding one); spoonerisms (exchanging the initial sounds of nearby words); anagrams (a word or phrase formed by rearranging the letters of another); and onomatopoeia (a word coined to approximate some non-speech or non-language sound). The structure inventory can be used to further specify the entities involved in certain types of ambiguity-based wordplay; its labels include homophony (words with the same pronunciation but different spelling), homography (words with the same spelling but different pronunciation), homonymy (words with the same spelling and pronunciation), paronymy (words with different spelling and pronunciation), lexical structure (the figurative and literal readings of an idiom), morphological structure (different morphological analyses of the same word), and syntactic structure (when a word takes on different meanings according to how the wider phrasal context is parsed).

The examples in Table 1 give some idea of our annotated data, about half of which consists of proper nouns and neologisms.

Table 1. Examples of annotated instances of wordplay.

Instance	Type	Structure	Interpretation
Why is music so painful? Because it HERTZ	pun	paronymy	hurts/hertz
Weasleys' Wildfire Whiz-bangs	alliteration, assonance, and consonance	—	alliteration in 'w'
She was my secretariat	malapropism	—	secretariat/secretary

Evaluation Metrics. Pilot Task 1 includes both classification and interpretation components. Classification performance will be evaluated with respect to accuracy, while interpretation performance will be measured by exact-match comparison to the gold-standard annotations. Accuracy is preferable over precision, recall, and F_1 as the latter are designed for binary classification. For the same reason, they are not appropriate to evaluate translation quality.

For the wordplay translation tasks (Pilot Tasks 2 and 3), there do not yet exist any accepted metrics of translation quality. MT is traditionally measured with the BLEU (Bilingual Evaluation Understudy) metric, which calculates vocabulary overlap between the candidate translation and a reference translation [22]. However, this metric is clearly inappropriate for use with wordplay, where a wide variety of translation strategies (and solutions implementing those strategies) are permissible. And as our *Wonderland* example from Sect. 1 demonstrates, many of these strategies require metalexical awareness and preservation of features such as lexical ambiguity and phonetic similarity. (Consider how substituting the synonymous *leçons* for *cours* in Parisot's translation would lose the wordplay, and indeed render the translation nonsensical, yet still result in a near-perfect BLEU score with the original translation.) Furthermore, overlap measures operate only on larger text spans and not on individual words, the morphological analysis of which can be crucial for neologisms.

Evaluation of human translation quality is similarly problematic, with past studies on wordplay translation (e.g., [16]) favouring qualitative rather than quantitative analyses, or else employing only subjective metrics such as "acceptability" or "successfulness". Part of the goal of the JOKER workshop is to work towards the development of evaluation metrics for the automated translation of wordplay. To this end, human evaluators will manually annotate the submitted translations according to both subjective measures and according to more concrete features such as whether wordplay exists in the target text, whether it corresponds to the type used in the source text, whether the target text preserves the semantic field, etc. At the end of the workshop, we will look for correspondences between the concrete and subjective measures, and consider how the concrete measures that best correlate with subjective translation quality might be automated.

4   Conclusion

The JOKER project addresses the issue of European identity through the study of humour in a cross-cultural perspective. Its main objective is to study the strategies of localization of humour and wordplay and to create a multilingual parallel corpus annotated according to these strategies, as well as to rethink evaluation metrics. To this end, we are organizing the CLEF 2022 JOKER track, consisting of a workshop and associated pilot tasks on automatic wordplay analysis and translation. A further goal of the workshop is to unify the scientific community interested in automatic localization of humour and wordplay and to facilitate future work in this area. Our multilingual corpus will be made freely available to the research community (to the extent permitted by third-party copyrights), and this data and evaluation framework will be a step forward to MT models adapted for creative language.

Further details on the pilot tasks and on how to participate in the track can be found in the call for papers and guidelines on the JOKER website.[1] Please join this effort and contribute by working on our challenges!

Acknowledgements. This work has been funded in part by the National Research Agency under the program *Investissements d'avenir* (Reference ANR-19-GURE-0001) and by the Austrian Science Fund under project M 2625-N31. JOKER is supported by *La Maison des sciences de l'homme en Bretagne*. We thank Adrien Couaillet and Ludivine Grégoire for data collection and the Master's students in translation at Université de Bretagne Occidentale for maintaining the JOKER website.

---

[1] http://www.joker-project.com/.

References

1. Blinov, V., Bolotova-Baranova, V., Braslavski, P.: Large dataset and language model fun-tuning for humor recognition. In: Proceedings of the 57th Annual Meeting of the Association for Computational Linguistics, pp. 4027–4032. Association for Computational Linguistics, Florence, Italy (2019). https://doi.org/10.18653/v1/P19-1394, https://www.aclweb.org/anthology/P19-1394
2. Castro, S., Chiruzzo, L., Rosá, A., Garat, D., Moncecchi, G.: A Crowd-annotated Spanish corpus for humor analysis. In: Proceedings of the Sixth International Workshop on Natural Language Processing for Social Media, pp. 7–11. Association for Computational Linguistics, Melbourne, Australia, July 2018. https://doi.org/10.18653/v1/W18-3502, https://www.aclweb.org/anthology/W18-3502
3. Cattle, A., Ma, X.: Recognizing humour using word associations and humour anchor extraction. In: Proceedings of the 27th International Conference on Computational Linguistics, pp. 1849–1858. Association for Computational Linguistics, Santa Fe, New Mexico, USA, August 2018. https://www.aclweb.org/anthology/C18-1157
4. Delabastita, D.: There's a Double Tongue: An Investigation Into the Translation of Shakespeare's Wordplay, with Special Reference to Hamlet. Rodopi, Amsterdam (1993)
5. Delabastita, D.: Wordplay as a translation problem: a linguistic perspective. In: 1. Teilband: Ein internationales Handbuch zur Übersetzungsforschung, pp. 600–606. De Gruyter Mouton, July 2008. https://doi.org/10.1515/9783110137088.1.6.600, https://www.degruyter.com/document/doi/10.1515/9783110137088.1.6.600/html
6. Ermilov, A., Murashkina, N., Goryacheva, V., Braslavski, P.: Stierlitz meets SVM: humor detection in Russian. In: Ustalov, D., Filchenkov, A., Pivovarova, L., Žižka, J. (eds.) Artificial Intelligence and Natural Language, pp. 178–184. Communications in Computer and Information Science, Springer International Publishing, Cham (2018). https://doi.org/10.1007/978-3-030-01204-5_17
7. Farwell, D., Helmreich, S.: Pragmatics-based MT and the translation of puns. In: Proceedings of the 11th Annual Conference of the European Association for Machine Translation. pp. 187–194, June 2006. http://www.mt-archive.info/EAMT-2006-Farwell.pdf
8. Francesconi, C., Bosco, C., Poletto, F., Sanguinetti, M.: Error analysis in a hate speech detection task: the case of HaSpeeDe-TW at EVALITA 2018. In: Bernardi, R., Navigli, R., Semeraro*, G. (eds.) Proceedings of the 6th Italian Conference on Computational Linguistics (CLiC-it 2019) (2018). http://ceur-ws.org/Vol-2481/paper32.pdf
9. Ghanem, B., Karoui, J., Benamara, F., Moriceau, V., Rosso, P.: IDAT@FIRE2019: overview of the track on irony detection in Arabic tweets. In: Mehta, P., Rosso, P., Majumder, P., Mitra, M. (eds.) Working Notes of FIRE 2019 - Forum for Information Retrieval Evaluation (2019)
10. Guibon, G., Ermakova, L., Seffih, H., Firsov, A., Noé-Bienvenu, G.L.: Multilingual fake news detection with satire. In: Proceedings of the 20th International Conference on Computational Linguistics and Intelligent Text Processing (CICLing 2019), April 2019. https://halshs.archives-ouvertes.fr/halshs-02391141
11. Hong, B.A., Ong, E.: Automatically extracting word relationships as templates for pun generation. In: Proceedings of the Workshop on Computational Approaches to Linguistic Creativity, pp. 24–31. Association for Computational Linguistics, Boulder, Colorado, June 2009. https://aclanthology.org/W09-2004

12. Hossain, N., Krumm, J., Gamon, M., Kautz, H.: SemEval-2020 task 7: assessing humor in edited news headlines. In: Proceedings of the Fourteenth Workshop on Semantic Evaluation. pp. 746–758. International Committee for Computational Linguistics, December 2020. https://aclanthology.org/2020.semeval-1.98

13. Karoui, J., Benamara, F., Moriceau, V., Patti, V., Bosco, C., Aussenac-Gilles, N.: Exploring the impact of pragmatic phenomena on irony detection in tweets: a multilingual corpus study. In: 15th European Chapter of the Association for Computational Linguistics (EACL 2017), vol. 1 - long pap, pp. 262–272. Association for Computational Linguistics (ACL), Valencia, ES (2017). https://oatao.univ-toulouse.fr/18921/

14. Karoui, J., Farah, B., Moriceau, V., Aussenac-Gilles, N., Hadrich-Belguith, L.: Towards a contextual pragmatic model to detect irony in tweets. In: Proceedings of the 53rd Annual Meeting of the Association for Computational Linguistics and the 7th International Joint Conference on Natural Language Processing (Volume 2: Short Papers), pp. 644–650. Association for Computational Linguistics, Beijing, China (2015). https://doi.org/10.3115/v1/P15-2106, http://aclweb.org/anthology/P15-2106

15. Klakow, D., Peters, J.: Testing the correlation of word error rate and perplexity. Speech Commun. **38**(1), 19–28 (2002). https://doi.org/10.1016/S0167-6393(01)00041-3, https://www.sciencedirect.com/science/article/pii/S0167639301000413

16. Kolb, W., Miller, T.: Human-computer interaction in pun translation. In: Hadley, J., Taivalkoski-Shilov, K., Teixeira, C.S.C., Toral, A. (eds.) Using Technologies for Creative-Text Translation. Routledge (2022), to appear

17. Meaney, J.A., Wilson, S., Chiruzzo, L., Lopez, A., Magdy, W.: SemEval-2021 task 7: HaHackathon, detecting and rating humor and offense. In: Proceedings of the 15th International Workshop on Semantic Evaluation, pp. 105–119. Association for Computational Linguistics (2021). https://doi.org/10.18653/v1/2021.semeval-1.9, https://aclanthology.org/2021.semeval-1.9

18. Mihalcea, R., Strapparava, C.: Making computers laugh: investigations in automatic humor recognition. In: Human Language Technology Conference and Conference on Empirical Methods in Natural Language Processing: Proceedings of the Conference, pp. 531–538. Association for Computational Linguistics, Stroudsburg, PA, October 2005. https://doi.org/10.3115/1220575.1220642, http://www.aclweb.org/anthology/H/H05/H05-1067

19. Miller, T.: The Punster's amanuensis: the proper place of humans and machines in the translation of wordplay. In: Proceedings of the Second Workshop Human-Informed Translation and Interpreting Technology associated with RANLP 2019, pp. 57–65. Incoma Ltd., Shoumen, Bulgaria, October 2019. https://doi.org/10.26615/issn.2683-0078.2019_007, https://acl-bg.org/proceedings/2019/RANLP_W1%202019/pdf/HiT-IT2019007.pdf

20. Miller, T., Hempelmann, C., Gurevych, I.: SemEval-2017 Task 7: detection and interpretation of English puns. In: proceedings of the 11th International Workshop on Semantic Evaluation (semeval-2017), pp. 58–68. Association for Computational Linguistics, Vancouver, Canada (2017). https://doi.org/10.18653/v1/S17-2005, http://aclweb.org/anthology/S17-2005

21. Nijholt, A., Niculescu, A., Valitutti, A., Banchs, R.E.: Humor in human-computer interaction: a short survey. In: INTERACT 2017 (2017)

22. Papineni, K., Roukos, S., Ward, T., Zhu, W.J.: BLEU: a method for automatic evaluation of machine translation. In: Proceedings of the 40th Annual Meeting on Association for Computational Linguistics, pp. 311–318. Association for Computational Linguistics (2002)

23. Potash, P., Romanov, A., Rumshisky, A.: SemEval-2017 Task 6: #HashtagWars: learning a sense of humor. In: Proceedings of the 11th International Workshop on Semantic Evaluation. pp. 49–57, August 2017. https://doi.org/10.18653/v1/S17-2004, http://www.aclweb.org/anthology/S17-2004

24. Reyes, A., Buscaldi, D., Rosso, P.: An analysis of the impact of ambiguity on automatic humour recognition. In: Matoušek, V., Mautner, P. (eds.) Text, Speech and Dialogue, pp. 162–169. Lecture Notes in Computer Science, Springer, Berlin, Heidelberg (2009). https://doi.org/10.1007/978-3-642-04208-9_25

25. Reyes, A., Rosso, P., Buscaldi, D.: From humor recognition to irony detection: The figurative language of social media. Data & Knowledge Engineering **74**, 1–12 (2012). https://doi.org/10.1016/j.datak.2012.02.005, https://doi.org/10.1016/j.datak.2012.02.005

26. Valitutti, A., Toivonen, H., Doucet, A., Toivanen, J.M.: Let everything turn well in your wife: generation of adult humor using lexical constraints. In: Proceedings of the 51st Annual Meeting of the Association for Computational Linguistics (Volume 2: Short Papers), pp. 243–248. Association for Computational Linguistics, Sofia, Bulgaria, August 2013. https://aclanthology.org/P13-2044

27. Vrticka, P., Black, J.M., Reiss, A.L.: The neural basis of humour processing. Nat. Rev. Neurosci. **14**(12), 860–868 (2013). https://doi.org/10.1038/nrn3566, https://www.nature.com/articles/nrn3566

28. Yang, D., Lavie, A., Dyer, C., Hovy, E.: Humor recognition and humor anchor extraction. In: Proceedings of the 2015 Conference on Empirical Methods in Natural Language Processing, pp. 2367–2376. Association for Computational Linguistics, Lisbon, Portugal, September 2015. https://doi.org/10.18653/v1/D15-1284, https://www.aclweb.org/anthology/D15-1284

29. Yu, Z., Tan, J., Wan, X.: A neural approach to pun generation. In: Proceedings of the 56th Annual Meeting of the Association for Computational Linguistics (Volume 1: Long Papers), pp. 1650–1660. Association for Computational Linguistics, Melbourne, Australia, July 2018. https://doi.org/10.18653/v1/P18-1153, https://aclanthology.org/P18-1153

Automatic Simplification of Scientific Texts: SimpleText Lab at CLEF-2022

Liana Ermakova[1]([⊠]) [iD], Patrice Bellot[2], Jaap Kamps[3], Diana Nurbakova[4],
Irina Ovchinnikova[5], Eric SanJuan[6], Elise Mathurin[1], Sílvia Araújo[7],
Radia Hannachi[8], Stéphane Huet[6], and Nicolas Poinsu[1]

[1] Université de Bretagne Occidentale, HCTI, EA 4249 Bretagne, France
`liana.ermakova@univ-brest.fr`
[2] Aix Marseille Univ, Université de Toulon, CNRS, LIS, Marseille, France
[3] University of Amsterdam, Amsterdam, The Netherlands
[4] Institut National des Sciences Appliquées de Lyon, LIRIS UMR 5205 CNRS, Lyon, France
[5] Sechenov University, Moscow, Russia
[6] Avignon Université, LIA, Avignon, France
[7] University of Minho, Braga, Portugal
[8] Université de Bretagne Sud, HCTI, EA 4249 Bretagne, France

Abstract. The Web and social media have become the main source of information for citizens, with the risk that users rely on shallow information in sources prioritizing commercial or political incentives rather than the correctness and informational value. Non-experts tend to avoid scientific literature due to its complex language or their lack of prior background knowledge. Text simplification promises to remove some of these barriers. The CLEF 2022 SimpleText track addresses the challenges of text simplification approaches in the context of promoting scientific information access, by providing appropriate data and benchmarks, and creating a community of NLP and IR researchers working together to resolve one of the greatest challenges of today. The track will use a corpus of scientific literature abstracts and popular science requests. It features three tasks. First, *content selection* (what is in, or out?) challenges systems to select passages to include in a simplified summary in response to a query. Second, *complexity spotting* (what is unclear?) given a passage and a query, aims to rank terms/concepts that are required to be explained for understanding this passage (definitions, context, applications). Third, *text simplification* (rewrite this!) given a query, asks to simplify passages from scientific abstracts while preserving the main content.

Keywords: Scientific text simplification · (Multi-document) summarization · Contextualization · Background knowledge

1 Introduction

Being science literate is an important ability for people. It is one of the keys for critical thinking, objective decision-making and judgment of the validity and significance of findings and arguments, which allows discerning facts from fiction. Thus, having a basic scientific knowledge may also help maintain one's health, both physiological and mental. The COVID-19 pandemic provides a good example of such a matter. Understanding the issue itself, being aware of and applying social distancing rules and sanitary policies, choosing to use or avoid particular treatment or prevention procedures can become crucial. In the context of a pandemic, the qualified and timely information should reach everyone and be accessible. That is what motivates projects such as EasyCovid.[1]

However, scientific texts are often hard to understand as they require solid background knowledge and use tricky terminology. Although there were some recent efforts on text simplification (e.g. [25]), removing such understanding barriers between scientific texts and general public in an automatic manner is still an open challenge. **SimpleText Lab** brings together researchers and practitioners working on the generation of simplified summaries of scientific texts. It is a new evaluation lab that follows up the SimpleText-2021 Workshop [9]. All perspectives on automatic science popularisation are welcome, including but not limited to: Natural Language Processing (NLP), Information Retrieval (IR), Linguistics, Scientific Journalism, etc.

SimpleText provides data and benchmarks for discussion of challenges of automatic text simplification by bringing in the following tasks:

- **TASK 1: What is in (or out)?** Select passages to include in a simplified summary, given a query.
- **TASK 2: What is unclear?** Given a passage and a query, rank terms/concepts that are required to be explained for understanding this passage (definitions, context, applications,..).
- **TASK 3: Rewrite this!** Given a query, simplify passages from scientific abstracts.
- **UNSHARED TASK** We welcome any submission that uses our data!

2   Background

2.1   Content Selection

We observe an accelerating growth of scientific publication and their major impact on the society, especially in medicine (e.g. the COVID-19 pandemic) and in computer science with unprecedented use of machine learning algorithms and their societal issues (biases, explainability, etc.). Numerous initiatives try to make science understandable for everyone. Efforts have been made by scientific journalism (Nature, The Guardian, ScienceX) researchers (Papier-Maché project[2]), and internet forums (*Explain Like I'm 5*[3]). The ScienceBites[4] platform publishes short simple posts about individual research papers, making state-of-the-art science accessible to a wide audience. While structured abstracts are an emerging trend since they tend to be informative [10, 12], non-experts are usually interested in other types of information. Popular science articles are generally much shorter than scientific publications. Thus, information selection is a crucial

---

[1] https://easycovid19.org/.

[2] https://papiermachesciences.org/.

[3] https://www.reddit.com/r/explainlikeimfive.

[4] https://sciencebites.org/.

but understudied task in document simplification especially with regard to the target audience [39]. In many cases the information in a summary designed for an expert in scientific domain is drastically different from that from a popularized version. Moreover, different levels of simplification, details, and explanation can be applied, e.g. for a given scientific article the Papier-Maché platform publishes two level of simplification: curiosity and advanced. Zhong et al. analyzed discourse factors related to sentence deletion on the Newsela corpus made of manually simplified sentences from news articles [39]. They found that professional editors utilize different strategies to meet the readability standards of elementary and middle schools.

The state-of-the-art in automatic summarization is achieved by deep learning models, in particular by pretrained Bidirectional Encoder Representations from Transformers (BERT) which can be used for both extractive and abstractive models [24]. It is important to study the limits of existing AI models, like GPT-2 [30] for English and CamemBERT for French [27], and how it is possible to overcome those limits. Recently, AI21 released the Jurassic-1 suite of language models, with 178B parameters for J1-Jumbo [23]. Jurassic-1 is a large AI model able to transform an existing text, e.g. in case of summarization. Multilingual T5 (mT5) is a large multilingual pretrained text-to-text transformer model developed by Google, covering 101 languages [36]. mT5 can be fine-tuned for any text-to-text generation, e.g. by using the SimpleT5 library.

2.2 Complexity Spotting

Our analysis of the queries from different sources revealed the gap between the actual interest of the wide readership and the expectations of the journalists [28]. People are interested in biology or modern technologies as long as there is a connection with their everyday life. Thus, a simplified scientific text needs to contain references to the daily experience of people. On the one hand, the *subjective complexity of terminology* is involved when readers face concepts that go beyond their area of expertise and general knowledge, and need additional definitions or clarifications. On the other hand, the *objective complexity of terminology* is a systematic feature caused by complexity of research areas, research traditions and socio-cultural diversity. The complexity of a scientific area depends on peculiar attributes and conditions [34]. Ladyman et al. [22] suggest five such conditions: numerosity of elements, numerosity of interactions, disorder, openness, feedback. The complexity of terminology is also associated with a formal representation (*signifier*) of a term. Apart from borrowings, scientific text is rich in symbols and abbreviations (acronyms, backronyms, syllabic abbreviations, etc.) that are meant to optimize content transferring, standardize the naming of numerous elements, allow frequent interaction among them, and facilitate data processing. But readers of popularized publications expect explanations of the symbols and abbreviations.

One of the technical challenges here is thus term recognition. Robertson provided theoretical justifications of the term-weighting function IDF (inverse document frequency) in the traditional probabilistic model of information retrieval [31]. IDF shows term specificity and can be used for difficult term extraction as it is connected to the Zipf's law. WordNet [11] distance to the basic terms can be used as a measure of the term difficulty. Task-independent AI models, like GPT-2 [30], Jurassic-1 [23], Multilingual T5 [36], can be fine-tuned for the terminology extraction. It should be noted that there are tools available online such as OneClick Terms or TermoStat Web that allow us to extract intuitively mono and multiwords.

2.3 Text Simplification

Existing works mainly focus on word/phrase-level (simplification of difficult words and constructions) [4, 15, 26, 29, 32, 38] or sentence-level simplifications [5, 7, 33, 35, 40, 41]. Koptient and Grabar analyzed the text transformation topology during simplification [21]. Among the most frequent transformations, they found synonymy, specification (insertion of information), generalization (deletion of information), pronominalization, substitution of adjectives by their corresponding nouns, and substitutions between singular and plural. In their further work, they proposed a rule-based system in French that combines lexical and syntactic simplification, for example, by transforming passive sentences into active sentences [19], and rating a lexicon [20]. Approaches based on rated lexicons are neither scalable nor robust to neologisms, which are frequent in scientific texts. Recent deep learning models with a large number of training parameters, like GPT-2 [30], Jurassic-1 [23], Multilingual T5 [36], can be applied for text simplification. Jurassic-1 Jumbo is the largest model publicly available with no waitlist. The AI21 studio's playground provides ready-to-use prompts for text simplification (see De-Jargonizer). However, as Jiang et al. showed, a text simplification system depends on the quality and quantity of training data [18]. Therefore, a major step in training artificial intelligence (AI) text simplification models is the creation of high quality data.

Researchers have proposed various approaches based on expert judgment [6], readability levels [13, 14], crowdsourcing [2, 6, 35], eye-tracking [16, 37], manual annotation [17]. Traditional evaluation like comparison to the reference data by standard evaluation measures is difficult to apply as one should consider the end user (young readers, foreigners, non-experts, people with different literacy levels, people with cognitive disabilities etc.) as well as source document content.

3 Data Set

In 2022, SimpleText's data is two-fold: *Medicine* and *Computer Science*, as these two domains are the most popular on forums like ELI5 [28]. For both domains, we provide datasets according to our shared tasks:

- content selection relevant for non-experts;
- terminology complexity spotting in a given passage;
- simplified passages.

As in 2021, we use the Citation Network Dataset: DBLP+Citation, ACM Citation network (12th version) [1] as source of scientific documents to be simplified [8]. It contains: 4.894.083 bibliographic references published before 2020, 4.232.520 abstracts in English, 3.058.315 authors with their affiliations, and 45.565.790 ACM citations. Scientific textual content about any topic related to computer science can be extracted from this corpus together with authorship. Although we manually preselected abstracts for topics, participants also have access to use an ElasticSearch index . This Index is adequate to passage retrieval using BM25. Additional datasets have been extracted to generate Latent Dirichlet Allocation models for query expansion or train Graph Neural Networks for citation recommendation as carried out in StellarGraph[5] for example. The shared datasets are: document abstract content for LDA or Word Embedding; document author relation for coauthoring analysis; document citation relation for co citation analysis; author citation relation for author impact factor analysis. These extra datasets are intended to be used to select passages by authors who are experts on the topic (highly cited by the community).

We propose 13 topics on *computer science* based on the recent *n* press titles from *The Guardian* enriched with keywords manually extracted from the content of the article (see Table 1). It has been checked that each keyword allows participants to extract at least 5 relevant abstracts. The use of these keywords is optional.

Table 1. Query examples

Query 12: Patient data from GP surgeries sold to US companies
Topic 12.1: patient data
Query 13: Baffled by digital marketing? Find your way out of the maze
Topic 13.1: digital marketing
Topic 13.2: advertising

We selected passages that are adequate to be inserted as plain citations in the original journalistic article. The comparison of the journalistic articles with the scientific ones, as well as the analysis we carried out to choose topics, demonstrated that for non-experts the most important information is the application of an object (which problem can be solved? how to use this information/object? what are the examples?).

Text passages issued from abstracts on computer science were simplified by either a master student in Technical Writing and Translation or a pair of experts: (1) a computer scientist and (2) a professional translator , English native speaker but not specialist in computer science [8]. Each passage was discussed and rewritten multiple times until it became clear for non-computer scientists. Sentences were shortened, excluding every detail that was irrelevant or unnecessary to the comprehension of the study, and rephrased, using simpler vocabulary. If necessary, concepts were explained.

In 2022, we introduce new data based on Google Scholar and PubMed articles on muscle hypertrophy and health annotated by a master student in Technical Writing and Translation, specializing in these domains. The selected abstracts included the objectives of the study, the results and sometimes the methodology. The abstracts including only the topic of the study were excluded because of lack of information. To avoid the curse of knowledge, another master student in Technical Writing and Translation not familiar with the domain was solicited for complexity spotting.

---

[5] https://stellargraph.readthedocs.io/.

4   Tasks

In 2022, SimpleText was transformed into a CLEF lab. We propose three shared tasks to help better understand the challenges as well as discuss these challenges and the way to evaluate solutions. Contributions should not exclusively rely on these shared tasks. We also welcome manual runs and runs within the unshared task.

Details on the tasks, guideline and call for contributions can be found at the SimpleText website. In this paper we just briefly introduce the planned shared tasks.

TASK 1: What is in (or out)? Select passages to include in a simplified summary, given a query. Based on an article from a major international newspaper general audience, this shared task aims to retrieve, from a large scientific bibliographic database with abstracts, all relevant passages to illustrate this article. Extracted passages should be adequate to be inserted as plain citations in the original paper. Sentence pooling and automatic metrics can be used to evaluate these results. The relevance of the source document can be evaluated as well as potential unresolved anaphora issues.

TASK 2: What is unclear? Given a passage and a query, rank terms/concepts that are required to be explained for understanding this passage (definitions, context, applications,..). The goal of this shared task is to decide which terms (up to 10) require explanation and contextualization to help a reader understand a complex scientific text—for example, with regard to a query, terms that need to be contextualized (with a definition, example and/or use-case). Terms should be ranked from 1 to 10 according to their complexity. *1* corresponds to the most difficult term, while lower ranks show that the term might be explained if there is space. Term pooling and automatic metrics (e.g. accuracy, NDCG, MSE, etc.) can be used to evaluate these results.

TASK 3: Rewrite this! Given a query, simplify passages from scientific abstracts. The goal of this shared task is to provide a simplified version of text passages. Participants are provided with queries and abstracts of scientific papers. The abstracts can be split into sentences as in the example. The simplified passages will be evaluated manually with use of aggregating metrics.

UNSHARED TASK. We welcome any submission that uses our data! This task is aimed at (but not limited to) Humanities, Social Science and Technical Communication. We encourage here manual and statistical analysis of content selection, readability and comprehensibility of simplified texts, terminology complexity analysis.

5   Conclusion and Future Work

The paper introduced the CLEF 2022 SimpleText track, containing three shared tasks and one unshared task on scientific text simplification. The created collection of simplified texts makes it possible to apply overlap metrics like ROUGE to text simplification. However, we will work on a new evaluation metric that can take into account unresolved anaphora [3] and information types. For the pilot task 2, participants will be asked to provide context for difficult terms. This context should provide a definition and take into account ordinary readers' needs to associate their particular problems with the opportunities that science provides them to solve the problems [28]. Full details about the lab can be found at the SimpleText website: http://simpletext-project.com. Help us to make scientific results understandable!

Acknowledgments. We thank Alain Kerhervé, University Translation Office, master students in Translation from the Université de Bretagne Occidentale, and the MaDICS research group.

References

1. AMiner. https://www.aminer.org/citation
2. Alva-Manchego, F., Martin, L., Bordes, A., Scarton, C., Sagot, B., Specia, L.: Asset: a dataset for tuning and evaluation of sentence simplification models with multiple rewriting transformations. arXiv preprint arXiv:2005.00481 (2020)
3. Bellot, P., Moriceau, V., Mothe, J., SanJuan, E., Tannier, X.: INEX tweet contextualization task: evaluation, results and lesson learned. Inf. Process. Manage. **52**(5), 801–819 (2016). https://doi.org/10.1016/j.ipm.2016.03.002
4. Biran, O., Brody, S., Elhadad, N.: Putting it simply: a context-aware approach to lexical simplification. In: Proceedings of the 49th Annual Meeting of the Association for Computational Linguistics: Human Language Technologies, pp. 496–501. Association for Computational Linguistics, Portland, Oregon, USA, June 2011. https://www.aclweb.org/anthology/P11-2087
5. Chen, P., Rochford, J., Kennedy, D.N., Djamasbi, S., Fay, P., Scott, W.: Automatic text simplification for people with intellectual disabilities. In: Artificial Intelligence Science and Technology, pp. 725–731. WORLD SCIENTIFIC, November 2016. https://doi.org/10.1142/9789813206823_0091, https://www.worldscientific.com/doi/abs/10.1142/9789813206823_0091
6. Orphée, D.: Using the crowd for readability prediction. Nat. Lang. Eng. **20**(3), 293–325 (2014), http://dx.doi.org/10.1017/S1351324912000344
7. Dong, Y., Li, Z., Rezagholizadeh, M., Cheung, J.C.K.: EditNTS: an neural programmer-interpreter model for sentence simplification through explicit editing. In: Proceedings of the 57th Annual Meeting of the Association for Computational Linguistics, pp. 3393–3402. Association for Computational Linguistics, Florence, Italy, Jul 2019. https://doi.org/10.18653/v1/P19-1331, https://www.aclweb.org/anthology/P19-1331
8. Ermakova, L., et al.: Overview of simpletext 2021 - CLEF workshop on text simplification for scientific information access. In: Candan, K.S., et al (eds.) Experimental IR Meets Multilinguality, Multimodality, and Interaction, pp. 432–449. LNCS, Springer International Publishing, Cham (2021). https://doi.org/10.1007/978-3-030-85251-1_27
9. Ermakova, L., et al.: Text simplification for scientific information access. In: Hiemstra, D., Moens, M.-F., Mothe, J., Perego, R., Potthast, M., Sebastiani, F. (eds.) ECIR 2021. LNCS, vol. 12657, pp. 583–592. Springer, Cham (2021). https://doi.org/10.1007/978-3-030-72240-1_68
10. Ermakova, L., Bordignon, F., Turenne, N., Noel, M.: Is the abstract a mere teaser? evaluating generosity of article abstracts in the environmental sciences. Front. Res. Metr. Anal. **3** (2018). https://doi.org/10.3389/frma.2018.00016, https://www.frontiersin.org/articles/10.3389/frma.2018.00016/full
11. Fellbaum, C. (ed.): WordNet: An Electronic Lexical Database. Language, Speech, and Communication, MIT Press, Cambridge, MA (1998)
12. Fontelo, P., Gavino, A., Sarmiento, R.F.: Comparing data accuracy between structured abstracts and full-text journal articles: implications in their use for informing clinical decisions. Evidence-Based Med. **18**(6), 207–11 (2013). https://doi.org/10.1136/eb-2013-101272, http://www.researchgate.net/publication/240308203_Comparing_data_accuracy_between_structured_abstracts_and_full-text_journal_articles_implications_in_their_use_for_informing_clinical_decisions

13. François, T., Fairon, C.: Les apports du tal à la lisibilité du français langue étrangère. Trait. Autom. des Langues **54**, 171–202 (2013)
14. Gala, N., François, T., Fairon, C.: Towards a french lexicon with difficulty measures: NLP helping to bridge the gap between traditional dictionaries and specialized lexicons. In: eLex-Electronic Lexicography (2013)
15. Glavaš, G., Štajner, S.: Simplifying lexical simplification: do we need simplified corpora? In: Proceedings of the 53rd Annual Meeting of the Association for Computational Linguistics and the 7th International Joint Conference on Natural Language Processing (Volume 2: Short Papers), pp. 63–68. Association for Computational Linguistics, Beijing, China, July 2015. https://doi.org/10.3115/v1/P15-2011, https://www.aclweb.org/anthology/P15-2011
16. Grabar, N., Farce, E., Sparrow, L.: Study of readability of health documents with eye-tracking approaches. In: 1st Workshop on Automatic Text Adaptation (ATA) (2018)
17. Grabar, N., Hamon, T.: A large rated lexicon with French medical words. In: LREC (Language Resources and Evaluation Conference) 2016 (2016)
18. Jiang, C., Maddela, M., Lan, W., Zhong, Y., Xu, W.: Neural CRF Model for Sentence Alignment in Text Simplification. arXiv:2005.02324 [cs] (June 2020)
19. Koptient, A., Grabar, N.: Fine-grained text simplification in French: steps towards a better grammaticality. In: ISHIMR Proceedings of the 18th International Symposium on Health Information Management Research. Kalmar, Sweden, September 2020. https://doi.org/10.15626/ishimr.2020.xxx, https://hal.archives-ouvertes.fr/hal-03095247
20. Koptient, A., Grabar, N.: Rated lexicon for the simplification of medical texts. In: The Fifth International Conference on Informatics and Assistive Technologies for Health-Care, Medical Support and Wellbeing HEALTHINFO 2020. Porto, Portugal, October 2020. https://hal.archives-ouvertes.fr/hal-03095275
21. Koptient, A., Grabar, N.: Typologie de transformations dans la simplification de textes. In: Congrès mondial de la linguistique française. Montpellier, France, July 2020. https://hal.archives-ouvertes.fr/hal-03095235
22. Ladyman, J., Lambert, J., Wiesner, K.: What is a complex system? European J. Philos. Sci. **3**(1), 33–67 (2013). https://doi.org/10.1007/s13194-012-0056-8
23. Lieber, O., Sharir, O., Lentz, B., Shoham, Y.: Jurassic-1: Technical Details and Evaluation, p. 9 (2021)
24. Liu, Y., Lapata, M.: Text Summarization with Pretrained Encoders. arXiv:1908.08345 [cs] (2019)
25. Maddela, M., Alva-Manchego, F., Xu, W.: Controllable Text Simplification with Explicit Paraphrasing. arXiv:2010.11004 [cs], April 2021
26. Maddela, M., Xu, W.: A word-complexity lexicon and a neural readability ranking model for lexical simplification. In: Proceedings of the 2018 Conference on Empirical Methods in Natural Language Processing, pp. 3749–3760. Association for Computational Linguistics, Brussels, Belgium (2018). https://doi.org/10.18653/v1/D18-1410, https://www.aclweb.org/anthology/D18-1410
27. Martin, L., et al.: CamemBERT: a tasty French language model. In: Proceedings of the 58th Annual Meeting of the Association for Computational Linguistics, pp. 7203–7219. Association for Computational Linguistics, Online (2020). https://doi.org/10.18653/v1/2020.acl-main.645, https://www.aclweb.org/anthology/2020.acl-main.645

28. Ovchinnikova, I., Nurbakova, D., Ermakova, L.: What science-related topics need to be popularized? a comparative study. In: Faggioli, G., Ferro, N., Joly, A., Maistro, M., Piroi, F. (eds.) Proceedings of the Working Notes of CLEF 2021 - Conference and Labs of the Evaluation Forum, Bucharest, Romania, September 21st - to - 24th, 2021. CEUR Workshop Proceedings, vol. 2936, pp. 2242–2255. CEUR-WS.org (2021). http://ceur-ws.org/Vol-2936/paper-203.pdf

29. Paetzold, G., Specia, L.: Lexical simplification with neural ranking. In: Proceedings of the 15th Conference of the European Chapter of the Association for Computational Linguistics: vol. 2, Short Papers, pp. 34–40. Association for Computational Linguistics, Valencia, Spain, April 2017. https://www.aclweb.org/anthology/E17-2006

30. Radford, A., Wu, J., Child, R., Luan, D., Amodei, D., Sutskever, I.: Language Models are Unsupervised Multitask Learners, p. 24 (2019)

31. Robertson, S.: Understanding inverse document frequency: on theoretical arguments for IDF. J. Doc. **60**(5), 503–520 (2004). https://doi.org/10.1108/00220410410560582, publisher: Emerald Group Publishing Limited

32. Specia, L., Jauhar, S.K., Mihalcea, R.: SemEval-2012 task 1: English lexical simplification. In: *SEM 2012: The First Joint Conference on Lexical and Computational Semantics - Volume 1: Proceedings of the main conference and the shared task, and Volume 2: Proceedings of the Sixth International Workshop on Semantic Evaluation (SemEval 2012), pp. 347–355. Association for Computational Linguistics, Montréal, Canada (2012). https://www.aclweb.org/anthology/S12-1046

33. Wang, T., Chen, P., Rochford, J., Qiang, J.: Text simplification using neural machine translation. In: Proceedings of the AAAI Conference on Artificial Intelligence, vol. 30, no. 1, March 2016. https://ojs.aaai.org/index.php/AAAI/article/view/9933, number: 1

34. Wiesner, K., Ladyman, J.: Measuring complexity. arXiv:1909.13243 [nlin], September 2020

35. Xu, W., Napoles, C., Pavlick, E., Chen, Q., Callison-Burch, C.: Optimizing statistical machine translation for text simplification. Trans. Assoc. Comput. Linguist. **4**, 401–415. MIT Press (2016)

36. Xue, L., et al.: mT5: a massively multilingual pre-trained text-to-text transformer. In: Proceedings of the 2021 Conference of the North American Chapter of the Association for Computational Linguistics: Human Language Technologies, pp. 483–498. Association for Computational Linguistics, Online, June 2021. https://doi.org/10.18653/v1/2021.naacl-main.41, https://aclanthology.org/2021.naacl-main.41

37. Yaneva, V., Temnikova, I., Mitkov, R.: Accessible texts for autism: an eye-tracking study. In: Proceedings of the 17th International ACM SIGACCESS Conference on Computers & Accessibility, pp. 49–57 (2015)

38. Yatskar, M., Pang, B., Danescu-Niculescu-Mizil, C., Lee, L.: For the sake of simplicity: unsupervised extraction of lexical simplifications from Wikipedia. In: Human Language Technologies: The 2010 Annual Conference of the North American Chapter of the Association for Computational Linguistics, pp. 365–368. Association for Computational Linguistics, Los Angeles, California, June 2010. https://www.aclweb.org/anthology/N10-1056

39. Zhong, Y., Jiang, C., Xu, W., Li, J.J.: Discourse level factors for sentence deletion in text simplification. In: Proceedings of the AAAI Conference on Artificial Intelligence, vol. 34, no. 05, pp. 9709–9716, April 2020. https://doi.org/10.1609/aaai.v34i05.6520, https://ojs.aaai.org/index.php/AAAI/article/view/6520, number: 05

40. Zhu, Z., Bernhard, D., Gurevych, I.: A monolingual tree-based translation model for sentence simplification. In: Proceedings of the 23rd International Conference on Computational Linguistics (Coling 2010), pp. 1353–1361. Coling 2010 Organizing Committee, Beijing, China, August 2010. https://www.aclweb.org/anthology/C10-1152

41. Štajner, S., Nisioi, S.: A detailed evaluation of neural sequence-to-sequence models for in-domain and cross-domain text simplification. In: Proceedings of the Eleventh International Conference on Language Resources and Evaluation (LREC 2018). European Language Resources Association (ELRA), Miyazaki, Japan, May 2018. https://www.aclweb.org/anthology/L18-1479

LeQua@CLEF2022: Learning to Quantify

Andrea Esuli[ID], Alejandro Moreo[ID], and Fabrizio Sebastiani[(✉)][ID]

Istituto di Scienza e Tecnologie dell'Informazione Consiglio Nazionale delle Ricerche,
56124 Pisa, Italy
{andrea.esuli,alejandro.moreo,fabrizio.sebastiani}@isti.cnr.it

Abstract. LeQua 2022 is a new lab for the evaluation of methods for "learning to quantify" in textual datasets, i.e., for training predictors of the relative frequencies of the classes of interest in sets of unlabelled textual documents. While these predictions could be easily achieved by first classifying all documents via a text classifier and then counting the numbers of documents assigned to the classes, a growing body of literature has shown this approach to be suboptimal, and has proposed better methods. The goal of this lab is to provide a setting for the comparative evaluation of methods for learning to quantify, both in the binary setting and in the single-label multiclass setting. For each such setting we provide data either in ready-made vector form or in raw document form.

1 Learning to Quantify

In a number of applications involving classification, the final goal is not determining which class (or classes) individual unlabelled items belong to, but estimating the *prevalence* (or "relative frequency", or "prior probability", or "prior") of each class in the unlabelled data. Estimating class prevalence values for unlabelled data via supervised learning is known as *learning to quantify* (LQ) (or *quantification*, or *supervised prevalence estimation*) [4,10].

LQ has several applications in fields (such as the social sciences, political science, market research, epidemiology, and ecological modelling) which are inherently interested in characterising *aggregations* of individuals, rather than the individuals themselves; disciplines like the ones above are usually *not* interested in finding the needle in the haystack, but in characterising the haystack. For instance, in most applications of tweet sentiment classification we are not concerned with estimating the true class (e.g., Positive, or Negative, or Neutral) of individual tweets. Rather, we are concerned with estimating the relative frequency of these classes in the set of unlabelled tweets under study; or, put in another way, we are interested in estimating as accurately as possible the true distribution of tweets across the classes.

It is by now well known that performing quantification by classifying each unlabelled instance and then counting the instances that have been attributed to the class (the "classify and count" method) usually leads to suboptimal quantification accuracy; this may be seen as a direct consequence of "Vapnik's principle" [21], which states

© The Author(s), under exclusive license to Springer Nature Switzerland AG 2022
M. Hagen et al. (Eds.): ECIR 2022, LNCS 13186, pp. 374–381, 2022.
https://doi.org/10.1007/978-3-030-99739-7_47

If you possess a restricted amount of information for solving some problem, try to solve the problem directly and never solve a more general problem as an intermediate step. It is possible that the available information is sufficient for a direct solution but is insufficient for solving a more general intermediate problem.

In our case, the problem to be solved directly is quantification, while the more general intermediate problem is classification.

Another reason why "classify and count" is suboptimal is that many application scenarios suffer from *distribution shift*, the phenomenon according to which the distribution across the classes in the sample (i.e., set) σ of *unlabelled* documents may substantially differ from the distribution across the classes in the labelled *training* set L; distribution shift is one example of *dataset shift* [15,18], the phenomenon according to which the joint distributions $p_L(\mathbf{x}, y)$ and $p_\sigma(\mathbf{x}, y)$ differ. The presence of distribution shift means that the well-known IID assumption, on which most learning algorithms for training classifiers hinge, does not hold. In turn, this means that "classify and count" will perform suboptimally on sets of unlabelled items that exhibit distribution shift with respect to the training set, and that the higher the amount of shift, the worse we can expect "classify and count" to perform.

As a result of the suboptimality of the "classify and count" method, LQ has slowly evolved as a task in its own right, different (in goals, methods, techniques, and evaluation measures) from classification. The research community has investigated methods to correct the biased prevalence estimates of general-purpose classifiers, supervised learning methods specially tailored to quantification, evaluation measures for quantification, and protocols for carrying out this evaluation. Specific applications of LQ have also been investigated, such as sentiment quantification, quantification in networked environments, or quantification for data streams. For the near future it is easy to foresee that the interest in LQ will increase, due (a) to the increased awareness that "classify and count" is a suboptimal solution when it comes to prevalence estimation, and (b) to the fact that, with larger and larger quantities of data becoming available and requiring interpretation, in more and more scenarios we will only be able to afford to analyse these data at the aggregate level rather than individually.

2   The Rationale for LeQua 2022

The LeQua 2022 lab (https://lequa2022.github.io/) at CLEF 2022 has a "shared task" format; it is a new lab, in two important senses:

- No labs on LQ have been organized before at CLEF conferences.
- Even outside the CLEF conference series, quantification has surfaced only episodically in previous shared tasks. The first such shared task was SemEval 2016 Task 4 "Sentiment Analysis in Twitter" [17], which comprised a *binary quantification* subtask and an *ordinal quantification* subtask (these two subtasks were offered again in the 2017 edition). Quantification also featured

in the Dialogue Breakdown Detection Challenge [11], in the Dialogue Quality subtasks of the NTCIR-14 Short Text Conversation task [22], and in the NTCIR-15 Dialogue Evaluation task [23]. However, quantification was never the real focus of these tasks. For instance, the real focus of the tasks described in [17] was sentiment analysis on Twitter data, to the point that almost all participants in the quantification subtasks used the trivial "classify and count" method, and focused, instead of optimising the quantification component, on optimising the sentiment analysis component, or on picking the best-performing learner for training the classifiers used by "classify and count". Similar considerations hold for the tasks discussed in [11,22,23].

This is the first time that a shared task whose explicit focus is quantification is organized. A lab on this topic was thus sorely needed, because the topic has great applicative potential, and because a lot of research on this topic has been carried out without the benefit of the systematic experimental comparisons that only shared tasks allow.

We expect the quantification community to benefit significantly from this lab. One of the reasons is that this community is spread across different fields, as also witnessed by the fact that work on LQ has been published in a scattered way across different areas, e.g., information retrieval [3,6,14], data mining [7,8], machine learning [1,5], statistics [13], or in the areas to which these techniques get applied [2,9,12]. In their own papers, authors often use as baselines only the algorithms from their own fields; we thus expect this lab to pull together different sub-communities, and to generate cross-fertilisation among them.

While quantification is a general-purpose machine learning/data mining task that can be applied to any type of data, in this lab we focus on its application to data consisting of textual documents.

3   Structure of the Lab

3.1   Tasks

Two tasks (T1 and T2) are offered within LeQua 2022, each admitting two subtasks (A and B).

In Task T1 (the *vector task*) participant teams are provided with vectorial representations of the (training/development/test) documents. This task has been offered so as to appeal to those participants who are not into *text* learning, since participants in this task do not need to deal with text preprocessing issues. Additionally, this task allows the participants to concentrate on optimising their quantification methods, rather than spending time on optimising the process for producing vectorial representations of the documents.

In Task T2 (the *raw documents task*), participant teams are provided with the raw (training/development/test) documents. This task has been offered so as to appeal to those participants who want to deploy end-to-end systems, or to those who want to also optimise the process for producing vectorial representations of the documents (possibly tailored to the quantification task).

The two subtasks of both tasks are the *binary quantification subtask* (T1A and T2A) and the *single-label multiclass quantification subtask* (T1B and T2B); in both subtasks each document belongs only to one class $y \in \mathcal{Y} = \{y_1, ..., y_n\}$, with $n = 2$ in T1A and T2A and $n > 2$ in T1B and T2B.

For each subtask in $\{T1A, T1B, T2A, T2B\}$, *participant teams are not supposed to use (training/development/test) documents other than those provided for that subtask.* In particular, participants are not supposed to use any document from either T2A or T2B in order to solve either T1A or T1B.

3.2 Evaluation Measures and Protocols

In a recent theoretical study on the adequacy of evaluation measures for the quantification task [19], *absolute error* (AE) and *relative absolute error* (RAE) have been found to be the most satisfactory, and are thus the only measures used in LeQua 2022. In particular, as a measure we do not use the once widely used Kullback-Leibler Divergence (KLD), since the same study has found it to be unsuitable for evaluating quantifiers.[1] AE and RAE are defined as

$$\mathrm{AE}(p_\sigma, \hat{p}_\sigma) = \frac{1}{n} \sum_{y \in \mathcal{Y}} |\hat{p}_\sigma(y) - p_\sigma(y)| \tag{1}$$

$$\mathrm{RAE}(p_\sigma, \hat{p}_\sigma) = \frac{1}{n} \sum_{y \in \mathcal{Y}} \frac{|\hat{p}_\sigma(y) - p_\sigma(y)|}{p_\sigma(y)} \tag{2}$$

where p_σ is the true distribution on sample σ, $\hat{p}_\sigma$ is the predicted distribution, $\mathcal{Y}$ is the set of classes of interest, and $n = |\mathcal{Y}|$. Note that RAE is undefined when at least one of the classes $y \in \mathcal{Y}$ is such that its prevalence in the sample σ of unlabelled items is 0. To solve this problem, in computing RAE we smooth all $p_\sigma(y)$'s and $\hat{p}_\sigma(y)$'s via additive smoothing, i.e., we take $\underline{p_\sigma}(y) = (\epsilon + p_\sigma(y))/(\epsilon \cdot n + \sum_{y \in \mathcal{Y}} p_\sigma(y))$, where $\underline{p_\sigma}(y)$ denotes the smoothed version of $p_\sigma(y)$ and the denominator is just a normalising factor (same for the $\hat{p_\sigma}(y)$'s); following [8], we use the quantity $\epsilon = 1/(2|\sigma|)$ as the smoothing factor. We then use the smoothed versions of $p_\sigma(y)$ and $\hat{p}_\sigma(y)$ in place of their original non-smoothed versions of Eq. 2; as a result, RAE is now always defined.

As the official measure according to which systems are ranked, we use AE; we also compute RAE results, but we do not use them for ranking the systems.

As the protocol for generating the test samples we adopt the so-called *artificial prevalence protocol* (APP), which is by now a standard protocol for the evaluation of quantifiers. In the APP we take the test set U of unlabelled items, and extract from it a number of subsets (the *test samples*), each characterised by a predetermined vector $(p_\sigma(y_1), ..., p_\sigma(y_n))$ of prevalence values: for extracting a

---

[1] One reason why KLD is undesirable is that it penalizes differently underestimation and overestimation; another is that it is very little robust to outliers. See [19, §4.7 and §5.2] for a detailed discussion of these and other reasons.

test sample σ, we generate a vector of prevalence values, and randomly select documents from U accordingly.[2]

The goal of the APP is to generate samples characterised by widely different vectors of prevalence values; this is meant to test the robustness of a *quantifier* (i.e., of an estimator of class prevalence values) in confronting class prevalence values possibly different (or very different) from the ones of the training set. For doing this we draw the vectors of class prevalence values uniformly at random from the set of all legitimate such vectors, i.e., from the *unit $(n-1)$-simplex* of all vectors $(p_\sigma(y_1), ..., p_\sigma(y_n))$ such that $p_\sigma(y_i) \in [0,1]$ for all $y_i \in \mathcal{Y}$ and $\sum_{y_i \in \mathcal{Y}} p_\sigma(y_i) = 1$. For this we use the Kraemer algorithm [20], whose goal is that of sampling in such a way that all legitimate class distributions are picked with equal probability.[3] For each vector thus picked we randomly generate a test sample. We use this method for both the binary case and the multiclass case.

The official score obtained by a given quantifier is the average value across all test samples of RAE, which we use as the official evaluation measure; for each system we also compute and report the value of AE. We use the non-parametric Wilcoxon signed-rank test on related paired samples in order to assess the statistical significance of the differences in performance between pairs of methods.

3.3   Data

The data we use are Amazon product reviews from a large crawl of such reviews. From the result of this crawl we remove (a) all reviews shorter than 200 characters and (b) all reviews that have not been recognized as "useful" by any users; this yields the dataset Ω that we will use for our experimentation. As for the class labels, (i) for the two binary tasks (T1A and T2A) we use two *sentiment* labels, i.e., Positive (which encompasses 4-stars and 5-stars reviews) and Negative (which encompasses 1-star and 2-stars reviews), while for the two multiclass tasks (T1B and T2B) we use 28 *topic* labels, representing the merchandise class the product belongs to (e.g., Automotive, Baby, Beauty).[4]

We use the same data (training/development/test sets) for the binary vector task (T1A) and for the binary raw document task (T2A); i.e., the former are the vectorized versions of the latter. Same for T1B and T2B.

The L_B (binary) training set and the L_M (multiclass) training set consist of 5,000 documents and 20,000 documents, respectively, sampled from the dataset Ω via *stratified sampling* so as to have "natural" prevalence values for all the class labels. (When doing stratified sampling for the binary "sentiment-based"

---

[2] Everything we say here on how we generate the test samples also applies to how we generate the development samples.

[3] Other seemingly correct methods, such as drawing n random values uniformly at random from the interval $[0,1]$ and then normalizing them so that they sum up to 1, tends to produce a set of samples that is biased towards the centre of the unit $(n-1)$-simplex, for reasons discussed in [20].

[4] The set of 28 topic classes is flat, i.e., there is no hierarchy defined upon it.

task, we ignore the topic dimension; and when doing stratified sampling for the multiclass "topic-based" task, we ignore the sentiment dimension).

The development (validation) sets D_B (binary) and D_M (multiclass) consist of 1,000 development samples of 250 documents each (D_B) and 1,000 development samples of 1,000 documents each (D_M) generated from $\Omega \setminus L_B$ and $\Omega \setminus L_M$ via the Kraemer algorithm.

The test sets U_B and U_M consist of 5,000 test samples of 250 documents each (U_B) and 5,000 test samples of 1,000 documents each (U_M), generated from $\Omega \setminus (L_B \cup D_B)$ and $\Omega \setminus (L_M \cup D_M)$ via the Kraemer algorithm. A submission for a given subtask will consist of prevalence estimations for the relevant classes (topic or sentiment) for each sample in the test set of that subtask.

3.4 Baselines

We have recently developed (and made publicly available) QuaPy, an open-source, Python-based framework that implements several learning methods, evaluation measures, parameter optimisation routines, and evaluation protocols, for LQ [16].[5] Among other things, QuaPy contains implementations of the baseline methods and evaluation measures officially adopted in LeQua 2022.[6]

Participant teams have been informed of the existence of QuaPy, so that they could use the resources contained in it; the goal was to guarantee a high average performance level of the participant teams, since everybody (a) had access to implementations of advanced quantification methods and (b) was able to test them according to the same evaluation standards as employed in LeQua 2022.

4   The LeQua Session at CLEF 2022

The LeQua 2022 session at the CLEF 2022 conference will host (a) one invited talk by a prominent scientist, (b) a detailed presentation by the organisers, overviewing the lab and the results of the participants, (c) oral presentations by selected participants, and (d) poster presentations by other participants.

Depending on how successful LeQua 2022 is, we plan to propose a LeQua edition for CLEF 2023; in that lab we would like to include a cross-lingual task.

Acknowledgments. This work has been supported by the SoBigdata++ project, funded by the European Commission (Grant 871042) under the H2020 Programme INFRAIA-2019-1, and by the AI4Media project, funded by the European Commission (Grant 951911) under the H2020 Programme ICT-48-2020. The authors' opinions do not necessarily reflect those of the European Commission. We thank Alberto Barron Cedeño, Juan José del Coz, Preslav Nakov, and Paolo Rosso, for advice on how to best set up this lab.

---

[5] https://github.com/HLT-ISTI/QuaPy.
[6] Check the branch https://github.com/HLT-ISTI/QuaPy/tree/lequa2022.

References

1. Alaíz-Rodríguez, R., Guerrero-Curieses, A., Cid-Sueiro, J.: Class and subclass probability re-estimation to adapt a classifier in the presence of concept drift. Neurocomputing **74**(16), 2614–2623 (2011)
2. Card, D., Smith, N.A.: The importance of calibration for estimating proportions from annotations. In: Proceedings of the 2018 Conference of the North American Chapter of the Association for Computational Linguistics (HLT-NAACL 2018), New Orleans, US, pp. 1636–1646 (2018)
3. Da San Martino, G., Gao, W., Sebastiani, F.: Ordinal text quantification. In: Proceedings of the 39th ACM Conference on Research and Development in Information Retrieval (SIGIR 2016), Pisa, IT, pp. 937–940 (2016)
4. José del Coz, J., González, P., Moreo, A., Sebastiani, F.: Learning to quantify: Methods and applications (LQ 2021). In: Proceedings of the 30th ACM International Conference on Knowledge Management (CIKM 2021), Gold Coast, AU (2021). Forthcoming
5. du Plessis, M.C., Niu, G., Sugiyama, M.: Class-prior estimation for learning from positive and unlabeled data. Mach. Learn. **106**(4), 463–492 (2016). https://doi.org/10.1007/s10994-016-5604-6
6. Esuli, A., Moreo, A., Sebastiani, F.: A recurrent neural network for sentiment quantification. In: Proceedings of the 27th ACM International Conference on Information and Knowledge Management (CIKM 2018), Torino, IT, pp. 1775–1778 (2018)
7. Esuli, A., Sebastiani, F.: Optimizing text quantifiers for multivariate loss functions. ACM Trans. Knowl. Discov. Data **9**(4), Article 27 (2015)
8. Forman, G.: Quantifying counts and costs via classification. Data Min. Knowl. Disc. **17**(2), 164–206 (2008)
9. Gao, W., Sebastiani, F.: From classification to quantification in tweet sentiment analysis. Soc. Netw. Anal. Min. **6**(1), 1–22 (2016). https://doi.org/10.1007/s13278-016-0327-z
10. González, P., Castaño, A., Chawla, N.V., José del Coz, J.: A review on quantification learning. ACM Comput. Surv. **50**(5), 74:1–74:40 (2017)
11. Higashinaka, R., Funakoshi, K., Inaba, M., Tsunomori, Y., Takahashi, T., Kaji, N.: Overview of the 3rd dialogue breakdown detection challenge. In: Proceedings of the 6th Dialog System Technology Challenge (2017)
12. Hopkins, D.J., King, G.: A method of automated nonparametric content analysis for social science. Am. J. Polit. Sci. **54**(1), 229–247 (2010)
13. King, G., Ying, L.: Verbal autopsy methods with multiple causes of death. Stat. Sci. **23**(1), 78–91 (2008)
14. Levin, R., Roitman, H.: Enhanced probabilistic classify and count methods for multi-label text quantification. In: Proceedings of the 7th ACM International Conference on the Theory of Information Retrieval (ICTIR 2017), Amsterdam, NL, pp. 229–232 (2017)
15. Moreno-Torres, J.G., Raeder, T., Alaíz-Rodríguez, R., Chawla, N.V., Herrera, F.: A unifying view on dataset shift in classification. Pattern Recogn. **45**(1), 521–530 (2012)
16. Moreo, A., Esuli, A., Sebastiani, F.: QuaPy: a python-based framework for quantification. In: Proceedings of the 30th ACM International Conference on Knowledge Management (CIKM 2021), Gold Coast, AU (2021). Forthcoming
17. Nakov, P., Ritter, A., Rosenthal, S., Sebastiani, F., Stoyanov, V.: SemEval-2016 Task 4: sentiment analysis in Twitter. In Proceedings of the 10th International Workshop on Semantic Evaluation (SemEval 2016), San Diego, US, pp. 1–18 (2016)

18. Quiñonero-Candela, J., Sugiyama, M., Schwaighofer, A., Lawrence, N.D. (eds.): Dataset Shift in Machine Learning. The MIT Press, Cambridge (2009)
19. Sebastiani, F.: Evaluation measures for quantification: an axiomatic approach. Inf. Retrieval J. **23**(3), 255–288 (2020)
20. Smith, N.A., Tromble, R.W.: Sampling uniformly from the unit simplex (2004). Unpublished manuscript. https://www.cs.cmu.edu/~nasmith/papers/smith+tromble.tr04.pdf
21. Vapnik, V.: Statistical Learning Theory. Wiley, New York (1998)
22. Zeng, Z., Kato, S., Sakai, T.: Overview of the NTCIR-14 short text conversation task: dialogue quality and nugget detection subtasks. In: Proceedings of NTCIR-14, pp. 289–315 (2019)
23. Zeng, Z., Kato, S., Sakai, T., Kang, I.: Overview of the NTCIR-15 dialogue evaluation task (DialEval-1). In: Proceedings of NTCIR-15, pp. 13–34 (2020)

ImageCLEF 2022: Multimedia Retrieval in Medical, Nature, Fusion, and Internet Applications

Alba G. Seco de Herrera[1(✉)], Bogdan Ionescu[2], Henning Müller[3],
Renaud Péteri[4], Asma Ben Abacha[5], Christoph M. Friedrich[6],
Johannes Rückert[6], Louise Bloch[6], Raphael Brüngel[6], Ahmad Idrissi-Yaghir[6],
Henning Schäfer[6], Serge Kozlovski[7], Yashin Dicente Cid[8], Vassili Kovalev[7],
Jon Chamberlain[1], Adrian Clark[1], Antonio Campello[9], Hugo Schindler[10],
Jérôme Deshayes[10], Adrian Popescu[10], Liviu-Daniel Ştefan[2],
Mihai Gabriel Constantin[2], and Mihai Dogariu[2]

[1] University of Essex, Colchester, UK
alba.garcia@essex.ac.uk
[2] University Politehnica of Bucharest, Bucharest, Romania
[3] University of Applied Sciences Western Switzerland (HES-SO), Delémont,
Switzerland
[4] University of La Rochelle, La Rochelle, France
[5] National Library of Medicine, Bethesda, USA
[6] University of Applied Sciences and Arts Dortmund (FH Dortmund),
Dortmund, Germany
[7] Belarussian Academy of Sciences, Minsk, Belarus
[8] University of Warwick, Coventry, UK
[9] Wellcome Trust, London, UK
[10] CEA LIST, Palaiseau, France

Abstract. ImageCLEF s part of the Conference and Labs of the Evaluation Forum (CLEF) since 2003. CLEF 2022 will take place in Bologna, Italy. ImageCLEF is an ongoing evaluation initiative which promotes the evaluation of technologies for annotation, indexing, and retrieval of visual data with the aim of providing information access to large collections of images in various usage scenarios and domains. In its 20th edition, Image-CLEF will have four main tasks: (i) a *Medical* task addressing concept annotation, caption prediction, and tuberculosis detection; (ii) a *Coral* task addressing the annotation and localisation of substrates in coral reef images; (iii) an *Aware* task addressing the prediction of real-life consequences of online photo sharing; and (iv) a new *Fusion* task addressing late fusion techniques based on the expertise of the pool of classifiers. In 2021, over 100 research groups registered at ImageCLEF with 42 groups submitting more than 250 runs. These numbers show that, despite the COVID-19 pandemic, there is strong interest in the evaluation campaign.

Keywords: User awareness · Medical image classification · Medical image understanding · Coral image annotation and classification · Fusion · ImageCLEF benchmarking · Annotated data

M. Hagen et al. (Eds.): ECIR 2022, LNCS 13186, pp. 382–389, 2022.
https://doi.org/10.1007/978-3-030-99739-7_48

1   Introduction

ImageCLEF is a benchmarking activity on the cross-language annotation and retrieval of images in the Conference and Labs of the Evaluation Forum (CLEF) [19,20]. The 20th anniversary of ImageCLEF will take place in Bologna, Italy, in September 2022[1]. The main goal of ImageCLEF is to promote research in the fields of multi-lingual and multi-modal information access evaluation. Hence, a set of benchmarking activities was designed to test different aspects of mono and cross-language information retrieval systems [16,19,20]. Both Image-CLEF [28] and also the overall CLEF campaign have important scholarly impact [28,29].

Since 2018, the AIcrowd[2] platform (previously crowdAI) is used to distribute the data and receive the submitted results. The platform provides access to the data beyond the competition and allows having an online leaderboard.

The following sections introduce the four tasks that are planned for 2022, namely: ImageCLEFmedical, ImageCLEFcoral, ImageCLEFaware, and the new ImageCLEFfusion. Figure 1 captures a few images the specificities of some tasks.

2   ImageCLEFmedical

The ImageCLEFmedical task has been carried out every year since 2004 [20]. The 2022 edition will include two tasks: the caption task, and the tuberculosis task. The *caption* task focuses on interpreting and summarising the insights gained from radiology images. In the 6th edition [12,13,22–24] of the task, there will be two subtasks: concept detection and caption prediction. The *concept detection* subtask aims to develop competent systems that are able to predict the Unified Medical Language System (UMLS®) Concept Unique Identifiers (CUIs) based on the visual image content. The F1-Score [15] will be used to evaluate the participating systems in this subtask. The *caption prediction* subtask focuses on implementing models to predict captions for given radiology images. The BLEU [21] score will be used for evaluating this subtask. In 2022, a subset of the Radiology Objects in Context (ROCO) [25] dataset will be used. As in the previous editions, the dataset will be manually curated after using multiple concept extraction methods to retrieve accurate CUIs.

The *tuberculosis* task will be extended from the previous TB-case and/or lesion classification problems [9–11,17,18] to the more advanced lesion detection problem. As in previous editions the task will use chest 3D Computed Tomography (CT) scans as source data, but this time participants are expected to detect cavern regions localisation rather than simply provide a label for the CT image. This problem is important because even after successful TB treatment, which satisfies the existing criteria of cavern recovery, patients may still contain colonies of Mycobacterium tuberculosis that could lead to unpredictable disease relapse. In addition, the subtask of predicting four binary features of caverns suggested by experienced radiologists will be available.

---

[1] https://clef2022.clef-initiative.eu/.

[2] https://www.aicrowd.com/.

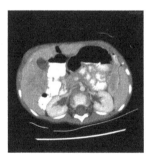
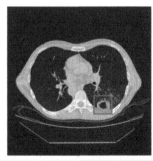

Concepts

Large Mass
Right lobe of liver
Tomography, Emission-
Computed

Captions

heterogenous mass with
well-defined
psudocapsule arises from
the right hepatic lobe.

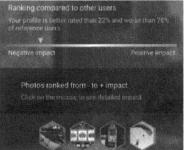

Fig. 1. Sample images from (left to right, top to bottom): ImageCLEFmedical caption with an image with the corresponding CUIs and caption, ImageCLEFmedical tuberculosis with a slice of a chest CT with tuberculosis cavern region, ImageCLEFcoral with 3D reconstruction of a coral reef and ImageCLEFaware with an example of user photos and predicted influence when searching for a bank loan.

3   ImageCLEFcoral

The increasing use of structure-from-motion photogrammetry for modelling large-scale environments from action cameras attached to drones has driven the next-generation of visualisation techniques that can be used in augmented and virtual reality headsets. Since 2019, the ImageCLEFcoral task addresses the issue automatically annotating these images for monitoring coral reef structure and composition, in support of their conservation. The 4th edition of the task will follow a similar format to previous editions [2–4] where participants automatically segment and label a collection of images that can be used in combination to create three-dimensional models of an underwater environment. Figure 1 shows the 3D reconstruction of a coral reef (approx. 4 × 6 m). To create this model, each image in the subset is represented by a blue rectangle in the image, with the track of multi-camera array clearly visible across the environment. In 2022, this task will contain the same two subtasks as in previous years: *coral reef image annotation and localisation* and *coral reef image pixel-wise parsing.*

The *coral reef image annotation and localisation* subtask requires participants to label the images with types of benthic substrate together with their bounding box. The *coral reef image pixel-wise parsing* subtask requires the participants to segment and parse each coral image into different image regions associated with benthic substrate types. As in previous editions, the performance of the submitted algorithms will be evaluated using the PASCAL VOC style metric of intersection over union (IoU) and the mean of pixel-wise accuracy per class.

Previous editions of ImageCLEFcoral in 2019 and 2020 showed improvements in task performance and promising results on cross-learning between images from different geographical regions. The 3rd edition in 2021 increased the task complexity and size of data available to participants through supplemental data, resulting in lower performance than in previous years. The 4th edition plans to address these issues by targeting algorithms for geographical regions and raising the benchmark performance. As with the 3rd edition, training and test data will form the complete set of images required for 3D reconstruction of the marine environment. This will allow participants to explore novel probabilistic computer vision techniques based on image overlap and transposition of data points.

4 ImageCLEFaware

The online disclosure of personal data often has effects which go beyond the initial context in which data was shared. Content which seems innocuous initially can be interpreted to the users' disadvantage by third parties which have access to the data. For instance, it is now common for prospective employers to search online information about candidates. This process can be done by humans or be based on automatic inferences. Users are entitled to be aware about the potential effects of online data sharing and this task hypothesises that feedback about these effects can be efficiently provided by simulating impactful real-life situations. Since images constitute a large part of the content shared online, the objective of the ImageCLEFaware task is to automatically rate user photographic profiles in four situations in which the users would search for, e.g., a bank loan, an accommodation, a waiter job, or a job in IT.

In the 2nd edition of the task, the dataset will be enriched to include 1000 profiles instead of the 500 included in the 1st edition. Each profile is labelled with an appeal score per situation by several annotators. Participants will be provided with the profiles along with the associated rankings. The objective of the task is to produce an automatic ranking which is as closely correlated as possible to the manual ranking. Correlation will be measured using a classical measure such as the Pearson correlation coefficient.

Task-related resources will be provided by the organisers to encourage participation of different communities. These resources include: (i) visual object ratings per situation obtained through crowdsourcing; (ii) automatically extracted visual object detection for over 350 objects (versus 270 in 2021) which have non-null rating in at least one situation.

In accordance with General Data Protection Regulation, data minimisation is applied and participants receive only the information necessary to carry out the task in an anonymised form. This resources include (i) anonymised visual concept ratings for each situation modelled; (ii) automatically extracted predictions for the images that compose the profiles.

5   ImageCLEFfusion

While the advent of deep learning systems greatly increased the overall performance of computer vision methods in general, there are still some tasks where system performance is low, thus impeding the adoption of automated computer vision methods in the industry. One representative example for this type of tasks is the prediction of subjective properties of multimedia samples. While these tasks may be harder to solve, given the inherent lower annotator agreement present in the datasets associated with such concepts, this task hypothesises that it is possible to significantly improve the current results by using late fusion approaches. *Late fusion*, also knows as *ensembling* or *decision-level fusion*, consists of a set of initial predictors, called *inducers*, that are trained and tested on the dataset, whose prediction outputs are combined in the final step via an *ensembling method* in order to create a new and improved set of predictions. In the current literature late fusion systems are sometimes successfully used even in traditional tasks such as video action recognition [27], and more often in subjective and multimodal tasks like memorability [1], violence detection [7] and media interestingness [30]. Furthermore, latest developments in this field, using deep neural networks as the primary ensembling method show major improvements over traditional ensembling methods, by greatly increasing the performance of individual inducers [5,6,26].

In this context, the 1st edition of the ImageCLEFfusion task is proposed. The organisers will provide several task-related resources such as: (i) the datasets that will be used throughout the task; (ii) a large set (more than 15) of pre-computed inducer prediction outputs for the corresponding datasets; and (iii) metrics and data-splits. Participants are tasked with creating novel ensembling methods to significantly increase the performance of the pre-computed inducers. The targeted datasets for the ImageCLEFfusion task will be composed of ground truth data extracted from several subjective multimedia processing tasks like interestingness [8], memorability [14] or result diversification [31].

While the metrics and data-splits will be used to measure ensembling method performance, this task will also look to provide answers to interesting theoretical questions, such as: (i) how does inducer correlation affect ensemble performance; (ii) how does inducer diversity affect ensemble performance; (iii) are there selection methods for inclusion and exclusion of inducers regarding an ensemble; (iv) are deep learning ensembling methods better than other types of approaches? Answering these questions may provide valuable insights for future research, not only with regards to the best performing ensemble methods, but also into the reduction of hardware requirements by inducer selection.

6   Conclusion

This paper presents an overview of the upcoming ImageCLEF at the CLEF 2022. ImageCLEF has been organising many tasks in a variety of domains in the field of visual media analysis, indexing, classification, and retrieval. The 20th anniversary of the task includes a variety of tasks in the fields of medical imaging, nature, system fusion, and internet applications. All the tasks will provide a set of new test collections simulating real-world situations. Such collections are important to enable researchers to assess the performance of their systems and to compare their results with others following a common evaluation framework.

Acknowledgement. Part of this work is supported under the H2020 AI4Media "A European Excellence Centre for Media, Society and Democracy" project, contract #951911. The work of Louise Bloch and Raphael Brüngel was partially funded by a PhD grant from the University of Applied Sciences and Arts Dortmund (FH Dortmund), Germany. The work of Ahmad Idrissi-Yaghir and Henning Schäfer was funded by a PhD grant from the DFG Research Training Group 2535 Knowledge- and data-based personalisation of medicine at the point of care (WisPerMed).

References

1. Azcona, D., Moreu, E., Hu, F., Ward, T.E., Smeaton, A.F.: Predicting media memorability using ensemble models. In: Working Notes Proceedings of the MediaEval 2019 Workshop. CEUR Workshop Proceedings, vol. 2670. CEUR-WS.org (2019)
2. Chamberlain, J., Campello, A., Wright, J.P., Clift, L.G., Clark, A., Seco de Herrera, A.G.: Overview of ImageCLEF coral 2019 task. In: Working Notes of Conference and Labs of the Evaluation Forum (CLEF 2019). CEUR Workshop Proceedings, vol. 2380. CEUR-WS.org (2019)
3. Chamberlain, J., Campello, A., Wright, J.P., Clift, L.G., Clark, A., Seco de Herrera, A.G.: Overview of the ImageCLEF coral 2020 task: automated coral reef image annotation. In: Working Notes of Conference and Labs of the Evaluation Forum (CLEF 2020). CEUR Workshop Proceedings, vol. 2696. CEUR-WS.org (2020)
4. Chamberlain, J., Seco de Herrera, A.G., Campello, A., Clark, A., Oliver, T.A., Moustahfid, H.: Overview of the ImageCLEF coral 2021 task: coral reef image annotation of a 3D environment. In: Working Notes of Conference and Labs of the Evaluation Forum (CLEF 2021). CEUR Workshop Proceedings, vol. 2936. CEUR-WS.org (2021)
5. Constantin, M.G., Ştefan, L.-D., Ionescu, B.: DeepFusion: deep ensembles for domain independent system fusion. In: Lokoč, J., et al. (eds.) MMM 2021. LNCS, vol. 12572, pp. 240–252. Springer, Cham (2021). https://doi.org/10.1007/978-3-030-67832-6_20
6. Constantin, M.G., Ştefan, L.D., Ionescu, B., Duong, N.Q., Demarty, C.H., Sjöberg, M.: Visual interestingness prediction: a benchmark framework and literature review. Int. J. Comput. Vis. 1–25 (2021)
7. Dai, Q., et al.: Fudan-Huawei at MediaEval 2015: detecting violent scenes and affective impact in movies with deep learning. In: Working Notes Proceedings of the MediaEval 2015 Workshop. CEUR Workshop Proceedings, vol. 1436. CEUR-WS.org (2015)
8. Demarty, C.H., Sjöberg, M., Ionescu, B., Do, T.T., Gygli, M., Duong, N.: MediaEval 2017 predicting media interestingness task. In: Working Notes Proceedings of the MediaEval 2017 Workshop. CEUR Workshop Proceedings, vol. 1984. CEUR-WS.org (2017)

9. Cid, Y.D., Kalinovsky, A., Liauchuk, V., Kovalev, V., Müller, H.: Overview of Image CLEF tuberculosis 2017 - predicting tuberculosis type and drug resistances. In: Working Notes of Conference and Labs of the Evaluation Forum (CLEF 2017). CEUR Workshop Proceedings, vol. 1866. CEUR-WS.org (2017)

10. Cid, Y.D., Liauchuk, V., Klimuk, D., Tarasau, A., Kovalev, V., Müller, H.: Overview of image CLEF tuberculosis 2019 - automatic CT-based report generation and tuberculosis severity assessment. In: Working Notes of Conference and Labs of the Evaluation Forum (CLEF 2019). CEUR Workshop Proceedings, vol. 2380. CEUR-WS.org (2019)

11. Cid, Y.D., Liauchuk, V., Kovalev, V., Müller, H.: Overview of Image CLEF tuberculosis 2018 - detecting multi-drug resistance, classifying tuberculosis type, and assessing severity score. In: Working Notes of Conference and Labs of the Evaluation Forum (CLEF 2018). CEUR Workshop Proceedings, vol. 2125. CEUR-WS.org (2018)

12. Eickhoff, C., Schwall, I., Seco de Herrera, A.G., Müller, H.: Overview of Image CLEF caption 2017 - the image caption prediction and concept extraction tasks to understand biomedical images. In: Working Notes of Conference and Labs of the Evaluation Forum (CLEF 2017). CEUR Workshop Proceedings, vol. 1866. CEUR-WS.org (2017)

13. Seco De Herrera, A.G., Eickhof, C., Andrearczyk, V., Müller, H.: Overview of the Image CLEF 2018 caption prediction tasks. In: Working Notes of Conference and Labs of the Evaluation Forum (CLEF 2018). CEUR Workshop Proceedings, vol. 2125. CEUR-WS.org (2018)

14. Seco De Herrera, A.G., et al.: Overview of MediaEval 2020 predicting media memorability task: what makes a video memorable? Working Notes Proceedings of the MediaEval 2020 Workshop (2020)

15. Goutte, C., Gaussier, E.: A probabilistic interpretation of precision, recall and F-score, with implication for evaluation. In: Losada, D.E., Fernández-Luna, J.M. (eds.) ECIR 2005. LNCS, vol. 3408, pp. 345–359. Springer, Heidelberg (2005). https://doi.org/10.1007/978-3-540-31865-1_25

16. Kalpathy-Cramer, J., Seco de Herrera, A.G., Demner-Fushman, D., Antani, S., Bedrick, S., Müller, H.: Evaluating performance of biomedical image retrieval systems: overview of the medical image retrieval task at image CLEF 2004–2014. Comput. Med. Imaging Graph. **39**, 55–61 (2015)

17. Kozlovski, S., Liauchuk, V., Cid, Y.D., Kovalev, V., Müller, H.: Overview of Image CLEF tuberculosis 2021 - CT-based tuberculosis type classification. In: Working Notes of Conference and Labs of the Evaluation Forum (CLEF 2021). CEUR Workshop Proceedings, vol. 2936. CEUR-WS.org (2021)

18. Kozlovski, S., Liauchuk, V., Cid, Y.D., Tarasau, A., Kovalev, V., Müller, H.: Overview of Image CLEF tuberculosis 2020 - automatic CT-based report generation. In: Working Notes of Conference and Labs of the Evaluation Forum (CLEF 2020). CEUR Workshop Proceedings, vol. 2696. CEUR-WS.org (2020)

19. Müller, H., Clough, P., Deselaers, T., Caputo, B. (eds.): Image CLEF - Experimental Evaluation in Visual Information Retrieval, The Springer International Series On Information Retrieval, vol. 32. Springer, Berlin Heidelberg (2010). https://doi.org/10.1007/978-3-642-15181-1

20. Müller, H., Kalpathy-Cramer, J., Seco de Herrera, A.G.: Experiences from the image CLEF medical retrieval and annotation tasks. In: Information Retrieval Evaluation in a Changing World, pp. 231–250. Springer (2019). https://doi.org/10.1007/978-3-030-22948-1_10

21. Papineni, K., Roukos, S., Ward, T., Zhu, W.J.: BLEU: a method for automatic evaluation of machine translation. In: Proceedings of the 40th Annual Meeting of the Association for Computational Linguistics (ACL 2002), pp. 311–318 (2002)
22. Pelka, O., et al.: Overview of the Image CLEFmed 2021 concept & caption prediction task. In: Working Notes of Conference and Labs of the Evaluation Forum (CLEF 2021). CEUR Workshop Proceedings, vol. 2936. CEUR-WS.org (2021)
23. Pelka, O., Friedrich, C.M., Seco de Herrera, A.G., Müller, H.: Overview of the Image CLEFmed 2019 concept detection task. In: Working Notes of Conference and Labs of the Evaluation Forum (CLEF 2019). CEUR Workshop Proceedings, vol. 2380. CEUR-WS.org (2019)
24. Pelka, O., Friedrich, C.M., Seco de Herrera, A.G., Müller, H.: Overview of the Image CLEFmed 2020 concept prediction task: medical image understanding. In: Working Notes of Conference and Labs of the Evaluation Forum (CLEF 2020). CEUR Workshop Proceedings, vol. 2696. CEUR-WS.org (2020)
25. Pelka, O., Koitka, S., Rückert, J., Nensa, F., Friedrich, C.M.: Radiology objects in COntext (ROCO): a multimodal image dataset. In: Intravascular Imaging and Computer Assisted Stenting and Large-Scale Annotation of Biomedical Data and Expert Label Synthesis, pp. 180–189. Springer (2018). https://doi.org/10.1007/978-3-030-01364-6_20
26. Ştefan, L.D., Constantin, M.G., Ionescu, B.: System fusion with deep ensembles. In: Proceedings of the 2020 International Conference on Multimedia Retrieval (ICMR 2020), pp. 256–260. Association for Computing Machinery (ACM) (2020)
27. Sudhakaran, S., Escalera, S., Lanz, O.: Gate-shift networks for video action recognition. In: Proceedings of the IEEE/CVF Conference on Computer Vision and Pattern Recognition (CVPR 2020), pp. 1102–1111 (2020)
28. Tsikrika, T., de Herrera, A.G.S., Müller, H.: Assessing the scholarly impact of ImageCLEF. In: Forner, P., Gonzalo, J., Kekäläinen, J., Lalmas, M., de Rijke, M. (eds.) CLEF 2011. LNCS, vol. 6941, pp. 95–106. Springer, Heidelberg (2011). https://doi.org/10.1007/978-3-642-23708-9_12
29. Tsikrika, T., Larsen, B., Müller, H., Endrullis, S., Rahm, E.: The scholarly impact of CLEF (2000–2009). In: Information Access Evaluation. Multilinguality, Multimodality, and Visualization, pp. 1–12. Springer (2013). https://doi.org/10.1007/978-3-642-40802-1_1
30. Wang, S., Chen, S., Zhao, J., Jin, Q.: Video interestingness prediction based on ranking model. In: Proceedings of the Joint Workshop of the 4th Workshop on Affective Social Multimedia Computing and First Multi-Modal Affective Computing of Large-Scale Multimedia Data (ASMMC-MMAC 2018), pp. 55–61. Association for Computing Machinery (ACM) (2018)
31. Zaharieva, M., Ionescu, B., Gînsca, A.L., Santos, R.L., Müller, H.: Retrieving diverse social images at MediaEval 2017: challenges, dataset and evaluation. In: Working Notes Proceedings of the MediaEval 2017 Workshop. CEUR Workshop Proceedings, vol. 1984. CEUR-WS.org (2017)

LifeCLEF 2022 Teaser: An Evaluation of Machine-Learning Based Species Identification and Species Distribution Prediction

Alexis Joly[1]([✉]), Hervé Goëau[2], Stefan Kahl[6], Lukáš Picek[10],
Titouan Lorieul[1], Elijah Cole[9], Benjamin Deneu[1],
Maximilien Servajean[7], Andrew Durso[11], Isabelle Bolon[8],
Hervé Glotin[3], Robert Planqué[4], Willem-Pier Vellinga[4], Holger Klinck[6],
Tom Denton[12], Ivan Eggel[5], Pierre Bonnet[2], Henning Müller[5],
and Milan Šulc[13]

[1] Inria, LIRMM, Univ Montpellier, CNRS, Montpellier, France
alexis.joly@inria.fr
[2] CIRAD, UMR AMAP, Montpellier, Occitanie, France
[3] Univ. Toulon, Aix Marseille Univ., CNRS, LIS, DYNI Team,
Marseille, France
[4] Xeno-canto Foundation, Amsterdam, The Netherlands
[5] HES-SO, Sierre, Switzerland
[6] KLYCCB, Cornell Lab of Ornithology, Cornell University, Ithaca, USA
[7] LIRMM, AMIS, Univ Paul Valéry Montpellier, University Montpellier, CNRS,
Montpellier, France
[8] ISG, Department of Community Health and Medicine,
UNIGE, Geneva, Switzerland
[9] Department of Computing and Mathematical Sciences, Caltech, Pasadena, USA
[10] Department of Cybernetics, FAV, University of West Bohemia, Plzen, Czechia
[11] Department of Biological Sciences, Florida Gulf Coast University,
Fort Myers, USA
[12] Google LLC, San Francisco, USA
[13] Department of Cybernetics, FEE, CTU in Prague, Prague, Czech Republic

Abstract. Building accurate knowledge of the identity, the geographic distribution and the evolution of species is essential for the sustainable development of humanity, as well as for biodiversity conservation. However, the difficulty of identifying plants, animals and fungi is hindering the aggregation of new data and knowledge. Identifying and naming living organisms is almost impossible for the general public and is often difficult even for professionals and naturalists. Bridging this gap is a key step towards enabling effective biodiversity monitoring systems. The LifeCLEF campaign, presented in this paper, has been promoting and evaluating advances in this domain since 2011. The 2022 edition proposes five data-oriented challenges related to the identification and

M. Hagen et al. (Eds.): ECIR 2022, LNCS 13186, pp. 390–399, 2022.
https://doi.org/10.1007/978-3-030-99739-7_49

prediction of biodiversity: (i) PlantCLEF: very large-scale plant identification, (ii) BirdCLEF: bird species recognition in audio soundscapes, (iii) GeoLifeCLEF: remote sensing based prediction of species, (iv) SnakeCLEF: Snake Species Identification in Medically Important scenarios, and (v) FungiCLEF: Fungi recognition from images and metadata.

1   LifeCLEF Lab Overview

Accurately identifying organisms observed in the wild is an essential step in ecological studies. Unfortunately, observing and identifying living organisms requires high levels of expertise. For instance, vascular plants alone account for more than 300,000 different species and the distinctions between them can be quite subtle. The world-wide shortage shortage of trained taxonomists and curators capable of identifying organisms has come to be known as the *taxonomic impediment*. Since the Rio Conference of 1992, it has been recognized as one of the major obstacles to the global implementation of the Convention on Biological Diversity[1]. In 2004, Gaston and O'Neill [7] discussed the potential of automated approaches for species identification. They suggested that, if the scientific community were able to (i) produce large training datasets, (ii) precisely evaluate error rates, (iii) scale up automated approaches, and (iv) detect novel species, then it would be possible to develop a generic automated species identification system that would open up new vistas for research in biology and related fields.

Since the publication of [7], automated species identification has been studied in many contexts [6,9,13,23,25,30,31,36]. This area continues to expand rapidly, particularly due to advances in deep learning [5,8,24,26,32–35]. In order to measure progress in a sustainable and repeatable way, the LifeCLEF[2] research platform was created in 2014 as a continuation and extension of the plant identification task that had been run within the ImageCLEF lab[3] since 2011 [10–12]. Since 2014, LifeCLEF expanded the challenge by considering animals in addition to plants, and including audio and video content in addition to images [14–21]. About 100-500 research groups annually register to LifeCLEF in order to either download the data, register to the mailing list or benefit from the shared evaluation tools. The number of participants who finally crossed the finish line by submitting runs was respectively: 22 in 2014, 18 in 2015, 17 in 2016, 18 in 2017, 13 in 2018, 16 in 2019, 16 in 2020, 1,022 in 2021 (including the 1,004 participants of the BirdCLEF Kaggle challenge). The 2022 edition proposes five data-oriented challenges: three in the continuity of the 2021 edition (BirdCLEF, GeoLifeCLEF and SnakeCLEF), one new challenge related to fungi recognition with a focus on the combination of visual information with meta-data and on edible vs. poisonous species (FungiCLEF), and a considerable expansion of the PlantCLEF

---

[1] https://www.cbd.int/.

[2] http://www.lifeclef.org/.

[3] http://www.imageclef.org/.

challenge towards the identification of the world's flora (about 300K species). In the following sections, we describe for each of the five challenges the motivation, the used data collection and the evaluated task.

2   PlantCLEF Challenge: Identify the World's Flora

Motivation: It is estimated that there are more than 300,000 species of vascular plants in the world. Increasing our knowledge of these species is of paramount importance for the development of human civilization (agriculture, construction, pharmacopoeia, etc.), especially in the context of the biodiversity crisis [22]. However, the burden of systematic plant identification by human experts strongly penalizes the aggregation of new data and knowledge. Since then, automatic identification has made considerable progress in recent years as highlighted during all previous editions of PlantCLEF. Deep learning techniques now seem mature enough to address the ultimate but realistic problem of global identification of plant biodiversity in spite of many problems that the data may present (a huge number of classes, very strongly unbalanced classes, partially erroneous identifications, duplications, variable visual quality, diversity of visual contents such as photos or herbarium sheets, etc.).

Data Collection: the training dataset that will be used this year can be distinguished in 2 main categories: labeled and unlabeled (i.e. with or without species labels provided and checked by humans). The labeled training dataset will be based on a dataset of more than 5M images covering more than 290k plant species based on a web crawl with Google and Bing search engines and the Encyclopedia of Life webportal. All datasets provided in previous editions of PlantCLEF can also be used and the use of external data will be encouraged, notably via the gbif-dl[4] package which facilitates the download of media data from the world's largest biodiversity database (GBIF[5]) by wrapping its public API. The unlabeled training dataset will be based on more than 9 million pictures coming from the Pl@ntNet platform [4] (associated with a pseudo-label but without human verification). Finally, the test set will be a set of tens of thousands pictures verified by world class experts related to various regions of the world and taxonomic groups.

Task Description: the task will be evaluated as a plant species retrieval task based on multi-image plant observations from the test set. The goal will be to retrieve the correct plant species among the top results of a ranked list of species returned by the evaluated system. The participants will first have access to the training set and a few months later, they will be provided with the whole test set. Semi-supervised or unsupervised approaches will be strongly encouraged and a starter package with a pre-trained model based on this type of method exploiting the unlabeled training dataset will be provided.

---

[4] https://github.com/plantnet/gbif-dl.
[5] https://www.gbif.org/.

3   BirdCLEF Challenge: Bird Species Identification in Soundscape Recordings

Motivation: Recognizing bird sounds in complex soundscapes is an important sampling tool that often helps to reduce the limitations of point counts[6]. In the future, archives of recorded soundscapes will become increasingly valuable as the habitats in which they were recorded will be lost in the near future. It is imperative to develop new technologies that can cope with the increasing amount of audio data and that can help to accelerate the process of species diversity assessments. In the past few years, deep learning approaches have transformed the field of automated soundscape analysis. Yet, when training data is sparse, detection systems struggle with the recognition of rare species. The goal of this competition is to establish training and test datasets that can serve as real-world applicable evaluation scenarios for endangered habitats and help the scientific community to advance their conservation efforts through automated bird sound recognition.

Data Collection: We will build on the experience from previous editions and adjust the overall task to encourage participants to focus on few-shot learning and task-specific model designs. We will select training and test data to suit this demand. As for previous years, Xeno-canto will be the primary source for training data, expertly annotated soundscape recordings will be used for testing. We will focus on bird species for which there is limited training data, but we will also include common species so that participants can train good recognition systems. In search of suitable test data, we will consider different data sources with varying complexity (call density, chorus, signal-to-noise ratio, anthropophony), and quality (mono and stereo recordings). We also want to focus on very specific real-world use cases (e.g., conservation efforts in Hawaii) and frame the competition based on the demand of the particular use case. Additionally, we are considering including unlabeled data to encourage self-supervised learning regimes.

Task Description: The competition will be held on Kaggle and the evaluation mode will resemble the 2021 test mode (i.e., hidden test data, code competition). We will use established metrics like F1 score and LwLRAP which reflect use cases for which precision is key and also allow organizers to assess system performance independent of fine-tuned confidence thresholds. Participants will be asked to return a list of species for short audio segments extracted from labeled soundscape data. In the past, we used 5-second segments, and we will consider increasing the duration of these context windows to better reflect the overall ground truth label distribution. However, the overall structure of the task will remain unchanged, as it provides a well-established base that has resulted in significant participation in past editions (e.g., 1,004 participants and 9,307 submissions in 2021). Again, we will strive to keep the dataset size reasonably

---

[6] e.g. some species might be oversampled or undersampled.

small (<50 GB) and easy to process, and we will also provide introductory code repositories and write-ups to lower the entry level of the competition.

4 GeoLifeCLEF Challenge: Species Prediction Based on Occurrence Data, Environmental Data and Remote Sensing Data

Motivation: Automatically predicting the list of species that are the most likely to be observed at a given location is useful for many scenarios in biodiversity conservation, ecotourism, land management, etc. First of all, it allows improve species identification tools by reducing the list of candidate species that are observable at a given location (be they automated, semi-automated or based on classical field guides or flora). More generally, it facilitates biodiversity inventories through the development of location-based recommendation services (typically on mobile phones), it favours the involvement of non-expert nature observers, as well as accelerate the annotation or validation of species observed by non-experts to produce high quality datasets. Last but not least, it might serve educational purposes thanks to biodiversity discovery applications providing functionalities such as contextualized educational pathways.

Data Collection: The GeoLifeCLEF dataset (already used in 2020 and 2021) contains about 2 million observations of around 30 thousand plant and animal species. Each observation is paired with very high-resolution covariates (aerial imagery, land cover, altitude) and environmental rasters (bioclimatic variables, soil type, etc.). The dataset took months to build in its raw format ($\sim$850 GB) and we reformatted it in a more convenient and memory efficient format ($\sim$100 GB). Indeed, it has not yet been used to its full potential due (i) to the computing power required to train models on it, and, (ii) to the complexity and the wide variety of challenges of the tackled task. In 2021, the challenge focused on measuring the efficiency of remote sensing imagery to predict the presence of species at a given location. In 2022, the objective is to make this competition more realistic by changing the evaluation protocol: the models will be evaluated on new presence/absence observation data. This means that the 2022 challenge will tackle two main issues in species presence prediction: (i) taking into consideration the sampling bias due to the presence-only nature of the training observation data, and, (ii) predicting relevant sets of species present at the given location.

Task Description: Given the test set of locations (i.e. geo-coordinates) and corresponding high-resolution and environmental covariates, the goal of the task will be to return for each location a ranked list of species sorted according to the likelihood that they might have been observed at that location. The metric used will be a multi-label metric such as mean average precision (mAP).

5   SnakeCLEF Challenge: Automated Snake Species Identification with Country-Level Focus

Motivation: Developing a robust system for identifying snake species from photographs is an important goal in biodiversity and global health. With over half a million of deaths and disability from venomous snakebite annually, understanding the global distribution of more than 3,900 snake species and differentiating them from images (particularly images of low quality) will significantly improve epidemiology data and treatment outcomes. From previous editions, we learned that "machines" are capable of accurate recognition (Macro F_{1_c} >90%, Accuracy ~95%) even in the scenarios with long-tailed class distributions and ~800 species. Thus, testing over real Medically Important Scenarios and integrating information on species toxicity is the next step to provide a more reliable "machine" prediction.

Data Collection: The dataset used in previous editions [27, 29] will be extended with new and rare species as well as with images from countries with no or just a few samples, reducing the uneven species distributions across all the countries included in the data. For testing, we tailored two sets, one for a machine evaluation and the second for the HUMAN vs AI comparison. The SnakeCLEF 2022 dataset covers 1,000 snake species on more than 500,000 images and from approximately 200 countries – adding 224 new species. In addition, we include: (i) snake species toxicity level, allowing us to research methods for lowering the possibility of medically-critical mis-prediction, i.e., confusion of venomous species with non-venomous. (ii) country-species mapping file describing species-country presence based on the The Reptile Database and allowing better worldwide regularization.

Task Description: Given the set of images and corresponding geographic locality information, the goal of the task is to return for each image a ranked list of species sorted according to the likelihood that they are in the image and might have been observed at that location and minimising the venomous/non-venomous confusion.

6   FungiCLEF Challenge: Fungi Recognition from Images and Metadata

Motivation: Automatic recognition of fungi species assists mycologists, citizen scientists and nature enthusiasts in species identification in the wild. Its availability supports the collection of valuable biodiversity data. In practice, species identification typically does not depend solely on the visual observation of the specimen but also on other information available to the observer – such as habitat, substrate, location and time. Thanks to rich metadata, precise annotations, and baselines available to all competitors, the challenge provides a benchmark for image recognition with the use of additional information. Moreover, similarly to SnakeCLEF, the toxicity of a mushroom can be crucial for the decision of a mushroom picker. The task will explore the decision process beyond the commonly assumed 0/1 cost function.

Data Collection: The challenge dataset is based on the DF20 dataset [28], contains 295,938 training images belonging to 1,604 species observed mostly in Denmark. All training samples passed an expert validation process, guaranteeing high quality labels. Rich observation metadata about habitat, substrate, time, location, EXIF etc. are provided. The challenge comes with two different test sets: (i) The first is unique in its annotation process, as all test images belong to physical samples sent for DNA sequencing. (ii) The second, with approximately 60k images, covers the whole year and includes observations collected across all substrate and habitat types.

Task Description: Given the set of images and corresponding metadata, the goal of the task is to return for each image a ranked list of species sorted according to the likelihood of the species appearing in the image. A baseline procedure to include meta-data in the decision problem, as well as pre-trained baseline image classifiers, will be provided as part of the task description to all participants.

7   Timeline and Registration Instructions

All information about the timeline and participation in the challenges is provided on the LifeCLEF 2022 web pages [3]. The challenges themselves are ran on the AIcrowd platform [1] and the Kaggle platform [2] for the registration, the submission of runs, the display of the leaderboard, etc.

8   Conclusions and Perspectives

To fully reach its objective, an evaluation campaign such as LifeCLEF requires a long-term research effort so as to (i) encourage non-incremental contributions, (ii) measure consistent performance gaps, (iii) progressively scale-up the problem and (iv) enable the emergence of a strong community. The 2022 edition of the lab supports this vision and also includes the following innovations:

- A new task on fungi recognition from images and metadata.
- A widening of the plant task at the scale of the world flora (100K-300K species).
- The inclusion of new data for the bird task with a focus on unsupervised training, stereo audio recordings and concrete conservation use cases.
- The inclusion of presence-absence test data for the GeoLifeCLEF challenge.
- The evaluation of decision problems for poisonous and venomous species identification represents a task beyond 0/1 cost function, not represented in computer vision benchmarks.

Acknowledgements. This project has received funding from the European Union's Horizon 2020 research and innovation programme under grant agreement No° 863463 (Cos4Cloud project), and the support of #DigitAG.

References

1. AICrowd. https://www.aicrowd.com/
2. Kaggle. https://www.kaggle.com/
3. LifeCLEF (2022). https://www.imageclef.org/LifeCLEF2022
4. Affouard, A., Goeau, H., Bonnet, P., Lombardo, J.C., Joly, A.: Pl@ntnet app in the era of deep learning. In: 5th International Conference on Learning Representations (ICLR 2017), 24–26 April 2017, Toulon, France (2017)
5. Goëau, H., et al.: Plant Identification: Experts vs. Machines in the Era of Deep Learning. In: Joly, A., Vrochidis, S., Karatzas, K., Karppinen, A., Bonnet, P. (eds.) Multimedia Tools and Applications for Environmental & Biodiversity Informatics. MSA, pp. 131–149. Springer, Cham (2018). https://doi.org/10.1007/978-3-319-76445-0_8
6. Cai, J., Ee, D., Pham, B., Roe, P., Zhang, J.: Sensor network for the monitoring of ecosystem: Bird species recognition. In: Intelligent Sensors, Sensor Networks and Information, 2007. ISSNIP 2007. 3rd International Conference on (2007). https://doi.org/10.1109/ISSNIP.2007.4496859
7. Gaston, K.J., O'Neill, M.A.: Automated species identification: why not? Philos. Trans. Royal Soc. London B: Biol. Sci. **359**(1444), 655–667 (2004)
8. Ghazi, M.M., Yanikoglu, B., Aptoula, E.: Plant identification using deep neural networks via optimization of transfer learning parameters. Neurocomputing **235**, 228–235 (2017)
9. Glotin, H., Clark, C., LeCun, Y., Dugan, P., Halkias, X., Sueur, J.: Proceeding 1st workshop on Machine Learning for Bioacoustics - ICML4B. ICML, Atlanta USA (2013). http://sabiod.org/ICML4B2013_book.pdf
10. Goëau, H., et al.: The imageclef 2013 plant identification task. In: CLEF task Overview 2013, CLEF: Conference and Labs of the Evaluation Forum, Sep. 2013, Valencia, Spain. Valencia (2013)
11. Goëau, H., et al.: The imageclef 2011 plant images classification task. In: CLEF task Overview 2011, CLEF: Conference and Labs of the Evaluation Forum, Sep. 2011, Amsterdam, Netherlands. (2011)
12. Goëau, H., et al.: Imageclef 2012 plant images identification task. In: CLEF Task Overview 2012, CLEF: Conference and Labs of the Evaluation Forum, Sep. 2012, Rome, Italy. Rome (2012)
13. Joly, A., et al.: Interactive plant identification based on social image data. Ecol. Inf. **23**, 22–34 (2014)
14. Joly, A., et al.: Overview of LifeCLEF 2018: a large-scale evaluation of species identification and recommendation algorithms in the era of ai. In: Jones, G.J., et al. (eds.) CLEF: Cross-Language Evaluation Forum for European Languages. Experimental IR Meets Multilinguality, Multimodality, and Interaction, vol. LNCS. Springer, Avigon, France (Sep 2018)
15. Joly, A., et al.: Overview of LifeCLEF 2019: Identification of Amazonian Plants, South & North American Birds, and Niche Prediction. In: Crestani, F., et al. (eds.) CLEF 2019 - Conference and Labs of the Evaluation Forum. Experimental IR Meets Multilinguality, Multimodality, and Interaction, vol. LNCS, pp. 387–401. Lugano, Switzerland (Sep 2019). https://doi.org/10.1007/978-3-030-28577-7_29, https://hal.umontpellier.fr/hal-02281455

16. Joly, A., et al.: LifeCLEF 2016: Multimedia Life Species Identification Challenges. In: Fuhr, N., et al. (eds.) CLEF: Cross-Language Evaluation Forum. Experimental IR Meets Multilinguality, Multimodality, and Interaction, vol. LNCS, pp. 286–310. Springer, Évora, Portugal (Sep 2016). https://doi.org/10.1007/978-3-319-44564-9_26, https://hal.archives-ouvertes.fr/hal-01373781

17. Joly, A., et al.: LifeCLEF 2017 Lab Overview: Multimedia Species Identification Challenges. In: Jones, G.J., et al. (eds.) CLEF: Cross-Language Evaluation Forum. Experimental IR Meets Multilinguality, Multimodality, and Interaction, vol. LNCS, pp. 255–274. Springer, Dublin, Ireland (Sep 2017). https://doi.org/10.1007/978-3-319-65813-1_24, https://hal.archives-ouvertes.fr/hal-01629191

18. Joly, A., et al.: LifeCLEF 2014: Multimedia Life Species Identification Challenges. In: CLEF: Cross-Language Evaluation Forum. Information Access Evaluation. Multilinguality, Multimodality, and Interaction, vol. LNCS, pp. 229–249. Springer International Publishing, Sheffield, United Kingdom (Sep 2014). https://doi.org/10.1007/978-3-319-11382-1_20, https://hal.inria.fr/hal-01075770

19. Joly, A., et al.: Lifeclef 2015: multimedia life species identification challenges. In: Experimental IR Meets Multilinguality, Multimodality, and Interaction, pp. 462–483. Springe, Chem (2015)

20. Joly, A., et al.: Overview of lifeclef 2020: a system-oriented evaluation of automated species identification and species distribution prediction. In: International Conference of the Cross-Language Evaluation Forum for European Languages, pp. 342–363. Springer, Chem (2020)

21. Joly, A., et al.: Overview of lifeclef 2021: an evaluation of machine-learning based species identification and species distribution prediction. In: International Conference of the Cross-Language Evaluation Forum for European Languages, pp. 371–393. Springer, Chem (2021)

22. Koh, L.P., Dunn, R.R., Sodhi, N.S., Colwell, R.K., Proctor, H.C., Smith, V.S.: Species coextinctions and the biodiversity crisis. Science **305**(5690), 1632–1634 (2004)

23. Lee, D.J., Schoenberger, R.B., Shiozawa, D., Xu, X., Zhan, P.: Contour matching for a fish recognition and migration-monitoring system. In: Optics East, pp. 37–48. International Society for Optics and Photonics (2004)

24. Lee, S.H., Chan, C.S., Remagnino, P.: Multi-organ plant classification based on convolutional and recurrent neural networks. IEEE Trans. Image Process. **27**(9), 4287–4301 (2018)

25. NIPS International Conference on Neural Information Processing Scaled for Bioacoustics, from Neurons to Big Data (2013). http://sabiod.org/nips4b

26. Norouzzadeh, M.S., Morris, D., Beery, S., Joshi, N., Jojic, N., Clune, J.: A deep active learning system for species identification and counting in camera trap images. Methods Ecol. Evol. **12**(1), 150–161 (2021)

27. Picek, L., Ruiz De Castañeda, R., Durso, A.M., Sharada, P.M.: Overview of the snakeclef 2020: Automatic snake species identification challenge. In: CLEF task overview 2020, CLEF: Conference and Labs of the Evaluation Forum, Sep. 2020, Thessaloniki, Greece (2020)

28. Picek, L., Sulc, M., Matas, J., Heilmann-Clausen, J., Jeppesen, T., Læssøe, T., Frøslev, T.: Danish fungi 2020 - not just another image recognition dataset. In: Proceedings of the IEEE/CVF Winter Conference on Applications of Computer Vision (WACV) (2022)

29. Picek, L., Durso, A.M., Ruiz De Castañeda, R., Bolon, I.: Overview of SnakeCLEF 2021: Automatic snake species identification with country-level focus. In: Working Notes of CLEF 2021 - Conference and Labs of the Evaluation Forum (2021)

30. Towsey, M., Planitz, B., Nantes, A., Wimmer, J., Roe, P.: A toolbox for animal call recognition. Bioacoustics **21**(2), 107–125 (2012)
31. Trifa, V.M., Kirschel, A.N., Taylor, C.E., Vallejo, E.E.: Automated species recognition of antbirds in a Mexican rainforest using hidden Markov models. J. Acoust. Soc. Am. **123**, 2424 (2008)
32. Van Horn, G., Mac Aodha, O., Song, Y., Cui, Y., Sun, C., Shepard, A., Adam, H., Perona, P., Belongie, S.: The inaturalist species classification and detection dataset. CVPR (2018)
33. Villon, S., Mouillot, D., Chaumont, M., Subsol, G., Claverie, T., Villéger, S.: A new method to control error rates in automated species identification with deep learning algorithms. Sci. Reports **10**(1), 1–13 (2020)
34. Wäldchen, J., Mäder, P.: Machine learning for image based species identification. Methods Ecol. Evol. **9**(11), 2216–2225 (2018)
35. Wäldchen, J., Rzanny, M., Seeland, M., Mäder, P.: Automated plant species identification-trends and future directions. PLoS Comput. Biol. **14**(4), e1005993 (2018)
36. Yu, X., Wang, J., Kays, R., Jansen, P.A., Wang, T., Huang, T.: Automated identification of animal species in camera trap images. EURASIP J. Image Video Process. **2013**(1), 1–10 (2013). https://doi.org/10.1186/1687-5281-2013-52

The ChEMU 2022 Evaluation Campaign: Information Extraction in Chemical Patents

Yuan Li[1], Biaoyan Fang[1], Jiayuan He[1,2], Hiyori Yoshikawa[1,3],
Saber A. Akhondi[4], Christian Druckenbrodt[5], Camilo Thorne[5], Zenan Zhai[1],
Zubair Afzal[4], Trevor Cohn[1], Timothy Baldwin[1], and Karin Verspoor[1,2(✉)]

[1] The University of Melbourne, Melbourne, Australia
karin.verspoor@rmit.edu.au
[2] RMIT University, Melbourne, Australia
[3] Fujitsu Limited, Kawasaki, Japan
[4] Elsevier BV, Amsterdam, The Netherlands
[5] Elsevier Information Systems GmbH, Frankfurt, Germany

Abstract. The discovery of new chemical compounds is a key driver of the chemistry and pharmaceutical industries, and many other industrial sectors. Patents serve as a critical source of information about new chemical compounds. The ChEMU (Cheminformatics Elsevier Melbourne Universities) lab addresses information extraction over chemical patents and aims to advance the state of the art on this topic. ChEMU lab 2022, as part of the 13th Conference and Labs of the Evaluation Forum (CLEF-2022), will be the third ChEMU lab. The ChEMU 2020 lab provided two information extraction tasks, named entity recognition and event extraction. The ChEMU 2021 lab introduced two more tasks, chemical reaction reference resolution and anaphora resolution. For ChEMU 2022, we plan to re-run all the four tasks with a new task on semantic classification for tables as the fifth one. In this paper, we introduce ChEMU 2022, including its motivation, goals, tasks, resources, and evaluation framework.

Keywords: Named entity recognition · Event extraction · Anaphora resolution · Reaction reference resolution · Table classification · Chemical patents · Text mining

1 Overview

The ChEMU campaign focuses on information extraction tasks over chemical reactions in patents. The ChEMU2020 lab [5,6,12] provided two information extraction tasks, named entity recognition and event extraction. The ChEMU 2021 lab [4,9,10] introduced two more tasks, chemical reaction reference resolution and anaphora resolution. This year, we plan to re-run all the four tasks with a new task on semantic classification for tables as the fifth one. Together, the tasks support comprehensive automatic chemical patent analysis.

M. Hagen et al. (Eds.): ECIR 2022, LNCS 13186, pp. 400–407, 2022.
https://doi.org/10.1007/978-3-030-99739-7_50

1.1   Why Is This Campaign Needed?

The discovery of new chemical compounds is a key driver of the chemistry and pharmaceutical industries, and many other industrial sectors. Patents serve as a critical source of information about new chemical compounds. Compared with journal publications, patents provide more timely and comprehensive information about new chemical compounds [1,2,13], since they are usually the first venues where new chemical compounds are disclosed. Despite the significant commercial and research value of the information in patents, manual effort is still the primary mechanism for extracting and organising this information. This is costly, considering the large volume of patents available [7,11]. Development of automatic natural language processing (NLP) systems for chemical patents, which aim to convert text corpora into structured knowledge about chemical compounds, has become a focus of recent research [6,8].

1.2   How Would the Community Benefit from the Campaign?

There are three key benefits of this campaign to our community. First, our tasks provide a unique chance for NLP experts to develop information extraction models for chemical patents and gain experience in analysing the linguistic properties of patent documents. Second, several high-quality data sets will be released for a range of complex information extraction tasks that have applicability beyond the chemical domain. Finally, the tasks provided in this campaign focus on the field of information extraction over chemical literature, which is an active research area. The campaign will provide strong baselines as well as a useful resource for future research in this area.

1.3   Usage Scenarios

The details of chemical synthesis are critical for tasks including drug design and analysis of environmental or health impacts of material manufacturing. A key usage scenario for ChEMU is population of databases collecting detailed information about chemicals such as Reaxys®,[1] The tasks within the ChEMU 2022 lab will lead towards detailed understanding of complex descriptions of chemicals, chemical properties, and chemical reactions in chemical patents, addressing a number of natural language processing challenges involving both local and longer-distance relations and table analysis.

2   Tasks

We first briefly introduce the tasks from previous years, then describe the new table classification task. For more details about previous tasks, please refer to the corresponding overview paper ChEMU 2020 [6,12], ChEMU 2021 [4,9], and our website hosting the shared tasks[2].

---

[1] Reaxys® Copyright ©2021 Elsevier Life Sciences IP Limited. Reaxys is a trademark of Elsevier Life Sciences IP Limited, used under license. https://www.reaxys.com.

[2] http://chemu.eng.unimelb.edu.au/.

2.1   Task 1 Expression-Level Information Extraction

Task 1 consists of three sub-tasks, i.e. named entity recognition, event extraction, and anaphora resolution, since they only consider entities or relations between them within a few consecutive sentences.

In our ChEMU corpus, every snippet has been annotated for all three tasks, which opens the opportunity to explore multi-task learning since the input data is the same for all three tasks, as illustrated in Table 1. Fang et al. [3] extended coreference resolution with four other bridging relations as the anaphora resolution task. Results show that the performance of coreference resolution model can be further improved if bridging relation annotations are also available on the same data and the model is jointly trained for 5 relations instead of just coreference. One possible explanation for this is that a large part of the jointly trained model is shared for both coreference resolution and bridging relation tasks so effectively the jointly trained model is making use of more data which reduces the risk of overfitting and improves its ability to generalization. We expect more exploration towards this direction.

Task 1a Named Entity Recognition. This task aims to identify chemical compounds and their specific types. In addition, this task also requires identification of the temperatures and reaction times at which the chemical reaction is carried out, as well as yields obtained for the final chemical product and the label of the reaction. In total, the participants need to find 10 types of named entities.

Task 1b Event Extraction. A chemical reaction leading to an end product often consists of a sequence of individual event steps. This task is to identify those steps which involve chemical entities recognized from Task 1a. It requires identification of event trigger words (e.g. "added" and "stirred") and then determination of the chemical entity arguments of these events.

Task 1c Anaphora Resolution. This task requires the resolution of anaphoric dependencies between expressions in chemical patents. The participants are required to find five types of anaphoric relationships in chemical patents, i.e. coreference, transformed, reaction-associated, work-up and contained.

2.2   Task 2 Document-Level Information Extraction

Tasks 2 groups together the two tasks chemical reaction reference resolution and table semantic classification, since both of these tasks take a complete patent document as input rather than the short snippet extracts of Task 1. This increases the complexity of the task from a language processing perspective. The reaction references can relate reaction descriptions that are far apart, and the semantics of a table may depend on linguistic context from the document structure or content (Table 2).

Table 1. Illustration of three tasks performed on the same snippet, namely, Task 1a Named Entity Recognition (NER), Task 1b Event Extraction (EE), and Task 1c Anaphora Resolution (AR).

Text	The title compound was used without purification (1.180 g, 95.2%) as yellow solid
NER	The **title compound** was used without purification (**1.180 g**, **95.2%**) as yellow solid
	REACTION_PRODUCT: **title compound**
	YIELD_OTHER: **1.180 g**
	YIELD_PERCENT: **95.2%**
EE	The **title compound** was *used* without purification (**1.180 g**, **95.2%**) as yellow solid
	REACTION_STEP: *used* → REACTION_PRODUCT: **title compound**
	REACTION_STEP: *used* → YIELD_OTHER: **1.180 g**
	REACTION_STEP: *used* → YIELD_PERCENT: **95.2%**
AR	**The title compound** was used without purification (**1.180 g, 95.2%**) as *yellow solid*
	COREFERENCE: *yellow solid* → **The title compound (1.180 g, 95.2%)**

Table 2. An example for Task 2a chemical reaction reference resolution, where reaction 2 (RX2) is producing Compound B13 following the procedure that reaction 1 (RX1) produces Compound B11.

	Text
RX1	A mixture of the obtained ester, ... was stirred under argon and heated at 110 °C. for 24 h. ... Column chromatography of the residue (silica gel-hexane/ethyl acetate, 9:1) gave **Compound B11**, ...
	...
RX2	Using 2-ethoxyethanol and following the procedure for **Compound B11** gave Compound B13, bis(2-ethoxyethyl) 3,3'-((2-(bromomethyl)-2-((3-((2-ethoxyethoxy)carbonyl)phenoxy)methyl)propane-1,3-diyl)bis(oxy))dibenzoate,

Some preliminary results on these tasks show that traditional machine learning models perform reasonably well and can sometimes do better than neural network models, especially on minority classes. It would be interesting to see if there exists a combined model that has the best of two worlds.

Task 2a Chemical Reaction Reference Resolution. Given a reaction description, this task requires identifying references to other reactions that the reaction relates to, and to the general conditions that it depends on. The participants are required to find pairs of reactions where one of them is the general condition for or is analogous to the other reaction.

Task 2b Table Semantic Classification. This task is about categorising tables in chemical patents based on their contents, which supports identification of tables containing key information. We define 8 types of tables as shown in Table 3. Figure 1 shows an example SPECT table. Please refer to Zhai et al. [15] for the dataset and Zhai et al. [16] for more details on the settings of this task.

Table 3. 8 labels defined for Task 2b semantic classification on tables, and examples of expected content.

Label	Description	Examples
SPECT	Spectroscopic data	Mass spectrometry, IR/NMR spectroscopy
PHYS	Physical data	Melting point, quantum chemical calculations
IDE	Identification of compounds	Chemical name, structure, formula, label
RX	All properties of reactions	Starting materials, products, yields
PHARM	Pharmacological data	Pharmacological usage of chemicals
COMPOSITION	Compositions of mixtures	Compositions made up by multiple ingredients
PROPERTY	Properties of chemicals	The time of resistance of a photoresist
OTHER	Other tables	–

Ex.	Structure	Purification, Physical properties
3	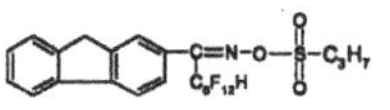	Recrystallization from 2-propanol ^{1}H-NMR and ^{19}F-NMR ($CDCl_3$) δ [ppm]: 1.10 (t, 3H), 1.87-1.98 (m, 2H), 3.39 (t, 2H), 3.98 (s, 2H), 6.05 (tt, 1 H), 7.33-7.43 (m, 3H), 7.54-7.62 (m, 2H), 7.84 (d, 1 H), 7.88 (d, 1H), -137.40 (d, 2F), - 129.74 (s, 2F), -123.80 (s, 2F), -121.43 (s, 2F), -120.55 (s, 2F), -109.83 (s, 2F), tentatively assigned as E-configuration White solid, mp: 66-68°C
4	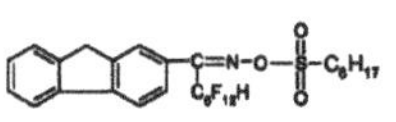	Recrystallization from 2-propanol ^{1}H-NMR and ^{19}F-NMR ($CDCl_3$). δ [ppm]: 0.89 (t, 3H), 1.20-1.50 (m, 10H), 1.83-1.96 (m, 2H), 3.40 (t, 2H), 3.98 (s, 2H), 6.05 (tt, 1H), 7.33-7.48 (m, 3H), 7.53-7.63 (m, 2H), 7.88 (d, 1H), 7.88 (d, 1H), - 137.47 (d, 2F), -129.75 (s, 2F), -123.81 (s, 2F), -121.45 (s, 2F), -120.02 (s, 2F), - 109.81 (s, 2F), tentatively assigned as E-configuration White solid, mp: 78-79°C

Fig. 1. An example table in SPECT category.

2.3 Changes Proposed for Rerunning Previous Tasks

The number of participating teams in ChEMU 2021 was much lower than that in ChEMU 2020 (2 vs. 11 teams). We believe the primary reason for this was

Table 4. A summary of the information about participation, data, and baseline models for all tasks. NER is short for Named Entity Recognition, EE for Event Extraction, AR for Anaphora Resolution, CR3 for Chemical Reaction Reference Resolution, TSC for Table Semantic Classification.

Task	Continued?	Data	Baseline models
1a NER	2020 task 1	Existing 1500 snippets as train and	He et al. [6]
1b EE	2020 task 2	dev sets. 500 new snippets will be	
1c AR	2021 task 2	annotated and used as the test set	Fang et al. [3]
2a CR3	2021 task 1	Data for ChEMU 2021 will be reused	Yoshikawa et al. [14]
2b TSC	New task	All the data is ready for release	Zhai et al. [16]

that the time given to participants was too short. The data for both tasks of ChEMU 2021 was released in early April, while the deadline for submitting the final predictions on test set was in mid-May, which left only 6 weeks to the participants to build and test their models. Additionally, the pandemic is not over yet, and one team mentioned that they faced several related challenges. Both teams that participated in ChEMU 2021 Task 2 asked for extensions to the various deadlines. This year, we will release the data for ChEMU 2022 by the end of this year, so that the participants will have a few months instead of a few weeks to work on them.

Furthermore, we will simplify the two tasks from ChEMU 2021 (2022 Tasks 1c and 2a), by providing the gold spans of mentions and chemical reactions, respectively. Since both teams have proposed a few potential directions for improving their relation extraction component, we hope to support exploration of more ideas on the this part. The simplification will also make it easier for participants to build models, and could potentially attract more people.

2.4   Data and Evaluation

A new corpus for Task 2b of 788 patents containing annotated tables will be first split into training, development, and test sets according to 60%/15%/25% portion. The training and development sets will be released in December, and the test set without annotations will be released one week before the evaluation deadline.

Data for other tasks will be released following the same schedule. For the three tasks of Task 1, the data released for ChEMU 2020 and 2021 (1500 snippets) will serve as the training and development sets, while 500 new snippets will be annotated for all three tasks and used as the test set. Since no one participated in Task 2a (ChEMU 2021 Task 1), its test set is untouched. Therefore, the data for this task will be reused as is for ChEMU 2022.

For evaluation, standard precision, recall, and F1 score will be used. For each task, we will take the model from our published papers as strong baselines and make them available to all participants, as shown in Table 4.

3   Conclusion

In this paper, we have presented a brief description of the upcoming ChEMU lab at CLEF-2022 including the re-run of all four tasks from ChEMU 2020/2021 and a new table semantic classification task.

We expect participants from both academia and industry and will advertise our tasks via social media and NLP-related mailing lists. In addition, we will invite previous participants and authors who have submitted to Frontiers In Research Metrics and Analytics special issue (Information Extraction from Bio-Chemical Text) to join ChEMU 2022.

References

1. Akhondi, S.A., et al.: Automatic identification of relevant chemical compounds from patents. Database **2019**, baz001 (2019)
2. Bregonje, M.: Patents: a unique source for scientific technical information in chemistry related industry? World Patent Inf. **27**(4), 309–315 (2005)
3. Fang, B., Druckenbrodt, C., Akhondi, S.A., He, J., Baldwin, T., Verspoor, K.M.: ChEMU-Ref: a corpus for modeling anaphora resolution in the chemical domain. In: Merlo, P., Tiedemann, J., Tsarfaty, R. (eds.) Proceedings of the 16th Conference of the European Chapter of the Association for Computational Linguistics: Main Volume, EACL 2021, Online, 19–23 April 2021, pp. 1362–1375. Association for Computational Linguistics (2021). https://www.aclweb.org/anthology/2021.eacl-main.116/
4. He, J., et al.: ChEMU 2021: reaction reference resolution and Anaphora resolution in chemical patents. In: Hiemstra, D., Moens, M.-F., Mothe, J., Perego, R., Potthast, M., Sebastiani, F. (eds.) ECIR 2021. LNCS, vol. 12657, pp. 608–615. Springer, Cham (2021). https://doi.org/10.1007/978-3-030-72240-1_71
5. He, J., et al.: Overview of ChEMU 2020: named entity recognition and event extraction of chemical reactions from patents. In: Arampatzis, A., et al. (eds.) CLEF 2020. LNCS, vol. 12260, pp. 237–254. Springer, Cham (2020). https://doi.org/10.1007/978-3-030-58219-7_18
6. He, J., et al.: ChEMU 2020: natural language processing methods are effective for information extraction from chemical patents. Frontiers Res. Metrics Anal. **6**, 654438 (2021). https://doi.org/10.3389/frma.2021.654438
7. Hu, M., Cinciruk, D., Walsh, J.M.: Improving automated patent claim parsing: dataset, system, and experiments. arXiv preprint arXiv:1605.01744 (2016)
8. Krallinger, M., Leitner, F., Rabal, O., Vazquez, M., Oyarzabal, J., Valencia, A.: CHEMDNER: the drugs and chemical names extraction challenge. J. Cheminform. **7**(1), 1–11 (2015)
9. Li, Y., et al.: Overview of ChEMU 2021: reaction reference resolution and Anaphora resolution in chemical patents. In: Candan, K.S., et al. (eds.) CLEF 2021. LNCS, vol. 12880, pp. 292–307. Springer, Cham (2021). https://doi.org/10.1007/978-3-030-85251-1_20
10. Li, Y., et al.: Extended overview of ChEMU 2021: reaction reference resolution and anaphora resolution in chemical patents. In: Faggioli, G., Ferro, N., Joly, A., Maistro, M., Piroi, F. (eds.) Proceedings of the Working Notes of CLEF 2021 - Conference and Labs of the Evaluation Forum, Bucharest, Romania, 21st–24th September 2021. CEUR Workshop Proceedings, vol. 2936, pp. 693–709. CEUR-WS.org (2021). http://ceur-ws.org/Vol-2936/paper-58.pdf

11. Muresan, S., et al.: Making every SAR point count: the development of chemistry connect for the large-scale integration of structure and bioactivity data. Drug Discovery Today **16**(23–24), 1019–1030 (2011)
12. Nguyen, D.Q., et al.: ChEMU: named entity recognition and event extraction of chemical reactions from patents. In: Jose, J.M., et al. (eds.) ECIR 2020. LNCS, vol. 12036, pp. 572–579. Springer, Cham (2020). https://doi.org/10.1007/978-3-030-45442-5_74
13. Senger, S., Bartek, L., Papadatos, G., Gaulton, A.: Managing expectations: assessment of chemistry databases generated by automated extraction of chemical structures from patents. J. Cheminform. **7**(1), 1–12 (2015). https://doi.org/10.1186/s13321-015-0097-z
14. Yoshikawa, H., et al.: Chemical reaction reference resolution in patents. In: Proceedings of the 2nd Workshop on on Patent Text Mining and Semantic Technologies (2021)
15. Zhai, Z., et al.: ChemTables: dataset for table classification in chemical patents (2021). https://doi.org/10.17632/g7tjh7tbrj.3
16. Zhai, Z., et al.: ChemTables: a dataset for semantic classification on tables in chemical patents. J. Cheminform. **13**(1), 97 (2021). https://doi.org/10.1186/s13321-021-00568-2

Advancing Math-Aware Search: The ARQMath-3 Lab at CLEF 2022

Behrooz Mansouri[1(✉)], Anurag Agarwal[1], Douglas W. Oard[2], and Richard Zanibbi[1]

[1] Rochester Institute of Technology, Rochester, NY, USA
{bm3302,axasma,rxzvcs}@rit.edu
[2] University of Maryland, College Park, MD, USA
oard@umd.edu

Abstract. ARQMath-3 is the third edition of the Answer Retrieval for Questions on Math lab at CLEF. In addition to the two main tasks from previous years, an interesting new pilot task will also be run. The main tasks include: (1) Answer Retrieval, returning posted answers to mathematical questions taken from a community question answering site (Math Stack Exchange (MSE)), and (2) Formula Retrieval, returning formulas and their associated question/answer posts in response to a query formula taken from a question. The previous ARQMath labs created a large new test collection, new evaluation protocols for formula retrieval, and established baselines for both main tasks. This year we will pilot a new *open domain* question answering task as Task 3, where questions from Task 1 may be answered using passages from documents from outside of the ARQMath collection, and/or that are generated automatically.

Keywords: Community question answering · Formula retrieval · Mathematical Information Retrieval · Math-aware search · Open domain QA

1 Introduction

Effective question answering systems for math would be valuable for math Community Question Answering (CQA) forums, and more broadly for the Web at large. Community Question Answering sites for mathematics such as Math Stack Exchange[1] (MSE) and Math Overflow [12] are widely-used resources. This indicates that there is great interest in finding answers to mathematical questions posed in natural language, using *both* text and mathematical notation.

The ARQMath lab [6,17] was established to support research into retrieval models that incorporate mathematical notation. With a number of Math Information Retrieval (MIR) systems having been introduced recently [1,5,8,9,15], a standard MIR benchmark is essential for understanding the behavior of retrieval

---

[1] https://math.stackexchange.com.

M. Hagen et al. (Eds.): ECIR 2022, LNCS 13186, pp. 408–415, 2022.
https://doi.org/10.1007/978-3-030-99739-7_51

Table 1. Example ARQMath-2 queries and results.

Question Answering (Task 1)	Formula Retrieval (Task 2)
QUESTION (TOPIC A.220) I'm having a difficult time understanding how to give a combinatorics proof of the identity $$\sum_{k=0}^{n} \binom{x+k}{k} = \binom{x+n+1}{n}$$	FORMULA QUERY (TOPIC B.220) I'm having a difficult time understanding how to give a combinatorics proof of the identity $$\sum_{k=0}^{n} \binom{x+k}{k} = \binom{x+n+1}{n}$$
RELEVANT The right side is the number of ways of choosing n elements from $\{1, 2, 3, ..., 2n\}$. The number of ways of choosing n elements from that set that starting with $1, 2, ..., n - k$ and not containing $n - k + 1$ is $\binom{n+k-1}{k}$.	RELEVANT Question: prove by induction on n+m the combinatoric identity: $$\sum_{k=0}^{n} \binom{m+k}{k} = \binom{m+n+1}{n}$$ I've tried to do on both n and m ...
NON-RELEVANT Hint: Find a combinatorial argument for which $$\sum_{k=0}^{n} \binom{n}{k}\binom{k}{2} = \binom{n}{2} 2^{n-2}$$ then use the previous identity. ...	NON-RELEVANT Hint $$\sum_{k=0}^{n} \binom{n}{k} x^k = (1+x)^n$$ Integrate twice both rhs and lhs with respect to x and when finished, plug $x = 1$ in your result.

models and implementations. To that end, the previous ARQMath collections produced a new collection, assessment protocols, parsing and evaluation tools, and a benchmark containing over 140 annotated topics for each of two tasks: math question answer retrieval, and formula retrieval.[2]

ARQMath is the first shared-task evaluation on question answering for math. Using formulae and text in posts from Math Stack Exchange (MSE), participating systems are given a question and asked to return potential answers. Relevance is determined by how well returned posts answer the provided question. The left column of Table 1 shows an example topic from Task 1 (ARQMath-2 Topic A.220), showing one answer assessed as relevant, and another assessed as non-relevant. The goal of Task 2 is retrieving relevant formulae for a formula query taken from a question post (e.g., the formula in the question post shown at top-right in Table 1), where relevance is determined *in-context*, based on the question post for the query formula and the question/answer posts in which retrieved formulae appears. This task is illustrated in the right column of Table 1 (ARQMath-2 Topic B.220).

Before ARQMath, early benchmarks for math-aware search were developed through the National Institute of Informatics (NII) Testbeds and Community for Information Access Research (at NTCIR-10 [2], NTCIR-11 [3] and NTCIR-12 [16]). The Mathematical Information Retrieval (MathIR) evaluations at NTCIR included tasks for both structured "text + math" queries and isolated formula retrieval, using collections created from arXiv and Wikipedia. ARQMath complements the NTCIR test collections by introducing additional test collections based on naturally occurring questions, by assessing formula relevance in context, and by substantially increasing the number of topics.

---

[2] https://www.cs.rit.edu/~dprl/ARQMath.

In this paper, we summarize existing data and tools, the second edition of the ARQMath lab, and planned changes for ARQMath-3. Briefly, ARQMath-3 will reuse the ARQMath collection, which consists of MSE posts from 2010 to 2018. The most substantial change is the addition of a new open domain question answering task as a pilot task.[3] Unlike our ongoing answer retrieval task, in which the goal is to return existing answers, for the new open domain question answering task systems may retrieve and/or generate answers, such as was done previously for the Question Answering tracks at TREC-8 [13] through TREC-13 [14], and the Conversational Question Answering Challenge [10].

2   ARQMath Tasks

ARQMath-3 will include the same two tasks as ARQMath-1 and -2, and it introduces a pilot task on open domain question answering for math, where external knowledge sources may be used to find, filter, and even generate answers.

2.1   Task 1: Answer Retrieval

The primary task for the ARQMath labs is answer retrieval, in which participants are presented with a question posted on MSE **after** 2018, and are asked to return a ranked list of up to 1,000 answers from **prior years** (2010–2018). In each lab, the participating teams ranked answer posts for 100 topics. In ARQMath-1 77 and in ARQMath-2 71 topics were assessed and used for the evaluation. In ARQMath-1, for primary runs the pooling depth was 50 and 20 for other runs. In ARQMath-2, these values were adjusted to 45 and 15 because the number of runs doubled, and participating teams also nearly doubled.

Table 2 summarizes the graded relevance scale used for assessment. System results ('runs') were evaluated using the nDCG' measure (read as "nDCG-prime"), introduced by Sakai [11] as the primary measure. nDCG' is simply normalized Discounted Cumulative Gain (nDCG), but with unjudged documents removed before scoring. Two additional measures, mAP' and P'@10, were also reported using binarized relevance judgments. In both labs, participants were allowed to submit up to 5 runs, with at least one designated as primary.

2.2   Task 2: Formula Retrieval

The ARQMath formula retrieval task has some similarity to the Wikipedia Formula Browsing Task from NTCIR-12 [16]. In the NTCIR-12 task, given a single query formula, similar formulae in a collection were to be returned. The NTCIR-12 formula browsing task test collection had only 20 formula queries (plus 20 modified versions with wildcards added), whereas in ARQMath-1, 74 queries (45 for evaluation + 29 additional for future training) and in ARQMath-2, 70 queries (58 for evaluation + 12 additional) were assessed.

---

[3] As proposed by Vít Novotný at CLEF 2021.

ARQMath has introduced two innovations for formula search evaluation. First, in ARQMath, relevance is decided by context, whereas in NTCIR-12, formula queries were compared by assessors with retrieved formula instances without consideration of the context for either. Second, in NTCIR-12 systems could receive credit for finding formula instances, whereas in ARQMath systems receive credit for finding *visually distinct* formulae. In other words, an NTCIR-12 system that found identical formulae in two different documents and returned that formula twice would get credit twice, whereas an ARQMath system would receive credit only once for each visually distinct formula. Deduplication of visually identical/near-identical formulae was done using Symbol Layout Trees produced from Presentation MathML by Tangent-S [4] where possible, and by comparing LATEX strings otherwise. In ARQMath-1, this clustering was done *post hoc* on submitted runs; for ARQMath-2 this clustering was done *a priori* on the full collection and shared with participating teams. In ARQMath-3 the cluster ids will again be provided with the collection. For efficiency reasons, we have limited the number of instances of any visually distinct formula that were assessed to 5 in ARQMath-1 and -2, and expect the same for ARQMath-3.

The relevance of a *visually distinct* formula is defined by the maximum relevance for any of its pooled instances, based on the associated question/answer post for each instance. Table 2 summarizes the graded relevance scale used for assessment. Here relevance is interpreted as the likelihood of a retrieved formula being *associated with* information that helps answer the question in which a formula query appeared. There is an important difference in relevance assessment for Task 2 in ARQMath-1 and -2: although the relevance scale shown in Table 2 was unchanged between ARQMath-1 and ARQMath-2, we did change how the table was interpreted for ARQMath-2. In ARQMath-1, only the context in the question post associated with the query formula was considered, with ARQMath-1 assessors instructed to mark exact matches as relevant. This was changed when we noticed that visually identical formulas at times had no bearing on the information represented by a query formula within its associated question post. As an example, we can have two visually identical formulae, but where one represents operations on sets, and the other operations on integers.

2.3   Task 3 (Pilot): Open Domain QA for Math

In this pilot task, participants are given Task 1 topics and asked to provide a single answer for each question that must not exceed a fixed maximum length. Unlike Task 1 where answers are taken from the MSE collection, answers may be produced using any technique, and any available knowledge sources (with the exception of MSE answers from 2019 to the present). For example, responses may be produced using any technique, and any available knowledge sources (with the exception of MSE answers from 2019 to the present). For example, responses may be a new machine-generated response, a single passage or complete answer post from MSE or another CQA platform (e.g., Math Overflow), or some combination of generated and existing content. For relevance assessment, responses from open domain QA systems will be included in the Task 1 pools. Rankings obtained from the Task 1 relevance measures will be compared with rankings produced by automated answer quality measures (e.g., derived from BLEU [7]) to assess

Table 2. Relevance scores, ratings, and definitions for tasks 1 and 2.

SCORE	RATING	DEFINITION
TASK 1: QUESTION ANSWERING		
3	High	Sufficient to answer the complete question on its own
2	Medium	Provides some path towards the solution. This path might come from clarifying the question, or identifying steps towards a solution
1	Low	Provides information that could be useful for finding or interpreting an answer, or interpreting the question
0	Not Relevant	Provides no information pertinent to the question or its answers. A post that restates the question without providing any new information is considered non-relevant
TASK 2: FORMULA RETRIEVAL		
3	High	Just as good as finding an exact match to the query formula would be
2	Medium	Useful but not as good as the original formula would be
1	Low	There is some chance of finding something useful
0	Not Relevant	Not expected to be useful

whether these measures may be used reliably to evaluate future systems. Task 3 answers will be further assessed separately for aspects such as fluency, and whether answers appear to be human-generated or machine-generated (for this, we may include MSE posts alongside Task 3 submissions).

3   The ARQMath Test Collection

ARQMath uses Math Stack Exchange (MSE) as its collection, which is freely available through the Internet Archive. The ARQMath collection contains MSE posts published from 2010 to 2018, with a total of 1 million questions and 1.4 million answers. In ARQMath-1, posts from 2019, and in ARQMath-2 posts from 2020 were used for topic construction. For ARQMath-3, posts from 2021 will be used.[4] Topic questions must contain at least one formula; with this constraint, 89,905 questions are available for ARQMath-3 topic development.

Topics. In previous ARQMath labs, topics were annotated with three categories: complexity, dependency, and type. In ARQMath-1, more than half of the Task 1 topics were categorized as questions seeking a proof. We aimed to better balance across question categories in ARQMath-2, but when category combinations are considered the Task 1 topic set still exhibited considerable skew towards a few combinations. In ARQMath-3, we introduce a fourth category, *parts*, which indicates whether a topic question calls for an answer that has a

---

[4] From a September 7, 2021 snapshot.

single part, or whether it contains sub-questions that each call for answers.[5] We do see that different systems seem to be doing better on different ARQMath-1 and ARQMath-2 question categories, so in ARQMath-3 we continue to aim to balance the topic selection process across combinations of question categories as best we can, including the new *parts* category.

Formulae. In the Internet Archive version of the collection, formulae appear between two '$' or '$$' signs, or inside a 'math-container' tag. For ARQMath, all posts (and all MSE comments on those posts) have been processed to extract formulae, assigning a unique identifier to each formula instance. Each formula is provided in three encodings: (a) as LaTeX strings, (b) as (appearance-based) Presentation MathML, and (c) as (operator tree) Content MathML.

The open source LaTeXML[6] tool we use for converting LaTeX to MathML fails for some MSE formulae. Moreover, producing Content MathML from LaTeX requires inference, and is thus potentially errorful. As a result, the coverage of Presentation MathML for detected formulae in the ARQMath-1 collection was 92%, and the coverage for Content MathML was 90%. For ARQMath-2, after LaTeXML updates the error rate was reduced to less than one percent for both representations, reducing the need for participating systems to fall back to using LaTeX. However, there are some remaining MathML encoding issues and formula parsing/clustering failures in the ARQMath-2 collection that we plan to correct in ARQMath-3.

Files. As with any CQA task, the ARQMath collection contains more than just question and answer posts. We distribute the collection as four main files:

- **Posts.** The post file contains a unique identifier for each question or answer post, along with information such as creation date and creator. Question posts contain a title and a body (with the body being the question), while answer posts have a body and the unique identifier of the associated question.
- **Comments.** Any post can have one or more comments, each having a unique id and the unique identifier of the associated post.
- **Votes.** This file records positive and negative votes for posts, along with additional annotations such as 'offensive' or 'spam.'
- **Users.** Posters of questions and answers have a unique User ID and a reputation score.

4   Conclusion

For ARQMath-3, we will continue our focus on answering math questions (Tasks 1 and 3), with formula search as the secondary task (Task 2). For question answering, we are adding a new pilot task for open domain QA (Task 3) alongside the answer retrieval task (Task 1). A single Math Stack Exchange (MSE) collection will again be used. This is both because MSE models an actual usage

---

[5] This is based on a suggestion at CLEF 2021 from Frank Tompa.

[6] https://dlmf.nist.gov/LaTeXML/.

scenario, and because we expect that reusing MSE will facilitate training and refinement of increasingly capable systems.

Acknowledgement. This material is based upon work supported by the Alfred P. Sloan Foundation under Grant No. G-2017-9827 and the National Science Foundation (USA) under Grant No. IIS-1717997. We thank Vítek Novotný for providing details for Task 3.

References

1. Ahmed, S., Davila, K., Setlur, S., Govindaraju, V.: Equation attention relationship network (EARN): A geometric deep metric framework for learning similar math expression embedding. In: 2020 25th International Conference on Pattern Recognition (ICPR), pp. 6282–6289. IEEE (2021)
2. Aizawa, A., Kohlhase, M., Ounis, I.: NTCIR-10 math pilot task overview. In: Proceedings of the 10th NTCIR Conference, pp. 654–661 (2013)
3. Aizawa, A., Kohlhase, M., Ounis, I., Schubotz, M.: NTCIR-11 Math-2 task overview. In: Proceedings of the 11th NTCIR Conference, pp. 88–98 (2014)
4. Davila, K., Zanibbi, R.: Layout and semantics: combining representations for mathematical formula search. In: Proceedings of the 40th International ACM SIGIR Conference on Research and Development in Information Retrieval, pp. 1165–1168 (2017)
5. Mansouri, B., Zanibbi, R., Oard, D.W.: Learning to rank for mathematical formula retrieval. In: Proceedings of the 44th International ACM SIGIR Conference on Research and Development in Information Retrieval, pp. 952–961 (2021)
6. Mansouri, B., Zanibbi, R., Oard, D.W., Agarwal, A.: Overview of ARQMath-2 (2021): second CLEF lab on answer retrieval for questions on math. In: Candan, K.S., et al. (eds.) CLEF 2021. LNCS, vol. 12880, pp. 215–238. Springer, Cham (2021). https://doi.org/10.1007/978-3-030-85251-1_17
7. Papineni, K., Roukos, S., Ward, T., Zhu, W.J.: Bleu: A method for automatic evaluation of machine translation. In: Proceedings of the 40th Annual Meeting of the Association for Computational Linguistics, pp. 311–318 (2002)
8. Peng, S., Yuan, K., Gao, L., Tang, Z.: MathBERT: A pre-trained model for mathematical formula understanding. arXiv preprint arXiv:2105.00377 (2021)
9. Pfahler, L., Morik, K.: Semantic search in millions of equations. In: Proceedings of the 26th ACM SIGKDD International Conference on Knowledge Discovery & Data Mining, pp. 135–143 (2020)
10. Reddy, S., Chen, D., Manning, C.D.: CoQA: A conversational question answering challenge. Trans. Assoc. Comput. Linguist. **7**, 249–266 (2019). https://doi.org/10.1162/tacl_a_00266
11. Sakai, T.: Alternatives to BPREF. In: Proceedings of the 30th Annual International ACM SIGIR Conference on Research and Development in Information Retrieval, pp. 71–78 (2007)
12. Tausczik, Y.R., Kittur, A., Kraut, R.E.: Collaborative problem solving: a study of MathOverflow. In: Proceedings of the 17th ACM conference on Computer Supported Cooperative Work & Social Computing, pp. 355–367 (2014)
13. Voorhees, E.M.: The TREC-8 question answering track report. In: Proceedings of the Eighth Text REtrieval Conference (TREC-8), pp. 77–82 (1999)

14. Voorhees, E.M.: Overview of the TREC 2004 question answering track. In: Proceedings of the Thirteenth Text REtrieval Conference (TREC 2004), pp. 52–62 (2005)
15. Wang, Z., Lan, A., Baraniuk, R.: Mathematical formula representation via tree embeddings (2021). https://people.umass.edu/~andrewlan/papers/preprint-forte.pdf
16. Zanibbi, R., Aizawa, A., Kohlhase, M., Ounis, I., Topic, G., Davila, K.: NTCIR-12 MathIR task overview. In: Proceedings of the 12th NTCIR Conference, pp. 299–308 (2016)
17. Zanibbi, R., Oard, D.W., Agarwal, A., Mansouri, B.: Overview of ARQMath 2020: CLEF lab on answer retrieval for questions on math. In: Arampatzis, A., et al. (eds.) CLEF 2020. LNCS, vol. 12260, pp. 169–193. Springer, Cham (2020). https://doi.org/10.1007/978-3-030-58219-7_15

The CLEF-2022 CheckThat! Lab on Fighting the COVID-19 Infodemic and Fake News Detection

Preslav Nakov[1]([⊠]), Alberto Barrón-Cedeño[2], Giovanni Da San Martino[3], Firoj Alam[1], Julia Maria Struß[4], Thomas Mandl[5], Rubén Míguez[6], Tommaso Caselli[7], Mucahid Kutlu[8], Wajdi Zaghouani[9], Chengkai Li[10], Shaden Shaar[11], Gautam Kishore Shahi[12], Hamdy Mubarak[1], Alex Nikolov[13], Nikolay Babulkov[14], Yavuz Selim Kartal[15], and Javier Beltrán[6]

[1] Qatar Computing Research Institute, HBKU, Doha, Qatar
{pnakov,fialam,hmubarak}@hbku.edu.qa
[2] DIT, Università di Bologna, Forlì, Italy
a.barron@unibo.it
[3] University of Padova, Padua, Italy
dasan@math.unipd.it
[4] University of Applied Sciences Potsdam, Potsdam, Germany
struss@fh-potsdam.de
[5] University of Hildesheim, Hildesheim, Germany
mandl@uni-hildesheim.de
[6] Newtral Media Audiovisual, Madrid, Spain
{ruben.miguez,javier.beltran}@newtral.es
[7] University of Groningen, Groningen, The Netherlands
t.caselli@rug.nl
[8] TOBB University of Economics and Technology, Ankara, Turkey
m.kutlu@etu.edu.tr
[9] Hamad Bin Khalifa University, Doha, Qatar
wzaghouani@hbku.edu.qa
[10] University of Texas at Arlington, Arlington, TX, USA
cli@cse.uta.edu
[11] Cornell University, Ithaca, NY, USA
ss2753@cornell.edu
[12] University of Duisburg-Essen, Duisburg, Germany
gautam.shahi@uni-due.de
[13] CheckStep, London, UK
alex.nikolov@checkstep.com
[14] Sofia University, Sofia, Bulgaria
[15] GESIS - Leibniz Institute for the Social Sciences, Mannheim, Germany
yavuzselim.kartal@gesis.org

Abstract. The fifth edition of the CheckThat! Lab is held as part of the 2022 Conference and Labs of the Evaluation Forum (CLEF). The lab evaluates technology supporting various factuality tasks in seven languages: Arabic, Bulgarian, Dutch, English, German, Spanish, and Turkish. Task 1 focuses on disinformation related to the ongoing COVID-19

© The Author(s), under exclusive license to Springer Nature Switzerland AG 2022
M. Hagen et al. (Eds.): ECIR 2022, LNCS 13186, pp. 416–428, 2022.
https://doi.org/10.1007/978-3-030-99739-7_52

infodemic and politics, and asks to predict whether a tweet is worth fact-checking, contains a verifiable factual claim, is harmful to the society, or is of interest to policy makers and why. Task 2 asks to retrieve claims that have been previously fact-checked and that could be useful to verify the claim in a tweet. Task 3 is to predict the veracity of a news article. Tasks 1 and 3 are classification problems, while Task 2 is a ranking one.

Keywords: Fact-checking · Disinformation · Misinformation · Check-worthiness · Verified claim retrieval · Fake news · Factuality · COVID-19

1 Introduction

The mission of the `CheckThat!` lab is to foster the development of technology to assist in the process of fact-checking news articles, political debates, and social media posts. Four editions of the lab have been held previously, targeting various natural language processing and information retrieval tasks related to factuality.

The 2018 edition of the lab focused on check-worthiness and fact-checking of claims in political debates [54]. The 2019 edition covered the various modules necessary to verify a claim: from check-worthiness, to ranking and classification of evidence in the form of Web pages, to actual fact-checking of claims against specific text snippets [24,25]. The 2020 edition featured three main tasks: detecting previously fact-checked claims, evidence retrieval, and actual fact-checking of claims [9,11]. Similarly, the 2021 edition focused on detecting check-worthy claims, previously fact-checked claims, and fake news [56,57]. Whereas the first editions focused mostly on political debates and speeches, and eventually tweets, the 2021 edition added the verification of news articles. Notably, all editions covered one of the most important initial stages in the fact-checking process: the identification of check-worthy claims—in debates, speeches, press conferences, and tweets. Finally, over the years, `CheckThat!` has witnessed an expansion in terms of language coverage, going from two (Arabic and English) to seven languages now (Arabic, Bulgarian, Dutch, English, German, Spanish, and Turkish).

The 2022 edition of the lab features three tasks to foster the technology on three timely problems in multiple languages.[1] Task 1 asks to detect relevant tweets: check-worthy, verifiable, harmful, and attention-worthy. Task 2 aims at detecting previously fact-checking claims. Task 3 focuses on checking the factuality of news articles. Automated systems to detect and to verify such multifaceted aspects can be very useful as supportive technology for investigative journalism, as they could provide help and guidance, thus saving time [3,27,37,39,55,81]. For example, a system could automatically identify check-worthy claims, could make sure they have not been fact-checked before by a reputable fact-checking organization, and can then present them to a journalist for further analysis in a ranked list. Similarly, a system can identify harmful and attention-worthy social

---

[1] http://sites.google.com/view/clef2022-checkthat/.

media content to support different stakeholders in their day-to-day decision-making process.

2   Description of the Tasks

The lab is organized around three tasks, each of which in turn has several subtasks. Figure 1 shows the full CheckThat! verification pipeline, with the three tasks we target this year highlighted.

2.1   Task 1: Identifying Relevant Claims in Tweets

Task 1 has four subtasks, three binary and one multi-class; Table 1 shows the class labels for each task. More detail about the original dataset on which this task is based can be found in [2,3].

Subtask 1A: Check-Worthiness Estimation Given a tweet, predict whether it is worth fact-checking by professional fact-checkers.

Subtask 1B: Verifiable Factual Claims Detection. Given a tweet, predict whether it contains a verifiable factual claim.

Subtask 1C: Harmful Tweet Detection. Given a tweet, predict whether it is harmful to the society.

Subtask 1D: Attention-Worthy Tweet Detection. Given a tweet, predict whether it should get the attention of policy makers and why.

2.2   Task 2: Detecting Previously Fact-Checked Claims

Given a check-worthy claim, and a set of previously-checked claims, determine whether the claim has been previously fact-checked with respect to a collection of fact-checked claims [67,69].

Subtask 2A: Detecting Previously Fact-checked Claims From Tweets. Given a tweet, detect whether the claim the tweet makes was previously fact-checked with respect to a collection of previously fact-checked claims. This is a ranking task, where the systems are asked to produce a list of top-n candidates.

Subtask 2B: Detecting Previously Fact-Checked Claims in Political Debates/Speeches. Within the context of a political debate or a speech, detect whether a claim has been previously fact-checked with respect to a collection of previously fact-checked claims. This is a ranking task, where systems are asked to produce a list of top-n candidates. It is offered in English only.

2.3   Task 3: Fake News Detection

This task targets news articles. Given the text and the title of an article, determine whether the main claim made in the article is true, partially true, false, or other (e.g., articles in dispute and unproven articles) [75,76]. This task is offered as a monolingual task in English and an English-German cross-lingual task.

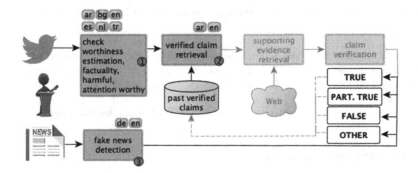

Fig. 1. The full verification pipeline. The lab covers three tasks from this pipeline: 1. identifying relevant claims in tweets, 2. verified claim retrieval, and 3. fake news detection. The languages covered are Arabic (ar), Bulgarian (bg), Dutch (nl), English (en), German (de), Spanish (es), and Turkish (tr).

Table 1. Overview of the classes for Subtasks 1A, 1B, 1C, and 1D.

Subtask 1A	Subtask 1C	Subtask 1D	
1. No	1. No	1. No	6. Yes, contains advice
2. Yes	2. Yes	2. Yes, asks question	7. Yes, discusses action taken
Subtask 1B		3. Yes, blame authorities	8. Yes, discusses cure
1. No		4. Yes, calls for action	9. Yes, other
2. Yes		5. Yes, Harmful	

3   Data

Table 2 summarizes the data available for each task and for each language.

Task 1: Identifying Relevant Claims in Tweets. We have more than 34K annotations about several topics, including COVID-19 and politics, which cover all subtasks 1A-1D [10,57].

Task 2: Detecting Previously Fact-Checked Claims. For Subtask 2A, we have 1,400 annotated examples ranging from 2016 till 2021. The test set comes from Snopes. For more details, refer to [71]. For Subtask 2B, we have 800 claims available for training [70].

Task 3: Fake News Detection We have 1,254 annotated examples in English on various topics including COVID-19, climate change, and politics from the 2021 edition of the lab [77]. We provide a new test set for English and a similarly annotated test data set in German.

Table 2. Data for all tasks. *VC*: Verified claim, *Input-VC*: Input–verified claim pair.

Task	ar	bl	en	nl	es	tr	Task	ar	en	Task	en	de
1A	3.4k	2.6k	2.9k	1.2k	15.0k	3.3k	2A: Input-claim	858	1.4k	2B: Input-claim	669	
1B	5.0k	3.7k	4.5k	2.7k			2A: Input-VC	1.0k	1.4k	2B: Input-VC	804	
1C	5.0k	3.7k	4.5k	2.7k			2A: VC	30.3k	13.8k	2B: VC	19.2k	
1D	5.0k	3.7k	4.5k	2.7k						3	1.3k	400*

*The process of crawling and annotating the data is not finished, and the final number will vary.

4 Evaluation Measures

For the classification tasks 1A and 1C, we use the F_1 measure with respect to the positive class, for Task 1B, we use accuracy, and for Task 1D, we use weighted-F_1. For the ranking problems in Tasks 2A and 2B, the official evaluation measure is *Mean Average Precision* (MAP), as in the two previous editions of these tasks. We also report reciprocal rank, and $P@k$ for $k \in \{1, 3, 5, 10, 20, 30\}$. Finally, for Task 3, we use macro-average F_1-measure as the official evaluation measure.

5 Previously in the CheckThat! Lab

Four editions of CheckThat! have been held so far, and some of the tasks in the 2022 edition are closely related to tasks from previous editions. Thus, considering the most successful approaches applied in the past is a good starting point to address the 2022 tasks. Below, we briefly discuss the tasks from previous years.

5.1 CheckThat! 2021

Task 1$_{2021}$. Given a topic and a set of potentially related tweets (or political debates/speeches), rank the tweets (or the sentences in the political debate/speech) according to their check-worthiness for the topic. BERT, AraBERT, and RoBERTa were by far the most popular large-scale pre-trained language models for the task [72,83]. Other approaches used WordNet [85] and LIWC [66].

Task 2$_{2021}$. Given a tweet, a political debate, or a speech, detect whether the claim it makes was previously fact-checked with respect to a collection of fact-checked claims. The most successful approaches were based on AraBERT, RoBERTa, and Sentence-BERT [18,47,65].

Task 3$_{2021}$. Given the text and the title of a news article, determine whether the main claim it makes is true, partially true, false, or other. Also, identify the domain of the article: health, crime, climate, elections, or education. The most successful pre-trained language model was RoBERTa [7,19,44]. Ensembles were also popular, with components using BERT [44] and LSTMs [19,44].

5.2 CheckThat! 2020

Task 1_{2020}. Given a topic and a stream of potentially-related tweets, rank the tweets by check-worthiness for the topic [35,73]. The most successful runs adopted state-of-the-art transformers. The top-ranked teams for the English version of this task used BERT [17] and RoBERTa [60,82]. For the Arabic version, the top systems used AraBERT [42,82] and a multilingual BERT [34].

Task 2_{2020}. Given a check-worthy claim and a collection of previously verified claims, rank these verified claims, so that those that verify the input claim (or a sub-claim in it) are ranked on top [73]. The most effective approaches fine-tuned BERT or RoBERTa [13].

Task 3_{2020}. Given a check-worthy claim in a tweet on a specific topic and a set of text snippets extracted from potentially relevant Web pages, return a ranked list of evidence snippets for the claim.

Task 4_{2020}. Given a check-worthy claim on a specific topic and a set of potentially-relevant Web pages, predict the veracity of the claim [35]. The top model used a scoring function that computes the degree of concordance and negation between a claim and all input text snippets for that claim [80].

Task 5_{2020}. Given a debate segmented into sentences, together with speaker information, prioritize sentences for fact-checking [73]. Only one out of eight runs outperformed a strong bi-LSTM baseline [46].

5.3 CheckThat! 2019

Task 1_{2019}. Given a political debate, an interview, or a speech, rank its sentences by the priority with which they should be fact-checked [5]. The most successful approaches used neural networks for the individual classification of the instances, e.g., based on domain-specific word embeddings, syntactic dependencies, and LSTMs [33].

Task 2_{2019}. Given a claim and a set of potentially relevant Web pages, identify which of these pages (and passages thereof) are useful for a human to fact-check the claim. Finally, determine the factuality of the claim [36]. The best approach used textual entailment and external data [28].

5.4 CheckThat! 2018

Task 1_{2018}. It was identical to Task 1_{2019} [4]. The best approaches used *pseudo-speeches* as a concatenation of all interventions by a debater [87], and represented the entries with embeddings, POS tags, and syntactic dependencies [32].

Task 2_{2018}. Given a check-worthy claim in the form of a (transcribed) sentence, determine whether the claim is true, half-true, or false [12]. The best approach grabbed information from the Web and fed the claim with the most similar Web text into a CNN [32].

6   Related Work

There has been a lot of research on checking the factuality of a claim, of a news article, or of an information source [6,8,41,45,51,59,64,68,86]. Special attention has been paid to disinformation and misinformation in social media [30,43,49, 74,78,84], more recently with focus on fighting the COVID-19 infodemic [2,3, 52,53]. Check-worthiness estimation is still an understudied problem, especially in social media [27,37–40,81], and fake news detection for news articles is mostly approached as a binary classification problem [59,61].

CheckThat! is related to several tasks at SemEval: on determining rumour veracity [22,29], on stance detection [50], on fact-checking in community question answering forums [48], on propaganda detection [21,23], and on semantic textual similarity [1,58]. It is also related to the FEVER task [79] on fact extraction and verification, to the Fake News Challenge [31,63], to the FakeNews task at MediaEval [62], as well as to the NLP4IF tasks on propaganda detection [20] and on fighting the COVID-19 infodemic in social media [68].

7   Conclusion

We presented the 2022 edition of the CheckThat! lab, which features tasks that span the full fact-checking pipeline: from spotting check-worthy claims to checking whether an input claim has been fact-checked before. We further have a fake news detection task. Last but not least, in line with one of the main missions of CLEF, we promote multi-linguality by offering tasks in seven languages: Arabic, Bulgarian, Dutch, English, German, Spanish, and Turkish.

Acknowledgments. This research is part of the Tanbih mega-project, developed at the Qatar Computing Research Institute, HBKU, which aims to limit the impact of "fake news", propaganda, and media bias, thus promoting media literacy and critical thinking. The Arabic annotation effort was partially made possible by NPRP grant NPRP13S-0206-200281 from the Qatar National Research Fund (a member of Qatar Foundation).

References

1. Agirre, E., et al.: SemEval-2016 task 1: semantic textual similarity, monolingual and cross-lingual evaluation. In: Proceedings of the 10th International Workshop on Semantic Evaluation. SemEval 2016, pp. 497–511 (2016)
2. Alam, F., et al.: Fighting the COVID-19 infodemic in social media: a holistic perspective and a call to arms. In: Proceedings of the International AAAI Conference on Web and Social Media, ICWSM 2021, pp. 913–922 (2021)
3. Alam, F., et al.: Fighting the COVID-19 infodemic: Modeling the perspective of journalists, fact-checkers, social media platforms, policy makers, and the society. In: Findings of EMNLP 2021, pp. 611–649 (2021)
4. Atanasova, P., et al.: Overview of the CLEF-2018 CheckThat! lab on automatic identification and verification of political claims. Task 1: Check-worthiness. In: Cappellato et al. [16]

5. Atanasova, P., Nakov, P., Karadzhov, G., Mohtarami, M., Da San Martino, G.: Overview of the CLEF-2019 CheckThat! lab on automatic identification and verification of claims. Task 1: Check-worthiness. In: Cappellato et al. [15]
6. Ba, M.L., Berti-Equille, L., Shah, K., Hammady, H.M.: VERA: a platform for veracity estimation over web data. In: Proceedings of the 25th International Conference on World Wide Web. WWW 2016, pp. 159–162 (2016)
7. Balouchzahi, F., Shashirekha, H., Sidorov, G.: MUCIC at CheckThat! 2021:FaDofake news detection and domain identification using transformersensembling. In: Faggioli et al. [26], pp. 455–464D
8. Baly, R., et al.: What was written vs. who read it: news media profiling using text analysis and social media context. In: Proceedings of the 58th Annual Meeting of the Association for Computational Linguistics. ACL 2020, pp. 3364–3374 (2020)
9. Barrón-Cedeño, A., et al.: **CheckThat!** at CLEF 2020: enabling the automatic identification and verification of claims in social media. In: Jose, J.M., et al. (eds.) ECIR 2020. LNCS, vol. 12036, pp. 499–507. Springer, Cham (2020). https://doi.org/10.1007/978-3-030-45442-5_65
10. Barrón-Cedeño, A., et al.: Overview of CheckThat! 2020 – automatic identification and verification of claims in social media. In: Proceedings of the 11th International Conference of the CLEF Association: Experimental IR Meets Multilinguality, Multimodality, and Interaction. CLEF 2020, pp. 215–236 (2020)
11. Barrón-Cedeño, A., et al.: Overview of CheckThat! 2020: automatic identification and verification of claims in social media. In: Arampatzis, A., et al. (eds.) CLEF 2020. LNCS, vol. 12260, pp. 215–236. Springer, Cham (2020). https://doi.org/10.1007/978-3-030-58219-7_17
12. Barrón-Cedeño, A., et al.: Overview of the CLEF-2018 CheckThat! lab on automatic identification and verification of political claims. Task 2: factuality. In: Cappellato et al. [16]
13. Bouziane, M., Perrin, H., Cluzeau, A., Mardas, J., Sadeq, A.: Buster.AI at CheckThat! 2020: Insights and recommendations to improve fact-checking. In: Cappellato et al. [14]
14. Cappellato, L., Eickhoff, C., Ferro, N., Névéol, A. (eds.): CLEF 2020 Working Notes. CEUR Workshop Proceedings (2020)
15. Cappellato, L., Ferro, N., Losada, D., Müller, H. (eds.): Working Notes of CLEF 2019 Conference and Labs of the Evaluation Forum. CEUR Workshop Proceedings (2019)
16. Cappellato, L., Ferro, N., Nie, J.Y., Soulier, L. (eds.): Working Notes of CLEF 2018-Conference and Labs of the Evaluation Forum. CEUR Workshop Proceedings (2018)
17. Cheema, G.S., Hakimov, S., Ewerth, R.: Check_square at CheckThat! 2020: claim detection in social media via fusion of transformer and syntacticfeatures. In: Cappellato et al. [14]
18. Chernyavskiy, A., Ilvovsky, D., Nakov, P.: Aschern at CLEF CheckThat! 2021: lambda-calculus of fact-checked claims. In: Faggioli et al. [26]
19. Cusmuliuc, C.G., Amarandei, M.A., Pelin, I., Cociorva, V.I., Iftene, A.: UAICS at CheckThat! 2021: fake news detection. In: Faggioli et al. [26]
20. Da San Martino, G., Barrón-Cedeño, A., Nakov, P.: Findings of the NLP4IF-2019 shared task on fine-grained propaganda detection. In: Proceedings of the Second Workshop on Natural Language Processing for Internet Freedom: Censorship, Disinformation, and Propaganda. NLP4IF 2019, pp. 162–170 (2019)

21. Da San Martino, G., Barrón-Cedeno, A., Wachsmuth, H., Petrov, R., Nakov, P.: SemEval-2020 task 11: detection of propaganda techniques in news articles. In: Proceedings of the 14th Workshop on Semantic Evaluation. SemEval 2020, pp. 1377–1414 (2020)

22. Derczynski, L., Bontcheva, K., Liakata, M., Procter, R., Wong Sak Hoi, G., Zubiaga, A.: SemEval-2017 task 8: RumourEval: determining rumour veracity and support for rumours. In: Proceedings of the 11th International Workshop on Semantic Evaluation. SemEval 2017, pp. 69–76 (2017)

23. Dimitrov, D., et al.: SemEval-2021 task 6: detection of persuasion techniques in texts and images. In: Proceedings of the International Workshop on Semantic Evaluation. SemEval 2021, pp. 70–98 (2021)

24. Elsayed, T., et al.: CheckThat! at CLEF 2019: Automatic identification and verification of claims. In: Azzopardi, L., Stein, B., Fuhr, N., Mayr, P., Hauff, C., Hiemstra, D. (eds.) Advances in Information Retrieval, pp. 309–315. Springer International Publishing, Cham (2019)

25. Elsayed, T., et al.: Overview of the CLEF-2019 CheckThat! lab: automatic identification and verification of claims. In: Crestani, F., et al. (eds.) CLEF 2019. LNCS, vol. 11696, pp. 301–321. Springer, Cham (2019). https://doi.org/10.1007/978-3-030-28577-7_25

26. Faggioli, G., Ferro, N., Joly, A., Maistro, M., Piroi, F. (eds.): CLEF 2021 Working Notes. Working Notes of CLEF 2021-Conference and Labs of the Evaluation Forum (2021)

27. Gencheva, P., Nakov, P., Màrquez, L., Barrón-Cedeño, A., Koychev, I.: A context-aware approach for detecting worth-checking claims in political debates. In: Proceedings of the International Conference Recent Advances in Natural Language Processing. RANLP 2017, pp. 267–276 (2017)

28. Ghanem, B., Glavaš, G., Giachanou, A., Ponzetto, S., Rosso, P., Rangel, F.: UPV-UMA at CheckThat! lab: verifying Arabic claims using cross lingual approach. In: Cappellato et al. [15]

29. Gorrell, G., et al.: SemEval-2019 task 7: RumourEval, determining rumour veracity and support for rumours. In: Proceedings of the 13th International Workshop on Semantic Evaluation. SemEval 2019, pp. 845–854 (2019)

30. Gupta, A., Kumaraguru, P., Castillo, C., Meier, P.: TweetCred: real-time credibility assessment of content on Twitter. In: Aiello, L.M., McFarland, D. (eds.) SocInfo 2014. LNCS, vol. 8851, pp. 228–243. Springer, Cham (2014). https://doi.org/10.1007/978-3-319-13734-6_16

31. Hanselowski, A., et al.: A retrospective analysis of the fake news challenge stance-detection task. In: Proceedings of the 27th International Conference on Computational Linguistics. COLING 2018, pp. 1859–1874 (2018)

32. Hansen, C., Hansen, C., Simonsen, J., Lioma, C.: The Copenhagen team participation in the check-worthiness task of the competition of automatic identification and verification of claims in political debates of the CLEF-2018 fact checking lab. In: Cappellato et al. [16]

33. Hansen, C., Hansen, C., Simonsen, J., Lioma, C.: Neural weakly supervised fact check-worthiness detection with contrastive sampling-based ranking loss. In: Cappellato et al. [15]

34. Hasanain, M., Elsayed, T.: bigIR at CheckThat! 2020: multilingual BERT for ranking Arabic tweets by check-worthiness. In: Cappellato et al. [14]

35. Hasanain, M., et al.: Overview of CheckThat! 2020 Arabic: automatic identification and verification of claims in social media. In: Cappellato et al. [14]

36. Hasanain, M., Suwaileh, R., Elsayed, T., Barrón-Cedeño, A., Nakov, P.: Overview of the CLEF-2019 CheckThat! lab on automatic identification and verification of claims. Task 2: evidence and factuality. In: Cappellato et al. [15]
37. Hassan, N., Li, C., Tremayne, M.: Detecting check-worthy factual claims in presidential debates. In: Proceedings of the 24th ACM International on Conference on Information and Knowledge Management. CIKM 2015, pp. 1835–1838 (2015)
38. Hassan, N., Tremayne, M., Arslan, F., Li, C.: Comparing automated factual claim detection against judgments of journalism organizations. In: Computation+Journalism Symposium, pp. 1–5 (2016)
39. Hassan, N., et al.: ClaimBuster: the first-ever end-to-end fact-checking system. Proc. VLDB Endowment 10(12), 1945–1948 (2017)
40. Jaradat, I., Gencheva, P., Barrón-Cedeño, A., Màrquez, L., Nakov, P.: ClaimRank: detecting check-worthy claims in Arabic and English. In: Proceedings of the 2018 Conference of the North American Chapter of the Association for Computational Linguistics: Demonstrations. NAACL-HLT 2018, pp. 26–30 (2018)
41. Karadzhov, G., Nakov, P., Màrquez, L., Barrón-Cedeño, A., Koychev, I.: Fully automated fact checking using external sources. In: Proceedings of the International Conference Recent Advances in Natural Language Processing. RANLP 2017, pp. 344–353 (2017)
42. Kartal, Y.S., Kutlu, M.: TOBB ETU at CheckThat! 2020: prioritizing English and Arabic claims based on check-worthiness. In: Cappellato et al. [14]
43. Kazemi, A., Garimella, K., Gaffney, D., Hale, S.: Claim matching beyond English to scale global fact-checking. In: Proceedings of the 59th Annual Meeting of the Association for Computational Linguistics and the 11th International Joint Conference on Natural Language Processing. ACL-IJCNLP 202, pp. 4504–45171 (2021)
44. Kovachevich, N.: BERT fine-tuning approach to CLEF CheckThat! fake news detection. In: Faggioli et al. [26]
45. Ma, J., et al.: Detecting rumors from microblogs with recurrent neural networks. In: Proceedings of the International Joint Conference on Artificial Intelligence. IJCAI 2016, pp. 3818–3824 (2016)
46. Martinez-Rico, J., Araujo, L., Martinez-Romo, J.: NLP&IR@UNED at CheckThat! 2020: a preliminary approach for check-worthiness and claim retrieval tasks using neural networks and graphs. In: Cappellato et al. [14]
47. Mihaylova, S., Borisova, I., Chemishanov, D., Hadzhitsanev, P., Hardalov, M., Nakov, P.: DIPS at CheckThat! 2021: verified claim retrieval. In: Faggioli et al. [26]
48. Mihaylova, T., Karadzhov, G., Atanasova, P., Baly, R., Mohtarami, M., Nakov, P.: SemEval-2019 task 8: fact checking in community question answering forums. In: Proceedings of the 13th International Workshop on Semantic Evaluation. SemEval 2019, pp. 860–869 (2019)
49. Mitra, T., Gilbert, E.: CREDBANK: a large-scale social media corpus with associated credibility annotations. In: Proceedings of the Ninth International AAAI Conference on Web and Social Media. ICWSM 2015, pp. 258–267 (2015)
50. Mohammad, S., Kiritchenko, S., Sobhani, P., Zhu, X., Cherry, C.: SemEval-2016 task 6: detecting stance in tweets. In: Proceedings of the 10th International Workshop on Semantic Evaluation. SemEval 2016, pp. 31–41 (2016)
51. Mukherjee, S., Weikum, G.: Leveraging joint interactions for credibility analysis in news communities. In: Proceedings of the 24th ACM International Conference on Information and Knowledge Management. CIKM 2015, pp. 353–362 (2015)

52. Nakov, P., Alam, F., Shaar, S., Da San Martino, G., Zhang, Y.: COVID-19 in Bulgarian social media: factuality, harmfulness, propaganda, and framing. In: Proceedings of the International Conference on Recent Advances in Natural Language Processing. RANLP 2021, pp. 997–1009 (2021)

53. Nakov, P., Alam, F., Shaar, S., Da San Martino, G., Zhang, Y.: A second pandemic? Analysis of fake news about COVID-19 vaccines in Qatar. In: Proceedings of Conference on Recent Advances in Natural Language Processing, pp. 1010–1021 (2021)

54. Nakov, P., et al.: Overview of the CLEF-2018 lab on automatic identification and verification of claims in political debates. In: Working Notes of CLEF 2018 - Conference and Labs of the Evaluation Forum. CLEF 2018 (2018)

55. Nakov, P., et al.: Automated fact-checking for assisting human fact-checkers. In: Proceedings of the 30th International Joint Conference on Artificial Intelligence. IJCAI 2021, pp. 4551–4558 (2021)

56. Nakov, P., et al.: The CLEF-2021 CheckThat! lab on detecting check-worthy claims, previously fact-checked claims, and fake news. In: Hiemstra, D., Moens, M.-F., Mothe, J., Perego, R., Potthast, M., Sebastiani, F. (eds.) ECIR 2021. LNCS, vol. 12657, pp. 639–649. Springer, Cham (2021). https://doi.org/10.1007/978-3-030-72240-1_75

57. Nakov, P., et al.: Overview of the CLEF–2021 CheckThat! lab on detecting check-worthy claims, previously fact-checked claims, and fake news. In: Candan, K.S., et al. (eds.) CLEF 2021. LNCS, vol. 12880, pp. 264–291. Springer, Cham (2021). https://doi.org/10.1007/978-3-030-85251-1_19

58. Nakov, P., et al.: SemEval-2016 Task 3: community question answering. In: Proceedings of the 10th International Workshop on Semantic Evaluation. SemEval 2015, pp. 525–545 (2016)

59. Nguyen, V.H., Sugiyama, K., Nakov, P., Kan, M.Y.: FANG: leveraging social context for fake news detection using graph representation. In: Proceedings of the 29th ACM International Conference on Information & Knowledge Management, CIKM 2020, pp. 1165–1174 (2020)

60. Nikolov, A., Da San Martino, G., Koychev, I., Nakov, P.: Team_Alex at CheckThat! 2020: identifying check-worthy tweets with transformer models. In: Cappellato et al. [14]

61. Oshikawa, R., Qian, J., Wang, W.Y.: A survey on natural language processing for fake news detection. In: Proceedings of the 12th Language Resources and Evaluation Conference. LREC 2020, pp. 6086–6093 (2020)

62. Pogorelov, K., et al.: FakeNews: corona virus and 5G conspiracy task at MediaEval 2020. In: Proceedings of the MediaEval 2020 Workshop. MediaEval 2020 (2020)

63. Pomerleau, D., Rao, D.: The fake news challenge: exploring how artificial intelligence technologies could be leveraged to combat fake news (2017). http://www.fakenewschallenge

64. Popat, K., Mukherjee, S., Strötgen, J., Weikum, G.: Credibility assessment of textual claims on the web. In: Proceedings of the 25th ACM International Conference on Information and Knowledge Management. CIKM 2016, pp. 2173–2178 (2016)

65. Pritzkau, A.: NLytics at CheckThat! 2021: check-worthiness estimation as a regression problem on transformers. In: Faggioli et al. [26]

66. Sepúlveda-Torres, R., Saquete, E.: GPLSI team at CLEF CheckThat! 2021: fine-tuning BETO and RoBERTa. In: Faggioli et al. [26]

67. Shaar, S., Alam, F., Da San Martino, G., Nakov, P.: The role of context in detecting previously fact-checked claims. arXiv:2104.07423 (2021)

68. Shaar, S., et al.: Findings of the NLP4IF-2021 shared tasks on fighting the COVID-19 infodemic and censorship detection. In: Proceedings of the Fourth Workshop on NLP for Internet Freedom: Censorship, Disinformation, and Propaganda. NLP4IF 2021, pp. 82–92 (2021)

69. Shaar, S., Alam, F., Martino, G.D.S., Nakov, P.: Assisting the human fact-checkers: detecting all previously fact-checked claims in a document. arXiv preprint arXiv:2109.07410 (2021)

70. Shaar, S., Babulkov, N., Da San Martino, G., Nakov, P.: That is a known lie: Detecting previously fact-checked claims. In: Proceedings of the 58th Annual Meeting of the Association for Computational Linguistics. ACL 2020, pp. 3607–3618 (2020)

71. Shaar, S., et al.: Overview of the CLEF-2021 CheckThat! lab task 2 on detecting previously fact-checked claims in tweets and political debates. In: Faggioli et al. [26]

72. Shaar, S., et al.: Overview of the CLEF-2021 CheckThat! lab task 1 on check-worthiness estimation in tweets and political debates. In: Faggioli et al. [26]

73. Shaar, S., et al.: Overview of CheckThat! 2020 English: automatic identification and verification of claims in social media. In: Cappellato et al. [14]

74. Shahi, G.K.: AMUSED: an annotation framework of multi-modal social media data. arXiv:2010.00502 (2020)

75. Shahi, G.K., Dirkson, A., Majchrzak, T.A.: An exploratory study of COVID-19 misinformation on Twitter. Online Social Networks Media 22, 100104 (2021)

76. Shahi, G.K., Nandini, D.: FakeCovid - a multilingual cross-domain fact check news dataset for COVID-19. In: Workshop Proceedings of the 14th International AAAI Conference on Web and Social Media (2020)

77. Shahi, G.K., Struß, J.M., Mandl, T.: Overview of the CLEF-2021 CheckThat! lab: task 3 on fake news detection. In: Faggioli et al. [26]

78. Shu, K., Sliva, A., Wang, S., Tang, J., Liu, H.: Fake news detection on social media: a data mining perspective. SIGKDD Explor. Newsl. 19(1), 22–36 (2017)

79. Thorne, J., Vlachos, A., Christodoulopoulos, C., Mittal, A.: FEVER: a large-scale dataset for fact extraction and VERification. In: Proceedings of the Conference of the North American Chapter of the Association for Computational Linguistics: Human Language Technologies. NAACL 2018, pp. 809–819 (2018)

80. Touahri, I., Mazroui, A.: EvolutionTeam at CheckThat! 2020: integration of linguistic and sentimental features in a fake news detection approach. In: Cappellato et al. [14]

81. Vasileva, S., Atanasova, P., Màrquez, L., Barrón-Cedeño, A., Nakov, P.: It takes nine to smell a rat: neural multi-task learning for check-worthiness prediction. In: Proceedings of the International Conference on Recent Advances in Natural Language Processing. RANLP 2019, pp. 1229–1239 (2019)

82. Williams, E., Rodrigues, P., Tran, S.: Accenture at CheckThat! 2021: interesting claim identification and ranking with contextually sensitive lexical training data augmentation. In: Faggioli et al. [14]

83. Williams, E., Rodrigues, P., Tran, S.: Accenture at CheckThat! 2021: interesting claim identification and ranking with contextually sensitive lexical training data augmentation. In: Faggioli et al. [26]

84. Zhao, Z., Resnick, P., Mei, Q.: Enquiring minds: early detection of rumors in social media from enquiry posts. In: Proceedings of the 24th International Conference on World Wide Web. WWW 2015, pp. 1395–1405 (2015)

85. Zhou, X., Wu, B., Fung, P.: Fight for 4230 at CLEF CheckThat! 2021: domain-specific preprocessing and pretrained model for ranking claims by check-worthiness. In: Faggioli et al. [26]
86. Zubiaga, A., Liakata, M., Procter, R., Hoi, G.W.S., Tolmie, P.: Analysing how people orient to and spread rumours in social media by looking at conversational threads. PLoS ONE **11**(3), e0150989 (2016)
87. Zuo, C., Karakas, A., Banerjee, R.: A hybrid recognition system for check-worthy claims using heuristics and supervised learning. In: Cappellato et al. [16]

BioASQ at CLEF2022: The Tenth Edition of the Large-scale Biomedical Semantic Indexing and Question Answering Challenge

Anastasios Nentidis[1,2]([✉]), Anastasia Krithara[1], Georgios Paliouras[1], Luis Gasco[3], and Martin Krallinger[3]

[1] National Center for Scientific Research "Demokritos", Athens, Greece
{tasosnent,akrithara,paliourg}@iit.demokritos.gr
[2] Aristotle University of Thessaloniki, Thessaloniki, Greece
nentidis@csd.auth.gr
[3] Barcelona Supercomputing Center, Barcelona, Spain
{luis.gasco,martin.krallinger}@bsc.es

Abstract. The tenth version of the BioASQ Challenge will be held as an evaluation Lab within CLEF2022. The motivation driving BioASQ is the continuous advancement of approaches and tools to meet the need for efficient and precise access to the ever-increasing biomedical knowledge. In this direction, a series of annual challenges are organized, in the fields of large-scale biomedical semantic indexing and question answering, formulating specific shared-tasks in alignment with the real needs of the biomedical experts. These shared-tasks and their accompanying benchmark datasets provide an unique common testbed for investigating and comparing new approaches developed by distinct teams around the world for identifying and accessing biomedical information. In particular, the BioASQ Challenge consists of shared-tasks in two complementary directions: (a) the automated indexing of large volumes of unlabelled biomedical documents, primarily scientific publications, with biomedical concepts, (b) the automated retrieval of relevant material for biomedical questions and the generation of comprehensible answers. In the first direction on semantic indexing, two shared-tasks are organized for English and Spanish content respectively, the latter considering human-interpretable evidence extraction (NER and concept linking) as well. In the second direction, two shared-tasks are organized as well, one for biomedical question answering and one particularly focusing on the developing issue of COVID-19. As BioASQ rewards the approaches that manage to outperform the state of the art in these shared-tasks, the research frontier is pushed towards ensuring that the valuable biomedical knowledge will be identifiable and accessible by the biomedical experts.

Keywords: Biomedical information · Semantic indexing · Question answering

M. Hagen et al. (Eds.): ECIR 2022, LNCS 13186, pp. 429–435, 2022.
https://doi.org/10.1007/978-3-030-99739-7_53

1 Introduction

BioASQ[1] [15] is a series of international challenges and workshops on biomedical semantic indexing and question answering. Each edition of BioASQ is structured into distinct but complementary tasks and sub-tasks relevant to biomedical information access. As a result, the participating teams can focus on particular tasks that are relevant to their specific area of expertise, including but not limited to hierarchical text classification, machine learning, information retrieval and multi-document query-focused summarization. The BioASQ challenge has been running annually since 2012, with more than 80 teams from 20 countries participating in its tasks. The BioASQ workshop has been taking place in the CLEF conference till 2015. In 2016 and 2017 it took place in ACL, in conjunction with BioNLP. In 2018, it took place in EMNLP as an independent workshop. In 2019 the workshop was again an independent workshop in ECML conference. Since 2020 the BioASQ workshop is again part of CLEF.

BioASQ allows multiple teams around the world that work on biomedical information access systems, to compete in the same realistic benchmark datasets and share, evaluate, and compare their ideas and approaches. As BioASQ consistently rewards the most successful approaches in each task and sub-task, it eventually pushes towards systems that outperform previous approaches. Such successful approaches for semantic indexing and question answering, can eventually lead to the development of tools to support more precise access to valuable biomedical knowledge and allow biomedical experts to provide high quality health services. A key contribution of BioASQ is the benchmark datasets developed for its tasks, as well as the corresponding open-source infrastructure developed for running the challenges. The impact of BioASQ is reportedly large, both in research and in industry, aiding the advancement of text mining in bioinformatics and enabling the development of novel computational models for life and health sciences.

2 BioASQ Evaluation Lab 2022

The tenth BioASQ challenge (BioASQ10) will consist of four tasks: *Task a* and *Task DisTEMIST* on indexing of large volumes of unlabeled documents with biomedical concepts, in English and Spanish respectively, *Task b* and *Task Synergy* on providing answers and supporting material to biomedical questions, under two distinct scenarios, as discussed in this section. As *Task a*, *Task b* and *Task Synergy* have also been organized in the context of previous editions of the BioASQ challenge, we refer to the current version of these tasks, in the context of BioASQ10, as *Task 10a*, *Task 10b* and *Task Synergy 10* respectively.

2.1 Task 10a: Large-scale Biomedical Semantic Indexing

BioASQ *task 10a* requires systems to assist the indexing of biomedical literature by automatically assigning MeSH [9] terms to biomedical articles added

---

[1] http://www.bioasq.org.

to the MEDLINE database[2]. In effect, this is a classification task that requires documents to be classified into a hierarchy of classes. Systems participating in *task 10a* are given newly published MEDLINE articles, before experts in NLM (curators) have assigned MeSH terms to them. The systems assign MeSH terms to the documents, which are then compared against the terms assigned by the NLM curators. In this manner, the evaluation of the systems is fully automated on the side of BioASQ and thus can run on a weekly basis throughout the year.

The performance of the systems taking part in *task 10a* is assessed with a range of different measures. Some of them are variants of standard information retrieval measures for multi-label classification problems (e.g. precision, recall, f-measure, accuracy). Additionally, measures that use the MeSH hierarchy to provide a more refined estimate of the systems' performance are used. The official measures for identifying the winners of the task are micro-averaged F-measure (MiF) and the Lowest Common Ancestor F-measure (LCA-F) [5]. In addition, as this task can be considered an extreme multi-label classification problem, rank-based evaluation metrics, such as precision@k and nDCG@k [2], are also being examined.

2.2 Task DisTEMIST: Disease Text Mining and Indexing Shared Task

Diseases are one of the top used semantic types when performing medical literature searches, with estimations that 20% of PubMed queries are related to disorders, diseases, anomalies or syndromes. In the corpus developed for the BioASQ MESINESP Task [4,10], 44% records (108,945 out of 250,539) had at least one disease-related MeSH descriptor, with disease-related MeSH descriptors being 58% for the clinical trials subset. Correct indexing of disease terms is critical for medical information retrieval systems. The novel DisTEMIST task will focus on the recognition and indexing of diseases in medical documents, by posing subtasks on (1) indexing medical documents with controlled terminologies (2) automatic detection indexing textual evidence, i.e. disease entity mentions in text and (3) normalization of these disease mentions to terminologies.

The BioASQ DisTEMIST track will rely primarily on 1,000 clinical case report publications in Spanish (SciELO [12] full text articles) for indexing diseases with concept identifiers from SNOMED-CT [3], MeSH and ICD10-CM[3]. A large silver standard collection of additional case reports and medical abstracts will also be provided. The evaluation of systems for this task will use flat evaluation measures following the *task 10a* [5] track (mainly micro-averaged F-measure, MiF).

2.3 Task 10b: Biomedical Question Answering

BioASQ *task 10b* takes place in two phases. In the first phase (Phase A), the participants are given English questions formulated by biomedical experts. For

---

[2] https://www.nlm.nih.gov/medline/medline_overview.html.

[3] https://www.cdc.gov/nchs/icd/icd10cm.htm.

each question, the participating systems have to retrieve relevant MEDLINE documents and snippets (passages) of these documents. Subsequently, in the second phase (Phase B) of *task 10b*, the participants are given some relevant documents and snippets that the experts themselves have identified (using tools developed in BioASQ [11]). In this phase, they are required to return 'exact' answers, such as names of particular diseases or genes, depending on the type of the question, and 'ideal' answers, which are paragraph-sized summaries of the most important information of the first phase for each question, regardless of its type. A training dataset of 4,239 biomedical questions will be available for participants of *task 10b* to train their systems and about 500 new biomedical questions, with corresponding golden annotations and answers, will be developed for testing the participating systems.

The evaluation of system responses is done both automatically and manually by the experts employing a variety of evaluation measures [6]. In phase A, the official evaluation for document retrieval is based on the Mean Average Precision (MAP) and for snippet retrieval with the F-measure. In phase B, for the exact answers the official evaluation measure depends on the type of the question. For yes/no questions the official measure is the macro-averaged F-Measure on questions with answers *yes* and *no*. The Mean Reciprocal Rank (MRR) is used for factoid questions, where the participants are allowed to return up to five candidate answers. For List questions, the official measure is the mean F-Measure. Finally, for ideal answers, even though automatic evaluation measures are provided and semi-automatic measures [14] are also considered, the official evaluation is still based on manual scores assigned by experts estimating the readability, recall, precision and repetition of each response.

2.4   BioASQ Synergy 10: Question Answering for Developing Issues

The original BioASQ *task b* is structured in a sequence of phases where the experts and the participating systems have a minimal interaction. This is acceptable for research questions that have a clear, undisputed answer. However, for questions on developing topics, such as the COVID-19 pandemic, that may remain open for some time and where new information and evidence appears every day, a more interactive model is needed, aiming at a synergy between the automated question answering systems and the biomedical experts.

In this direction, last year we introduced the BioASQ *task Synergy* which is designed as a continuous dialog, that allows biomedical experts to pose unanswered questions for developing problems, such as COVID-19 and receive the system responses to these questions, including relevant material (documents and snippets) and potential answers. Then, the experts assess these responses, and provide feedback to the systems, in order to improve their responses. This process repeats iteratively with new feedback and new system responses for the same questions, as well as with new additional questions that may arise meanwhile. In each round of this task, new material is also considered based on the current version of the COVID-19 Open Research Dataset (CORD-19) [16]. This year, in the new edition of *task Synergy* in the context of BioASQ10 (*task Synergy*

10), only documents that come from PubMed, PubMed Central or ArXiv are considered. Additionally, the questions are not required to have definite answers and the answers to the questions can be more volatile.

The same evaluation measures used in *task 10b* are also employed in *task Synergy 10* for comparison. However, in order to capture the iterative nature of the task, only new material is considered for the evaluation of a question in each round, an approach known as *residual collection evaluation* [13]. In parallel, additional evaluation metrics are also examined in this direction. Through this task, we aim to facilitate the incremental understanding of COVID-19 and contribute to the discovery of new solutions. For this purpose, the BioASQ infrastructure was adapted and the BioASQ expert community was expanded to address information needs on the COVID-19 pandemic, as well as new developing public health issues in the future.

2.5   BioASQ Datasets and Tools

During the nine years of BioASQ hundreds of systems from research teams around the world have been evaluated on the indexing and the retrieval of hundreds of thousands of fresh biomedical publications. In this direction, BioASQ has developed a lively ecosystem of tools that facilitate research, such as the BioASQ Annotation Tool [11] for question-answering dataset development and a range of evaluation measures for automated assessment of system performance in all tasks. All BioASQ software and datasets are publicly available[4].

In particular, for *task a* on semantic indexing, BioASQ uses the real stream of articles provided by MEDLINE, developing a training dataset of more that 15.6 millions articles and fifteen weekly test sets of around 6,000 articles each.

For the novel *DisTEMIST* task, following the previous MESINESP tracks on medical semantic indexing in Spanish, a dataset of 1,000 semantically annotated medical documents in Spanish labelled with text-bound evidence mentions of diseases together with concept identifiers for entity linking and semantic indexing will be released. Additionally, a set of disease-relevant mentions from over 200,000 biomedical articles in Spanish will be provided. At the same time, for *task b* on biomedical question answering, BioASQ employs a team of trained biomedical experts who provide a set of about 500 questions on their specialized field of expertise annually. For *task 10b* in particular, a set of 4,239 realistic questions accompanied with answers, and supporting evidence (documents and snippets) is already available as a unique resource for the development of question answering systems. In addition, from the two introductory versions of *Synergy 9 task* that took place last year, a dataset of 202 questions on COVID-19 is already available. These questions, are incrementally annotated with different versions of exact and ideal answers, as well as documents and snippets assessed by the experts as relevant or irrelevant. During the *Synergy 10 task* this set will be extended with more than fifty new open questions on COVID-19 and any exist-

---

[4] https://github.com/bioasq.

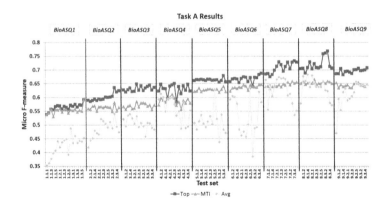

Fig. 1. Performance of the participating systems in *task a*, on semantic indexing. Each year, the participating systems push the state-of-the-art to higher levels

ing questions that remain relevant may be enriched with more updated answers and more recent evidence (documents and snippets).

3   The Impact of BioASQ Results

The impact of BioASQ is reportedly very large, both in research and in industry; it has vastly helped in the advancement of the field of text mining in bioinformatics and has enabled the development of novel computational models for life and health sciences. BioASQ significantly facilitates the exchange and fusion of ideas, as it brings people who work on the same benchmark data together. Therefore, it eventually accelerates progress in the field. For example, the Medical Text Indexer (MTI) [8], which is developed by the NLM to assist in the indexing of biomedical literature, has improved its performance by almost 10% in the last 8 years (Fig. 1). NLM has announced that improvement in MTI is largely due to the adoption of ideas from the systems that compete in the BioASQ challenge [7]. Recently, MTI has reached a performance level that allows it to be used in the fully automated indexing of articles of specific types [1].

Acknowledgments. Google was a proud sponsor of the BioASQ Challenge in 2021. The tenth edition of BioASQ is also sponsored by Atypon Systems inc. The DisTEMIST task is supported by the Spanish Plan for the Advancement of Language Technologies (Plan TL), the 2020 Proyectos de I+D+i-RTI Tipo A (Descifrando El Papel De Las Profesiones En La Salud De Los Pacientes A Traves De La Mineria De Textos, PID2020-119266RA-I00), and HORIZON-CL4-2021-RESILIENCE-01 (BIOMAT+, 101058779).

References

1. Incorporating values for indexing method in medline/pubmed xml. https://www.nlm.nih.gov/pubs/techbull/ja18/ja18_indexing_method.html. Accessed 01 Sep 2019

2. Bhatia, K., et al.: The extreme classification repository: multi-label datasets and code (2016). http://manikvarma.org/downloads/XC/XMLRepository.html
3. Donnelly, K., et al.: Snomed-ct: the advanced terminology and coding system for ehealth. Stud. Health Technol. Inform. **121**, 279 (2006)
4. Gasco, L., et al.: Overview of BioASQ 2021-MESINESP track. Evaluation of advance hierarchical classification techniques for scientific literature, patents and clinical trials. In: CEUR Workshop Proceedings (2021)
5. Kosmopoulos, A., Partalas, I., Gaussier, E., Paliouras, G., Androutsopoulos, I.: Evaluation measures for hierarchical classification: a unified view and novel approaches. Data Min. Knowl. Discov. **29**(3), 820–865 (2014). https://doi.org/10.1007/s10618-014-0382-x
6. Malakasiotis, P., Pavlopoulos, I., Androutsopoulos, I., Nentidis, A.: Evaluation measures for task b. Technical report. BioASQ (2018). http://participants-area.bioasq.org/Tasks/b/eval_meas_2018
7. Mork, J., Aronson, A., Demner-Fushman, D.: 12 years on-is the nlm medical text indexer still useful and relevant? J. Biomed. Semant. **8**(1), 8 (2017)
8. Mork, J., Jimeno-Yepes, A., Aronson, A.: The nlm medical text indexer system for indexing biomedical literature (2013)
9. National Library of Medicine (US): Medical subject headings, vol. 41. US Department of Health and Human Services, Public Health Service, National (2000)
10. Nentidis, A.A., et al.: Overview of BioASQ 2021: the ninth BioASQ challenge on large-scale biomedical semantic indexing and question answering. In: Candan, K.S., et al. (eds.) CLEF 2021. LNCS, vol. 12880, pp. 239–263. Springer, Cham (2021). https://doi.org/10.1007/978-3-030-85251-1_18
11. Ngomo, A.C.N., Heino, N., Speck, R., Ermilov, T., Tsatsaronis, G.: Annotation tool. Project deliverable D3.3 (February 2013). http://www.bioasq.org/sites/default/files/PublicDocuments/2013-D3.3-AnnotationTool.pdf
12. Packer, A.L., et al.: Scielo: uma metodologia para publicação eletrônica. Ciência da informação 27, nd-nd (1998)
13. Salton, G., Buckley, C.: Improving retrieval performance by relevance feedback. J. Am. Soc. Inf. Sci. **41**(4), 288–297 (1990). https://doi.org/10.1002/(SICI)1097-4571(199006)41:4⟨288::AID-ASI8⟩3.0.CO;2-H
14. ShafieiBavani, E., Ebrahimi, M., Wong, R., Chen, F.: Summarization evaluation in the absence of human model summaries using the compositionality of word embeddings. In: Proceedings of the 27th International Conference on Computational Linguistics, pp. 905–914. Association for Computational Linguistics, Santa Fe, New Mexico, USA (August 2018). https://www.aclweb.org/anthology/C18-1077
15. Tsatsaronis, G., et al.: An overview of the bioasq large-scale biomedical semantic indexing and question answering competition. BMC Bioinform. **16**, 138 (2015). https://doi.org/10.1186/s12859-015-0564-6
16. Wang, L.L., et al.: Cord-19: The COVID-19 open research dataset. ArXiv (2020). https://arxiv.org/abs/2004.10706v2

eRisk 2022: Pathological Gambling, Depression, and Eating Disorder Challenges

Javier Parapar[1](✉) , Patricia Martín-Rodilla[1] , David E. Losada[2] ,
and Fabio Crestani[3]

[1] Information Retrieval Lab, Centro de Investigación en Tecnoloxías da Información
e as Comunicacións (CITIC), Universidade da Coruña, A Coruña, Spain
{javierparapar,patricia.martin.rodilla}@udc.es
[2] Centro Singular de Investigación en Tecnoloxías Intelixentes (CiTIUS),
Universidade de Santiago de Compostela, Santiago de Compostela, Spain
david.losada@usc.es
[3] Faculty of Informatics, Università della Svizzera italiana (USI),
Lugano, Switzerland
fabio.crestani@usi.ch

Abstract. In 2017, we launched eRisk as a CLEF Lab to encourage
research on early risk detection on the Internet. The eRisk 2021 was the
fifth edition of the Lab. Since then, we have created a large number of
collections for early detection addressing different problems (e.g., depression, anorexia or self-harm). This paper outlines the work that we have
done to date (2017, 2018, 2019, 2020, and 2021), discusses key lessons
learned in previous editions, and presents our plans for eRisk 2022, which
introduces a new challenge to assess the severity of eating disorders.

1 Introduction

As a part of CLEF (Conference and Labs of the Evaluation Forum), the eRisk
Lab[1] is a forum for exploring evaluation methodologies and effectiveness metrics
related to early risk detection on the Internet (with past challenges particularly
focused on health and safety). Over the previous editions [6–9,11], we have presented a number of testbeds and tools under the eRisk's umbrella. Our dataset
construction methodology and evaluation strategies are general and, thus, potentially applicable to different application domains.

Our Lab brings together researchers from various fields (such as information retrieval, computational linguistics, machine learning, and psychology) to
interdisciplinary address the presented tasks. Furthermore, participants develop
models for solving the eRisk defined challenges that may play a critical role in
helping in solving socially worrying problems. For example, when an individual
begins broadcasting suicidal thoughts or threats of self-harm on social networks,

---

[1] https://erisk.irlab.org.

M. Hagen et al. (Eds.): ECIR 2022, LNCS 13186, pp. 436–442, 2022.
https://doi.org/10.1007/978-3-030-99739-7_54

systems may send warning alerts. Previous eRisk editions proposed shared tasks centred on specific health and security issues, such as depression, anorexia, or self-harm detection.

eRisk proposes two types of challenges: early alert and severity estimation tasks. On early risk tasks, risk prediction is viewed as a sequential process of evidence accumulation. Participant systems must automatically analyse the continuous data flow in a given source (e.g., social media entries). Those algorithms must estimate when and if there is enough aggregated evidence about a specific type of risk during this process. On each shared task, participants have access to temporally organised writing histories and must balance making early alerts (e.g., based on a few social media entries) versus making not-so-early (late) alerts (e.g., evaluating a wider range of entries and only emitting alerts after analysing a larger number of pieces of evidence). On the other hand, severity estimation tasks are concerned with computing a fine-grained estimate of the symptoms of a specific risk based on the entire set of user writings. Participants are challenged to create models that fill out a standard questionnaire the same way that a real user would.

2 Five Years of eRisk

eRisk, a CLEF lab for research on early risk prediction on the Internet, began in 2017 as a forum to lay the experimental groundwork for early risk detection. Our fifth anniversary was last year. Since the Lab's inception, we have created numerous reference collections in the field of risk prediction and organised a number of challenges based on those datasets. Each challenge centred on a specific risk detection issue, such as depression, anorexia, or self-harm.

eRisk participants only addressed the detection of early risk of depression in its first edition (2017) [6]. This resulted in the first proposals for exploiting the relationship between the use of language in social networks and early signs of depression. Because it was the first edition of such an innovative evaluation scheme, eRisk 2017 was extremely demanding for both participants and organisers. Temporal data chunks were released in sequential order (one chunk per week). Following each release, participants were required to send their predictions about the users in the collection. Only eight of the thirty participating groups completed the tasks by the deadline. More than 30 different approaches to the problem were proposed by these teams (variants or runs). The evaluation methodology and metrics were those defined in [5].

eRisk [7] included two shared tasks in 2018: 1) a continuation of the task on early detection of depression from 2017 and 2) a task on early detection of signs of anorexia. Both tasks were organised similarly and used the same eRisk 2017 evaluation methods. eRisk 2018 had 11 final participants (out of 41 registered). Participants submitted 45 systems for Task 1 (depression) and 35 for Task 2 (anorexia).

In 2019, we organised three tasks [8]. Task 1 was a continuation of the 2018 task on early detection of indicators of anorexia. Task 2 was a new one on early detection of signs of self-harm. Furthermore, and more novel, a new activity, Task 3, was developed to automatically fill out a depression questionnaire based on user interactions in social media. It should be noted that this new task does not address early detection but rather another complex task (fine-grained depression level estimation). For eRisk 2019, 14 participants (out of 62 registered teams) actively participated in the three challenges and submitted 54, 33, and 33 system versions (runs) for each one, respectively.

We opted to continue the tasks of early detection of self-harm (Task 1) and estimating the severity of depression symptoms (depression level estimation, Task 2) for the 2020 edition [9]. Task 1 had 12 final participants who submitted 46 possible system variants, whilst Task 2 had six active participants who presented 17 different system variants.

Finally, in 2021 we proposed three tasks to the participants. Following our three-year-per-task cycle, we closed the early detection of signs of self-harm challenge (Task 2) and the estimation of the severity of the symptoms of depression (Task 3). Additionally, we presented a new domain for early detection, in this case, pathological gambling (Task 1) [11]. We received 115 runs from 18 teams out of 75 registered. Those are distributed as follows: 26 systems for Task 1, 55 for Task 2 and 34 for Task 3.

Over these five years, eRisk has received a steady number of active participants, slowly placing the Lab as a reference forum for early risk research. For celebrating those five years of efforts by participants, we are just finishing the edition of a book on the topic of eRisk [3].

2.1 Early Risk Prediction Tasks

The majority of the proposed challenges were geared toward early risk prediction in various domains (depression, anorexia, self-harm). They were all organised in the same way: the teams had to analyse social media writings (posts or comments) sequentially (in chronological order) to detect risk signals as soon as possible. The main objective was to produce useful algorithms and models for monitoring social network activity.

The social media platform Reddit was used as a source for all shared tasks in the various versions. It is vital to note that Reddit's terms of service allow for data extraction for research purposes. Except as provided by the notion of fair use, Reddit does not enable unauthorised commercial use or redistribution of its material. The research activities of eRisk are an example of fair usage.

Reddit users frequently portray a highly active profile with a large thread of submissions (covering several years). In terms of mental health problems, there are subcommunities (e.g. subreddits) dedicated to depression, anorexia, self-harm, and pathological gambling, to name a few. We used these valuable sources to create collections of writings (posts or comments) made by *redditors* for the eRisk test collections (as mentioned in [5]). In our datasets, *redditors* are divided into positive class (e.g., depressed) and negative class (control group).

When building the datasets, we followed the same approach as Coppersmith and colleagues [2]. We generated the positive class by employing a retrieval strategy for identifying *redditors* who were diagnosed with the disease at hand (e.g. depressed). This was determined through searches for self-expressions associated with medical diagnosis (e.g. "Today, I was diagnosed with depression"). Many *redditors* are active on subreddits about mental health, and they are frequently open about their medical condition. Following that, we carefully checked the collected results to ensure that the expressions about diagnosis were genuine. For example, *"I am anorexic"*, *"I have anorexia"*, or *"I believe I have anorexia"* were not deemed explicit affirmations of a diagnosis. We only included a user in the positive set where there was a clear and explicit indication of a diagnosis (e.g., *"Last month, I was diagnosed with anorexia nervosa"*, *"After struggling with anorexia for a long time, last week I was diagnosed"*). We have a high level of confidence in the reliability of these labels. This semi-automatic extraction method has successfully extracted information about patients who have been diagnosed with a particular disease. Since 2020, we have used Beaver [10], a new tool for labelling positive and negative instances, for aiding us in this task.

The first edition of eRisk presented a new measure called ERDE (Early Risk Detection Error) for measuring early detection [5]. This metric served as a supplement to normal classification measures, which neglect prediction delay. ERDE considers the correctness of the (binary) decision as well as the latency, which is calculated by counting the number (k) of texts seen prior to reaching the decision.

Later on, eRisk added a ranking-based way to evaluate participation. Since 2019, after each round of writings, a user ranking has been generated (ranked by decreasing estimated risk). These ranks are assessed using common information retrieval metrics (for example, P@10 or nDCG). The ranking-based evaluation is described in detail in [8] We have also embraced $F_{latency}$ in 2019, an alternative assessment metric for early risk prediction proposed by Sadeque et al. [12].

2.2 Severity Level Estimation Task

In 2019 we introduced a new task on estimating the severity level of depression that we continued in 2020 and 2021. The Depression Level Estimation Task investigates the feasibility and possible ways for automatically measuring the occurrence and intensity of numerous well-known depression symptoms. In this task, participants had access to the whole history of writings of some *redditors*. With that in hand, participants had to devise models that fill out a standard depression questionnaire using each user's history. Models have to capture evidence from users texts to decide on the answer to each questionnaire item. The questionnaire presents 21 questions regarding the severity of depression signs and symptoms (with four alternative responses corresponding to different severity levels) (e.g., loss of energy, sadness, and sleeping problems). The questionnaire is based on the Beck's Depression Inventory (BDI) [1].

To produce the ground truth, we compiled a series of surveys filled out by social media users together with their writing history. Because of the unique nature of the task, new evaluation measures for evaluating the participants' estimations were required. We defined four metrics: Average Closeness Rate

(ACR), Average Hit Rate (AHR), Average DODL (ADODL), and Depression Category Hit Rate (DCHR), details can be found in [8].

2.3 Results

According to the CLEF tradition, Labs' Overview and Extended Overview papers compile the summaries and critical analysis of the participants' systems results [6–9,11].

So far, we have presented eight editions of early risk tasks on four mental health issues. Participants contributing to those past editions have presented a wide variety of models and approaches. The majority of the methods are based on standard classification techniques. That is, most of our competitors were centred on optimising classification accuracy on the training data. In general, participants were less concerned with the accuracy-delay trade-off. In terms of performance, the results over the years demonstrate some variances between challenges, with anorexia detection yielding better results than depression detection. These differences may be attributable to the amount and quality of the released training data and the very nature of the disorder. We hypothesise that, depending on the condition, patients are more or less prone to leave traces of the language used in social media. The performance figures show a trend on how participants improved detection accuracy edition over edition. This trend motivates us to continue supporting research on text-based early risk screening in social media. Furthermore, given the success of some participants, it appears that automatic or semi-automatic screening systems that predict the commencement of specific hazards are within reach.

In terms of estimating depression levels, the results show that automatic analysis of the user's writings could be a complementary strategy for extracting some signs or symptoms associated with depression. Some participants had a 40 per cent hit rate (the systems answered, i.e., 40% of the BDI questions with the exact same response given by the real user). This still has a lot of room for improvement, but it shows that the participants were able to extract some signals from the chaotic social media data.

The difficulties in locating and adapting measures for these novel jobs has also prompted us to develop new metrics for eRisk.Some eRisk participants [12,13], were also engaged in proposing novel modes of evaluation, which is yet another beneficial outcome of the Lab. We are also planning to incorporate new metrics for automatic risk estimation tasks. Mean Absolute Error (MAE) and Root Mean Square Error (RMSE) were two widely metrics used in rating prediction for users in recommendation systems [4]. We think that they may be suitable metrics for the problem.

3 eRisk 2022

The results of past editions have inspired us to continue with the Lab in 2022 and further investigate the relationship between text-based screening from social media and risk prediction and estimation. The scheme of tasks for eRisk 2022 is as follows:

- Task 1: Early detection of pathological gambling. We will continue the new task from 2021. This is the second edition of the task, so following our new three-year cycle, in 2022, participants will have training data.
- Task 2: Early detection of depression. After the second edition in 2018, we will close the cycle for this task with its edition next year. Moreover, and different from previous editions for the disease, the participants will have a post-by-post release of user history through our web service.
- Task 3: Measuring the severity of the signs of eating disorders. This is a new severity estimation task in the field of eating disorders. Eating disorders (ICD-10-CM code F50) affect up to 5% of the population, most often developing in adolescence and young adulthood. The task consists of estimating the severity of the eating disorder from a thread of user submissions. The participants will be given a history of postings for each user, and the participants will have to fill a standard questionnaire (based on the evidence found in the history of postings). The questionnaire that we will use is the Eating Disorder Examination Questionnaire (EDE-Q). EDE-Q assesses the range and severity of features associated with the diagnosis of eating disorders. It is a 28-item questionnaire with four subscales (restrain, eating, concern, shape concern, and weight concern).

4 Conclusions

The results achieved so far under eRisk and the engagement of the research community motivate us to continue with the proposal of new shared-tasks related to early risk detection. We are truly thankful to participants for their contribution to the success of eRisk. We want to encourage the research teams working on the early risk to keep improving and creating new models for future tasks and risks. Even if generating new resources is tedious, we are convinced that the societal benefits outweigh the costs.

Acknowledgements. This work was supported by projects RTI2018-093336-B-C21, RTI2018-093336-B-C22 (Ministerio de Ciencia e Innvovación & ERDF). The first and second authors thank the financial support supplied by the Consellería de Educación, Universidade e Formación Profesional (accreditation 2019-2022 ED431G/01, ED431B 2019/03) and the European Regional Development Fund, which acknowledges the CITIC Research Center in ICT of the University of A Coruña as a Research Center of the Galician University System. The third author also thanks the financial support supplied by the Consellería de Educación, Universidade e Formación Profesional (accreditation 2019-2022 ED431G-2019/04, ED431C 2018/29) and the European Regional Development Fund, which acknowledges the CiTIUS-Research Center in Intelligent Technologies of the University of Santiago de Compostela as a Research Center of the Galician University System. The first, second, and third author also thank the funding of project PLEC2021-007662 (MCIN/AEI/10.13039/501100011033, Ministerio de Ciencia e Innovación, Agencia Estatal de Investigación, Plan de Recuperación, Transformación y Resiliencia, Unión Europea-Next Generation EU).

References

1. Beck, A.T., Ward, C.H., Mendelson, M., Mock, J., Erbaugh, J.: An inventory for measuring depression. JAMA Psychiatry **4**(6), 561–571 (1961)
2. Coppersmith, G., Dredze, M., Harman, C.: Quantifying mental health signals in Twitter. In: ACL Workshop on Computational Linguistics and Clinical Psychology (2014)
3. Crestani, F., Losada, D.E., Parapar, J.: Early Detection of Mental Health Disorders by Social Media Monitoring. Springer, Berlin (2021)
4. Herlocker, J.L., Konstan, J.A., Terveen, L.G., Riedl, J.T.: Evaluating collaborative filtering recommender systems. ACM Trans. Inf. Syst. **22**(1), 5–53 (2004)
5. Losada, D.E., Crestani, F.: A test collection for research on depression and language use. In: Proceedings Conference and Labs of the Evaluation Forum CLEF 2016. Evora, Portugal (2016)
6. Losada, D.E., Crestani, F., Parapar, J.: eRISK 2017: CLEF lab on early risk prediction on the internet: experimental foundations. In: Jones, G.J.F., et al. (eds.) CLEF 2017. LNCS, vol. 10456, pp. 346–360. Springer, Cham (2017). https://doi.org/10.1007/978-3-319-65813-1_30
7. Losada, D.E., Crestani, F., Parapar, J.: Overview of erisk: early risk prediction on the internet. In: Bellot, P., et al. (eds.) CLEF 2018. LNCS, vol. 11018, pp. 343–361. Springer, Cham (2018). https://doi.org/10.1007/978-3-319-98932-7_30
8. Losada, D.E., Crestani, F., Parapar, J.: Overview of erisk 2019 early risk prediction on the internet. In: Crestani, F., et al. (eds.) CLEF 2019. LNCS, vol. 11696, pp. 340–357. Springer, Cham (2019). https://doi.org/10.1007/978-3-030-28577-7_27
9. Losada, D.E., Crestani, F., Parapar, J.: Overview of erisk 2020: early risk prediction on the internet. In: Arampatzis, A., et al. (eds.) CLEF 2020. LNCS, vol. 12260, pp. 272–287. Springer, Cham (2020). https://doi.org/10.1007/978-3-030-58219-7_20
10. Otero, D., Parapar, J., Barreiro, Á.: Beaver: efficiently building test collections for novel tasks. In: Proceedings of the Joint Conference of the Information Retrieval Communities in Europe (CIRCLE 2020), Samatan, Gers, France, 6–9 July 2020. http://ceur-ws.org/Vol-2621/CIRCLE20_23.pdf
11. Parapar, J., Martín-Rodilla, P., Losada, D.E., Crestani, F.: Overview of erisk 2021: early risk prediction on the internet. In: Candan, K.S., et al. (eds.) CLEF 2021. LNCS, vol. 12880, pp. 324–344. Springer, Cham (2021). https://doi.org/10.1007/978-3-030-85251-1_22
12. Sadeque, F., Xu, D., Bethard, S.: Measuring the latency of depression detection in social media. In: Proceedings of the Eleventh ACM International Conference on Web Search and Data Mining, pp. 495–503. WSDM 2018, ACM, New York, NY, USA (2018)
13. Trotzek, M., Koitka, S., Friedrich, C.M.: Utilizing neural networks and linguistic metadata for early detection of depression indications in text sequences. IEEE Trans. Knowl. Data Eng. **32**(3), 588–601 (2018)

Doctoral Consortium

Continually Adaptive Neural Retrieval Across the Legal, Patent and Health Domain

Sophia Althammer[(✉)]

Institute for Information Systems Engineering, TU Wien, Vienna, Austria
sophia.althammer@tuwien.ac.at

1 Motivation

In the past years neural retrieval approaches using contextualized language models have driven advancements in information retrieval (IR) and demonstrated great effectiveness gains for retrieval, primarily in the web domain [1,10,33]. This is enabled by the availability of large-scale, open-domain labelled collections [6].

Besides retrieval in the web domain there are numerous IR tasks in other domains [13,41], we focus on the legal, patent and health domain. In these domains there are different search tasks and the task setting can differ from the web domain. The user of the retrieval system might be a professional working in the respective domain, the query might contain domain-specific terminology and the query might be a long document. Furthermore for domain-specific retrieval tasks smaller labelled collections are available compared to the size of training collections in the web domain. This setting holds common challenges over these three domains for neural retrieval methods, which differ from the challenges in the web domain. So far neural approaches remain understudied in the legal, patent and health domain, although we reason that neural retrieval models would benefit the retrieval tasks in those domains by taking into account the semantic context of the text, by learning relevance signals from labelled collections, by domain adaptation of neural models from the web domain to other domains, and by transferring knowledge across the domains in a cross-domain retrieval setting. We define the concept of domain adaptation of a model from a source to a target domain by further training the model, which is pre-trained on the source domain, on target domain data.

In the first research questions we investigate how we can adapt neural retrieval models from the web domain for retrieval tasks in the legal, patent and health domain and how well the findings for the web domain generalize for the retrieval tasks not only solely in a specific domain, but across those specific domains.

The second emerging open research question for neural retrieval models for real-world applications is how to continually adapt these models in production systems in the web domain as well as in domain-specific retrieval. For now the contextualized neural retrieval models contain language models which are pre-trained on a static collection and the retrieval model itself is trained on a static training collection with static relevance judgements. But in a real-world scenario, the content of the corpus as well as the notion of relevance changes over time. For example with the Covid-19 crisis emerging in 2020, every day new scientific articles were published and the common knowledge about the crisis changed over time. To be able to find high-quality, relevant, and recent

© The Author(s), under exclusive license to Springer Nature Switzerland AG 2022
M. Hagen et al. (Eds.): ECIR 2022, LNCS 13186, pp. 445–454, 2022.
https://doi.org/10.1007/978-3-030-99739-7_55

results, novel content not only needs to be included in production systems, but the content shift needs to be accounted for in the contextualized retrieval model. With traditional inverted indices, the problem reduces to adding new words in the vocabulary, but as contextualized retrieval models learn to embed the semantic meaning of words and passages for retrieval, there is the need to study how to continually learn and update neural contextualized retrieval models with a constant stream of emerging data. It is yet to be understood how to continually adapt and learn the neural retrieval model in order to take changing content in the corpus as well as changing relevance notions into account. For production systems it is also an open question how to update the index of neural retrieval models in real-time without having to re-index the whole corpus.

We see these research questions as crucial and emerging questions for the applicability of neural retrieval models across domains as well as for the use of the neural retrieval models in a continually adaptive scenarios.

In parallel to our main research questions, we see it as an important part of our work to make the research findings accessible and reproducible through open-sourcing the code of the experiments, which we release on https://github.com/sophiaalthammer/.

2 Research Questions

Our motivation above leads us to our research questions, which we divide into two parts: the first one addresses neural retrieval across the three domains and the second part tackles continually adapting neural retrieval models for in-production systems.

RQ-1 How can we use neural retrieval and ranking models for retrieval in the legal, patent and health domains?

RQ-1.1 How can we adapt dense retrieval for domain-specific retrieval?

RQ-1.2 How well do approaches for neural re-ranking generalize across different domains?

RQ-1.3 How well do approaches for neural retrieval in the first stage retrieval generalize across different domains?

RQ-2 How can we continually adapt neural retrieval models and systems for in-production systems?

RQ-2.1 How can we continually adapt dense retrieval models to a continual content shift?

RQ-2.2 How can we continually update dense retrieval indexes for in-production systems in an efficient and effective way?

3 Background and Related Work

In this section we give an overview of the related work as well as our preliminary work.

Related Work. The advent of neural methods for retrieval has demonstrated enormous effectiveness gains for re-ranking retrieved results [2, 10, 19, 29, 33, 34] as well as for first stage retrieval with methods like dense retrieval [12, 15, 18, 25, 27, 39, 54, 56], mainly focusing on the web domain.

In order to create and provide queries and relevance judgements for domain-specific retrieval and with this promote research in medical, legal and patent IR, there are numerous evaluation campaigns for medical [31,32,41,46,49,50], legal [7,13,37], and patent [36] IR. Related work for domain-specific neural retrieval also addresses the problem of domain specific language modelling so far only in the medical domain [23,38,53]. Long documents in the legal domain are handled with either automatically creating summaries [5,43,48] or by determining relevance on the paragraph-level [45]. The challenge of little labelled data is tackled in the medical domain by synthetic training data generation [38] or by sampling medical subsets from existing larger collections [28]. Another line of research for domain specific retrieval investigates cross-domain retrieval [1,21] or multi-task learning [9,20,52].

Continually learning contextualized models so far has only been addressed in the area of Natural Language Processing. Here recent advances investigate how to continually pre-train contextualized language models with a constant stream of information in order to address the time-wise or topic-wise content shift [17,30,44] or how to learn to rank in an online manner with a changed user intent [58].

Our Preliminary Work. In our recent work, which is currently under review, we addressed **RQ-1.1** "How can we adapt dense retrieval for domain-specific retrieval?". For the medical domain, we investigated the effect of domain specific language modelling for dense retrieval on TripClick [40]. In another work we studied dense retrieval for legal case retrieval and proposed a paragraph-level aggregation model for dense document-to-document retrieval. We investigated how to train a dense retrieval model with little labelled data, we studied the effect of domain specific language modelling. Furthermore we proposed to handle long documents by determining relevance on paragraph-level as contextualized language models have a limited input length and cannot take into account the whole long document.

Furthermore we answer **RQ-1.2** "How well do approaches for neural re-ranking generalize across different domains?" in our paper "Cross-domain Retrieval in the Legal and Patent Domains: a Reproducibility Study" [4]. In this paper, we investigated findings of Shao et al. [45] for paragraph-level re-ranking with BERT for the task of legal case retrieval. We studied the generalization capability of Shao et al.'s proposed model BERT-PLI for the task of prior art retrieval in the patent domain, as prior art retrieval poses similar challenges as legal case retrieval like long documents and little training data at hand. We found that BERT-PLI does not outperform the BM25 baseline [42] for prior art retrieval on the CLEF-IP dataset [36]. Furthermore, we evaluated the BERT-PLI model for cross-domain retrieval between the legal and patent domain on individual components, both on a paragraph and document-level. We found that the transfer of the BERT-PLI model on the paragraph-level leads to comparable results between both domains as well as first promising results for the cross-domain transfer on the document-level.

4 Methods and Experiments

In this section we give an overview of how we plan to tackle the remaining research questions and which experiments we are planning for this.

4.1   RQ-1.3: How Well Do Approaches for Neural Retrieval in the First Stage Retrieval Generalize Across Different Domains?

In order to study the generalizability of dense retrieval methods across different domains, we plan to study the generalization capabilities of trained dense retrieval methods across different domains in a zero-shot setting.

Previous work has shown that training dense retrieval models jointly on similar retrieval collections leads to an inferior retrieval effectiveness compared to solely in-domain training [18,20,21,24] and that the generalization capabilities across different domains underperform unsupervised lexical matching methods [47]. Furthermore recent work in compressing dense representations [26,55] shows that the information contained in the 768-dimensional embeddings can be compressed to much fewer dimensions, suggesting that the dense embeddings of 768 dimensions do not exploit fully all dimensions to store information signals. If we view lexical methods also as sparse representations with a dimension of the size of the vocabulary (typically much larger than 768), we suggest that the robust retrieval effectiveness of lexical retrieval models come from the uniform covering of the sparse representation space. We reason that increasing the coverage and diversification of the embeddings of dense retrieval models increases robustness and generalization capabilities of dense retrieval models as well as making joint training with multiple datasets beneficial for dense retrieval models.

In order to gain more insight we plan to analyze and compare the coverage and diversification of lexical and dense embedding spaces with methods from linear algebra and cluster analysis [11]. Furthermore we plan to compare dense retrieval embedding spaces trained with single or multiple retrieval collections. With this intuition of sparse and dense embeddings we suggest that training dense retrieval models jointly on multiple data sources "squeezes" the embeddings leading to less information encoded in the space and therefore an inferior retrieval performance. We plan to investigate dense retrieval training with joint collections by introducing a training loss to diversify the dense representations and to maximize embedding space coverage. This study would give insights on how dense retrieval embeddings are learned in a joint training setting and how these embeddings can be learned so that they are more robust and generalizable.

We plan to jointly train the dense retrieval models on the MS Marco dataset [6], the TREC-COVID collection [49] and the TripClick dataset [40] and evaluate the models in a zero-shot setting on the BEIR benchmark [47] for domain specific retrieval.

With the findings of the study, future directions could include transfer learning and domain adaptation for dense retrieval models.

4.2   RQ-2: How Can We Continually Adapt Neural Retrieval Models for In-Production Systems?

In order to answer **RQ-2** we divide the question into the two subquestions. The first one is how to continually learn and update dense retrieval methods, the second one is concerned with updating the dense retrieval index for in-production systems once we have an updated retrieval model at hand.

RQ-2.1: How Can We Continually Adapt Dense Retrieval Models to a Continual Content Shift? We will investigate how we can continually adapt dense retrieval models for a time-wise data stream, inspired by recent work in life-long language modelling [17]. The retrieval model should continually learn with the goal that the previous knowledge of the retrieval model is not lost, but the model is continually adapted to new content.

We will experiment with different continual learning approaches like adapter-based [35,51], memory-replay [8] or distillation-based approaches [14,16,22]. For training dense retrieval models, it is an active research question how to choose negative samples in the training to achieve the highest retrieval effectiveness [15,54,57]. Therefore it would be also an open question how to do effective negative sampling for continually learning the dense retrieval methods.

For modeling the constant data stream, we divide the data stream in different time-steps. Here we need for every time-step a different training collection. Different to a static test collection, we also need different test subsets for each time- or topic-step in order to evaluate how the updated retrieval model adapts and generalizes to new data and in order to evaluate if and how the updated retrieval model forgets previous knowledge. For that we plan to divide the training and test collection of large scale collections into disjunct subsets.

In order to evaluate the forgetting of the retrieval model, we plan to evaluate the retrieval effectiveness of the adapted retrieval model for the previous test collection and compare it to the retrieval effectiveness before adapting the model. For the evaluation of the adaptation capability, the retrieval effectiveness of the previous retrieval model for the new test subset is compared to the retrieval effectiveness of the adapted model.

RQ-2.2: How Can We Continually Update Dense Retrieval Indexes for In-Production Systems in an Efficient and Effective Way? Once we investigated how to continually adapt the retrieval model, we see another open challenge in integrating the continually updated retrieval model in production systems. Especially we see a challenge in continually updating the dense retrieval index in real-time. In production systems the search indexes have a size up to 100 millions of terabytes, thus re-indexing the whole corpus is computationally expensive and not feasible in real-time scenarios.

Therefore we proposed in Althammer [3] the concept RUDI for Real-time learning to Update Dense retrieval Indices with simple transformations. In RUDI a computationally lightweight vector space transformation function $T : \mathbb{V} \rightarrow \mathbb{V}_u$ between the vector embedding space of the previous retrieval model $\mathbb{V}$ and of the updated dense retrieval model $\mathbb{V}_u$ is used to transform the vector embeddings of the previous index to the embeddings of the updated indexing model. The advantage of RUDI is that the index embedding does not need to be fully re-indexed with the updated dense retrieval model, but the index is updated with a learned, computationally lightweight transformation function. This allows updating the dense retrieval index in real-time.

To approximate the embeddings in $\mathbb{V}_u$, the transformation function T takes the embedding $v^d \in \mathbb{V}$ of the document d from the previous embedding space as input and outputs the approximated vector space embedding v_a^d. The approximated vector space embedding of document d v_a^d is then the updated embedding of vector space $\mathbb{V}_r$. The transformation function T is learned in real-time on a small, sampled fraction $\mathbb{D}$

of the documents in the corpus by minimizing the distance between the approximated vector space embedding v_a^d and the embedding of d in the updated vector space v_u^d.

Here we see several open challenges. For learning the transformation function, we plan to investigate different, lightweight transformations like a fully connected layer or an exponential transformation and compare their approximation performance. Also we plan to investigate how the overall retrieval effectiveness is influenced by updating the retrieval index with RUDI compared to re-indexing the whole index. Furthermore we plan to analyze the trade-off between the number of training documents for learning the transformation and overall retrieval quality on the updated index. We plan to study different sampling strategies for sampling the training documents from the overall index like random sampling or sampling strategies aiming to sample documents from the index with maximal orthogonal embeddings. Furthermore we plan to do speed comparisons between updating the dense retrieval index with different size of training samples and between re-indexing the whole index.

5   Research Issues

RQ-1.3: How Well Do Approaches for Neural Retrieval in the First Stage Retrieval Generalize Across Different Domains? The first issue of our proposed methodology is that our intuition of the problem could be misleading and that jointly training the dense retrieval model with multiple labelled collections does not lead to a lower coverage and diversity of the embedding space. Therefore we first want to analyze and compare the embeddings of different dense indices using methods from linear algebra and cluster analysis [11]. For answering this research question we also see a challenge in how to train the dense retrieval model in order to diversify the learned embeddings and with this uniformly cover the embedding space. The first ideas are to introduce an additional loss function which maximizes the distance of embeddings between batches while balancing the batches topic-wise [15]. Another issue is the choice of training collections, which should be somehow related to each other but still come from different domains, so that the dense retrieval model learns to embed those different domains in the same embedding space. With this training objective our intuition is that the dense retrieval model generalizes better to domains different from the training distribution.

RQ-2: How Can We Continually Adapt Neural Retrieval Models for In-Production Systems? For updating and continually adapting a neural retrieval model we see a major challenge in simulating the data stream for training and test collections with the collections at hand. We plan to model this data stream using large scale retrieval training collections like MS Marco dataset [6] and the TripClick dataset [40] and divide the training set into disjunct subsets of documents. These subsets could be separated to model the changing relevance judgements for certain queries over time, so that we can elaborate how the adapted model learns to incorporate this shift. Another collection we want to experiment with is the TREC-COVID collection [49], as it contains 5 different subsets with different topics and relevance judgements, which result from the constant updates of the collection during creation time. Nevertheless we are not sure if the test collections have the capability to reflect the changing relevance over time, therefore we also consider an in-production system evaluation with A/B testing.

References

1. Akkalyoncu Yilmaz, Z., Wang, S., Yang, W., Zhang, H., Lin, J.: Applying BERT to document retrieval with birch. In: Proceedings of the 2019 Conference on Empirical Methods in Natural Language Processing and the 9th International Joint Conference on Natural Language Processing (EMNLP-IJCNLP): System Demonstrations, Hong Kong, China, pp. 19–24. Association for Computational Linguistics, November 2019. https://doi.org/10.18653/v1/D19-3004. https://aclanthology.org/D19-3004
2. Akkalyoncu Yilmaz, Z., Yang, W., Zhang, H., Lin, J.: Cross-domain modeling of sentence-level evidence for document retrieval. In: Proceedings of the 2019 Conference on Empirical Methods in Natural Language Processing and the 9th International Joint Conference on Natural Language Processing (EMNLP-IJCNLP), Hong Kong, China, pp. 3490–3496. Association for Computational Linguistics, November 2019. https://doi.org/10.18653/v1/D19-1352. https://aclanthology.org/D19-1352
3. Althammer, S.: RUDI: real-time learning to update dense retrieval indices. In: Proceedings of DESIRES 2021–2nd International Conference on Design of Experimental Search Information REtrieval Systems (2021)
4. Althammer, S., Hofstätter, S., Hanbury, A.: Cross-domain retrieval in the legal and patent domains: a reproducibility study. In: Advances in Information Retrieval, 43rd European Conference on IR Research, ECIR 2021 (2021)
5. Askari, A., Verberne, S.: Combining lexical and neural retrieval with longformer-based summarization for effective case law retrieval. In: Proceedings of DESIRES 2021–2nd International Conference on Design of Experimental Search Information REtrieval Systems (2021)
6. Bajaj, P., et al.: MS MARCO: a human generated MAchine reading COmprehension dataset. In: Proceedings of the NIPS (2016)
7. Bhattacharya, P., et al.: FIRE 2019 AILA track: artificial intelligence for legal assistance. In: Proceedings of the 11th Forum for Information Retrieval Evaluation, FIRE 2019, pp. 4–6. Association for Computing Machinery, New York (2019). https://doi.org/10.1145/3368567.3368587
8. Chaudhry, A., et al.: On tiny episodic memories in continual learning. In: arXiv (2019)
9. Cohan, A., Feldman, S., Beltagy, I., Downey, D., Weld, D.: SPECTER: document-level representation learning using citation-informed transformers. In: Proceedings of the 58th Annual Meeting of the Association for Computational Linguistics, pp. 2270–2282. Association for Computational Linguistics, July 2020. https://doi.org/10.18653/v1/2020.acl-main.207. https://aclanthology.org/2020.acl-main.207
10. Dai, Z., Callan, J.: Context-aware term weighting for first stage passage retrieval, pp. 1533–1536. Association for Computing Machinery, New York (2020). https://doi.org/10.1145/3397271.3401204
11. Feldbauer, R., Leodolter, M., Plant, C., Flexer, A.: Fast approximate hubness reduction for large high-dimensional data. In: 2018 IEEE International Conference on Big Knowledge (ICBK), pp. 358–367 (2018). https://doi.org/10.1109/ICBK.2018.00055
12. Gao, L., Dai, Z., Chen, T., Fan, Z., Durme, B.V., Callan, J.: Complementing lexical retrieval with semantic residual embedding, April 2020. http://arxiv.org/abs/2004.13969
13. Hedin, B., Zaresefat, S., Baron, J., Oard, D.: Overview of the TREC 2009 legal track. In: The Eighteenth Text REtrieval Conference (TREC 2009) Proceedings, January 2009
14. Hinton, G., Vinyals, O., Dean, J.: Distilling the knowledge in a neural network. In: arXiv (2015)
15. Hofstätter, S., Lin, S.C., Yang, J.H., Lin, J., Hanbury, A.: Efficiently teaching an effective dense retriever with balanced topic aware sampling, pp. 113–122. Association for Computing Machinery, New York (2021). https://doi.org/10.1145/3404835.3462891

16. Hou, S., Pan, X., Loy, C.C., Wang, Z., Lin, D.: Lifelong learning via progressive distillation and retrospection. In: Ferrari, V., Hebert, M., Sminchisescu, C., Weiss, Y. (eds.) ECCV 2018. LNCS, vol. 11207, pp. 452–467. Springer, Cham (2018). https://doi.org/10.1007/978-3-030-01219-9_27

17. Jin, X., et al.: Lifelong pretraining: continually adapting language models to emerging corpora. In: arXiv:2110.08534 (2021)

18. Karpukhin, V., et al.: Dense passage retrieval for open-domain question answering. In: Proceedings of the 2020 Conference on Empirical Methods in Natural Language Processing (EMNLP), pp. 6769–6781. Association for Computational Linguistics, November 2020. https://doi.org/10.18653/v1/2020.emnlp-main.550. https://www.aclweb.org/anthology/2020.emnlp-main.550

19. Khattab, O., Potts, C., Zaharia, M.: Relevance-guided supervision for openQA with colbert. arXiv preprint arXiv:2007.00814 (2020)

20. Li, M., Li, M., Xiong, K., Lin, J.: Multi-task dense retrieval via model uncertainty fusion for open-domain question answering. In: Findings of the 2021 Conference on Empirical Methods in Natural Language Processing (EMNLP). Association for Computational Linguistics (2021)

21. Li, M., Lin, J.: Encoder adaptation of dense passage retrieval for open-domain question answering. In: arXiv (2021)

22. Li, Z., Hoiem, D.: Learning without forgetting. IEEE Trans. Pattern Anal. Mach. Intell. **40**, 2935–2947 (2018). https://doi.org/10.1109/TPAMI.2017.2773081

23. Lima, L.C., et al.: Denmark's participation in the search engine TREC COVID-19 challenge: lessons learned about searching for precise biomedical scientific information on COVID-19. arXiv preprint arXiv:2011.12684 (2020)

24. Lin, J.: A proposed conceptual framework for a representational approach to information retrieval. arXiv preprint arXiv:2110.01529 (2021)

25. Lin, S.C., Yang, J.H., Lin, J.: Distilling dense representations for ranking using tightly-coupled teachers. ArXiv abs/2010.11386 (2020)

26. Ma, X., Li, M., Sun, K., Xin, J., Lin, J.: Simple and effective unsupervised redundancy elimination to compress dense vectors for passage retrieval. In: Findings of the Association for Computational Linguistics: EMNLP 2021. Association for Computational Linguistics, November 2021

27. Ma, X., Sun, K., Pradeep, R., Lin, J.: A replication study of dense passage retriever. In: arXiv (2021)

28. MacAvaney, S., Cohan, A., Goharian, N.: SLEDGE-Z: a zero-shot baseline for COVID-19 literature search. In: Proceedings of the 2020 Conference on Empirical Methods in Natural Language Processing (EMNLP), pp. 4171–4179. Association for Computational Linguistics, November 2020. https://doi.org/10.18653/v1/2020.emnlp-main.341. https://www.aclweb.org/anthology/2020.emnlp-main.341

29. MacAvaney, S., Yates, A., Cohan, A., Goharian, N.: CEDR: contextualized embeddings for document ranking. In: Proceedings of the 42nd International ACM SIGIR Conference on Research and Development in Information Retrieval, SIGIR 2019, pp. 1101–1104. Association for Computing Machinery, New York (2019). https://doi.org/10.1145/3331184.3331317

30. Maronikolakis, A., Schütze, H.: Multidomain pretrained language models for green NLP. In: Proceedings of the Second Workshop on Domain Adaptation for NLP, Kyiv, Ukraine, pp. 1–8. Association for Computational Linguistics, April 2021. https://aclanthology.org/2021.adaptnlp-1.1

31. Möller, T., Reina, A., Jayakumar, R., Pietsch, M.: COVID-QA: a question answering dataset for COVID-19. In: Proceedings of the 1st Workshop on NLP for COVID-19 at ACL 2020. Association for Computational Linguistics, July 2020. https://www.aclweb.org/anthology/2020.nlpcovid19-acl.18

32. Nentidis, A., et al.: Overview of BioASQ 2020: the eighth BioASQ challenge on large-scale biomedical semantic indexing and question answering, pp. 194–214, September 2020. https://doi.org/10.1007/978-3-030-58219-7_16
33. Nogueira, R., Cho, K.: Passage re-ranking with BERT. arXiv preprint arXiv:1901.04085 (2019)
34. Nogueira, R., Yang, W., Cho, K., Lin, J.: Multi-stage document ranking with BERT. In: arXiv (2019)
35. Pfeiffer, J., Kamath, A., Rücklé, A., Cho, K., Gurevych, I.: AdapterFusion: non-destructive task composition for transfer learning. In: Proceedings of the 16th Conference of the European Chapter of the Association for Computational Linguistics: Main Volume, pp. 487–503. Association for Computational Linguistics, April 2021. https://doi.org/10.18653/v1/2021.eacl-main.39. https://aclanthology.org/2021.eacl-main.39
36. Piroi, F., Lupu, M., Hanbury, A., Zenz, V.: CLEF-IP 2011: retrieval in the intellectual property domain, January 2011
37. Rabelo, J., et al.: A summary of the COLIEE 2019 competition. In: Sakamoto, M., Okazaki, N., Mineshima, K., Satoh, K. (eds.) JSAI-isAI 2019. LNCS (LNAI), vol. 12331, pp. 34–49. Springer, Cham (2020). https://doi.org/10.1007/978-3-030-58790-1_3
38. Reddy, R.G., et al.: End-to-end QA on COVID-19: domain adaptation with synthetic training. arXiv preprint arXiv:2012.01414 (2020)
39. Reimers, N., Gurevych, I.: Sentence-BERT: sentence embeddings using Siamese BERT-networks. In: Proceedings of the 2019 Conference on Empirical Methods in Natural Language Processing and the 9th International Joint Conference on Natural Language Processing (EMNLP-IJCNLP), Hong Kong, China, pp. 3982–3992. Association for Computational Linguistics, November 2019. https://doi.org/10.18653/v1/D19-1410. https://aclanthology.org/D19-1410
40. Rekabsaz, N., Lesota, O., Schedl, M., Brassey, J., Eickhoff, C.: TripClick: the log files of a large health web search engine. In: Proceedings of the 44th International ACM SIGIR Conference on Research and Development in Information Retrieval (SIGIR 2021), Virtual Event, Canada, 11–15 July 2021. https://doi.org/10.1145/3404835.3463242
41. Roberts, K., et al.: Overview of the TREC 2019 precision medicine track. In: TREC. Text REtrieval Conference 26 (2019)
42. Robertson, S., Zaragoza, H.: The probabilistic relevance framework: BM25 and beyond. Found. Trends Inf. Retr. 3(4), 333–389 (2009)
43. Rossi, J., Kanoulas, E.: Legal information retrieval with generalized language models. In: Proceedings of the 6th Competition on Legal Information Extraction/Entailment, COLIEE 2019 (2019)
44. Röttger, P., Pierrehumbert, J.B.: Temporal adaptation of BERT and performance on downstream document classification: insights from social media. In: arXiv (2021)
45. Shao, Y., et al.: BERT-PLI: modeling paragraph-level interactions for legal case retrieval. In: Bessiere, C. (ed.) Proceedings of the Twenty-Ninth International Joint Conference on Artificial Intelligence, IJCAI 2020, pp. 3501–3507. International Joint Conferences on Artificial Intelligence Organization, July 2020. https://doi.org/10.24963/ijcai.2020/484. Main track
46. Tang, R., et al.: Rapidly bootstrapping a question answering dataset for COVID-19. CoRR abs/2004.11339 (2020)
47. Thakur, N., Reimers, N., Rücklé, A., Srivastava, A., Gurevych, I.: BEIR: a heterogeneous benchmark for zero-shot evaluation of information retrieval models. In: Thirty-fifth Conference on Neural Information Processing Systems Datasets and Benchmarks Track (Round 2) (2021). https://openreview.net/forum?id=wCu6T5xFjeJ

48. Tran, V., Nguyen, M.L., Satoh, K.: Building legal case retrieval systems with lexical matching and summarization using a pre-trained phrase scoring model. In: Proceedings of the Seventeenth International Conference on Artificial Intelligence and Law, ICAIL 2019, pp. 275–282. Association for Computing Machinery, New York (2019). https://doi.org/10.1145/3322640.3326740

49. Voorhees, E., et al.: TREC-COVID: constructing a pandemic information retrieval test collection. ArXiv abs/2005.04474 (2020)

50. Wang, L.L., et al.: CORD-19: The COVID-19 open research dataset. arXiv preprint arXiv:2004.10706 (2020)

51. Wang, R., et al.: K-Adapter: infusing knowledge into pre-trained models with adapters. In: Findings of the Association for Computational Linguistics: ACL-IJCNLP 2021, pp. 1405–1418. Association for Computational Linguistics, August 2021. https://doi.org/10.18653/v1/2021.findings-acl.121

52. Xin, J., Xiong, C., Srinivasan, A., Sharma, A., Jose, D., Bennett, P.N.: Zero-shot dense retrieval with momentum adversarial domain invariant representations. In: arXiv (2021)

53. Xiong, C., et al.: CMT in TREC-COVID round 2: mitigating the generalization gaps from web to special domain search. In: arXiv (2020)

54. Xiong, L., et al.: Approximate nearest neighbor negative contrastive learning for dense text retrieval. In: International Conference on Learning Representations (2021), https://openreview.net/forum?id=zeFrfgyZln

55. Yamada, I., Asai, A., Hajishirzi, H.: Efficient passage retrieval with hashing for open-domain question answering. In: Proceedings of the 59th Annual Meeting of the Association for Computational Linguistics and the 11th International Joint Conference on Natural Language Processing (Volume 2: Short Papers), pp. 979–986. Association for Computational Linguistics, August 2021. https://doi.org/10.18653/v1/2021.acl-short.123. https://aclanthology.org/2021.acl-short.123

56. Yang, L., Zhang, M., Li, C., Bendersky, M., Najork, M.: Beyond 512 tokens: Siamese multi-depth transformer-based hierarchical encoder for long-form document matching. In: Proceedings of the 29th ACM International Conference on Information & Knowledge Management, CIKM 2020, pp. 1725–1734. Association for Computing Machinery, New York (2020). https://doi.org/10.1145/3340531.3411908

57. Zhan, J., Mao, J., Liu, Y., Guo, J., Zhang, M., Ma, S.: Optimizing dense retrieval model training with hard negatives. In: Proceedings of the 44th International ACM SIGIR Conference on Research and Development in Information Retrieval, SIGIR 2021, pp. 1503–1512. Association for Computing Machinery, New York (2021). https://doi.org/10.1145/3404835.3462880

58. Zhuang, S., Zuccon, G.: How do online learning to rank methods adapt to changes of intent?, pp. 911–920. Association for Computing Machinery, New York (2021). https://doi.org/10.1145/3404835.3462937

Understanding and Learning from User Behavior for Recommendation in Multi-channel Retail

Mozhdeh Ariannezhad[✉]

AIRLab, University of Amsterdam, Amsterdam, The Netherlands
m.ariannezhad@uva.nl

Abstract. Online shopping is gaining more and more popularity everyday. Traditional retailers with physical stores adjust to this trend by allowing their customers to shop online as well as offline, i.e., in-store. Increasingly, customers can browse and purchase products across multiple shopping channels. Understanding how customer behavior relates to the availability of multiple shopping channels is an important prerequisite for many downstream machine learning tasks, such as recommendation and purchase prediction. However, previous work in this domain is limited to analyzing single-channel behavior only. In this project, we first provide a better understanding of the similarities and differences between online and offline behavior. We further study the next basket recommendation task in a multi-channel context, where the goal is to build recommendation algorithms that can leverage the rich cross-channel user behavior data in order to enhance the customer experience.

1 Motivation

The emergence of e-commerce in recent years has encouraged traditional store-based business owners to provide the possibility of online shopping for their customers in addition to in-store (offline) shopping. The addition of an online channel does not isolate the offline channel. Instead, it creates a multi-channel shopping experience for customers in retail sectors like grocery, cosmetics, and apparel [2]. Multi-channel retail introduces interesting challenges for the business and at the same time promising scientific problems that can be tackled using artificial intelligence, and in particular machine learning.

In contemporary multi-channel retail, digital (i.e., online) and physical (i.e., offline) shopping channels typically operate under a different set of initiatives and approaches. However, treating these channels separately as distinct units has downsides from both customer and business points of view, leading to operational inefficiencies in supply chain and replenishment, and adding frictions for customers who seek a seamless shopping experience across channels [18].

My research is based on proprietary data from a large food retailer in Europe, with a number of physical stores, an online website, and a mobile application. The same product inventory is offered on all channels. Customers can get a loyalty card either in-store or online, and can use that card for their shopping across all channels. A customer can be tracked across offline and online channels if they use their loyalty card at the cashier in-store, and use the information of the same card to login to the mobile application or the website to browse or purchase online. A customer needs to be identified to fulfill the order in the online channel, while this is not the case for in-store shopping (offline). Nonetheless, most of the offline transactions are associated with a loyalty card, due to possibility of getting discounts that are only available when the loyalty card is used during payment.

Customers do not only buy goods via multiple shopping channels, but they can also leverage the online channel for exploring the product inventory, comparing products, and saving products for later purchases, before shopping offline. In addition, the data generated via online shopping provides a further opportunity for personalizing the shopping experience through recommendation [1,18], in addition to the transaction data that is typically used as the only source of information for recommendation [10,28].

While there are numerous studies examining user behavior in online shopping platforms, little is known about multi-channel customer behavior. Moreover, existing recommendation systems are tailored for single-channel scenarios, and leveraging cross-channel information remains under-explored. The goal of my research is to first provide an understanding of customer behavior in multi-channel retail, and further develop recommendation systems that take the multi-channel setting into account.

2   Related Work

Customer Behavior Understanding. Previous work on customer behavior understanding relies mainly on click stream data [5,8,17,22,26,27,30]. Other sources of customer behavior include transaction data [24], digital receipts of online purchases extracted from emails [13], transaction logs of a bank [25], or search logs of a commercial product search engine [20]. However, these studies utilize data from a single shopping channel only, and do not explore multi-channel customer behavior. So far, multi-channel customer behavior has mostly been studied in the marketing and retail research literature [1,4,7,11,12]. These studies rely on perceptions gathered via interviews and customer surveys, in order to model, e.g., lock-in effects or physical store surface needs. Yet, perceptions are often different from actions, and prior work does not consider actual transaction data from customers.

Next Basket Recommendation. The goal of the next basket recommendation (NBR) task is to recommend a full basket composed of a set of items to the user for their next basket, based on the history of the items that they have purchased in the past. One of the key challenges in NBR that makes it distinct from other types of recommendation is learning representations for the user history, which is a sequence of sets.

Learning representations of the sequence of past baskets with neural networks is the most dominant approach in the literature for NBR. Yu et al. [28]

use recurrent layers and inspired by word2vec in natural language processing, Wan et al. [23] introduce triple2vec for NBR. Correlation between items in baskets are used to recommend more coherent baskets in [14]. An encoder-decoder architecture using recurrent layers is proposed in [9]. Inspired by the transformer architecture, Sun et al. [21] leverage multi-head attention to learn a representation for a sequence of sets. Yu et al. [29] build a co-occurrence graph for items and use graph convolutions to learn the item relationships in baskets. A contrastive learning framework is introduced in [19] to denoise basket generation by considering only the relevant items in the history.

In another line of work, Hu et al. [10] propose TIFU-KNN, a nearest neighbor based model for NBR that directly models the personal history of users. Faggioli et al. [6] also propose a relatively simple model that combines personal popularity with collaborative filtering. In our recent study [15], we show that these models perform superior to strong neural network baselines designed for the task, demonstrating the importance of repeat behavior, i.e., recommending the items that a user has purchased before. In our follow up work [16], we propose a novel NBR algorithm that models the repeat behavior explicitly by separating the repeat item prediction task from recommending new items. However, none of the existing recommendation models is designed for the multi-channel setting, and only single-channel retail scenarios are considered.

3 Proposed Research and Methodology

I plan to write my thesis in two main parts. First, I provide an understanding of the customer behavior in multi-channel retail, through extensive studies of shopping transactions and click behavior data. The findings from this part will subsequently guide the design of recommendation algorithms for multi-channel retail in the second part of the thesis.

3.1 Understanding Customer Behavior

Understanding customer behavior in retail serves as a basis for many downstream machine learning tasks, such as recommending products and predicting purchases. In our recent work [3], we observed that the insights gained during user behavior analysis can lead to design of effective recommendation algorithms. In this part of my research, I provide a comprehensive picture of customer behavior through answering two main research questions:

RQ1.1 *How does the choice of shopping channel affect the behavior of customers?*

I started working on this research question by analyzing the transaction data from a food retailer with multiple physical stores and two online platforms. The initial findings are already published in [2]. In this work, three groups of customers are defined based on their choice of shopping channels, namely *online-only*, *offline-only*, and *multi-channel* customers. We find that the tendency to purchase previously bought products, defined as repeat behavior ratio, is higher for online-only customers. Zooming in on multi-channel customers, our analysis reveals that there is little overlap in online and offline baskets of multi-channel customers; they use each channel for different sets of items. Our analysis further indicates that online baskets are larger, and contain items from more diverse product categories.

RQ1.2 *How do the customers of a multi-channel retail organization make use of its online platforms?*

While there are numerous studies on understanding the online behavior in retail, they are focused on single-channel shopping settings [8,17,22]; the customers are not able to shop in-store in addition to the online platforms. To answer this research question, I will study the characteristics of the online behavior leading to online purchases vs. offline purchases. This research will be built up on prior studies on online user behavior understanding [3,8,13,17]. I will study the similarities and differences of online sessions from different aspects, such as type of interactions (e.g., search, product view, add to basket) and interaction duration.

My goal is to use the insights from the user behavior analysis in order to detect customer intention, given the interactions in an online session. The customer intention can further be classified as an online purchase, an offline purchase, or simply browsing. In addition to informing the design of the downstream recommendation algorithm, the insights from this study can further be adapted as online features that improve the customer experience with the online platforms. As an example, pointing the customer to the location of a product in a store can be useful if an offline purchase is predicted, while displaying the "add to basket" button might not be as practical for the user's need.

3.2   Recommendation in Multi-channel Retail

The focus in this part of the thesis is on the task of next basket recommendation, where the goal is to recommend a list of items for the next basket of a customer. Such recommendations would reduce the burden on users to proactively find the items of their interest every time they need to shop. I plan to answer two research questions in this part:

RQ2.1 *How can we design a recommendation algorithm for the multi-channel setting?*

Our customer behavior analysis in [2] revealed that the online and offline baskets have different characteristics. Furthermore, customers that use both channels for shopping purchase different products in different channels. While we have shown the importance of recommending previously purchased items in our previous studies on NBR [15,16], the repeat behavior differs significantly across channels. Our preliminary experiments with an NBR model showed that the recommendation performance is not consistent across channels, when using a model that is not designed for the multi-channel setting. This calls for recommendation algorithms that are tailored for multi-channel retail.

I propose to design a neural recommendation algorithm that has separate components for modeling online and offline baskets in the customers' history, as well as a component to predict the channel for the next basket. The final component that is responsible for generating the recommendation list will make use of the channel predictor. Following previous works [9, 14, 19], sequence modeling architectures such as recurrent layers and self-attention layers will be used to encode the baskets in a user history. To encode each basket, a.k.a., a set of items, I will experiment with graph neural networks [29]. Given the differences between online and offline baskets, a single architecture might not be optimal for encoding, and I will experiment with separate architectures for different type of baskets. The channel predictor component can be integrated with the recommendation algorithm, or can be designed as a separate model that operates beforehand. I will study both approaches in order to find the optimal solution.

Although the recommendation algorithm is designed for a multi-channel setting, it should perform consistently well for the customers that use a single channel for shopping. To this end, I aim to validate the proposed algorithm on publicly available NBR datasets as well, where the data comes from a single channel.

RQ2.2 *How can we leverage the online behavior of customers in a multi-channel recommendation algorithm?*

As mentioned in Sect. 1, loyalty cards enable us to connect the online behavior of customers to their offline behavior. In other words, when users are logged in on their mobile phones or on the website, their interactions can be linked to their offline baskets, assuming they use the loyalty card during checkout. The loyalty card further provides the opportunity to offer personal discounts to customers, in the form of a recommendation list. Currently, personal offers are generated solely based on the purchasing behavior, and click data is not considered. Using online signals to provide recommendation for offline shopping is under explored in the literature, with the exception of a probabilistic graphical model proposed in [18].

My goal in this section of the thesis is to utilize online click behavior data for providing recommendations not only for the online channel, but for in-store shopping as well. I will make use of the insights gained in answering RQ1.2, and integrate the online signals with the recommendation algorithm designed in answer to RQ2.1. To this end, the online interaction sessions will be encoded and fed into the neural recommendation model, with the indication of customer intention for each session. An intermediate neural component is then used to decide the extent of the contribution of each online session to the final recommendations.

4   Research Issues for Discussion

I seek suggestions and feedback on how to improve this proposal in general. In particular, there are two areas that I would like to discuss during the doctoral consortium. The first issue is the broader impact of this research. For example, are there other domains beside retail, or other settings beside multi-channel shopping that can benefit from the findings in this research? How can we transfer our knowledge and proposed models in that case? Second, I have not considered the societal concerns in studying the multi-channel shopping behavior, e.g., from a privacy preservation point of view. I would appreciate input on this matter.

References

1. Acquila-Natale, E., Iglesias-Pradas, S.: A matter of value? predicting channel preference and multichannel behaviors in retail. Technol. Forecast. Soc. Chang. **162**, 120401 (2021)
2. Ariannezhad, M., Jullien, S., Nauts, P., Fang, M., Schelter, S., de Rijke, M.: Understanding multi-channel customer behavior in retail. In: CIKM 2021: The 30th ACM International Conference on Information and Knowledge Management, Virtual Event, Queensland, Australia, 1–5 November 2021, pp. 2867–2871. ACM (2021)
3. Ariannezhad, M., Yahya, M., Meij, E., Schelter, S., de Rijke, M.: Understanding and learning from user interactions with financial company filings (2021, Under review)
4. Chatterjee, P.: Multiple-channel and cross-channel shopping behavior: role of consumer shopping orientations. Market. Intell. Plann. **28** (2010)
5. Chen, C., et al.: Predictive analysis by leveraging temporal user behavior and user embeddings. In: Proceedings of the 27th ACM International Conference on Information and Knowledge Management, CIKM 2018, Torino, Italy, 22–26 October 2018, pp. 2175–2182. ACM (2018)
6. Faggioli, G., Polato, M., Aiolli, F.: Recency aware collaborative filtering for next basket recommendation. In: Proceedings of the 28th ACM Conference on User Modeling, Adaptation and Personalization, UMAP 2020, Genoa, Italy, 12–18 July 2020, pp. 80–87. ACM (2020)
7. Gao, F., Agrawal, V., Cui, S.: The effect of multichannel and omnichannel retailing on physical stores. Manage. Sci. (2021)
8. Hendriksen, M., Kuiper, E., Nauts, P., Schelter, S., de Rijke, M.: Analyzing and predicting purchase intent in e-commerce: anonymous vs. identified customers. In: eCOM 2020: The 2020 SIGIIR Workshop on eCommerce. ACM (2020)
9. Hu, H., He, X.: Sets2sets: learning from sequential sets with neural networks. In: Proceedings of the 25th ACM SIGKDD International Conference on Knowledge Discovery & Data Mining, KDD 2019, Anchorage, AK, USA, 4–8 August 2019, pp. 1491–1499. ACM (2019)
10. Hu, H., He, X., Gao, J., Zhang, Z.: Modeling personalized item frequency information for next-basket recommendation. In: Proceedings of the 43rd International ACM SIGIR conference on research and development in Information Retrieval, SIGIR 2020, Virtual Event, China, 25–30 July 2020, pp. 1071–1080. ACM (2020)

11. Hult, G.T.M., Sharma, P.N., Morgeson, F.V., Zhang, Y.: Antecedents and consequences of customer satisfaction: do they differ across online and offline purchases? J. Retail. **95**(1), 10–23 (2019)

12. Hussein, R.S., Kais, A.: Multichannel behaviour in the retail industry: evidence from an emerging market. Int. J. Log. Res. Appl. **24**(3), 242–260 (2021)

13. Kooti, F., Lerman, K., Aiello, L.M., Grbovic, M., Djuric, N., Radosavljevic, V.: Portrait of an online shopper: understanding and predicting consumer behavior. In: Proceedings of the Ninth ACM International Conference on Web Search and Data Mining, San Francisco, CA, USA, 22–25 February 2016, pp. 205–214. ACM (2016)

14. Le, D., Lauw, H.W., Fang, Y.: Correlation-sensitive next-basket recommendation. In: Proceedings of the Twenty-Eighth International Joint Conference on Artificial Intelligence, IJCAI 2019, Macao, China, 10–16 August 2019, pp. 2808–2814. ijcai.org (2019)

15. Li, M., Jullien, S., Ariannezhad, M., de Rijke, M.: A next basket recommendation reality check. CoRR abs/2109.14233 (2021). https://arxiv.org/abs/2109.14233

16. Li, M., Jullien, S., Ariannezhad, M., de Rijke, M.: TREX: a flexible repetition and exploration framework for next basket recommendation (2021, Under review)

17. Lo, C., Frankowski, D., Leskovec, J.: Understanding behaviors that lead to purchasing: a case study of Pinterest. In: Proceedings of the 22nd ACM SIGKDD International Conference on Knowledge Discovery and Data Mining, San Francisco, CA, USA, 13–17 August 2016, pp. 531–540. ACM (2016)

18. Luo, P., Yan, S., Liu, Z., Shen, Z., Yang, S., He, Q.: From online behaviors to offline retailing. In: Proceedings of the 22nd ACM SIGKDD International Conference on Knowledge Discovery and Data Mining, San Francisco, CA, USA, 13–17 August 2016, pp. 175–184. ACM (2016)

19. Qin, Y., Wang, P., Li, C.: The world is binary: contrastive learning for denoising next basket recommendation. In: SIGIR 2021: The 44th International ACM SIGIR Conference on Research and Development in Information Retrieval, Virtual Event, Canada, 11–15 July 2021, pp. 859–868. ACM (2021)

20. Su, N., He, J., Liu, Y., Zhang, M., Ma, S.: User intent, behaviour, and perceived satisfaction in product search. In: Proceedings of the Eleventh ACM International Conference on Web Search and Data Mining, WSDM 2018, Marina Del Rey, CA, USA, 5–9 February 2018, pp. 547–555. ACM (2018)

21. Sun, L., Bai, Y., Du, B., Liu, C., Xiong, H., Lv, W.: Dual sequential network for temporal sets prediction. In: Proceedings of the 43rd International ACM SIGIR Conference on Research and Development in Information Retrieval, SIGIR 2020, Virtual Event, China, 25–30 July 2020, pp. 1439–1448. ACM (2020)

22. Toth, A., Tan, L., Fabbrizio, G.D., Datta, A.: Predicting shopping behavior with mixture of RNNs. In: Proceedings of the SIGIR 2017 Workshop On eCommerce co-located with the 40th International ACM SIGIR Conference on Research and Development in Information Retrieval, eCOM@SIGIR 2017, Tokyo, Japan, 11 August 2017, vol. 2311. CEUR-WS.org (2017)

23. Wan, M., Wang, D., Liu, J., Bennett, P., McAuley, J.J.: Representing and recommending shopping baskets with complementarity, compatibility and loyalty. In: Proceedings of the 27th ACM International Conference on Information and Knowledge Management, CIKM 2018, Torino, Italy, 22–26 October 2018, pp. 1133–1142. ACM (2018)

24. Wang, P., Guo, J., Lan, Y.: Modeling retail transaction data for personalized shopping recommendation. In: Proceedings of the 23rd ACM International Conference on Conference on Information and Knowledge Management, CIKM 2014, Shanghai, China, 3–7 November 2014, pp. 1979–1982. ACM (2014)
25. Wen, Y.T., Yeh, P., Tsai, T., Peng, W., Shuai, H.: Customer purchase behavior prediction from payment datasets. In: Proceedings of the Eleventh ACM International Conference on Web Search and Data Mining, WSDM 2018, Marina Del Rey, CA, USA, 5–9 February 2018, pp. 628–636. ACM (2018)
26. Xia, Q., Jiang, P., Sun, F., Zhang, Y., Wang, X., Sui, Z.: Modeling consumer buying decision for recommendation based on multi-task deep learning. In: Proceedings of the 27th ACM International Conference on Information and Knowledge Management, CIKM 2018, Torino, Italy, 22–26 October 2018, pp. 1703–1706. ACM (2018)
27. Yeo, J., Kim, S., Koh, E., Hwang, S., Lipka, N.: Browsing2purchase: online customer model for sales forecasting in an e-commerce site. In: Proceedings of the 25th International Conference on World Wide Web, WWW 2016, Montreal, Canada, 11–15 April 2016, Companion Volume, pp. 133–134. ACM (2016)
28. Yu, F., Liu, Q., Wu, S., Wang, L., Tan, T.: A dynamic recurrent model for next basket recommendation. In: Proceedings of the 39th International ACM SIGIR conference on Research and Development in Information Retrieval, SIGIR 2016, Pisa, Italy, 17–21 July 2016, pp. 729–732. ACM (2016)
29. Yu, L., Sun, L., Du, B., Liu, C., Xiong, H., Lv, W.: Predicting temporal sets with deep neural networks. In: KDD 2020: The 26th ACM SIGKDD Conference on Knowledge Discovery and Data Mining, Virtual Event, CA, USA, 23–27 August 2020, pp. 1083–1091. ACM (2020)
30. Zhou, M., Ding, Z., Tang, J., Yin, D.: Micro behaviors: a new perspective in e-commerce recommender systems. In: Proceedings of the Eleventh ACM International Conference on Web Search and Data Mining, WSDM 2018, Marina Del Rey, CA, USA, 5–9 February 2018, pp. 727–735. ACM (2018)

An Entity-Oriented Approach for Answering Topical Information Needs

Shubham Chatterjee[✉][iD]

University of New Hampshire, Durham, USA
shubham.chatterjee@unh.edu

Abstract. In this dissertation, we adopt an entity-oriented approach to identify relevant materials for answering a topical keyword query such as "Cholera". To this end, we study the interplay between text and entities by addressing three related prediction problems: (1) Identify knowledge base entities that are relevant for the query, (2) Understand an entity's meaning in the context of the query, and (3) Identify text passages that elaborate the connection between the query and an entity. Through this dissertation, we aim to study some overarching questions in entity-oriented research such as the importance of query-specific entity descriptions, and the importance of entity salience and context-dependent entity similarity for modeling the query-specific context of an entity.

1 Introduction

Wikipedia is useful for users seeking information on topics such as "Cholera"; however, it is mostly focused on recent and popular topics. Through this dissertation, we address the first step in answering topical queries: the retrieval of relevant materials (text and entities) that constitutes the answer. We envision a downstream system to utilize these relevant materials to automatically construct a Wikipedia-like article for such topical queries.

Humans usually think about topics in terms of entities, and the background stories and roles of these entities with respect to the topic. Motivated by this, we adopt an entity-oriented approach to identify relevant material for answering such topical queries by addressing three related prediction problems: (1) **Entity Retrieval:** Identify entities that are relevant for a discussion about the query, (2) **Entity Aspect Linking:** Link the mention of an entity in a given context (e.g., sentence) to the aspect (from a catalog) that best captures the meaning of the entity in that context, and (3) **Entity Support Passage Retrieval:** Find passages that explain why the entity is relevant for the query.

Research Questions. An entity such as "Oyster" may be referred to in multiple contexts in text, e.g., "Cultivation", "Ecosystem Services", etc. – each is called an *aspect* of the entity "Oyster". Often, features for entity ranking are derived from entity links found in query-relevant documents [10,16,20]. While entity linking can disambiguate the different mentions of "Oyster" in a text (animal

M. Hagen et al. (Eds.): ECIR 2022, LNCS 13186, pp. 463–472, 2022.
https://doi.org/10.1007/978-3-030-99739-7_57

Query: Cholera
Entity: Oyster

Most cholera cases in developed countries are a result of transmission by food. Food transmission can occur when people harvest seafood such as **oysters** in waters infected with sewage, as Vibrio cholerae accumulates in planktonic crustaceans and the **oysters** eat the zooplankton.	**Oyster** is the common name for a number of different families of salt-water bivalve molluscs that live in marine or brackish habitats. **Oysters** influence nutrient cycling, water filtration, habitat structure, biodiversity, and food web dynamics.	Cholera is an infection of the small intestine by some strains of the bacterium Vibrio cholerae. The classic symptom is large amounts of watery diarrhea that lasts a few days. Vomiting and muscle cramps may also occur. Cholera can be caused by eating **oysters**.

Fig. 1. Example query and entity with support passages. **Left:** The passage is relevant to the query and entity, and the entity is salient in the passage. The passage clarifies how oysters may cause cholera. Hence, this is a good support passage for the query-entity pair. **Middle:** The passage is relevant to the entity but not to the query. **Right:** The passage is relevant to the query but not to the entity as the entity is not salient in the passage. The passages in the middle and right are not good support passages.

versus place), it cannot identify the different aspects of "Oyster". Entity aspect linking [24] can remedy this by using a unique aspect id to resolve the different meanings (aspects) of an entity in the context of the query. Hence, we study the following research question: **(RQ1)** How can we leverage fine-grained entity aspects for entity retrieval, and to what extent are they useful?

Although we study the utility of entity aspects in the context of Web search, applications such as question-answering and recommender systems that aim to understand the subtleties in the human language would also benefit from research on entity aspect linking. Moreover, since entity aspect linking aims to match an entity's context to a candidate aspect, approaches to entity aspect linking would also be applicable to other text similarity problems. The current entity aspect linking system from Nanni et al. [24] has scope for improvement (see Sect. 3). Hence, we study the following research question: **(RQ2)** How can we improve the current entity aspect linking system?

We use entity aspect linking to obtain a fine-grained understanding of the entity in the context of the query. Alternatively, *entity support passages* may also be used for this purpose. Entity support passages [5,18] are paragraph-size text passages that explain why an entity, e.g., "Oyster", from the entity ranking for a query, e.g., "Cholera", is relevant to the query (Fig. 1). Entity support passages may serve as text to be summarized when generating an answer to a topical query. There are two challenges in finding a good support passage: (1) Support passages must be relevant to both, the query and the entity, and (2) The entity must be *salient*, i.e., *central* to the discussion in the text and not just mentioned as an aside. Hence, we study the following research questions: **(RQ3)** How can we model the joint relevance of a paragraph to an entity and a query? To what extent is entity salience helpful? **(RQ4)** How can we leverage support passages for entity retrieval and to what extent are they useful?

An important consideration in entity-oriented research is regarding the construction of entity descriptions. Often, entity descriptions are constructed without considering the query, by using knowledge bases [8,15,19], the entity's

Wikipedia article [10,21,22], or by collecting documents from a corpus that mention the entity [1,9,28]. As a result, such query-independent descriptions may contain information about the entity that is non-relevant in the context of the query. For example, although the entity "Oyster" is relevant to the query "Cholera", the Wikipedia page of "Oyster" does not even mention Cholera. We study the utility of entity aspects and entity support passages as *query-specific* entity descriptions for learning query-specific entity embeddings. In this regard, we study RQ1 and RQ4 in addition to the following research question: **(RQ5)** Is it sufficient to use the lead text of an entity's Wikipedia page as the entity's description?

2 Related Work

Entity Retrieval. The Sequential Dependence Model [23] assigns different weights to matching unigrams and bigarams of different types. Zhiltsov et al. [36] propose the Fielded Sequential Dependence Model (FSDM) that uses field-specific background models across multiple fields. Nikolaev et al. [25] estimate the probability of unigrams and bigrams being mapped onto a field in the FSDM dynamically. Hasibi et al. [16] leverage entity links in queries and propose a parameter-free estimation of the field weights in FSDM. Several models utilize information from Knowledge Bases. For example, Kaptein et al. [19] utilize the types of entities in the query, and Balog et al. [2] utilize category information about an entity obtained from a user. Learning-To-Rank (LTR) is another common approach. Schuhmacher et al. [30] utilize several features to re-rank entities in a LTR setting. Dietz [11] proposed ENT Rank, a LTR model that combines information about an entity, the entity's neighbors, and context using a hypergraph.

Entity Aspect Linking. Several works treat Wikipedia sections as entity aspects. For example, Fetahu et al. [14] enrich Wikipedia sections with news-article references. Banerjee et al. [3] seek to improve Wikipedia stubs by generating content for each section automatically. Nanni et al. [24] address the entity aspect linking task using the top-level sections from an entity's Wikipedia article as the entity's aspects. Their approach is based on LTR with lexical and semantic features derived from various contexts (e.g., sentence, paragraph, section) where the entity is mentioned in text. Ramsdell et al. [29] released a large dataset for entity aspect linking using the definition of aspects from Nanni et al.

Entity Support Passage Retrieval. Blanco et al. [5] rank entity support sentences using LTR with features based on named entity recognition, and term-based retrieval. Kadry et al. [18] study the importance of relation extraction for entity support passage retrieval. A related task is entity relationship explanation that aims to explain the relationship between two entities in a Knowledge Graph using a text passage. Pirro et al. [26] address the problem from a graph perspective by finding the sub-graph consisting of nodes and edges in the set of paths between the two input entities. Voskarides et al. [31,32] use textual, entity and relationship features within a LTR framework, whereas Bhatia et al. [4] address the problem from a probabilistic perspective.

3   Methodology

Datasets. The TREC Complex Answer Retrieval (CAR) [12] provides large and suitable benchmarks consisting of topical keyword queries. We use the CAR benchmarks for experiments on entity retrieval and entity support passage retrieval. Additionally, we also use the DBpedia-Entity v2 [17] dataset for our entity retrieval experiments. For experiments on entity aspect linking, we use the dataset from Ramsdell et al. [29]. We also use the aspect catalog and aspect linker implementation from Ramsdell et al. to entity aspect link the CAR corpus of English Wikipedia paragraphs.

Completed: Entity Support Passage Retrieval. We refer to the entity for which we want to find a support passage as *target entity*. Our approach [6] to entity support passage retrieval uses Learning-To-Rank with a rich set of features. The features are based on: (1) Modelling the joint relevance of a passage to the query and target entity, and (2) Salience of the target entity in a passage. Below, we describe our approach to study our research question **RQ3**. Our approach assumes that a high-precision entity ranking is available as input.

To model the relevance of a passage to a query, we retrieve a candidate set of passages $\mathcal{D}$ using BM25. We assume that a passage is relevant to the target entity if the passage contains many entities that are related to the target entity in the context of the query. We derive the query-relevant context $\mathcal{D}_e$ of the target entity from the candidate set $\mathcal{D}$ by retaining passages in $\mathcal{D}$ that contain a link to the target entity. We consider each passage $p \in \mathcal{D}_e$ as a candidate support passage. To identify entities from $\mathcal{D}_e$ which are related to the target entity, we treat $\mathcal{D}_e$ as a bag-of-entities [16,34]. We assume that entities $e_x \in \mathcal{D}_e$ which frequently co-occur with the target entity within $\mathcal{D}_e$ are related to the target entity. We then score a candidate support passage $p \in \mathcal{D}_e$ by the number of frequently co-occurring entities e_x linked to the passage. We also derive several features based on the salience of the target entity in the support passage.[1]

We outperform the state-of-the-art method from Blanco et al. [5] by a large margin on various benchmarks from TREC CAR. We find that salience is a strong indicator of support passages for entities that have a passage with a salient mention in the candidate set.

Completed: Entity Retrieval. We study **RQ1** based on the hypothesis that different mentions of an entity in a query-specific context contribute differently to determine the relevance of that entity for the query. For example, when determining the relevance of the entity "Oyster" for the query "Cholera", the aspect "Diseases" is more important than "Cultivation". Our approach [7] is based on Learning-To-Rank (LTR) with features derived from entity aspects: (1) Aspect Retrieval features: We rank aspects via their text[2] using the query, and (2)

---

[1] We use the salience detection system from Ponza et al. [27].
[2] Available from the aspect catalog from Ramsdell et al.

Search result context of entity "Oyster"

Entity: Oyster
Aspect: Ecosystem services

The *Nature Conservancy*, and the *Oyster Recovery Partnership*, *Maryland Department of Natural Resources*, the *National Oceanographic and Atmospheric Administration*, and the *U.S. Army Corps of Engineers* planted oyster spat on 350 underwater acres. Planting began in 2012. Water quality is measured with a vertical profiler and *water quality sondes* moored at the bottom. In 2013, 112,500 tons of fossilized oyster shell were transported from *Florida*, and 42,536 tons of the shell went into *Harris Creek* (the rest went to the *Little Choptank River*.

As an ecosystem engineer oysters provide "supporting" *ecosystem services*, along with "provisioning", "regulating" and "cultural" services. *Oysters* influence *nutrient cycling*, water filtration, habitat structure, biodiversity, and *food web dynamics*. [...]

Fig. 2. Depiction of our entity aspect linking approach for the entity "Oyster". Left: context from search results. Right: Correct aspect "Ecosystem services" of the entity "Oyster". The example text, entities, and aspects taken from Wikipedia. Entity links marked in bold italics. In objective 1, we address the issue that not all words are relevant for the decision – non-relevant words are depicted in grey. As described in objective 2, it is rare that identical entities are mentioned in both context and aspect content, hence we need to identify which entities are related in this context, such as entities related to ecosystems (green frame) and regarding water quality (orange frame). In objective 3, we study how integrating the prediction of relevant words and entities is helpful for most accurate predictions of entity aspect links. (Color figure online)

Aspect Link PRF features: After retrieving an initial candidate set of passages using BM25, the frequency distribution of entity aspect links in these passages are weighted by the retrieval score of the passages to obtain a distribution of relevant entity aspects. To study **RQ4**, we also build the candidate set of passages from an entity support passage ranking instead of BM25 when deriving Aspect Link PRF features. Furthermore, we study RQ1, RQ4, and **RQ5** by fine-tuning a BERT model for the entity ranking task using entity aspects and entity support passages as query-specific entity descriptions.[3]

We find that our LTR and BERT models trained using entity aspects and entity support passages significantly outperform both neural and non-neural baselines using both TREC CAR and DBpedia-Entity v2. Moreover, significant performance improvements are obtained by replacing a query-independent entity description (e.g., lead text of an entity's Wikipedia article) with a query-specific description (e.g., entity support passage).

Proposed: Entity Aspect Linking. Below, we identify three research objectives to study **RQ2**.

1. **Identify relevant words in context and aspect.** Nanni et al. [24] find that using all words from the entity's context leads to poor results. They alleviate this by considering only the sentence mentioning the entity. However, as shown in Fig. 2, to help us make the aspect linking decision, we need to consider the whole passage which mentions the entity "Oyster", and not

---

[3] Paper under review.

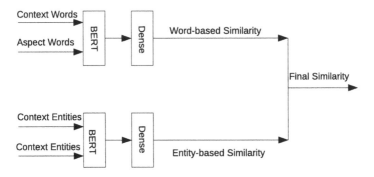

Fig. 3. Depiction of our proposed end-to-end entity aspect linking system. The top model takes the words from the aspect and context and learns a representation that pays attention to the words from the context that are important for the aspect linking decision. Similarly, the bottom model learns a context-dependent similarity between the entities from the context and aspect. The final similarity is obtained by combining the word-based and entity-based similarity.

just the sentence. Since the majority of words in the larger context are not relevant, we propose to use the attention mechanism in deep learning to select words from the context which are most beneficial for the aspect linking decision.

2. **Identify contextually related entities.** Nanni et al. base the aspect linking decisions on whether a direct relationship exists between an aspect-entity and a context-entity. However, as shown in Fig. 2, otherwise unrelated entities are related in the given context. Hence, we propose to base the aspect linking decisions on whether two entities are related in context, by learning embeddings of these entities using BERT, taking the context into account. Our preliminary work (See Footnote 3) in learning query-specific entity embeddings has shown promising results.

3. **Integrate information from words and entities.** Previous works which leverage entities for retrieving text [13, 20, 33–35] have found that combining indicators of relevance obtained using words and entities leads to better performance for distinguishing relevant from non-relevant text. In this light, I propose to integrate the information from relevant words and entities (from 1 and 2 above) by learning the similarity between the context and the aspect end-to-end using a Siamese Neural Network (Fig. 3).

4   Conclusion

In this research statement, we describe our entity-oriented approach to identify relevant text and entities for answering a topical keyword query such as "Cholera". We describe three related tasks that we address for this purpose, and identify several overarching research questions in entity-oriented research that we aim to answer with this work. We envision this work to serve as a stepping stone towards building more intelligent systems. Such systems would one day respond to a user's open-ended and complex information needs with a complete answer instead of a ranked list of results, thus transforming the "search" engine into an "answering" engine.

References

1. Balog, K., Azzopardi, L., de Rijke, M.: Formal models for expert finding in enterprise corpora. In: Proceedings of the 29th Annual International ACM SIGIR Conference on Research and Development in Information Retrieval, SIGIR 2006, pp. 43–50. Association for Computing Machinery, New York (2006). https://doi.org/10.1145/1148170.1148181
2. Balog, K., Bron, M., De Rijke, M.: Query modeling for entity search based on terms, categories, and examples. ACM Trans. Inf. Syst. **29**(4) (2011). https://doi.org/10.1145/2037661.2037667
3. Banerjee, S., Mitra, P.: WikiKreator: improving Wikipedia stubs automatically. In: Proceedings of the 53rd Annual Meeting of the Association for Computational Linguistics and the 7th International Joint Conference on Natural Language Processing (Volume 1: Long Papers), Beijing, China, pp. 867–877. Association for Computational Linguistics, July 2015. https://doi.org/10.3115/v1/P15-1084. https://aclanthology.org/P15-1084
4. Bhatia, S., Dwivedi, P., Kaur, A.: That's interesting, tell me more! Finding descriptive support passages for knowledge graph relationships. In: Vrandečić, D., et al. (eds.) ISWC 2018. LNCS, vol. 11136, pp. 250–267. Springer, Cham (2018). https://doi.org/10.1007/978-3-030-00671-6_15
5. Blanco, R., Zaragoza, H.: Finding support sentences for entities. In: Proceedings of the 33rd International ACM SIGIR Conference on Research and Development in Information Retrieval, SIGIR 2010, pp. 339–346. Association for Computing Machinery, New York (2010). https://doi.org/10.1145/1835449.1835507
6. Chatterjee, S., Dietz, L.: Why does this entity matter? Support passage retrieval for entity retrieval. In: Proceedings of the 2019 ACM SIGIR International Conference on Theory of Information Retrieval, ICTIR 2019, pp. 221–224. Association for Computing Machinery, New York (2019). https://doi.org/10.1145/3341981.3344243
7. Chatterjee, S., Dietz, L.: Entity retrieval using fine-grained entity aspects. In: Proceedings of the 44th International ACM SIGIR Conference on Research and Development in Information Retrieval, SIGIR 2021. Association for Computing Machinery, New York (2021). https://doi.org/10.1145/3404835.3463035
8. Chen, J., Xiong, C., Callan, J.: An empirical study of learning to rank for entity search. In: Proceedings of the 39th International ACM SIGIR Conference on Research and Development in Information Retrieval, SIGIR 2016, pp. 737–740. Association for Computing Machinery, New York (2016). https://doi.org/10.1145/2911451.2914725
9. Conrad, J.G., Utt, M.H.: A system for discovering relationships by feature extraction from text databases. In: Croft, B.W., van Rijsbergen, C.J. (eds.) SIGIR 1994, pp. 260–270. Springer, London (1994). https://doi.org/10.1007/978-1-4471-2099-5_27

10. Dalton, J., Dietz, L., Allan, J.: Entity query feature expansion using knowledge base links. In: Proceedings of the 37th International ACM SIGIR Conference on Research and Development in Information Retrieval, SIGIR 2014, pp. 365–374. Association for Computing Machinery, New York (2014). https://doi.org/10.1145/2600428.2609628

11. Dietz, L.: ENT rank: retrieving entities for topical information needs through entity-neighbor-text relations. In: Proceedings of the 42nd International ACM SIGIR Conference on Research and Development in Information Retrieval, SIGIR 2019, pp. 215–224. Association for Computing Machinery, New York (2019). https://doi.org/10.1145/3331184.3331257

12. Dietz, L., Foley, J.: TREC CAR Y3: complex answer retrieval overview. In: Proceedings of Text REtrieval Conference (TREC) (2019)

13. Ensan, F., Bagheri, E.: Document retrieval model through semantic linking. In: Proceedings of the 10th ACM International Conference on Web Search and Data Mining, WSDM 2017, pp. 181–190. Association for Computing Machinery, New York (2017). https://doi.org/10.1145/3018661.3018692

14. Fetahu, B., Markert, K., Anand, A.: Automated news suggestions for populating wikipedia entity pages. In: Proceedings of the 24th ACM International on Conference on Information and Knowledge Management, CIKM 2015, pp. 323–332. Association for Computing Machinery, New York (2015). https://doi.org/10.1145/2806416.2806531

15. Graus, D., Tsagkias, M., Weerkamp, W., Meij, E., de Rijke, M.: Dynamic collective entity representations for entity ranking. In: Proceedings of the Ninth ACM International Conference on Web Search and Data Mining, WSDM 2016, pp. 595–604. Association for Computing Machinery, New York (2016). https://doi.org/10.1145/2835776.2835819

16. Hasibi, F., Balog, K., Bratsberg, S.E.: Exploiting entity linking in queries for entity retrieval. In: Proceedings of the 2016 ACM International Conference on the Theory of Information Retrieval, ICTIR 2016, pp. 209–218. Association for Computing Machinery, New York (2016). https://doi.org/10.1145/2970398.2970406

17. Hasibi, F., et al.: DBpedia-entity V2: a test collection for entity search. In: Proceedings of the 40th International ACM SIGIR Conference on Research and Development in Information Retrieval, SIGIR 2017, pp. 1265–1268. Association for Computing Machinery, New York (2017). https://doi.org/10.1145/3077136.3080751

18. Kadry, A., Dietz, L.: Open relation extraction for support passage retrieval: merit and open issues. In: Proceedings of the 40th International ACM SIGIR Conference on Research and Development in Information Retrieval, SIGIR 2017, pp. 1149–1152. Association for Computing Machinery, New York (2017). https://doi.org/10.1145/3077136.3080744

19. Kaptein, R., Serdyukov, P., De Vries, A., Kamps, J.: Entity ranking using wikipedia as a pivot. In: Proceedings of the 19th ACM International Conference on Information and Knowledge Management, CIKM 2010, pp. 69–78. Association for Computing Machinery, New York (2010). https://doi.org/10.1145/1871437.1871451

20. Liu, X., Fang, H.: Latent entity space: a novel retrieval approach for entity-bearing queries. Inf. Retr. J. **18**(6), 473–503 (2015)

21. Manotumruksa, J., Dalton, J., Meij, E., Yilmaz, E.: CrossBERT: a triplet neural architecture for ranking entity properties. In: Proceedings of the 43rd International ACM SIGIR Conference on Research and Development in Information Retrieval, SIGIR 2020, pp. 2049–2052. Association for Computing Machinery, New York (2020). https://doi.org/10.1145/3397271.3401265

22. Meij, E., Bron, M., Hollink, L., Huurnink, B., de Rijke, M.: Mapping queries to the linking open data cloud: a case study using DBpedia. J. Web Semant. **9**(4), 418–433 (2011). https://doi.org/10.1016/j.websem.2011.04.001. http://www.sciencedirect.com/science/article/pii/S1570826811000187. jWS special issue on Semantic Search

23. Metzler, D., Croft, W.B.: A Markov random field model for term dependencies. In: Proceedings of the 28th Annual International ACM SIGIR Conference on Research and Development in Information Retrieval, SIGIR 2005, pp. 472–479. Association for Computing Machinery, New York (2005). https://doi.org/10.1145/1076034.1076115

24. Nanni, F., Ponzetto, S.P., Dietz, L.: Entity-aspect linking: providing fine-grained semantics of entities in context. In: Proceedings of the 18th ACM/IEEE on Joint Conference on Digital Libraries, JCDL 2018, pp. 49–58. Association for Computing Machinery, New York (2018). https://doi.org/10.1145/3197026.3197047

25. Nikolaev, F., Kotov, A., Zhiltsov, N.: Parameterized fielded term dependence models for ad-hoc entity retrieval from knowledge graph. In: Proceedings of the 39th International ACM SIGIR Conference on Research and Development in Information Retrieval, SIGIR 2016, pp. 435–444. Association for Computing Machinery, New York (2016). https://doi.org/10.1145/2911451.2911545

26. Pirrò, G.: Explaining and suggesting relatedness in knowledge graphs. In: Arenas, M., et al. (eds.) ISWC 2015. LNCS, vol. 9366, pp. 622–639. Springer, Cham (2015). https://doi.org/10.1007/978-3-319-25007-6_36

27. Ponza, M., Ferragina, P., Piccinno, F.: SWAT: a system for detecting salient Wikipedia entities in texts. Comput. Intell. (2018). https://doi.org/10.1111/coin.12216

28. Raghavan, H., Allan, J., McCallum, A.: An exploration of entity models, collective classification and relation description. In: KDD Workshop on Link Analysis and Group Detection, pp. 1–10 (2004)

29. Ramsdell, J., Dietz, L.: A large test collection for entity aspect linking. In: Proceedings of the 29th ACM International Conference on Information and Knowledge Management, CIKM 2020, pp. 3109–3116. Association for Computing Machinery, New York (2020). https://doi.org/10.1145/3340531.3412875

30. Schuhmacher, M., Dietz, L., Paolo Ponzetto, S.: Ranking entities for web queries through text and knowledge. In: Proceedings of the 24th ACM International on Conference on Information and Knowledge Management, CIKM 2015, pp. 1461–1470. Association for Computing Machinery, New York (2015). https://doi.org/10.1145/2806416.2806480

31. Voskarides, N., Meij, E., de Rijke, M.: Generating descriptions of entity relationships. In: Jose, J.M., et al. (eds.) ECIR 2017. LNCS, vol. 10193, pp. 317–330. Springer, Cham (2017). https://doi.org/10.1007/978-3-319-56608-5_25

32. Voskarides, N., Meij, E., Tsagkias, M., de Rijke, M., Weerkamp, W.: Learning to explain entity relationships in knowledge graphs. In: Proceedings of the 53rd Annual Meeting of the Association for Computational Linguistics and the 7th International Joint Conference on Natural Language Processing (Volume 1: Long Papers), Beijing, China, pp. 564–574. Association for Computational Linguistics, July 2015. https://doi.org/10.3115/v1/P15-1055. https://www.aclweb.org/anthology/P15-1055

33. Xiong, C., Callan, J.: Query expansion with freebase. In: Proceedings of the 2015 International Conference on The Theory of Information Retrieval, ICTIR 2015, pp. 111–120. Association for Computing Machinery, New York (2015). https://doi.org/10.1145/2808194.2809446

34. Xiong, C., Callan, J., Liu, T.Y.: Bag-of-entities representation for ranking. In: Proceedings of the 2016 ACM International Conference on the Theory of Information Retrieval, ICTIR 2016, pp. 181–184. Association for Computing Machinery, New York (2016). https://doi.org/10.1145/2970398.2970423
35. Xiong, C., Callan, J., Liu, T.Y.: Word-entity duet representations for document ranking. In: Proceedings of the 40th International ACM SIGIR Conference on Research and Development in Information Retrieval, SIGIR 2017, pp. 763–772. Association for Computing Machinery, New York (2017). https://doi.org/10.1145/3077136.3080768
36. Zhiltsov, N., Kotov, A., Nikolaev, F.: Fielded sequential dependence model for ad-hoc entity retrieval in the web of data. In: Proceedings of the 38th International ACM SIGIR Conference on Research and Development in Information Retrieval, SIGIR 2015, pp. 253–262. Association for Computing Machinery, New York (2015). https://doi.org/10.1145/2766462.2767756

Cognitive Information Retrieval

Dima El Zein[✉]

Université Côte d'Azur, CNRS, I3S, UMR 7271, Nice, France
elzein@i3s.unice.fr
http://www.i3s.unice.fr/∼elzein/

Abstract. Several existing search personalisation techniques tailor the returned results by using information about the user that often contains demographic data, query logs, or history of visited pages.

These techniques still lack awareness about the user's cognitive aspects like beliefs, knowledge, and search goals. They might return, for example, results that answer the query and fit the user's interests but contain information that the user already knows. Considering the user's cognitive components in the domain of Information Retrieval (IR) is still recognized as one of the "major challenges" by the IR community. This paper overviews my recent doctoral work on the exploration of the approaches to represent the user's cognitive aspects (especially knowledge and search goals) and on the investigation of incorporating them into information retrieval systems. Knowing that those aspects are subject to constant change, the thesis also aims to consider this dynamic characteristic. The research's objective is to better understand the knowledge acquisition process and the goal achievement task in an IR context. That will help search users find the information they seek for.

Keywords: Personalized search · User knowledge · User search goals

1 Introduction and Thesis Objectives

Users on the Web are believed to be cognitive agents [4] having their own beliefs and knowledge about the world. They try to fill their information gaps (missing or uncertain) by conducting information search activities. Users interact with search systems by submitting queries expecting to receive useful or relevant information that will help them reach their search goal and thus reduce their information gap [1]. In consequence, the search results must consider what has been already proposed to the user [9] to avoid information redundancy. Being aware of the user's search need/goal can also contribute in providing relevant information. Taking into account the cognitive components of the user in information search engines has been set as one of the "major challenges" by the Information Retrieval (IR) community [5] in 2018.

The common IR evaluation methods (i.e. nDCG, MAP, and bpref) measure a document's relevance against a query: a document is often judged in isolation independently of what has been already proposed to the user and to what

© The Author(s), under exclusive license to Springer Nature Switzerland AG 2022
M. Hagen et al. (Eds.): ECIR 2022, LNCS 13186, pp. 473–479, 2022.
https://doi.org/10.1007/978-3-030-99739-7_58

the user already knows. They often do not consider the user's search goal too. In other terms, a search system might return a document that is relevant to the query but not useful with respect to the user's knowledge (i.e. contains information already known). By studying the literature in information retrieval, knowledge understanding, and cognitive aspects domains, we could identify some important gaps that need to be bridged: (i) Existing cognitive retrieval frameworks need to consider the dynamic aspect of knowledge and information needs. They must go beyond the relevance to one query or one session. (ii) Evaluation measures need to include more user's cognitive aspects. (iii) Judgment files or test collections reflecting user's knowledge change and goal achievement progress are missing.

The objective of this thesis is to explore methodologies that make IR systems "aware" of the user's cognitive aspects (especially knowledge and goals). The proposed methodologies are put in a framework that must consider the evolution of the user's knowledge and the change of his/her search goals.

The mentioned methodologies can be applied in other domains like recommendation systems, search-as-learning SAL, MOOC courses, question-answering, etc. To sum up, the contribution will allow those systems to answer the user's need given his/her previous knowledge and what has been previously proposed; hence avoiding information redundancy. It will also help satisfy the user's search goal - fill his/her information need - in an efficient number of interaction iterations with the search system.

2   Research Methodology

We present the research questions, our performed preliminary work and our proposed solutions.

2.1   RQ1: How to Extract and Represent the User's Knowledge?

The user's knowledge studied in our research is mainly the information he/she acquires during search sessions. The studied information will be the unstructured texts - referred to as "document" here - (i.e. Web pages, documents, courses,...): the content of the texts is considered to be the acquired knowledge. A document brings a set of positive facts α (i.e. coronavirus Covid is an infectious disease) or negated facts $\neg\beta$ (i.e. the vaccine is not effective).

Our objective is to estimate this knowledge, represent it and employ it to customise the user's search results. In other terms, the intent is to learn what the user knows and what he/she does not. It is to note that the content read by the users might not necessarily be adopted as confirmed knowledge, because users also acquire knowledge outside the context of search systems (i.e. interpersonal exchanges, offline information like books, personal beliefs, etc.). Users can also earn information as a result of their reasoning process.

In real life, knowledge is gradual: individuals might have information more entrenched (certain or accepted) than others. We associate a qualitative order of preference *Degree of Certainty* for every piece of knowledge α that represents the extent to which a user is knowledgeable/certain about α.

The information inside a proposed document might be novel, redundant, or contradictory to the user's existing knowledge. To integrate this new information and "update" the representation of the user's knowledge, we apply *belief revision* algorithms. Those algorithms modify the knowledge representation to maintain its consistency whenever new information becomes available. A representation's consistency means that the user doesn't believe in contradictory facts α and $\neg\alpha$ (i.e. cannot believe that Covid is contagious and not contagious at the same time). The certainty degrees will play a role in deciding which facts will be kept in the representation.

As a preliminary work in this direction, we presented in [7] the knowledge in form of weighted keywords. The fact α is then a literal associated with a *degree of certainty* to form a pair (literal, degree) i.e. (delta_variant,0.7). The keywords are extracted from texts using Rapid Automatic Keyword Extraction RAKE [15]. This method detects the most representative words or phrases in a text and calculates a related score. Its advantage over traditional methods (i.e. TF-IDF) is that it can extract complex phrases – not just keywords – that might be more meaningful. To calculate the degree of certainty of an extracted keyword. To calculate the degree of certainty of an extracted keyword, we normalised its score then multiplied it by an adjustment factor.

By trying to simulate the user's reasoning process, we explored the possibility of using knowledge rules to derive new knowledge. For example, the knowledge rule *Covid* & *Pfizer* $\rightarrow$ *Covid_vaccine* means that if the user knows about Covid disease and the pharmaceutical company Pfizer then he/she probably knows about Covid vaccines. The knowledge *covid_vaccine* will be "derived" and added to the user's knowledge. Its certainty degree will depend on the degrees of the facts that derived it. The rules could originate, for example, from mining contextual knowledge from textual corpus, like the information flow.

The mentioned concepts are all currently developed as a prototype in Python code. First, as a proof of concept, we intend to evaluate the framework having the knowledge represented as weighted keywords.

The next step is to find better knowledge representation methods, taking into consideration the semantic aspects of the text for example. Two other representation methods are on the test road-map: (1) Triples relation: Recognising entities, linking them to existing ones in knowledge bases, and extracting the relation stated in the text. A fact α would be represented by *(head entity, relationship, tail entity)* with a degree associated with it. The knowledge will be represented by a graph embedding. (2) Vector representation using GloVe - Global Vectors for Word Representation [14] or representation using BERT - Bidirectional Transformers for Language Understanding - embedding [6]. The vector weights will represent the degree of certainty. Another planned future work is to compare the "performance" of the framework when using the three methods. The remaining challenge would be to update (using belief revision) the representation when the knowledge has the form of vectors.

One other challenge is the extraction and representation of the negated facts from the texts. Being aware of the facts the user knows are not true will allow us to perform real belief revision operations and hence better reflect the user's knowledge. Unfortunately, we had the confirmation from several NLP experts that the extraction of such information is a task under research in the field.

2.2   RQ2: How to Model the User's Search Goals?

Users interact with search systems by submitting queries expecting to receive information that helps them reach their search goal and thus reduce their information gaps. User queries might be ambiguous; also, different users could submit the same query but have different search goals. Getting to know the intention behind their queries will help return the relevant document(s).

Inferring the user's search goal(s) received attention in the last decades. The most common methods were tracking feedback sessions, click-through, and Web caching. The novel idea presented in this thesis is that goal achievement is not necessarily satisfied by a single document. A document can partially cover the information need. This relevance depends on the user's knowledge, his search goals (and the progress towards it), and the document(s) already presented to him/her. The notion of "missing information" presented in [2] represented the difference between the needed information to achieve a goal and the user's actual knowledge. Hence, to measure a document's usefulness to a user's goal, we evaluate the amount of "missing information" it covers.

The user's goal can be explicitly expressed, in a text form or selected from a suggested list for example. It can also be implicit and automatically predicted by analyzing the user's profile, previous queries, or the terms used during the search. The developed framework currently considers the first option where goals are fed to the system in form of a textual set of statements or tasks. Similar to the user's knowledge, the goal is represented by weighted keywords resulting from RAKE. The retrieval framework proposed in [17] represented, in the context of vocabulary learning, the user's need as a set of keywords. It demonstrated the effectiveness of the personalised results when accounted for individual user's learning goals, as well as the effort required to achieve those goals. Semantic meaning is planned to be considered in future work. For implicit and automatic identification of the user's goal, we refer to the set of proposed methods in the literature [10,12].

The user goals might change too: might need an update, become obsolete, inactive, or achieved. Therefore, the representation of the user's goals must be revised too: we are referring to "goal change". This latter aspect is on the roadmap of our research work and has not been explored yet.

2.3   RQ3: How to Design the Framework Architecture

An intelligent agent running on the client-side will manage the representation of the user's knowledge and search goals. That agent will be placed in an intermediary between the user and the search system and will act as a filter for the results returned by the system. Hence, the agent is not responsible for the retrieval process itself but for cognitive filtering.

The agent is constructed in BDI (Belief-Desire-Intention) architecture where the agent's beliefs represent what the agent knows about the user. A belief is an information (knowledge or goal) the agent believes to be true, but which is not necessarily true. When the agent has α in its belief base, it believes that the user knows that α is true. If the belief base contains $\neg\alpha$, then the agent believes the user knows that α is not true. When neither α nor $\neg\alpha$ is in the belief base, the agent believes neither the user knows α is true nor the user knows that α is false. The belief base also contains a set of user goals g. It has been earlier proved in [11] show that the belief-based symbolic IR model is more effective than a classical quantitative IR model.

The agent's beliefs about the user might not always be true. Therefore, beliefs and goals can be revised if the agent realizes that it has made a mistake.

The framework has 5 main modules: (1) Knowledge extractor: extracts knowledge from the texts examined by the user. (2) Knowledge reasoner: derives new beliefs using knowledge rules. It also revises the belief set to maintain consistency. (3) Goal extractor: extracts the user's search goal and represents them as agent beliefs. (4) Goal reasoner: updates beliefs about goals. (5) Result filtering: compares the content of the candidate document to the agent beliefs and selects the "useful" documents to be proposed to the user.

2.4   RQ3: How to Employ the Cognitive Aspects to Personalise Results?

The chosen representation for knowledge and goals (i.e. keywords, triples graph, embedding) will be also used for the candidate documents. The filtering decision will be taken after the comparison of three components: the agent's beliefs about the user knowledge, the agent's beliefs about the user's search goal, and the candidate document. These components are compared in pairs: distance (document, knowledge) and distance(document, goal).

First, we discuss the coverage criterion presented in [13] that quantifies the degree inclusion of one vector $v1$, representing the document into another $v2$ representing the user's knowledge or goal. The coverage function produces the maximum value of 1 when the non-null elements in $v2$'s vector also belong to $v1$'s vector. It produces the value zero when the two vectors have no common element. Moreover, the value of the function increases with the increase of the number of common elements.

On another side, the well-known "aboutness" measure is based on a weighted cosine similarity that accounts for the number and weight of common vector elements. It measures the similarity between two non-zero vectors of an inner product space. It is defined to be equal to the cosine of the angle between them, which is also the same as the inner product of the same vectors normalized to both have length 1.

We use those two measures to quantify the "distances" between the pair's components. They can be employed for different purposes depending on the intended application: when the purpose of the framework is to reinforce the user knowledge, then documents that are "close" to the agent beliefs will be returned. Contrarily, when the framework is employed for novelty purposes, those documents could be considered redundant with what the user already knows and will be excluded by the filter. Also, when the system is helping the user reach its search goals, then the documents that are "closer" to the goal are returned.

2.5   RQ4: How to Evaluate the Model?

Finding adequate datasets to evaluate the proposed idea(s) is challenging. The existing datasets dealing with users' knowledge usually measure the knowledge at the end of the search session. What we are trying to evaluate is the user's knowledge change after reading one single document, to finally prove that our framework is proposing documents providing relevant knowledge. Also, we could not find a dataset that will help evaluating the goal satisfaction change on a document by document level.

On another side, there is no benchmark to compare the performance of our framework. Our goal is to propose a framework that helps the user acquire new knowledge and reach its goal in a minimal number of interaction exchanges.

We adapted an existing dataset [8] that logged the search behaviour of 500 users and assessed their knowledge level (a score between 0 and 100) before and after the search session through knowledge tests. We proposed - in a work under review - a benchmark and collection that estimates the knowledge level brought by every document. We then tracked the evolution of every user's knowledge on a document-by-document level. Our planned future work is to submit the set of queries in the collection to the proposed framework, and compare the evolution of the users' knowledge to the one in the benchmark. The evaluation metrics will consider the number of user-system exchanges needed to reach the maximal knowledge gain. In addition, we intend to compare the overall knowledge gain brought by our framework to the one provided in the benchmark. The resulting benchmark can serve other researchers studying cognitive learning behaviour during the session. Another considerable evaluation measure could be the $\alpha - nDCG$ proposed in [3] that rewarded novelty and diversity in documents with respect to the information need. We aim to adapt it to consider the user's previous knowledge too.

In the direction of building our own dataset, we initiated an experiment design that presents, consecutively, a set of documents to users. We then ask them about the knowledge brought by every document, as well as the novelty it shows. The work has been put on hold until the framework is tested against the mentioned benchmark first.

3   Conclusion

We seek suggestions and comments on how to improve this proposal. We are specifically interested in discussing the experimentation designs to evaluate our framework.

References

1. Boyce, B.: Beyond topicality: a two stage view of relevance and the retrieval process. Inf. Process. Manage. **18**, 105–109 (1982)
2. Cholvy, L., Costa Pereira, C.: Usefulness of information for goal achievement. In: International Conference on Principles and Practice of Multi-Agent Systems, pp. 123–137 (2019)
3. Clarke, C., et al.: Novelty and diversity in information retrieval evaluation. In: Proceedings of the 31st Annual International ACM SIGIR Conference on Research and Development in Information Retrieval, pp. 659–666 (2008)
4. Costa Móra, M., Lopes, J., Vicari, R., Coelho, H.: BDI models and systems: bridging the gap. In: ATAL, pp. 11–27 (1998)
5. Culpepper, J., Diaz, F., Smucker, M.: Research frontiers in information retrieval: report from the third strategic workshop on information retrieval in Lorne (SWIRL 2018). In: ACM SIGIR Forum, vol. 52, pp. 34–90 (2018)
6. Devlin, J., Chang, M., Lee, K., Toutanova, K.: BERT: pre-training of deep bidirectional transformers for language understanding. ArXiv Preprint ArXiv:1810.04805 (2018)
7. El Zein, D., Costa Pereira, C.: A cognitive agent framework in information retrieval: using user beliefs to customize results. In: The 23rd International Conference on Principles and Practice of Multi-Agent Systems (2020)
8. Gadiraju, U., Yu, R., Dietze, S., Holtz, P.: Analyzing knowledge gain of users in informational search sessions on the web. In: Proceedings of the 2018 Conference on Human Information Interaction & Retrieval, pp. 2–11 (2018)
9. Goffman, W.: A searching procedure for information retrieval. Inf. Storage Retr. **2**, 73–78 (1964)
10. Jansen, B., Booth, D., Spink, A.: Determining the user intent of web search engine queries. In: Proceedings of the 16th International Conference on World Wide Web, pp. 1149–1150 (2007)
11. Lau, R., Bruza, P., Song, D.: Belief revision for adaptive information retrieval. In: Proceedings of the 27th Annual International ACM SIGIR Conference on Research and Development in Information Retrieval, pp. 130–137 (2004)
12. Lee, U., Liu, Z., Cho, J.: Automatic identification of user goals in web search. In: Proceedings of the 14th International Conference on World Wide Web, pp. 391–400 (2005)
13. Pasi, G., Bordogna, G., Villa, R.: A multi-criteria content-based filtering system. In: Proceedings of the 30th Annual International ACM SIGIR Conference on Research and Development in Information Retrieval, pp. 775–776 (2007)
14. Pennington, J., Socher, R., Manning, C.: GloVe: global vectors for word representation. In: Proceedings of the 2014 Conference on Empirical Methods in Natural Language Processing (EMNLP), pp. 1532–1543 (2014)
15. Rose, S., Engel, D., Cramer, N., Cowley, W.: Automatic keyword extraction from individual documents. Text Mining Appl. Theory **1**, 1–20 (2010)
16. Foundation, P. rake-nltk 1.0.4. https://pypi.org/project/rake-nltk/
17. Syed, R., Collins-Thompson, K.: Retrieval algorithms optimized for human learning. In: Proceedings of the 40th International ACM SIGIR Conference on Research and Development in Information Retrieval, pp. 555–564 (2017)

Graph-Enhanced Document Representation for Court Case Retrieval

Tobias Fink[✉]

TU Wien, Favoritenstr. 9-11, 1040 Vienna, Austria
tobias.fink@tuwien.ac.at

Abstract. To reach informed decisions, legal domain experts in Civil Law systems need to have knowledge not only about legal paragraphs, but also about related court cases. However, court case retrieval is challenging due to the domain-specific language and large document sizes. While modern transformer models such as BERT create dense text representations suitable for efficient retrieval in many domains, without domain specific adaptions they are outperformed by established lexical retrieval models in the legal domain. Although citations of court cases and codified law play an important role in the domain, there has been little research on utilizing a combination of text representations and citation graph data for court case retrieval. In other domains, attempts have been made to combine these two with methods such as concatenating graph embeddings to text embeddings. In the PhD research project, domain-specific challenges of legal retrieval systems will be tackled. To help with this task, a dataset of Austrian court cases, their document labels as well as their citations of other court cases and codified law on a document and paragraph level will be created and made public. Experiments in this project will include various ways of enhancing transformer-based text representations methods with citation graph data, such as graph based transformer re-training or graph embeddings.

Keywords: Information retrieval · Text representation · Citation graph

1 Introduction and Motivation

Reaching correct decisions for legal processes in Civil Law systems such as building permit issuance is a complex matter. Domain experts require an extensive amount of legal knowledge since just missing a single important detail could be the cause for a decision to be challenged in court. This means domain experts need not only be able to learn about sections of codified law ("legal paragraphs") but also of related court cases. However, retrieving related court cases is a challenging task, even for domain experts who are familiar with the subject matter.

When domain experts search for a court case to answer one of their questions, they often base their search around some legal paragraphs that they know will

© The Author(s), under exclusive license to Springer Nature Switzerland AG 2022
M. Hagen et al. (Eds.): ECIR 2022, LNCS 13186, pp. 480–487, 2022.
https://doi.org/10.1007/978-3-030-99739-7_59

be relevant. Questions could be for example "Based on § 81, is it permitted for a building to have a maximum height of 10 m in construction zone 1? What if the distance to the property border is less than 2 m?". Of the many cases that relate to a particular legal paragraph, only a small subset might be relevant for a particular question. Moreover, cases are often long documents that relate to many different legal paragraphs with varying degrees of importance.

Domain experts are typically familiar with a few exemplary court cases per legal paragraph. While similar court cases would provide relevant insights, using keyword search and checking court cases for relevance is generally very difficult and time consuming. Instead, court cases are typically retrieved based on document queries which are either whole or parts of related court cases (see for example Task 1a of the FIRE AILA 2020 workshop [4]).

In recent years, neural network based vector representations or embeddings of text have become increasingly widespread in many IR domains. Vector representations of individual words are now already a well researched area [5,20,22] and models dealing with sentence or passage level tasks such as transformer models are also commonly used [6,8,23]. Nonetheless, classical lexical methods such as BM25 or language models still outperform newer methods in many court case IR benchmarks (e.g. see COLIEE 2021[1]). While transformer models are much more flexible and adaptable than lexical methods, out-of-the-box they are incapable of dealing with the challenges of the legal domain such as long documents and required knowledge of legal context.

By developing a novel method for court case IR that overcomes the issues of out-of-the-box transformer models, it will be easier for domain experts to retrieve the court cases necessary to understand the legal context of their questions, causing them to make fewer mistakes.

To overcome the issue of document length, a variety of options are described in the literature. One is to use a specialized architecture to increase the number of tokens that can be processed by a transformer model [3]. Another is to split a long document into smaller passages and create passage level text representations for document level IR [1,11]. Since this approach is close to how domain experts compare documents, seeking to explore it further is natural.

The second issue is the required knowledge of legal context. Citation graph data is an important source of information in the legal domain and provides valuable legal context for a piece of text. Court cases usually cite related court cases or legal paragraphs, information that is often instrumental for domain experts to reach decisions. In the proposed research project we will investigate ways of representing citation graph data and develop a new way of adding this critically required knowledge to transformer models. To gain the required data for the project, a dataset containing court case and legal paragraph documents as well as their citation graph will be created based on data from the Austrian law information system (Rechtsinformationssystem - RIS), a source of high quality

---

[1] https://sites.ualberta.ca/~rabelo/COLIEE2021/.

Civil Law legal documents. The RIS is the official announcement platform for Austrian legal texts and is used as part of the E-government project BRISE-Vienna[2].

2   Related Work

Text Representation. Neural network based models have been used very successfully for a variety of tasks such as question answering, language inference, named entity recognition and also information retrieval and have seen much development in recent years. One such model is the transformer-based BERT language model [8], which is pre-trained on a large corpus and often fine-tuned for a specific task. By using a function like the dot product, text encoded by BERT can be compared with each other and a similarity score for ranking documents calculated [14]. However, due to quadratic memory requirements it does not scale very well to long documents [15]. Subsequent research has tried to address this problem in various ways. For example, Beltagy et al. [3] modify the transformer architecture to handle longer sequences of tokens. A different approach is taken by Hofstätter et al. [11] in an IR setting, who split long documents into smaller passages using a sliding window and only apply transformer computations for term matching inside one of these windows. Then a saturation function and a two-stage pooling strategy are used to identify which regions of the document are relevant in relation to a single passage query.

Graph Based Methods. If a document is also a node in a citation graph, position and structural context of a node also define its semantics. Algorithms such as SimRank [12] or Linked Data Semantic Distance (LSDS) [21] only require knowledge about the structure of the graph to measure the similarity between documents. In contrast, Hamilton et al. [10] make use of both the structure of the graph and the features of a node to map a node to vector space. They introduce a general neural network based method called GraphSAGE that is capable of generating node embeddings for previously unseen data with the help of node features, such as text attributes. In this method, feature information from neighboring nodes is aggregated with a learned aggregation function to create the node embedding. The performance of their method is evaluated by classifying papers using the Web of Science citation dataset, classifying Reddit posts and classifying protein-protein interactions. However, how well these embeddings can be used to express node similarity was not evaluated. Cohan et al. [7] introduce a method called SPECTER that fine-tunes BERT based on citation graph features. During training, title and abstract of a paper are encoded with SciBERT [2], a BERT model that is pre-trained on scientific text. The task is to minimize the L2 norm distance between the query document P^Q and a related document P^+ as well as maximize the distance between P^Q and an unrelated document P^-. Whether a document is considered related or unrelated to another document is

---

[2] https://digitales.wien.gv.at/projekt/brisevienna/.

decided based on the citation graph. While P^+ intuitively refers to documents that are cited by P^Q, P^- refers to random documents as well as documents that are cited by P^+ but not by P^Q. So although there might still be a relation between P^- and P^Q, in this negative sampling scheme the model is conditioned to learn a cutoff point at depth $= 2$ of the citation graph.

Jeong et al. [13] concatenate BERT embeddings and citation graph embeddings from a graph convolutional network for the task of citation recommendation. This combined representation is then passed through a feed-forward neural network to calculate probabilities of citation labels.

Court Case Retrieval. The topic of court case retrieval has been explored in various workshops such as the COLIEE 2020 and 2021 workshops or the FIRE AILA 2019 and 2020 workshops. In both of these workshops one of tasks was to retrieve court cases related to a query court case. While for COLIEE the task was to retrieve supporting cases from the Federal Court of Canada, FIRE AILA required to retrieve precedents from the Indian supreme court, both of which are long unstructured court case documents. In the COLIEE 2021 workshop[3], the best ranking submission by a large margin measured in F1 score was a language model [19], while the second best was a standard BM25 implementation [24]. To handle the length issue of the legal domain, many submissions performed document retrieval at the passage level. Althammer et al. [1] compared the performance of BM25, dense retrieval and a combination of the two. In the context of first stage retrieval, they found that passage level retrieval with an aggregation strategy outperforms document level retrieval for all of their methods.

In the FIRE AILA 2020 workshop [4], almost all submitted models were lexical models with query preprocessing/reduction methods [9,16–18]. Only few models based on embeddings were submitted but they were outperformed by the other models.

3   Research Description and Research Questions

We propose to develop a novel method for creating court case document representations by combining recent successful methods for representing text, such as transformer-based models, with methods for utilizing graph data. To successfully develop such a method, the following questions must be answered.

Which methods for text representation outperform previous methods for court case retrieval? Current transformer-based methods lack domain-specific adaptions to the legal domain of which the most important would be dealing with the length of court case documents. Splitting long documents into smaller passages has already shown some success in COLIEE 2021. Moreover, domain experts appear to also determine case relevancy on passage level. We expect that this question can be answered with further research of transformer-based methods on passage level for document level retrieval.

---

[3] https://sites.ualberta.ca/~rabelo/COLIEE2021/.

How does a combined approach of text representation and graph representation methods perform compared to individual methods for court case retrieval? Although of great relevance to the legal domain, in the COLIEE and AILA workshops little attention has been given to citation networks. In other domains, various forms of graph neural networks (GNN) are used to create document representations and to classify documents (modelled as graph nodes) using text-based input features. Furthermore, various graph algorithms exist to calculate a similarity measure between nodes. We expect that adding the information provided by graph based models to text based similarity models will yield a better performance than text based and graph based similarity models by themselves.

What is the most effective way of combining text representation and graph representation methods for court case retrieval? There are multiple ways of making use of graph based data in text representation models. One way is using an explicit component for graph representation and adding it to a neural network by ways such as concatenating it to the output of a different component. Further, a network can learn citation graph data implicitly based on sample selection. One part of this project is to answer this question by experimenting with multiple ways of adding passage level citation graph data to the most promising text representation methods and evaluating their court case retrieval performance.

Are the combined representations useful for classification and link prediction? Finally, it is useful to know whether the created combined document representations can be used for other tasks such as court case classification or legal paragraph link prediction and how they perform. We expect that although the representations will most likely be on a passage level, only few modifications will be necessary to perform document level predictions.

4   Methodology

To answer our research questions, we intend to create a suitable dataset of legal documents then use this dataset to develop a new approach for creating document representations.

Datasets. Since this PhD research project is part of the BRISE-Vienna[4] project, we can make use of the data and expert knowledge available to the BRISE-Vienna project. As part of the BRISE-Vienna project, domain experts handling building permit issuance will be supported with a search system. A function of this search system is to suggest similar court cases from the Austrian higher administrative court (Verwaltungsgerichtshof - Vwgh[5]). The RIS Vwgh data includes around 400.000 court case documents and related files in German and meta data including a list of citations of legal paragraphs as well as multi-label document category labels. We will enhance this dataset by building a citation graph that models how court case documents and legal paragraph

---

[4] https://digitales.wien.gv.at/projekt/brisevienna/.
[5] https://www.ris.bka.gv.at/Vwgh/.

documents are connected. We also plan to further enhance the dataset with text mining by extracting the exact locations of legal paragraph mentions and cited court case mentions from the case text. We intend to use cited court case mentions as ground truth data for court cases similarity. Since text mining can be a challenging task by itself, we do not aim to solve this problem completely and instead focus on the information that we can extract with simple rule based methods to create this dataset.

At a later stage, we also plan to investigate other datasets such as the COL-IEE 2021 Legal Case Retrieval dataset[6] which consists of query case documents as well as related supporting case documents. The selection of other datasets will depend on the effort that is necessary to create a similar citation graph of legal paragraphs (or equivalent) and court case documents.

Retrieval Experiments and Method Development. To understand what level of performance we can expect from our created dataset, we propose to implement two simple baseline methods to suggest similar court case documents: a lexical method such as BM25 and a citation graph based method such as LSDS [21]. We will evaluate how well these methods can retrieve the cases of our court case citation ground truth data. Further, we want integrate these methods into the BRISE search system at an early stage of our research. We hope that this way we can quickly gain insights into typical domain/dataset specific retrieval issues and gather some additional ground truth data on document similarity. We want to compare these methods with our later methods as part of an A/B Test.

Next, we plan to adapt the method described by Cohan et al. [7] to our data. In this method the citation graph data is represented only *implicitly* in the model and is used during training to determine which documents should or should not be similar. Since the English SciBERT is not suitable for our data, we instead propose to replace the underlying BERT model with a suitable German model, which will be used to create document representations on a document level and on a passage level. To train document similarity, we propose the following heuristic: if a document/passage A cites the same legal paragraph as document/passage B in the RIS citation graph, we would consider these documents/passages to be related. Following [7], if B is related to C but A is not related to C, we would consider A and C to be unrelated. To determine document similarity from passage level representations, an additional aggregation step will be employed.

We expect that further improvements will be possible with an *explicit* representation of the citation graph or the neighborhood of a document in the graph. This could be achieved by simply concatenating the output of a GNN node representation to that of a BERT model, but there are certainly more approaches that are worth trying to make use of GNN embeddings as part of a transformer model. Another way could be to encode passages with a transformer model and use the

---

[6] https://sites.ualberta.ca/~rabelo/COLIEE2021/.

produced vectors as feature vectors for GNN training and node embedding generation. Further literature research is necessary to determine which methods are the most promising.

Other Experiments. The RIS dataset will give us options to further evaluate the quality of the generated document representations besides document retrieval. One option is predicting legal paragraph relations instead of court cases for a test set of court cases. Since there are additional category labels available for each court case, we will use the document or passage representations to train a simple machine learning algorithm for the task of multi-label classification.

References

1. Althammer, S., Askari, A., Verberne, S., Hanbury, A.: DoSSIER@ COLIEE 2021: leveraging dense retrieval and summarization-based re-ranking for case law retrieval. arXiv preprint arXiv:2108.03937 (2021)
2. Beltagy, I., Lo, K., Cohan, A.: SciBERT: a pretrained language model for scientific text. arXiv preprint arXiv:1903.10676 (2019)
3. Beltagy, I., Peters, M.E., Cohan, A.: Longformer: the long-document transformer. arXiv preprint arXiv:2004.05150 (2020)
4. Bhattacharya, P., et al.: Overview of the FIRE 2020 AILA track: artificial intelligence for legal assistance. In: Proceedings of FIRE 2020 - Forum for Information Retrieval Evaluation, Hyderabad, India, December 2020
5. Bojanowski, P., Grave, E., Joulin, A., Mikolov, T.: Enriching word vectors with subword information. Trans. Assoc. Comput. Linguist. **5**, 135–146 (2017)
6. Brown, T.B., et al.: Language models are few-shot learners. arXiv preprint arXiv:2005.14165 (2020)
7. Cohan, A., Feldman, S., Beltagy, I., Downey, D., Weld, D.S.: Specter: document-level representation learning using citation-informed transformers. arXiv preprint arXiv:2004.07180 (2020)
8. Devlin, J., Chang, M.W., Lee, K., Toutanova, K.: BERT: pre-training of deep bidirectional transformers for language understanding. arXiv preprint arXiv:1810.04805 (2018)
9. Fink, T., Recski, G., Hanbury, A.: FIRE2020 AILA track: legal domain search with minimal domain knowledge. In: FIRE (Working Notes), pp. 76–81 (2020)
10. Hamilton, W.L., Ying, R., Leskovec, J.: Inductive representation learning on large graphs. In: Proceedings of the 31st International Conference on Neural Information Processing Systems, pp. 1025–1035 (2017)
11. Hofstätter, S., Zamani, H., Mitra, B., Craswell, N., Hanbury, A.: Local self-attention over long text for efficient document retrieval. In: Proceedings of the 43rd International ACM SIGIR Conference on Research and Development in Information Retrieval, pp. 2021–2024 (2020)
12. Jeh, G., Widom, J.: SimRank: a measure of structural-context similarity. In: Proceedings of the Eighth ACM SIGKDD International Conference on Knowledge Discovery and Data Mining, pp. 538–543 (2002)
13. Jeong, C., Jang, S., Park, E., Choi, S.: A context-aware citation recommendation model with BERT and graph convolutional networks. Scientometrics **124**(3), 1907–1922 (2020). https://doi.org/10.1007/s11192-020-03561-y

14. Karpukhin, V., et al.: Dense passage retrieval for open-domain question answering. arXiv preprint arXiv:2004.04906 (2020)
15. Kitaev, N., Kaiser, Ł., Levskaya, A.: Reformer: the efficient transformer. arXiv preprint arXiv:2001.04451 (2020)
16. Leburu-Dingalo, T., Motlogelwa, N.P., Thuma, E., Modongo, M.: UB at FIRE 2020 precedent and statute retrieval. In: FIRE (Working Notes), pp. 12–17 (2020)
17. Li, Z., Kong, L.: Language model-based approaches for legal assistance. In: FIRE (Working Notes), pp. 49–53 (2020)
18. Liu, L., Liu, L., Han, Z.: Query revaluation method for legal information retrieval. In: FIRE (Working Notes), pp. 18–21 (2020)
19. Ma, Y., Shao, Y., Liu, B., Liu, Y., Zhang, M., Ma, S.: Retrieving Legal Cases from a Large-scale Candidate Corpus (2021)
20. Mikolov, T., Sutskever, I., Chen, K., Corrado, G.S., Dean, J.: Distributed representations of words and phrases and their compositionality. In: Advances in Neural Information Processing Systems, pp. 3111–3119 (2013). http://papers.nips.cc/paper/5021-distributed-representations
21. Passant, A.: Measuring semantic distance on linking data and using it for resources recommendations. In: 2010 AAAI Spring Symposium Series (2010)
22. Pennington, J., Socher, R., Manning, C.: GloVe: global vectors for word representation. In: Proceedings of the 2014 Conference on Empirical Methods in Natural Language Processing (EMNLP), pp. 1532–1543 (2014)
23. Peters, M.E., et al.: Deep contextualized word representations. arXiv preprint arXiv:1802.05365 (2018)
24. Rosa, G.M., Rodrigues, R.C., Lotufo, R., Nogueira, R.: Yes, BM25 is a strong baseline for legal case retrieval. arXiv preprint arXiv:2105.05686 (2021)

Relevance Models Based on the Knowledge Gap

Yasin Ghafourian[1,2(✉)] 🆔

[1] Research Studios Austria FG, 1090 Vienna, Austria
`yasin.ghafourian@researchstudio.at`
[2] Vienna University of Technology (TU Wien), 1040 Vienna, Austria

Abstract. Search systems are increasingly used for gaining knowledge through accessing relevant resources from a vast volume of content. However, search systems provide only limited support to users in knowledge acquisition contexts. Specifically, they do not fully consider the knowledge gap which we define as the gap existing between what the user knows and what the user intends to learn. The effects of considering the knowledge gap for knowledge acquisition tasks remain largely unexplored in search systems. We propose to model and incorporate the knowledge gap into search algorithms. We plan to explore to what extent the incorporation of the knowledge gap leads to an improvement in the performance of search systems in knowledge acquisition tasks. Furthermore, we aim to investigate and design a metric for the evaluation of the search systems' performance in the context of knowledge acquisition tasks.

Keywords: Information retrieval · Knowledge delta · Knowledge acquisition

1 Motivation and Problem Statement

Web search is nowadays considered more to be a source for accessing information resources and exploration, and educational resources are also not an exception to this. Web search has become the most popular medium for education in schools, universities and for professional training [1,2]. In addition, web search is being used more often for the aim of gaining new knowledge [3]. There have been numerous studies done on the information-seeking behaviors of students and academic staff in different parts of the world (some examples are [4–7]) which also verify that the majority of individuals and students consider the internet to be the most useful information source for learning.

Marchionini [2] categorises the search activities into two broad categories of *look-up* and *exploratory* search activities. Look up search is the activity in which users know in particular what information they want and have a concrete expectation of what would the desired search results be, while in exploratory search users go through multiple iterations of searching the online resources. With regard to using search systems for learning purposes, look-up search is

M. Hagen et al. (Eds.): ECIR 2022, LNCS 13186, pp. 488–495, 2022.
https://doi.org/10.1007/978-3-030-99739-7_60

viewed as a type of search that users initiate based on their current knowledge of the subject of interest that leads them to the relevant neighborhood of information. From this neighborhood, users will then start their exploration into the learning resources and might reformulate their query multiple times according to their findings with the aim of reaching more satisfying resources.

Users typically have different levels of background knowledge on the topic in question. We define a user's background knowledge as the current level of the user's familiarity with the topic. Due to different levels of background knowledge, users might perceive the relevance of the documents on a topic differently.

Search engines currently do not take this diversity in the background knowledge levels of the users into account and assume that users' information needs are well represented in their queries [8]. This is a clear limitation. The optimal sequence of documents leading to satisfying a user's learning goal depends on the user's specific background knowledge. What if search engines could exploit information about the user's background knowledge to provide the suitable (sequence of) documents helping the user to satisfy their learning goal as fast as possible.

As search engines are being used by students and researchers, it is crucial for search engines to pursue more developments in this direction so as to be a better fit for learning tasks, especially *knowledge acquisition tasks*. We define *knowledge acquisition tasks* as tasks in which users aim to acquire knowledge with a learning goal as part of a sequence of interactions with an information retrieval (IR) system. Supposing that we have a search system specialized for knowledge acquisition tasks that takes the users' knowledge level into account, it will provide the users with resources that best fit their learning needs according to their knowledge level. This, in turn, saves effort from users in terms of spending longer search times and consuming documents that are returned as relevant but cannot be utilised by the users according to their knowledge level.

To incorporate the users' knowledge level into the search, one needs to enable the search system to have a model representing the users' knowledge within the topic of interest ("what the user knows"). Furthermore, one needs to define and represent the knowledge to be acquired ("what the user wants to know"). The goal is then to overcome the gap between these two representations. We call this gap the *knowledge gap*. Having defined the knowledge gap, the objective is to develop a retrieval method that provides users with an order list (a path) of resources that helps them to overcome the knowledge gap. A suitable order will be one where more complex resources requiring more background knowledge will be preceded by resources that are more easily approachable based on the user's background knowledge. In this research, we will investigate the means to measure the *knowledge gap* and understand how one applies it within a search system designed for acquiring knowledge so that the users can effectively overcome the knowledge gap. In the rest of this paper, we will first discuss the research questions that we seek to answer throughout this research. Later and in the section that follows, we will provide a brief overview of the research surrounding the concepts of the *knowledge gap* and the *knowledge delta*. Finally, we will explain our planned methodology to approach the research questions.

2 Research Questions

Our motivation is to investigate on how to improve the retrieval effectiveness of the search systems for knowledge acquisition by incorporating the knowledge gap into the ranking mechanisms. We propose our main research question as follows:

High Level Research Question: How can the search system help users to effectively overcome their knowledge gap in a knowledge acquisition task?

This high-level research question is comprised of three fine-grained research questions to be investigated:

- RQ1: How can we model users' background knowledge, target knowledge and the knowledge gap in knowledge acquisition tasks?
- RQ2: To what extent can the incorporation of the knowledge gap into a search system facilitate a more efficient journey for users in knowledge acquisition tasks?
- RQ3: How should search systems be evaluated in the context of knowledge acquisition tasks?

3 Background and Related Work

In this section we provide the result of our literature review done in an attempt to capture the viewpoint of the papers that discuss the concepts of *knowledge gap* and *knowledge delta* and how they approach and incorporate these concepts.

3.1 Knowledge Gap

What we defined earlier, i.e. the gap between the users' knowledge level and the level of knowledge in the field of interest for learning has been discussed in the literature under the title of *knowledge gap*. Knowledge gap is one of the causing factors of information need [9]. Another factor that causes information need is referred to as *Anomaly in the state of knowledge* by Belkin et al. [10] which is the phenomenon in the state of knowledge that causes the information need. In one of the early studies on information use, Dervin and Nilan [11] used the phrase *knowledge gap* to refer to a situation where a person's cognitive state has recognized an incompleteness in its currently possessed information. This incompleteness happens as a result of interaction with information sources or through thinking processes, and thereby later that incompleteness will turn into an information need. Additionally, Thellefsen et al. [12] use *knowledge gap* and *information need* interchangeably in a discussion where some of the definitions of *knowledge gap* are covered to argue the intricacy of the concept of information need and the importance of incorporating users' information need while developing a knowledge organization system. In a research done by Yu et al. [13] the *knowledge gap* also has a similar explanation, however, the research is more focused on knowledge predictive models for the knowledge state of users which are being calibrated through questionnaires. Considering the definition of *knowledge gap*, there are studies that use the same definition and provide search solutions that are adaptive to the *knowledge gap* [14–16]. In addition, there are similar studies that seek to model the knowledge gap by modeling the knowledge of the user and compare it against an existing knowledge level for the topic of interest [1,17–21]. Among these works, the work done by Zhao et al. [21] explores the knowledge paths between users' already acquired knowledge and the target knowledge that the user aims to obtain. Considering this knowledge path and in the context of recommender systems, the authors' method recommends a number of papers to the users so that their learning goals are achieved in the best satisfying way. Similarly, as our goal will be to assist users in knowledge acquisition tasks, we will take into consideration the target knowledge that users aim to obtain. Having modeled the knowledge from users' background knowledge and the target knowledge level, our goal is to estimate the knowledge gap between these levels. Thereafter, we will use the modeled knowledge gap to improve the retrieval effectiveness of the search systems for knowledge acquisition.

3.2 Knowledge Delta

Knowledge delta is another concept that is semantically closely related to knowledge gap. One definition of *knowledge delta* in the literature is the amount of knowledge change in a user's knowledge level which can be measured through questionnaires. As a case in point, we can refer to the work of Grunewald et al. [22], where the expertise gain is calculated through asking users of a "Massive Open Online Courses" system about their knowledge level in a field before and after taking an online course and denoting it as *knowledge delta*. Similar studies have also been done to measure the knowledge change of the users [22–26].

In all the aforementioned works, *knowledge delta* is used as a concept that demonstrates the change in users knowledge level. This concept is also associated with the name of *knowledge gain* in other works such as [27].

4   Research Methodology

The methodology begins with the investigation to find an answer to the first research question all the way to the third research question. The steps of the methodology are three-fold. Firstly, we will build a model for a user's knowledge and use it to build a model for the user's knowledge gap. Secondly, we will ask users to participate in knowledge acquisition tasks to gain knowledge on learning goals that will be defined for them. During this experiment, we will utilize the modeled knowledge gap for each user during the user's search session in order to provide the user with better results for learning. during this step, we will investigate the extent of improvement that incorporating the knowledge gap will bring about in a learning session. Thirdly, we will design a function whose output is a measure that will score the user's learning progress throughout the session.

RQ1: How can we model users' background knowledge, target knowledge and the knowledge gap in knowledge acquisition tasks?

To answer this research question, we need to design learning tasks for the users and we need to estimate the users' domain knowledge. Each learning task will set gaining knowledge on a sub-topic within a topic as a goal for the users. This goal will establish a desired level of knowledge as the target knowledge that a user wants to acquire. We will choose a fixed number of sub-topics and assign each user to a sub-topic for the learning task.

As it was mentioned under Sect. 3, several studies have modeled a user's knowledge in a variety of tasks and in different ways [1,16,21,28,29]. Extending the work done by Zhang et al. [29], we aim to represent the user's knowledge of a topic with a set of concepts within that topic.

We will assess the user's knowledge before and after the learning session using knowledge tests. We will design the knowledge tests in such a way that the user's knowledge of each concept can be attributed to a level of learning according to Bloom's taxonomy of educational objectives [30,31].

Having completed the knowledge test, each user's knowledge will be represented with its set of constituent concepts. There are a variety of options to represent the user's knowledge. One such way is a one-dimensional vector of concepts and the user's understanding of the concepts as weights. The same model will also be used to represent the target knowledge. The knowledge gap will then be computed between these two models.

Having modeled the knowledge gap, the methodology continues the investigation by moving to the second research question.

RQ2: To what extent can the incorporation of the knowledge gap into a search system facilitate a more efficient journey for users in knowledge acquisition tasks?

After the pre-task knowledge assessment, users will have access to a search interface connected to a search engine (here after collectively referred to as the search system) to carry out their learning tasks. The interface will allow for the recording of the users' interactions which will later be used as features to evaluate the quality of the learning sessions.

Each user will participate in at least two learning sessions during this experiment. In one learning session, the search system will not take the knowledge gap into account and original retrieved results by the system will be presented to the user. In the other learning session, the search system will take the knowledge gap into account and adapt the results before presenting them to the user. There are several ways to adapt the search results to the knowledge gap. One such way will be to re-rank the results of the user's each query submission based on a personalized understandability score of those results. For each of the learning sessions, we will design a different learning task. So as to ensure that learning about one topic doesn't affect the user's knowledge about other topics. After the learning session, the users will be asked to take a knowledge test again (Post-task knowledge tests). Comparing the result of the pre-task knowledge test with the post-task test for each user, we will have a score for the progress of the user's knowledge. On the other hand, for each user, we will compare the knowledge gained in the learning sessions to observe the effect of incorporating the knowledge gap on the quality of the learning session. We will evaluate this quality based on the interaction features recorded during the session.

RQ3: How should search systems be evaluated in the context of knowledge acquisition tasks?

In this research question, we will investigate how to define evaluation performance metrics for search systems in knowledge acquisition tasks. Previously in [32], we have explained why current retrieval metrics are insufficient in this context as they treat each query during the session independently and don't consider the users' knowledge factor in the evaluations. Correspondingly, we proposed three directions forward as well as their advantages and shortcomings: 1) online evaluation approach, 2) prerequisite-labeled relevance judgements approach, and 3) session-based evaluation approach. It's essential to mention at this point that what has been defined in this research proposal as *knowledge gap* was defined as *knowledge delta* in [32]. However, in order to maintain a greater consistency with the literature in this area considering the subtle distinction between the *knowledge delta* and the *knowledge gap*, we have adopted the term *knowledge gap* in this research proposal.

Our objective will be to extend this previous work and formalise the *session-based evaluation approach*. The goal is to design a session-based evaluation function that gauges the quality of learning sessions in knowledge acquisition tasks.

Up until this point in the methodology, we will have collected data on knowledge that the users have gained and information about learning sessions. In addition we will have recorded the users' interactions with the search system. As a result, a function that will serve as a new performance metric will be designed. This performance metric will use interaction features, such as time, number of queries used, etc., as cost indicators. It will combine these cost indicators such that the function's value aligns well with the experiences from the qualitative data (Knowledge tests). As a result, for the evaluation of future learning sessions, it will suffice to only have access to cost indicators and to use the designed function to evaluate the learning session.

There two main research challenges for the implementation of the discussed methodology:

1. The challenge that exists for the implementation for research question two is maintaining a balance between guiding a user to resources that are better suited for them and are adapted to their level of knowledge by the system, and just responding to the queries the user submitted.
2. In defining the learning goal for the experiments, the level of understanding of the user in the topic is a variable and the desired change in the level of understanding for the sake of experiments should be fixed. (e.g. Does it suffice if users are just familiarized with a concept, or should they be able to use it or should they be able to implement it?)

References

1. Aroyo, L., Denaux, R., Dimitrova, V., Pye, M.: Interactive ontology-based user knowledge acquisition: a case study. In: Sure, Y., Domingue, J. (eds.) ESWC 2006. LNCS, vol. 4011, pp. 560–574. Springer, Heidelberg (2006). https://doi.org/10.1007/11762256_41
2. Marchionini, G.: Exploratory search: from finding to understanding. Commun. ACM **49**(4), 41–46 (2006)
3. Gadiraju, U., Yu, R., Dietze, S., Holtz, P.: Analyzing knowledge gain of users in informational search sessions on the web. In: Proceedings of the 2018 Conference on Human Information Interaction & Retrieval, pp. 2–11 (2018)
4. Sulaiman, H.H.: Information needs and information seeking behaviour at the College of Medicine, The Sultan Qaboos University, Oman/Huda Sulaiman al-Haddabi. Ph.D. thesis, University of Malaya (2005)
5. Nkomo, N.: A comparative analysis of the web information seeking behaviour of students and staff at the University of Zululand and the Durban University of Technology. Ph.D. thesis (2009)
6. Igbinovia, M.O., Ikenwe, I.I.J.C.: Information seeking behavior of academic librarians for effective performance: a study of UNIBEN, AAU and AUCHI Poltechnic, Edo state, Nigeria (2014)
7. Gyesi, K.: Information seeking behaviour of graduate students of the university of professional studies, Accra (UPSA). Library Philos. Pract. (e-J.) 4155 (2020). https://digitalcommons.unl.edu/libphilprac/4155
8. Kelly, D., Belkin, N.J.: A user modeling system for personalized interaction and tailored retrieval in interactive IR. Proc. Am. Soc. Inf. Sci. Technol. **39**(1), 316–325 (2002)
9. Cosijn, E., Ingwersen, P.: Dimensions of relevance. Inf. Process. Manag. **36**(4), 533–550 (2000)
10. Belkin, N.J., Oddy, R.N., Brooks, H.M.: Ask for information retrieval: part I. background and theory. J. Doc. **38**(2), 61–71 (1982)
11. Dervin, B., Nilan, M.: Information needs and uses. In: Williams, M. (ed.) Annual Review of Information Science and Technology, vol. 21, pp. 3–33 (1986)
12. Thellefsen, M., Thellefsen, T., Sørenson, B.: A pragmatic semeiotic perspective on the concept of information need and its relevance for knowledge organization. KO Knowl. Organ. **40**(4), 213–224 (2014)
13. Yu, R., Tang, R., Rokicki, M., Gadiraju, U., Dietze, S.: Topic-independent modeling of user knowledge in informational search sessions. Inf. Retrieval J. **24**(3), 240–268 (2021). https://doi.org/10.1007/s10791-021-09391-7
14. Stojanovic, N.: On the role of a user's knowledge gap in an information retrieval process. In: Proceedings of the 3rd International Conference on Knowledge Capture, pp. 83–90 (2005)
15. Soldaini, L.: The knowledge and language gap in medical information seeking. In: ACM SIGIR Forum, vol. 52, pp. 178–179. ACM, New York (2019)
16. Zhang, Y., Liu, C.: Users' knowledge use and change during information searching process: a perspective of vocabulary usage. In: Proceedings of the ACM/IEEE Joint Conference on Digital Libraries in 2020, pp. 47–56 (2020)
17. Xiong, Y.: An automated feedback system to support student learning of conceptual knowledge in writing-to-learn activities. Ph.D. thesis, New Jersey Institute of Technology (2020)
18. Thaker, K., Zhang, L., He, D., Brusilovsky, P.: Recommending remedial readings using student knowledge state. International Educational Data Mining Society (2020)

19. Lindstaedt, S.N., Beham, G., Kump, B., Ley, T.: Getting to know your user – unobtrusive user model maintenance within work-integrated learning environments. In: Cress, U., Dimitrova, V., Specht, M. (eds.) EC-TEL 2009. LNCS, vol. 5794, pp. 73–87. Springer, Heidelberg (2009). https://doi.org/10.1007/978-3-642-04636-0_9

20. Syed, R.: Models and algorithms for understanding and supporting learning goals in information retrieval. Ph.D. thesis (2020)

21. Zhao, W., Ran, W., Liu, H.: Paper recommendation based on the knowledge gap between a researcher's background knowledge and research target. Inf. Process. Manag. **52**(5), 976–988 (2016)

22. Grünewald, F., Meinel, C., Totschnig, M., Willems, C.: Designing MOOCs for the support of multiple learning styles. In: Hernández-Leo, D., Ley, T., Klamma, R., Harrer, A. (eds.) EC-TEL 2013. LNCS, vol. 8095, pp. 371–382. Springer, Heidelberg (2013). https://doi.org/10.1007/978-3-642-40814-4_29

23. Daghio, M.M., Fattori, G., Ciardullo, A.V.: Assessment of readability and learning of easy-to-read educational health materials designed and written with the help of citizens by means of two non-alternative methods. Adv. Health Sci. Educ. **11**(2), 123–132 (2006)

24. Monica Daghio, M., Fattori, G., Ciardullo, A.V.: Evaluation of easy-to-read information material on healthy life-styles written with the help of citizens' collaboration through networking. Promot. Educ. **13**(3), 91–196 (2006)

25. Schaumberg, A.: Variation in closeness to reality of standardized resuscitation scenarios: effects on the success of cognitive learning of medical students. Anaesthesist **64**(4), 286–291 (2015)

26. de Resende, H.C., Slamnik-Krijestorac, N., Both, C.B., Marquez-Barja, J.: Introducing engineering undergraduate students to network management techniques: a hands-on approach using the Citylab Smart City. In: 2020 IEEE Global Engineering Education Conference (EDUCON), pp. 1316–1324. IEEE (2020)

27. Vakkari, P., Völske, M., Potthast, M., Hagen, M., Stein, B.: Modeling the usefulness of search results as measured by information use. Inf. Process. Manag. **56**(3), 879–894 (2019)

28. Yu, R., Gadiraju, U., Holtz, P., Rokicki, M., Kemkes, P., Dietze, S.: Predicting user knowledge gain in informational search sessions. In: The 41st International ACM SIGIR Conference on Research & Development in Information Retrieval, pp. 75–84 (2018)

29. Zhang, X., Liu, J., Cole, M., Belkin, N.: Predicting users' domain knowledge in information retrieval using multiple regression analysis of search behaviors. J. Am. Soc. Inf. Sci. **66**(5), 980–1000 (2015)

30. Bloom, B.S.: Taxonomy of educational objectives, handbook I: The cognitive domain. New York: David Mckay Co Inc. as cited in file. D:/bloom. html (1956)

31. Anderson, L.W., Bloom, B.S., et al.: A Taxonomy for Learning, Teaching, and Assessing: A Revision of Bloom's Taxonomy of Educational Objectives. Longman, New York (2001)

32. Ghafourian, Y., Knoth, P., Hanbury, A.: Information retrieval evaluation in knowledge acquisition tasks. In: Proceedings of the Third Workshop on Evaluation of Personalisation in Information Retrieval (WEPIR), pp. 88–95 (2021)

Evidence-Based Early Rumor Verification in Social Media

Fatima Haouari[(✉)]

Computer Science and Engineering Department, Qatar University, Doha, Qatar
200159617@qu.edu.qa

Abstract. A plethora of studies has been conducted in the past years on rumor verification in micro-blogging platforms. However, most of them exploit the propagation network, i.e., replies and retweets to verify rumors. We argue that first, subjective evidence from the propagation network is insufficient for users to understand, and reason the veracity of the rumor. Second, the full propagation network of the rumor can be sufficient, but for early detection when only part of the network is used, inadequate context for verification can be a major issue. As time is critical for early rumor verification, and sufficient evidence may not be available at the posting time, the objective of this thesis is to verify any tweet as soon as it is posted. Specifically, we are interested in exploiting evidence from Twitter as we believe it will be beneficial to 1) improve the veracity prediction 2) improve the user experience by providing convincing evidence 3) early verification, as waiting for subjective evidence may not be needed. We first aim to retrieve authority Twitter accounts that may help verify the rumor. Second, we aim to retrieve relevant tweets, i.e., tweets stating the same rumor, or tweets stating an evidence that contradicts or supports the rumor along with their propagation networks. Given the retrieved evidence from multiple sources namely evidence from authority accounts, evidence from relevant tweets, and their propagation networks, we intend to learn an effective model for rumor verification, and show rationales behind the decisions it makes.

Keywords: Claims · Veracity · Twitter · Social networks

1 Introduction

In the last decade, micro-blogging platforms (e.g., Twitter) have become a major source of information. As per Global Digital Report statistics, social media users have grown to a global total of 4.48 billion by the start of July 2021 which is almost 57% of the world's population.[1] This led news agencies and traditional newspapers to move to social media in order to cope with the societal change. However, social media have also become a medium to disseminate rumors and misinformation. Rumors are defined as circulating claims whose veracity is not

---

[1] https://datareportal.com/reports/digital-2021-july-global-statshot.

© The Author(s), under exclusive license to Springer Nature Switzerland AG 2022
M. Hagen et al. (Eds.): ECIR 2022, LNCS 13186, pp. 496–504, 2022.
https://doi.org/10.1007/978-3-030-99739-7_61

yet known at the time of posting [44]. Zubiaga et al. [45] found that unverified rumors spread quickly and largely at early stages. They also found that users tend to support unverified rumors during the first few minutes, and that the number of users supporting a rumor decreases after the rumor resolution. Previous studies showed that on average it takes 12 h between the start of a false rumor, and its debunking time [35,45]. These findings together motivate the importance of *early* detection and verification of rumors.

To solve the *early* rumor verification in social media problem, we propose retrieving and exploiting evidence from Twitter as we believe it will be beneficial to 1) improve the veracity prediction 2) improve the user experience by providing more convincing evidence 3) achieve the earliness objective, as waiting for subjective evidence, i.e., replies to the tweet to be verified, may not be needed. In our work, we are targeting Arabic social media, but our proposed system can be applied to any language. Specifically, we aim to tackle the following research questions: **(RQ1)** How can we find authoritative Twitter accounts that can help verify a tweet?, **(RQ2)** How can we effectively extract *evidence* from Twitter to verify a given tweet?, and **(RQ3)** Can we learn an effective and explainable model that can *early verify* tweets by exploiting evidence from other tweets?

The remainder of this paper is organized as follows: a literature review is presented in Sect. 2. We discuss our proposed methodology, and the evaluation setup in Sect. 3, and Sect. 4 respectively. Finally, we discuss the issues and challenges in Sect. 5.

2   Related Work

Previous studies on rumor verification at tweet-level exploited the propagation network of the post focusing on propagation structure [4,10,29,30,37], the information extracted from the profiles of the engaged users [26], or incorporating the stance of the replies [7,22,39,42,43]. Variant approaches were proposed ranging from adopting traditional machine learning models such as SVM [29] to neural networks [26,30], transformers [20,21], graph neural networks [4,10,37], and reinforcement learning [8,43]. Expert finding for a relevant piece of misinformation is under studied, the literature shows that only two studies addressed this problem [23,24]. Liang, Liu, and Sun [24] addressed identifying experts for Chinese rumors by searching a collection of Sina Weibo users. Their approach relies on expertise tags users assign to themselves. Li et al. [23] targeted only domain-specific misinformation which rarely mention specific people, places, or organizations as shown in Liang, Liu, and Sun [24] empirical analysis. Both studies adopted an unsupervised approach, and used the Bayes theorem to estimate the probability that a user is an expert given a piece of misinformation.

The limitations of existing studies are three-folds. First, existing studies exploit the propagation network, i.e., replies and retweets to detect and verify rumors. Using the full propagation network of the rumor can be sufficient, but for *early* verification when only part of the network is used, inadequate context for verification can be a major issue. Second, although many models were proposed in the literature and showed to perform well, the majority lacks showing evidence and rationales behind the model decision. A model that outputs, in addition to the veracity, other sources of evidence such as links to web pages, other social media posts, or authoritative accounts were this decision can be confirmed will be more robust as it allows the users to assess its reliability. Third, there are no datasets for evidence-based rumor verification in social media. The only datasets available whether in English [9,12,13,16,29,45], Chinese [29], or Arabic [19] contains subjective evidence, i.e., retweets and replies. To our knowledge, extracting evidence from relevant tweets, or from authoritative Twitter accounts for rumor verification was not addressed previously in the literature.

3   Proposed Methodology

In this section, we present our proposed methodology to address the research questions listed in Sect. 1.

3.1   RQ1: Authority Finding

Given a tweet stating a rumor, the task is to find the top k potential authoritative Twitter accounts that can help verify the given rumor. To address this problem, we first aim to experiment with existing topic expert finding techniques [5,15,23,32], but considering exploiting the recent advancement in information retrieval, and natural language processing. Specifically, we intend to study the effectiveness of these techniques to retrieve authoritative accounts for rumor verification, and adopting them as baselines to our proposed system. Second, as opposed to existing studies for expert finding for rumor verification [23,24] that adopted an unsupervised approach, we propose a *supervised approach* and address the problem as a ranking task by exploiting pre-trained transformers [25].

We propose to enrich the context of the rumor by adopting two stages of expansion 1) *pre-retrieval expansion:* by exploiting embedded information in the tweet such as, linked web pages and captions on images, 2) *post-retrieval:* by extracting keywords and topics from the top N users timeline tweets. For *Initial retrieval*, to get a list of candidates, we plan to experiment with different techniques such as retrieval using BM25 [33], which is based on keyword match, or topic similarity between query and users where each query and user are represented as topic vectors. To further improve the retrieval performance, we propose expanding the users representations with extra context, such as adopting entity linking [41]. We can then *rerank* the list of retrieved candidates by fine-tuning a pre-trained transformer model such as BERT [14], or ELECTRA [11].

3.2   RQ2: Evidence Retrieval from Twitter

Relevant Tweets Retrieval: We aim to retrieve from Twitter relevant tweets to the rumor. In our work, we distinguish relevant tweets into two categories 1) *relevant tweets stating the same rumor*, and 2) *relevant tweets that may contain evidence supporting or refuting the rumor*, i.e., those that are not explicitly supporting or denying the given rumor. We intend to retrieve and identify both categories as we believe this will be beneficial to improve the veracity prediction, and improve the user experience by providing more convincing evidence. The intuition is that we can extract extra signals from *relevant tweets stating the same rumor* propagation networks, and evidence from *relevant tweets that may contain evidence* to help verify the rumor, and act as rationals behind the model decisions. To retrieve relevant tweets, we plan to explore different queries construction techniques such as, searching with entities, regular expressions, or keyphrase extraction. To classify the relevant tweets into the mentioned categories, we propose detecting *relevant tweets stating the same rumor* using either clustering, near-duplicate detection [38], or fine-tuning a transformer-based model for relevance classification. We can then assume that all the retrieved tweets that are not discussing the rumor as tweets containing evidence.

Evidence Retrieval from Authoritative Accounts and Relevant Tweets: We propose to retrieve evidence from *authoritative Twitter accounts*, and *relevant tweets that may contain evidence* to select the best tweet, or set of tweets that can help verify a rumor. Specifically, we plan to experiment with the following techniques 1) similarity between claim and evidence representations, 2) classification task by fine-tuning transformer-based models.

3.3   RQ3: Effective Explainable Early Rumor Verification

To address RQ3, we aim to adopt the following earliness or detection deadline approaches:

1. **Time:** To experiment with either 1) considering tweets from authority accounts, relevant tweets and their replies posted *earlier* than the tweet to be verified, or 2) considering tweets from authority accounts, relevant tweets and their replies posted *after* the tweet to be verified but with variant detection deadlines.
2. **Number of tweets:** To experiment with either 1) pruning the number of relevant tweets, 2) pruning relevant tweets replies, or 3) pruning the tweet to be verified replies.

Exploiting Evidence from Authoritative Accounts and Relevant Tweets: We plan to exploit top k evidence tweets selected from authority accounts, and relevant tweets to *early* verify a rumor. To achieve this we intend to experiment with two approaches 1) transformer-based approach: given the tweet to be verified and an evidence tweet, we can fine-tune a pre-trained transformer model such as BERT, or ELECTRA to classify whether the tweet is refuted, supported, or has not enough information to be verified, 2) graph-based approach: we plan to experiment with SOTA approaches for evidence-based fact-checking [27,36], where evidence sentences are retrieved from Wikipedia pages.

In our work, we will consider the tweet to be verified as the claim, and top k tweets selected from authority accounts, and relevant tweets as evidence sentences.

Incorporating the Stance of Relevant Tweets Replies: We propose to incorporate the stance of relevant tweets and their replies for rumor verification. In our work, we will consider *the replies of all relevant tweets stating the same rumor* as replies to the tweet to be verified as we believe this will provide extra signals from users comments toward the rumor. To achieve this, we plan to experiment with the following 1) transformer-based approach: fine-tuning BERT, or ELECTRA to classify the stance of each reply towards the tweet to be verified, then consider the majority voting, 2) multi-task learning approach: to experiment with SOTA models for stance-aware rumor verification [7,17,42] to jointly predict the stance of replies, and verify the rumor.

Exploiting the Propagation Structure of Relevant Tweets and their Replies: We aim to exploit the propagation structure of relevant tweets and their replies. Specifically, we plan to experiment with adopting graph embeddings [3,31], or graph neural networks [40] to represent the propagation structure of relevant tweets and their replies.

Jointly Exploiting Multiple Sources of Evidence: We plan to incorporate all sources of evidence discussed previously to estimate the veracity of a given tweet. My proposed system is composed of different components namely 1) a graph representing evidence selected from authority accounts, 2) a graph representing evidence stated in relevant tweets, 3) a graph of relevant tweets replies, 4) the stance features, and 5) the representation of the tweet to be verified. As proposed by Lu and Li [28], we aim to experiment with incorporating multiple co-attention into our system to capture the correlation between the tweet to be verified, and each component in the system, and finally estimating the veracity of the tweets by concatenating all the representations.

Explainable Early Rumor Verification: Finally, we aim to provide the user with visual explanations that justify the model's decisions, and make them trust the system. As mentioned previously, our proposed system will select evidence from authority accounts, and from relevant tweets. These will be presented to the user as a kind of justification of the system decisions. We plan to experiment with either 1) abstractive summarization: to fine-tune GPT-3 [6] to generate explanations given the set of evidence tweets selected from authority accounts and relevant tweets, 2) evidence text extraction: we plan to experiment with the TokenMasker architecture proposed by Shah, Schuster, and Barzilay [34] to detect the crucial text spans connecting the claim and evidence.

Preliminary Work: Based on the literature review, we found that there is no study addressing *early* Arabic rumor verification in Twitter given the propagation network of the tweet to be verified. In order to evaluate our proposed system, we need to compare against SOTA models exploiting the propagation networks, and thus we need an Arabic rumor verification Twitter dataset, which includes the propagation networks of the tweets. We started by constructing *ArCOV-19* [18], the first Arabic COVID-19 Twitter dataset with propagation networks. We then worked on manually annotating a subset of it to construct *ArCOV19-Rumors* [19], that covers tweets spreading COVID-19 related rumors labeled as true, false. Moreover, we implemented some baselines [19] to our proposed system, we experimented with SOTA models that either exploit content only, user profile features, temporal features, or propagation structure of the replies. Specifically, we experimented with two models that exploit the propagation networks for tweet verification namely Bi-GCN [4], i.e., a bidirectional Graph Convolutional Networks model that leverages Graph Convolutional Networks to verify tweets given the target tweet and replies content, in addition to the replies tree structure, and PPC-RNN+CNN [26], i.e., a multivariate time series model that exploits user profiles to verify tweets. Moreover, due to proven effectiveness of BERT-based classifiers in versatile text classification tasks, we fine-tuned two Arabic pre-trained BERT models namely MARBERT [1], and AraBERT [2].

4   Experimental Evaluation

To evaluate our proposed system, we plan to extend *ArCOV19-Rumors* dataset to associate authority Twitter accounts, and relevant tweets to each tweet in the *tweet verification subset* which consists of 3,584 tweets labeled as true or false. Additionally, we aim to annotate the replies of the tweets with their stance towards the rumor. Moreover, we plan to extend the dataset to include topics other than COVID-19 for my models to be able to generalize to rumors from variant topics. To evaluate the effectiveness of the authority finding model, we will adopt Precision@k and nDCG scores. As we are addressing the *evidence retrieval*, and the *early* tweet verification as classification tasks, we will compute precision, recall, and F1 scores.

5   Issues for Discussion

I plan to discuss several points, namely 1) what are the limitations of restricting evidence retrieval to Twitter, 2) should I consider other sources for evidence retrieval such as Wikipedia and web pages?, 3) what are the potential features that can be extracted from users accounts to find authoritative accounts?, 4) whether to consider the authority finding problem a ranking, or a classification task, and what are the limitations of each approach?, 5) what are the possible approaches to model the propagation structure of relevant tweets replies?, Additionally, I would like to get feedback about the general limitations of the proposed system.

Acknowledgments. This work was made possible by GSRA grant# GSRA6-1-0611-19074 from the Qatar National Research Fund. The statements made herein are solely the responsibility of the authors.

References

1. Abdul-Mageed, M., Elmadany, A., Nagoudi, E.M.B.: ARBERT & MARBERT: deep bidirectional transformers for Arabic. In: Proceedings of the 59th Annual Meeting of the Association for Computational Linguistics and the 11th International Joint Conference on Natural Language Processing (Volume 1: Long Papers), pp. 7088–7105. Association for Computational Linguistics, August 2021
2. Baly, F., Hajj, H., et al.: AraBERT: transformer-based model for Arabic language understanding. In: Proceedings of the 4th Workshop on Open-Source Arabic Corpora and Processing Tools, with a Shared Task on Offensive Language Detection, pp. 9–15 (2020)
3. Beladev, M., Rokach, L., Katz, G., Guy, I., Radinsky, K.: tdGraphEmbed: temporal dynamic graph-level embedding. In: Proceedings of the 29th ACM International Conference on Information & Knowledge Management, pp. 55–64 (2020)
4. Bian, T., et al.: Rumor detection on social media with bi-directional graph convolutional networks. In: Proceedings of the AAAI Conference on Artificial Intelligence, pp. 549–556 (2020)
5. Bozzon, A., Brambilla, M., Ceri, S., Silvestri, M., Vesci, G.: Choosing the right crowd: expert finding in social networks. In: Proceedings of the 16th International Conference on Extending Database Technology, EDBT 2013, pp. 637–648. Association for Computing Machinery, New York (2013)
6. Brown, T.B., et al.: Language models are few-shot learners. arXiv preprint arXiv:2005.14165 (2020)
7. Chen, L., Wei, Z., Li, J., Zhou, B., Zhang, Q., Huang, X.J.: Modeling evolution of message interaction for rumor resolution. In: Proceedings of the 28th International Conference on Computational Linguistics, pp. 6377–6387 (2020)
8. Chen, Y., Yin, C., Zuo, W.: Multi-task learning for stance and early rumor detection. Opt. Memory Neural Netw. **30**(2), 131–139 (2021)
9. Cheng, M., et al.: A COVID-19 rumor dataset. Front. Psychol. **12** (2021)
10. Choi, J., Ko, T., Choi, Y., Byun, H., Kim, C.k.: Dynamic graph convolutional networks with attention mechanism for rumor detection on social media. Plos One **16**(8), e0256039 (2021)
11. Clark, K., Luong, M.T., Le, Q.V., Manning, C.D.: Electra: pre-training text encoders as discriminators rather than generators. arXiv preprint arXiv:2003.10555 (2020)
12. Cui, L., Lee, D.: CoAID: COVID-19 healthcare misinformation dataset. arXiv preprint arXiv:2006.00885 (2020)
13. Derczynski, L., Bontcheva, K., Liakata, M., Procter, R., Hoi, G.W.S., Zubiaga, A.: SemEval-2017 task 8: RumourEval: determining rumour veracity and support for rumours. arXiv preprint arXiv:1704.05972 (2017)
14. Devlin, J., Chang, M.W., Lee, K., Toutanova, K.: BERT: pre-training of deep bidirectional transformers for language understanding. arXiv preprint arXiv:1810.04805 (2018)

15. Ghosh, S., Sharma, N., Benevenuto, F., Ganguly, N., Gummadi, K.: Cognos: crowd-sourcing search for topic experts in microblogs. In: Proceedings of the 35th International ACM SIGIR Conference on Research and Development in Information Retrieval, pp. 575–590 (2012)

16. Gorrell, G., et al.: SemEval-2019 task 7: RumourEval, determining rumour veracity and support for rumours. In: Proceedings of the 13th International Workshop on Semantic Evaluation, pp. 845–854 (2019)

17. Han, X., Huang, Z., Lu, M., Li, D., Qiu, J.: Rumor verification on social media with stance-aware recursive tree. In: Qiu, H., Zhang, C., Fei, Z., Qiu, M., Kung, S.-Y. (eds.) KSEM 2021. LNCS (LNAI), vol. 12817, pp. 149–161. Springer, Cham (2021). https://doi.org/10.1007/978-3-030-82153-1_13

18. Haouari, F., Hasanain, M., Suwaileh, R., Elsayed, T.: ArCOV-19: the first Arabic COVID-19 Twitter dataset with propagation networks. In: Proceedings of the Sixth Arabic Natural Language Processing Workshop, pp. 82–91 (2021)

19. Haouari, F., Hasanain, M., Suwaileh, R., Elsayed, T.: ArCOV19-rumors: Arabic COVID-19 Twitter dataset for misinformation detection. In: Proceedings of the Sixth Arabic Natural Language Processing Workshop, pp. 72–81 (2021)

20. Khandelwal, A.: Fine-tune longformer for jointly predicting rumor stance and veracity. In: 8th ACM IKDD CODS and 26th COMAD, pp. 10–19 (2021)

21. Khoo, L.M.S., Chieu, H.L., Qian, Z., Jiang, J.: Interpretable rumor detection in microblogs by attending to user interactions. In: Proceedings of the AAAI Conference on Artificial Intelligence, vol. 34, pp. 8783–8790 (2020)

22. Kumar, S., Carley, K.: Tree LSTMs with convolution units to predict stance and rumor veracity in social media conversations. In: Proceedings of the 57th Annual Meeting of the Association for Computational Linguistics, Florence, Italy. Association for Computational Linguistics, July 2019

23. Li, G., et al.: Misinformation-oriented expert finding in social networks. World Wide Web 23(2), 693–714 (2019). https://doi.org/10.1007/s11280-019-00717-6

24. Liang, C., Liu, Z., Sun, M.: Expert finding for microblog misinformation identification. In: Proceedings of COLING 2012: Posters, pp. 703–712 (2012)

25. Lin, J., Nogueira, R., Yates, A.: Pretrained transformers for text ranking: BERT and beyond (2021)

26. Liu, Y., Wu, Y.F.B.: Early detection of fake news on social media through propagation path classification with recurrent and convolutional networks. In: Thirty-Second AAAI Conference on Artificial Intelligence (2018)

27. Liu, Z., Xiong, C., Sun, M., Liu, Z.: Fine-grained fact verification with kernel graph attention network. In: Proceedings of the 58th Annual Meeting of the Association for Computational Linguistics, pp. 7342–7351. Association for Computational Linguistics, July 2020

28. Lu, Y.J., Li, C.T.: GCAN: graph-aware co-attention networks for explainable fake news detection on social media. In: Proceedings of the 58th Annual Meeting of the Association for Computational Linguistics, pp. 505–514. Association for Computational Linguistics, July 2020

29. Ma, J., Gao, W., Wong, K.F.: Detect rumors in microblog posts using propagation structure via kernel learning. Association for Computational Linguistics (2017)

30. Ma, J., Gao, W., Wong, K.F.: Rumor detection on Twitter with tree-structured recursive neural networks. Association for Computational Linguistics (2018)

31. Narayanan, A., Chandramohan, M., Venkatesan, R., Chen, L., Liu, Y., Jaiswal, S.: graph2vec: learning distributed representations of graphs. arXiv preprint arXiv:1707.05005 (2017)

32. Pal, A., Counts, S.: Identifying topical authorities in microblogs. In: Proceedings of the Fourth ACM International Conference on Web Search and Data Mining, pp. 45–54 (2011)

33. Robertson, S., Zaragoza, H.: The probabilistic relevance framework: BM25 and beyond. Found. Trends Inf. Retr. **3**(4), 333–389. Now Publishers Inc, Hanover (2009). https://doi.org/10.1561/1500000019

34. Shah, D., Schuster, T., Barzilay, R.: Automatic fact-guided sentence modification. In: Proceedings of the AAAI Conference on Artificial Intelligence, vol. 34, pp. 8791–8798 (2020)

35. Shao, C., Ciampaglia, G.L., Flammini, A., Menczer, F.: Hoaxy: a platform for tracking online misinformation. In: Proceedings of the 25th International Conference Companion on World Wide Web, pp. 745–750 (2016)

36. Si, J., Zhou, D., Li, T., Shi, X., He, Y.: Topic-aware evidence reasoning and stance-aware aggregation for fact verification. In: Proceedings of the 59th Annual Meeting of the Association for Computational Linguistics and the 11th International Joint Conference on Natural Language Processing (Volume 1: Long Papers), pp. 1612–1622. Association for Computational Linguistics, August 2021

37. Song, C., Shu, K., Wu, B.: Temporally evolving graph neural network for fake news detection. Inf. Process. Manag. **58**(6), 102712 (2021)

38. Tao, K., Abel, F., Hauff, C., Houben, G.J., Gadiraju, U.: Groundhog day: near-duplicate detection on Twitter. In: Proceedings of the 22nd International Conference on World Wide Web, WWW 2013, pp. 1273–1284. Association for Computing Machinery, New York (2013)

39. Wu, L., Rao, Y., Jin, H., Nazir, A., Sun, L.: Different absorption from the same sharing: sifted multi-task learning for fake news detection. In: Proceedings of the 2019 Conference on Empirical Methods in Natural Language Processing and the 9th International Joint Conference on Natural Language Processing (EMNLP-IJCNLP), Hong Kong, China. Association for Computational Linguistics, November 2019

40. Wu, L., et al.: Graph neural networks for natural language processing: a survey. arXiv e-prints arXiv:2106 (2021)

41. Yamada, I., Takeda, H., Takefuji, Y.: Enhancing named entity recognition in Twitter messages using entity linking. In: ACL-IJCNLP 2015, p. 136 (2015)

42. Yu, J., Jiang, J., Khoo, L.M.S., Chieu, H.L., Xia, R.: Coupled hierarchical transformer for stance-aware rumor verification in social media conversations. In: Proceedings of the 2020 Conference on Empirical Methods in Natural Language Processing (EMNLP), pp. 1392–1401. Association for Computational Linguistics, November 2020

43. Yuan, C., Qian, W., Ma, Q., Zhou, W., Hu, S.: SRLF: a stance-aware reinforcement learning framework for content-based rumor detection on social media. arXiv preprint arXiv:2105.04098 (2021)

44. Zubiaga, A., Aker, A., Bontcheva, K., Liakata, M., Procter, R.: Detection and resolution of rumours in social media: a survey. ACM Comput. Surv. (CSUR) **51**(2), 1–36 (2018)

45. Zubiaga, A., Liakata, M., Procter, R., Hoi, G.W.S., Tolmie, P.: Analysing how people orient to and spread rumours in social media by looking at conversational threads. PloS One **11**(3), e0150989 (2016)

Multimodal Retrieval in E-Commerce

From Categories to Images, Text, and Back

Mariya Hendriksen[✉]

Informatics Institute, University of Amsterdam, Amsterdam, The Netherlands
m.hendriksen@uva.nl

Abstract. E-commerce provides rich multimodal data that is barely leveraged in practice. The majority of e-commerce search mechanisms are uni-modal, which are cumbersome and often fail to grasp the customer's needs. For the Ph.D. we conduct research aimed at combining information across multiple modalities to improve search and recommendations in e-commerce. The research plans are organized along the two principal lines. First, motivated by the mismatch between a textual and a visual representation of a given product category, we propose the task of category-to-image retrieval, i.e., the problem of retrieval of an image of a category expressed as a textual query. Besides, we propose a model for the task. The model leverages information from multiple modalities to create product representations. We explore how adding information from multiple modalities impacts the model's performance and compare our approach with state-of-the-art models. Second, we consider fine-grained text-image retrieval in e-commerce. We start off by considering the task in the context of reproducibility. Moreover, we address the problem of attribute granularity in e-commerce. We select two state-of-the-art (SOTA) models with distinct architectures, a CNN-RNN model and a Transformer-based model, and consider their performance on various e-commerce categories as well as on object-centric data from general domain. Next, based on the lessons learned from the reproducibility study, we propose the model for the fine-grained text-image retrieval.

1 Motivation

Multimodal retrieval is an important but understudied problem in e-commerce [48]. Even though e-commerce products are associated with rich multimodal information, research currently focuses mainly on textual and behavioral signals to support product search and recommendation [1,15,42]. The majority of prior work in multimodal retrieval for e-commerce focuses on applications in the fashion domain, such as recommendation of fashion items [34] and cross-modal fashion retrieval [13,25]. In the more general e-commerce domain, multimodal retrieval has not been explored that well yet [17,31]. Motivated by the knowledge gap, we lay out two directions for the research agenda: category-to-image retrieval, and fine-grained text-image retrieval (Fig. 1).

M. Hagen et al. (Eds.): ECIR 2022, LNCS 13186, pp. 505–512, 2022.
https://doi.org/10.1007/978-3-030-99739-7_62

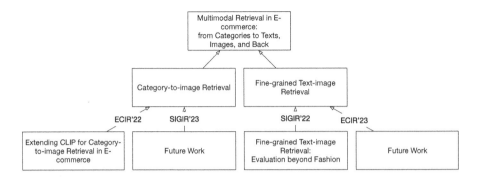

Fig. 1. Dissertation overview.

Category-to-Image Retrieval. First, we focus on the category information in e-commerce. Product category trees are a key component of modern e-commerce as they assist customers when navigating across large product catalogues [16, 24, 46, 50]. Yet, the ability to retrieve an image for a given product category remains a challenging task mainly due to noisy category and product data, and the size and dynamic character of product catalogues [28, 48]. Motivated by this challenge, we introduce the task of retrieving a ranked list of relevant images of products that belong to a given category, which we call the *category-to-image* (CtI) retrieval task. Unlike image classification tasks that operate on a predefined set of classes, in the CtI retrieval task we want to be able not only to understand which images belong to a given category but also to generalize towards unseen categories. Use cases that motivate the CtI retrieval task include (1) the need to showcase different categories in search and recommendation results [24, 46, 48]; (2) the task can be used to infer product categories in the cases when product categorical data is unavailable, noisy, or incomplete [52]; and (3) the design of cross-categorical promotions and product category landing pages [39].

Fine-Grained Text-Image Retrieval. Second, we address the problem of fine-grained text-image retrieval. Text-image retrieval is the task of finding similar items across textual and visual modalities. Successful performance on the task depends on the domain. In the general domain, where images typically depict complex scenes of objects in their natural contexts information across modalities is matched coarsely. Some examples of such datasets include MS COCO [33], and Flick30k [53]. By contrast, in the e-commerce domain, where there is typically one object per image, fine-grained matching is more important. Therefore, we focus on *fine-grained text-image retrieval*. We define the task as a combination of two subtasks: 1. *text-to-image retrieval*: given a noun phrase that describes an object, retrieve the image that depicts to the object; 2. *image-to-text retrieval*: given an image of an object, retrieve the noun phrase that describes an object.

We start off by examining the topic in the context of reproducibility. Reproducibility is one of the major pillars of the scientific method and is of utmost importance for Information Retrieval (IR) as a discipline rooted in experimentation [10]. One of the first works that touch upon reproducibility in IR is the study by Armstrong et al. [2] where the authors conducted a longitudinal analysis of papers published in proceedings of CIKM and SIGIR between 1998–2008 and discovered that the ad-hoc retrieval was not measurably improving. Later on, Yang et al. [51] provided a meta-analysis of results reported on the TREC Robust04 and found out that some of the more recent neural models were outperformed by strong baselines. Similar discoveries were made in the domain of recommender systems research [5,6]. Motivated by the findings, we explore the reproducibility of fine-grained text-image retrieval results. More specifically, we examine how SOTA models for fine-grained text-image fashion retrieval generalize towards other categories of e-commerce products. After analyzing SOTA models in the domain, we plan to improve upon them in a subsequent future work.

2 Related Work

Category-to-Image Retrieval. Early work in image retrieval grouped images into a restricted set of semantic categories and allowed users to retrieve images by using category labels as queries [44]. Later work allowed for a wider variety of queries ranging from natural language [20,49], to attributes [37], to combinations of multiple modalities (e.g., title, description, and tags) [47]. Across these multimodal image retrieval approaches we find three common components: (1) an image encoder, (2) a query encoder, and (3) a similarity function to match the query to images [14,40]. Depending on the focus of the work some components might be pre-trained, whereas the others are optimized for a specific task. In our work, we rely on pre-trained image and text encoders but learn a new multimodal composite of the query to perform CtI retrieval.

Fine-Grained Text-Image Retrieval. Early approaches to cross-modal mapping focused on correlation maximization through canonical correlation analysis [18,19,45]. Later approaches centered around convolutional and recurrent neural networks [11,22,23,29]. They were further expanded by adding attention on top of encoders [29,35,38]. More recently, inspired by the success of transformers [8], a line of work centered around creating a universal vision-language encoder emerged [4,30,32,36]. To address the problem of attribute granularity in the context of cross-modal retrieval, a line of work proposed to segment images into fragments [27], use attention mechanisms [26], combine image features across multiple levels [13], use pre-trained BERT as a backbone [12,54]. Unlike prior work in this domain that focused on fashion, we focus on the general e-commerce domain.

3 Research Description and Methodology

The dissertation comprises two parts. Below, we describe every part of the thesis and elaborate on the methodology.

Category-to-Image Retrieval. Product categories are used in various contexts in e-commerce. However, in practice, during a user's session, there is often a mismatch between a textual and a visual representation of a given category. Motivated by the problem, we introduce the task of category-to-image retrieval in e-commerce and propose a model for the task.

We use the XMarket dataset recently introduced by Bonab et al. [3] that contains textual, visual, and attribute information of e-commerce products as well as a category tree. Following [7,21,43] we use BM25, MPNet, CLIP as our baselines. To evaluate model performance, we use Precision@K where $K = \{1, 5, 10\}$, mAP@K where $K = \{5, 10\}$, and R-precision.

RQ1.1 *How do baseline models perform on the **CtI** retrieval task? Specifically, how do unimodal and bi-modal baseline models perform? How does the performance differ w.r.t. category granularity?*

To answer the question, we feed BM25 corpora that contain textual product information, i.e., product titles. We use an MPNet in a zero-shot manner. For all the products in the dataset, we pass the product title through the model. During the evaluation, we pass a category expressed as textual query through MPNet and retrieve top-k candidates ranked by cosine similarity w.r.t. the target category. We compare categories of the top-k retrieved candidates with the target category. Besides, we use pre-trained CLIP in a zero-shot manner with a text transformer and a vision transformer (ViT) [9] configuration. We pass the product image through the image encoder. For evaluation, we pass a category through the text encoder and retrieve top-k image candidates ranked by cosine similarity w.r.t. the target category. We compare categories of the top-k retrieved image candidates with the target category.

RQ1.2 *How does a model, named CLIP-I, that uses product image information for building product representations impact the performance on the CtI retrieval task?*

To answer the question, we build product representations by training on e-commerce data. We investigate how using product image data for building product representations impacts performance on the CtI retrieval task. To introduce visual information, we extend CLIP in two ways: (1) We use ViT from CLIP as an image encoder. We add a product projection head that takes as an input product visual information. (2) We use the text encoder from MPNet as category encoder; we add a category projection head on top of the category encoder. We name the resulting model CLIP-I. We train CLIP-I on category-product pairs from the training set. We only use visual information for building product representations.

RQ1.3 *How does CLIP-IA, which extends CLIP-I with product attribute information, perform on the CtI retrieval task?*

To answer the question, we extend CLIP-I by introducing attribute information to the product information encoding pipeline. We add an attribute encoder through which we obtain a representation of product attributes. We concatenate the resulting attribute representation with image representation and pass the resulting vector to the product projection head. Thus, the resulting product representation **p** is based on both visual and attribute product information. We name the resulting model CLIP-IA. We train CLIP-IA on category-product pairs and we use visual and attribute information for building product representation.

RQ1.4 *And finally, how does CLIP-ITA, which extends CLIP-IA with product text information, perform on the CtI task?*

To answer the question, we investigate how extending the product information processing pipeline with the textual modality impacts performance on the CtI retrieval task. We add a title encoder to the product information processing pipeline and use it to obtain title representation. We concatenate the resulting representation with product image and attribute representations. We pass the resulting vector to the product projection head. The resulting model is CLIP-ITA. We train and test CLIP-ITA on category-product pairs. We use visual, attribute, and textual information for building product representations. The results are to be published in ECIR'22 [16]. The follow-up work is planned to be published at SIGIR 2023.

Fine-Grained Text-Image Retrieval. The ongoing work is focused on fine-grained text-image retrieval in the context of reproducibility. For the experiments, we select two SOTA models for fine-grained cross-modal fashion retrieval, each model with distinctive architecture. One of them is based on Transformer while another one is CNN-RNN-based. The Transformer-based model is Kaleido-BERT [54], that extends BERT [8]. Another model is a Multi-level Feature approach (MLF) [13]. Both models claim to deliver SOTA performance by being able to learn image representations that can better represent fine-grained attributes. They were evaluated on Fashion-Gen dataset [41] but, to the best of our knowledge, were not compared against each other.

In the work, we aim to answer the following research questions:

RQ2.1 *How well Kaleido-BERT and MLF perform on data from an e-commerce category that is different from Fashion?*

RQ2.2 *How well both models generalize beyond e-commerce domain? More specifically, how do they perform on object-centric data from the general domain?*

RQ2.3 *How Kaleido-BERT and MLF compare to each other w.r.t performance?*

The results are planned to be published as a paper at SIGIR 2022. The follow-up work is planned to be published at ECIR 2023.

References

1. Ariannezhad, M., Jullien, S., Nauts, P., Fang, M., Schelter, S., de Rijke, M.: Understanding multi-channel customer behavior in retail. In: Proceedings of the 30th ACM International Conference on Information & Knowledge Management, pp. 2867–2871 (2021)
2. Armstrong, T.G., Moffat, A., Webber, W., Zobel, J.: Improvements that don't add up: ad-hoc retrieval results since 1998. In: Proceedings of the 18th ACM Conference on Information and Knowledge Management, pp. 601–610. Association for Computing Machinery (2009)

3. Bonab, H., Aliannejadi, M., Vardasbi, A., Kanoulas, E., Allan, J.: Cross-market product recommendation. In: CIKM. ACM (2021)

4. Chen, Y.C., et al.: UNITER: learning universal image-text representations. arXiv preprint arXiv:1909.11740 (2019)

5. Dacrema, M.F., Cremonesi, P., Jannach, D.: Are we really making much progress? A worrying analysis of recent neural recommendation approaches. In: Proceedings of the 13th ACM Conference on Recommender Systems, pp. 101–109 (2019)

6. Dacrema, M.F., Boglio, S., Cremonesi, P., Jannach, D.: A troubling analysis of reproducibility and progress in recommender systems research. ACM Trans. Inf. Syst. (TOIS) **39**(2), 1–49 (2021)

7. Dai, Z., Lai, G., Yang, Y., Le, Q.V.: Funnel-transformer: filtering out sequential redundancy for efficient language processing. arXiv preprint arXiv:2006.03236 (2020)

8. Devlin, J., Chang, M.W., Lee, K., Toutanova, K.: BERT: pre-training of deep bidirectional transformers for language understanding. arXiv preprint arXiv:1810.04805 (2018)

9. Dosovitskiy, A., et al.: An image is worth 16x16 words: transformers for image recognition at scale. arXiv preprint arXiv:2010.11929 (2020)

10. Ferro, N., Fuhr, N., Järvelin, K., Kando, N., Lippold, M., Zobel, J.: Increasing reproducibility in IR: findings from the Dagstuhl seminar on "reproducibility of data-oriented experiments in e-science". In: ACM SIGIR Forum, vol. 50, pp. 68–82. ACM New York (2016)

11. Frome, A., et al.: Devise: a deep visual-semantic embedding model. In: Burges, C.J.C., Bottou, L., Welling, M., Ghahramani, Z., Weinberger, K.Q. (eds.) Advances in Neural Information Processing Systems 26, pp. 2121–2129. Curran Associates Inc (2013)

12. Gao, D., et al.: FashionBERT: text and image matching with adaptive loss for cross-modal retrieval. In: Proceedings of the 43rd International ACM SIGIR Conference on Research and Development in Information Retrieval, pp. 2251–2260 (2020)

13. Goei, K., Hendriksen, M., de Rijke, M.: Tackling attribute fine-grainedness in cross-modal fashion search with multi-level features. In: SIGIR 2021 Workshop on eCommerce. ACM (2021)

14. Gupta, T., Vahdat, A., Chechik, G., Yang, X., Kautz, J., Hoiem, D.: Contrastive learning for weakly supervised phrase grounding. In: Vedaldi, A., Bischof, H., Brox, T., Frahm, J.-M. (eds.) ECCV 2020. LNCS, vol. 12348, pp. 752–768. Springer, Cham (2020). https://doi.org/10.1007/978-3-030-58580-8_44

15. Hendriksen, M., Kuiper, E., Nauts, P., Schelter, S., de Rijke, M.: Analyzing and predicting purchase intent in e-commerce: anonymous vs. identified customers. arXiv preprint arXiv:2012.08777 (2020)

16. Hendriksen, M., Bleeker, M., Vakulenko, S., van Noord, N., Kuiper, E., de Rijke, M.: Extending CLIP for category-to-image retrieval in e-commerce. In: Hagen, M., et al. (eds.) ECIR 2022. LNCS, vol. 13186, pp. 289–303. Springer, Cham (2022)

17. Hewawalpita, S., Perera, I.: Multimodal user interaction framework for e-commerce. In: 2019 International Research Conference on Smart Computing and Systems Engineering (SCSE), pp. 9–16. IEEE (2019)

18. Hodosh, M., Young, P., Hockenmaier, J.: Framing image description as a ranking task: data, models and evaluation metrics. J. Artif. Intell. Res. **47**, 853–899 (2013)

19. Hotelling, H.: Relations between two sets of variates. In: Kotz, S., Johnson, N.L. (eds.) Breakthroughs in Statistics. SSS, pp. 162–190. Springer, New York (1992). https://doi.org/10.1007/978-1-4612-4380-9_14

20. Hu, R., Xu, H., Rohrbach, M., Feng, J., Saenko, K., Darrell, T.: Natural language object retrieval. In: Proceedings of the IEEE Conference on Computer Vision and Pattern Recognition, pp. 4555–4564 (2016)
21. Jabeur, L.B., Soulier, L., Tamine, L., Mousset, P.: A product feature-based user-centric ranking model for e-commerce search. In: Fuhr, N., et al. (eds.) CLEF 2016. LNCS, vol. 9822, pp. 174–186. Springer, Cham (2016). https://doi.org/10.1007/978-3-319-44564-9_14
22. Karpathy, A., Fei-Fei, L.: Deep visual-semantic alignments for generating image descriptions. In: Proceedings of the IEEE Conference on Computer Vision and Pattern Recognition, pp. 3128–3137 (2015)
23. Kiros, R., Salakhutdinov, R., Zemel, R.S.: Unifying visual-semantic embeddings with multimodal neural language models. arXiv preprint arXiv:1411.2539 (2014)
24. Kondylidis, N., Zou, J., Kanoulas, E.: Category aware explainable conversational recommendation. arXiv preprint arXiv:2103.08733 (2021)
25. Laenen, K., Moens, M.-F.: Multimodal neural machine translation of fashion e-commerce descriptions. In: Kalbaska, N., Sádaba, T., Cominelli, F., Cantoni, L. (eds.) FACTUM 2019, pp. 46–57. Springer, Cham (2019). https://doi.org/10.1007/978-3-030-15436-3_4
26. Laenen, K., Moens, M.F.: A comparative study of outfit recommendation methods with a focus on attention-based fusion. Inf. Process. Manag. **57**(6), 102316 (2020)
27. Laenen, K., Zoghbi, S., Moens, M.F.: Cross-modal search for fashion attributes. In: Proceedings of the KDD 2017 Workshop on Machine Learning Meets Fashion, vol. 2017, pp. 1–10. ACM (2017)
28. Laenen, K., Zoghbi, S., Moens, M.F.: Web search of fashion items with multimodal querying. In: Proceedings of the Eleventh ACM International Conference on Web Search and Data Mining, pp. 342–350 (2018)
29. Lee, K.H., Chen, X., Hua, G., Hu, H., He, X.: Stacked cross attention for image-text matching. In: Proceedings of the European Conference on Computer Vision (ECCV), pp. 201–216 (2018)
30. Li, G., Duan, N., Fang, Y., Jiang, D., Zhou, M.: Unicoder-VL: a universal encoder for vision and language by cross-modal pre-training. arXiv preprint arXiv:1908.06066 (2019)
31. Li, H., Yuan, P., Xu, S., Wu, Y., He, X., Zhou, B.: Aspect-aware multimodal summarization for Chinese e-commerce products. In: Proceedings of the AAAI Conference on Artificial Intelligence, vol. 34, pp. 8188–8195 (2020)
32. Li, L.H., Yatskar, M., Yin, D., Hsieh, C.J., Chang, K.W.: VisualBERT: a simple and performant baseline for vision and language. arXiv preprint arXiv:1908.03557 (2019)
33. Lin, T.-Y., et al.: Microsoft COCO: common objects in context. In: Fleet, D., Pajdla, T., Schiele, B., Tuytelaars, T. (eds.) ECCV 2014. LNCS, vol. 8693, pp. 740–755. Springer, Cham (2014). https://doi.org/10.1007/978-3-319-10602-1_48
34. Lin, Y., Ren, P., Chen, Z., Ren, Z., Ma, J., de Rijke, M.: Improving outfit recommendation with co-supervision of fashion generation. In: The World Wide Web Conference, pp. 1095–1105 (2019)
35. Liu, C., Mao, Z., Liu, A.A., Zhang, T., Wang, B., Zhang, Y.: Focus your attention: a bidirectional focal attention network for image-text matching. In: Proceedings of the 27th ACM International Conference on Multimedia, pp. 3–11 (2019)
36. Lu, J., Batra, D., Parikh, D., Lee, S.: ViLBERT: pretraining task-agnostic visiolinguistic representations for vision-and-language tasks. In: Advances in Neural Information Processing Systems, pp. 13–23 (2019)

37. Nagarajan, T., Grauman, K.: Attributes as operators: factorizing unseen attribute-object compositions. In: Proceedings of the European Conference on Computer Vision (ECCV), pp. 169–185 (2018)
38. Nam, H., Ha, J.W., Kim, J.: Dual attention networks for multimodal reasoning and matching. In: Proceedings of the IEEE Conference on Computer Vision and Pattern Recognition, pp. 299–307 (2017)
39. Nielsen, J., Molich, R., Snyder, C., Farrell, S.: E-commerce user experience. Nielsen Norman Group (2000)
40. Radford, A., et al.: Learning transferable visual models from natural language supervision. arXiv preprint arXiv:2103.00020 (2021)
41. Rostamzadeh, N., et al.: Fashion-Gen: the generative fashion dataset and challenge. arXiv preprint arXiv:1806.08317 (2018)
42. Rowley, J.: Product search in e-shopping: a review and research propositions. J. Consum. Market. (2000)
43. Shen, S., et al.: How much can clip benefit vision-and-language tasks? arXiv preprint arXiv:2107.06383 (2021)
44. Smeulders, A., Worring, M., Santini, S., Gupta, A., Jain, R.: Content-based image retrieval at the end of the early years. IEEE Trans. Pattern Anal. Mach. Intell. **22**(12), 1349–1380 (2000)
45. Socher, R., Fei-Fei, L.: Connecting modalities: semi-supervised segmentation and annotation of images using unaligned text corpora. In: 2010 IEEE Computer Society Conference on Computer Vision and Pattern Recognition, pp. 966–973. IEEE (2010)
46. Tagliabue, J., Yu, B., Beaulieu, M.: How to grow a (product) tree: personalized category suggestions for ecommerce type-ahead. arXiv preprint arXiv:2005.12781 (2020)
47. Thomee, B., et al.: YFCC100M: the new data in multimedia research. Commun. ACM **59**(2), 64–73 (2016)
48. Tsagkias, M., King, T.H., Kallumadi, S., Murdock, V., de Rijke, M.: Challenges and research opportunities in ecommerce search and recommendations. In: SIGIR Forum, vol. 54, no. 1 (2020)
49. Vo, N., et al.: Composing text and image for image retrieval-an empirical odyssey. In: Proceedings of the IEEE/CVF Conference on Computer Vision and Pattern Recognition, pp. 6439–6448 (2019)
50. Wirojwatanakul, P., Wangperawong, A.: Multi-label product categorization using multi-modal fusion models. arXiv preprint arXiv:1907.00420 (2019)
51. Yang, W., Lu, K., Yang, P., Lin, J.: Critically examining the "neural hype": weak baselines and the additivity of effectiveness gains from neural ranking models. In: Proceedings of the 42nd International ACM SIGIR Conference on Research and Development in Information Retrieval, pp. 1129–1132. Association for Computing Machinery (2019)
52. Yashima, T., Okazaki, N., Inui, K., Yamaguchi, K., Okatani, T.: Learning to describe e-commerce images from noisy online data. In: Lai, S.-H., Lepetit, V., Nishino, K., Sato, Y. (eds.) ACCV 2016. LNCS, vol. 10115, pp. 85–100. Springer, Cham (2017). https://doi.org/10.1007/978-3-319-54193-8_6
53. Young, P., Lai, A., Hodosh, M., Hockenmaier, J.: From image descriptions to visual denotations: new similarity metrics for semantic inference over event descriptions. Trans. Assoc. Comput. Linguist. **2**, 67–78 (2014)
54. Zhuge, M., et al.: Kaleido-BERT: vision-language pre-training on fashion domain. In: Proceedings of the IEEE/CVF Conference on Computer Vision and Pattern Recognition, pp. 12647–12657 (2021)

Medical Entity Linking in Laypersons' Language

Annisa Maulida Ningtyas[1,2(✉)]

[1] Technische Universität Wien, Vienna, Austria
[2] Universitas Gadjah Mada, Yogyakarta, Indonesia
annisa.ningtyas@student.tuwien.ac.at

Abstract. Due to the vast amount of health-related data on the Internet, a trend toward digital health literacy is emerging among laypersons. We hypothesize that providing trustworthy explanations of informal medical terms in social media can improve information quality. Entity linking (EL) is the task of associating terms with concepts (entities) in the knowledge base. The challenge with EL in lay medical texts is that the source texts are often written in loose and informal language. We propose an end-to-end entity linking approach that involves identifying informal medical terms, normalizing medical concepts according to SNOMED-CT, and linking entities to Wikipedia to provide explanations for laypersons.

Keywords: Medical entity linking · Medical concept normalization · Named entity recognition

1 Motivation

Social media has become a platform for users to discuss various medical issues [12,16]. This trend brings an abundance of health-related information in the form of free text. Identifying formal medical concepts from free text is valuable for medical companies, such as drug manufacturers. Formal representation can help a drug manufacturer summarize the side effects of their product. The task is known as Medical Concept Normalization (MCN), which aims to link informal medical phrases to formal medical concepts in a knowledge base (KB), such as Unified Medical Language System (UMLS) [11]. Although the MCN proved to be useful for medical organizations, such as identifying adverse drug effects [11,16], but this capability has not yet been made accessible for laypersons. According to Eurobarometer [3], over three-quarters of EU citizens believe the internet can help them learn more about health issues. This is consistent with research showing that patients who go online have a desire to improve their functional health literacy [4]. The authors emphasize the importance of maintaining a lay language or patient-centered terminological level to keep the forum engaging as a learning environment [4]. In addition, Fage-Butler and Brøgger [5] point out that

M. Hagen et al. (Eds.): ECIR 2022, LNCS 13186, pp. 513–519, 2022.
https://doi.org/10.1007/978-3-030-99739-7_63

by incorporating medical terminology in online patient-patient communication could help broaden patient navigation skills in the medical community.

We therefore hypothesize that the accessibility of the MCN to laypersons may facilitate patient learning. The learning can be established by providing an explanation or definition of the terms based on Wikipedia articles. We chose Wikipedia as a source of explanations, due to its ability to increase laypersons the readability in expert medical terms [13].

2   Background and Related Work

The process of mapping words or phrases (mentions) in a text to concepts (entities) in a knowledge base [8] is known as Entity Linking (EL). Unlike MCN, Medical EL focuses on entity mentions associated with entity types, such as *drugs and disease*. MCN, on the other hand, works with phrases that may or may not be associated with an entity type [12]. Meanwhile, MCN works with phrases that may or may not have an entity type [12]. For instance, even though the phrase *cannot shut up for the whole day* is not recognized as a medical named entity, the MCN model will map it to *Hyperactive Behavior* [12]. Currently, the MCN model only maps towards a formal KB. However, Wikipedia articles may be a better alternative to the formal medical knowledge base in terms of increasing user comprehension [13]. Nevertheless, the formal medical KB can serve as a bridge for medical EL.

The source texts for informal medical EL are often written in layperson's language, which is more casual and descriptive than medical text. For example, *"I feel a **bit drowsy** & have a **little blurred vision** , so far ..."*. The phrase *bit drowsy* refers to **drowsy or sleepy** [7]. The language gap between formal medical concepts and laypersons' language can make medical entity recognition difficult [1]. For this reason, direct dictionary matching is not ideal for detecting medical terms. Most of the current research formulates medical entity linking as a classification problem [9–11,16]. However, as stated previously, laypeople find these medical concepts impractical. Thus, medical entity linking for laypeople is still a challenge. The recent approaches on MCN models [9–11,16] rely on annotated labeled data sets, such as Psychiatric Treatment Adverse Reaction (PsyTar) [18] and CSIRO Adverse Drug Event Corpus (CADEC) [7]. The most comprehensive data set is COMETA, which covers 20.015 informal medical phrases mapped to 7.648 concepts of 350.000 concepts of SNOMED-CT [2]. Thus, the problem of unseen medical concepts still occurs [1]. Increasing the coverage of the data sets by human annotation requires a lot of time and cost. Moreover, the advancement of medical science leads to more medical concepts. The scarcity of data sets, combined with the variation of free form informal language, makes it difficult to achieve good precision, which is critical in the medical domain.

The approach to solving data set scarcity is through the utilization of data sets collected for a low resource scenario [6]. The first method is called distant supervision, which automatically generates labeled data with existing knowledge bases or dictionaries. However, the existing distant supervision approach [12] suffers from language disparity between the informal medical phrases and the formal medical concept. For example the phrase *spinning sensation* refers to *vertigo*. Due to the similarity method they used [12], the disparity can't be detected. The second approach is data augmentation [17], which attempts to artificially add more data to boost the model performance and add various representations of informal phrases. However, the impact of augmented data on Medical Entity Linking (MEL) model performance has yet to be investigated. In this research, we present an approach for automatically generating additional data and we extend the work of Seiffe et al. [15] by including a general explanation for each medical term to aid in layperson reading comprehension.

3   Proposed Research

Based on the motivation, we raise the first research question:

RQ1: How effective is medical entity linking used in medical forums or social media to increase digital health literacy among lay people?

The effectiveness in the RQ1 refers to the usefulness of the linked entities and the description provided by the entity linking model. There are various medical entities, such as drug and disease. Additionally, we aim to investigate which entities or entity types should be linked to the KB. Then, based on the data scarcity problem that we explained in the background, we raise the second research question as:

RQ2: How effective are data augmentation and distant supervision in overcoming the problem of data scarcity in MCN tasks?

The effectiveness with respect to RQ2 refers to the impact of performance on supervised MCN tasks and the increase in concept coverage of the current MCN dataset. Data augmentation will be used to increase the diversity of lay medical terms. One of the challenges in data augmentation is a change in informal medical terms or phrases may lead to a different medical concept. For example, *weight gain* as a result of *obesity* can be transformed into *burden gain*, which may be associated with *struggle*. In contrast to data augmentation, distant supervision will be used to increase the coverage of the formal medical phrases from the available data. The automatic labeling by distant supervision can produce noise. Thus, it will be a research challenge to prevent this noise.

4   Research Methodology and Proposed Experiments

The goal of this research is to develop a model and framework for medical entity linking for texts written in lay language. The overall pipeline is intended to answer RQ1, while the data augmentation and remote monitoring parts are intended to answer RQ2. Figure 1 shows the general workflow for our approach. We divided each module into subsections for further explanation.

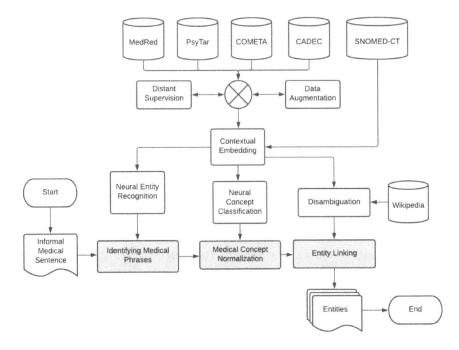

Fig. 1. The workflow of informal medical entity linking.

4.1   Data Sets Available

Various data sets are available for handling informal medical sentences, namely PsyTAR [18], COMETA [1], and CADEC [7]. The union of these data sets covers less than 10% of concepts from SNOMED-CT, with an average of 4–5 sentence representations for each concept. All of those data sets explain disease, symptoms, and drug-related issues. Most of the datasets are linked to SNOMED-CT. We use an additional data set called Medical Entities in Reddit (MedRed) [14], a medical named entity recognition corpus for identifying medical phrases.

4.2   Data Augmentation and Distant Supervision

The primary goal of these modules is to address the issue of data scarcity in MCN tasks. Data augmentation is a technique for generating additional data. We intend to add the additional data by imitating the writing style of laypeople in order to maintain the sentence's context and avoid concept shifting. We propose the following augmentation techniques: 1) Character augmentation (e.g. keyboard errors), Word augmentation (e.g. synonym replacement); 3) Paraphrase; 4) Semantic mention replacement. Based on our experiments on CADEC and PsyTar data sets, the augmentation could increase the variation of informal medical phrases, and improved the model performance on the MCN module compared to the original data sets.[1] Secondly, we intend to expand the concept coverage for distant supervision by extracting pairs of informal and formal concepts from a knowledge base. We will refer to Wikipedia articles for informal terms and SNOMED-CT for formal concepts. Informal medical terms are derived from Wikipedia components such as redirect names, wikilink, lay definitions, and aliases. Then we assign formal concepts to the informal terms by associating them with a SNOMED-CT concept.

4.3 Informal Medical Entity Linking

The first step of medical entity linking is recognizing the correct span of the medical phrases. We will use the Bidirectional LSTM-CRF sequence labeling architecture based on contextual embedding from the original and the augmented data sets. Within this step, we aim to detect the entity type, such as disease and drug.

The second step is to properly link a layperson's phrase to a medical concept. This task is called MCN, and we treat it as a multi-class classification task. The goal is to standardize informal medical phrases into formal medical concepts represented by SNOMED-CT codes as class labels.

Finally, the entity linking module aims to generate an explanation for informal medical phrases from Wikipedia articles. The most suitable explanation is chosen in three steps: 1) Take all formal medical concepts from the output of the MCN module as *candidate mentions*; 2) Retrieve the list of candidates of Wikipedia articles (called as *candidate entities*) for each formal term; 3) Candidate ranking based on commonness, semantic similarity, and link probability to select the correct Wikipedia article.

4.4 Evaluation

The evaluation will focus on RQ1 and RQ2. For RQ2, we intend to examine the effect of data augmentation and distant supervision on medical phrase identification and the MCN modules. The recent research from Basaldella et al. [1] will be the benchmark for medical entity linking. We extend the COMETA [1] and CADEC [7] data set to be linked to Wikipedia. We use this as the ground truth for our entity linking model. Additionally, to address RQ1, we intend to conduct controlled experiments to evaluate the correctness and usefulness of entity linking. The term correctness refers to the phrase's span being correctly associated with the target entity. The term usefulness refers to the span's ability to assist the user in comprehending the medical concept contained in the informal text. To determine the tool's functionality, we will conduct controlled experiments with respondents from various backgrounds. A layperson annotator will be given a set of sentences with connected entities and asked to answer a series of questions. We use Item Response Theory (IRT) to assess reading comprehension, which

---

[1] The paper was presented in Forum for Information Retrieval Evaluation (FIRE) 2021 conference.

is composed of latent properties of our entity linking model. The IRT is used to ascertain the amount of information or the degree of precision required on the reading comprehension skills test. Finally, RQ1 will be answered using the correctness and usefulness scores.

References

1. Basaldella, M., Liu, F., Shareghi, E., Collier, N.: COMETA: a corpus for medical entity linking in the social media. In: Proceedings of the 2020 Conference on EMNLP, pp. 3122–3137. ACL, November 2020
2. Donnelly, K.: Snomed-ct: the advanced terminology and coding system for ehealth. Stud. Health Technol. Inf. **121**, 279–290 (2006)
3. Eurobarometer: European citizens' digital health literacy. A report to the European Commission (2014)
4. Fage-Butler, A.M., Nisbeth Jensen, M.: Medical terminology in online patient-patient communication: evidence of high health literacy? Health Expect. **19**(3), 643–653 (2016)
5. Fage-Butler, A.M., Jensen, M.N.: The interpersonal dimension of online patient forums: How patients manage informational and relational aspects in response to posted questions. HERMES-J. Lang. Commun. Bus. **51**, 21–38 (2013)
6. Hedderich, M.A., Lange, L., Adel, H., Strötgen, J., Klakow, D.: A survey on recent approaches for natural language processing in low-resource scenarios. arXiv preprint arXiv:2010.12309 (2020)
7. Karimi, S., Metke-Jimenez, A., Kemp, M., Wang, C.: Cadec: a corpus of adverse drug event annotations. J. Biomed. Inform. **55**, 73–81 (2015)
8. Kolitsas, N., Ganea, O.E., Hofmann, T.: End-to-end neural entity linking. In: Proceedings of the 22nd Conference on CoNLL, pp. 519–529. ACL, Brussels, Belgium, October 2018
9. Limsopatham, N., Collier, N.: Adapting phrase-based machine translation to normalise medical terms in social media messages. In: Proceedings of the 2015 Conference on EMNLP, pp. 1675–1680. ACL, Lisbon, Portugal, September 2015
10. Limsopatham, N., Collier, N.: Normalising medical concepts in social media texts by learning semantic representation. In: Proceedings of the 54th Annual Meeting of the ACL, pp. 1014–1023. ACL, Berlin, Germany, August 2016
11. Miftahutdinov, Z., Tutubalina, E.: Deep neural models for medical concept normalization in user-generated texts. In: Proceedings of the 57th Annual Meeting of the ACL: Student Research Workshop, pp. 393–399. ACL, Florence, Italy, July 2019
12. Pattisapu, N., Anand, V., Patil, S., Palshikar, G., Varma, V.: Distant supervision for medical concept normalization. J. Biomed. Inform. **109**, 103522 (2020)
13. Polepalli Ramesh, B., Houston, T., Brandt, C., Fang, H., Yu, H.: Improving patients' electronic health record comprehension with noteaid. In: MEDINFO 2013, pp. 714–718. IOS Press (2013)
14. Scepanovic, S., Martin-Lopez, E., Quercia, D., Baykaner, K.: Extracting medical entities from social media. In: Proceedings of the ACM Conference on Health, Inference, and Learning, CHIL 2020, pp. 170–181. ACM, New York (2020)
15. Seiffe, L., Marten, O., Mikhailov, M., Schmeier, S., Möller, S., Roller, R.: From witch's shot to music making bones - resources for medical laymen to technical language and vice versa. In: Proceedings of the 12th LREC, pp. 6185–6192. ELRA, Marseille, France, May 2020

16. Tutubalina, E., Miftahutdinov, Z., Nikolenko, S., Malykh, V.: Medical concept normalization in social media posts with recurrent neural networks. J. Biomed. Inform. **84**, 93–102 (2018)
17. Wei, J., Zou, K.: EDA: easy data augmentation techniques for boosting performance on text classification tasks. In: Proceedings of the 2019 Conference on EMNLP and the 9th EMNLP-IJCNLP. pp. 6382–6388. ACL, Hong Kong, China, November 2019
18. Zolnoori, M., et al.: The psytar dataset: from patients generated narratives to a corpus of adverse drug events and effectiveness of psychiatric medications. Data Brief **24**, 103838 (2019)

A Topical Approach to Capturing Customer Insight Dynamics in Social Media

Miguel Palencia-Olivar[1,2]($\boxtimes$)

[1] Lizeo IT, 42 quai Rambaud, 69002 Lyon, France
miguel.palencia-olivar@lizeo-group.com
[2] Laboratoire ERIC, Université de Lyon,
5 Avenue Pierre Mendès France, 69500 Bron, France

Abstract. With the emergence of the internet, customers have become far more than mere consumers: they are now opinion makers. As such, they share their experience of goods, services, brands, and retailers. People interested in a certain product often reach for these opinions on all kinds of channels with different structures, from forums to microblogging platforms. On these platforms, topics about almost everything proliferate, and can become viral for a certain time before they begin stagnating, or extinguishing. The amount of data is massive, and the data acquisition processes frequently involve web scraping. Even if basic parsing, cleaning, and standardization exist, the variability of noise create the need for ad-hoc tools. All these elements make it difficult to extract customer insights from the internet. To address these issues, I propose to devise time-dynamic, nonparametric neural-based topic models that take topic, document and word linking into account. I also want to extract opinions accordingly with multilingual contexts, all the while making my tools relevant for pretreatment improvement. Last but not least, I want to devise a proper way of evaluating models so as to assess all their aspects.

Keywords: Topic modeling · Text mining · Social media mining · Business analytics · Natural language processing · Deep learning

1 Introduction

The age of social media has opened new opportunities for businesses. Customers are no longer the final link of a linear value chain; they have also become informants and influencers as they review goods and services, talk about their buying interests and share their opinion about brands, manufacturers and retailers. This flourishing wealth of information located outside traditional channels and frameworks of marketing research also poses many challenges. These challenges, data analysis practitioners must address when trying to test the viability of a business idea, or to capture the full picture and latest trends of consumers' opinion. In particular, social media constitute massive, heterogeneous and noisy document

© The Author(s), under exclusive license to Springer Nature Switzerland AG 2022
M. Hagen et al. (Eds.): ECIR 2022, LNCS 13186, pp. 520–527, 2022.
https://doi.org/10.1007/978-3-030-99739-7_64

sources that are often accessed through web scraping when no API is available. Additionally, customer trends tend to evolve through time, thus causing data drifts that create the need for Machine Learning models' updates. Last but not least, documents' structure is oftentimes more complex than in classical applications, as documents can exhibit some linking in the form of a graph (e.g.: Twitter) or in the form of a *nested* hierarchy (e.g.: Reddit). Linking between words and linking between topics are also important for efficient meaning extraction and better interpretation.

To come up with these challenges, I propose to devise time-dynamic, nonparametric neural-based topic models that take topic, document and word linking into account. I also want to address the issue of crosslingual topic modeling, as a single topic can have several versions in different languages. Whatever the modeling decisions, the settings must acknowledge that data is noisy in its founding assumptions, either by filtering it, or by isolating it.

2   Background and Research Questions

My research originates from industrial needs for customer insight extraction[1] from massive streams of texts regarding social media in a broad sense: technical reports, blogs, microblogs (tweets, etc.), and forum posts. Lizeo IT, the company that provides my experimental material, harvests data on a daily basis. This harvesting is mainly performed through web scraping on more than a thousand websites in 6 different languages[2]. The data acquisition pipeline includes parsing and basic data cleaning steps, but noise stills remains, e.g., markup languages, mispells, and documents in a given language that comes from a source supposedly in another language. Due to this noise and to its variability, off-the-shelf tools seldom work. Additionally, no tools are available on specific domains such as the tire industry, which is the main field Lizeo IT evolves in. In-house data dictionaries and ontologies about the tire industries do exist, but they rely on manual, expert knowledge-backed labeling. Lizeo IT's intent for this project is to be able to extract information without any kind of prior background information -objectively observable elements inherent to data set aside- so as to work with data related to other industries. The use case is purely exploratory data analysis (EDA), that is considerably hindered by data opaqueness in absence of background, by data heterogeneity, and by data noisiness.

Consequently, I pursue several goals in terms of both modeling and inference, under precise hypothesis and conditions:

RQ1: Nonparametric topic extraction As the data volumes are huge, one cannot reasonably know what the topics are, nor their number.

RQ2: Integrative extraction I need to extract customers' insights in a way that preserves document, topic and word structures and relationships while taking temporal dependencies, languages and noise into account.

---

[1] My work solely focuses on the insights per-se, not the emitters, and only includes corpus-related information.

[2] English, french, spanish, italian, german, and dutch.

RQ3: Data cleaning processes improvement I need my approaches to cope with noise directly, either by isolating it, or by filtering it.

RQ4: Scalability The algorithms have to adapt to Big Data-like settings, as the training datasets are massive and as the models need to apply quickly to newly available data.

RQ5: Annotator's bias avoidance Exposed simply, experts, on the one hand, tend to focus on product caracteristics, while on the other hand, customers tend to *focus on their experience of the product*. I want to extract customers' insights to their fullest possible extent.

To solve these problems, I propose fully unsupervised topic modeling approaches. Probabilistic settings in the vein of the Latent Dirichlet Allocation [2] appear as a go-to set of techniques for this task, as they are very flexible statistical models. However, probabilistic topic models are still unsupervised learning techniques that bear an additional -yet desirable- burden: that of *interpretable language modeling*. Analysts have two *simultaneous* expectations: from the purely statistical perspective, they need a model that efficiently extracts latent variables (the topics), that in turn are representative of -or coherent with-the dataset at hand; from the linguistic perspective, they also need *direct* interpretability of the topics. To comply with these expectations:

RQ6: Proper evaluation There is a need for a way of evaluating a topic model by simultaneously taking into account their statistical nature, whatever the modeling or inference settings, and the necessities for interpretability and coherence of the latent variables.

3   Research Methodology

As probabilistic topic models are Bayesian statistics to their very core, it is possible to use all the tools and perspectives the field offers. In particular, it permits to consider distributions as building blocks of a generative process to capture the desired aspects of the dataset at hand. I intend to use *Dirichlet Processes (DPs)* [24] *to automatically determine the number of topics (RQ1)*, and *survival analysis to model their birth, survival and death in time (RQ2)*. The usual Bayesian statistics' workhorse for inference is Markov Chain Monte Carlo (MCMC), but variational inference (VI) [5] has been gaining momentum in the last years due to its ability to scale to massive amounts of data. One of the most recent examples of modern VI is the *Variational Autoencoder framework* [16], *which I intend to use for both scalability (RQ4) and flexibility (RQ1, RQ2)*. The use of black-box variational inference [22] enables fluidifying going through iterative modeling processes such as George Box's modeling loop.

Overall, George Box's modeling loop [3] is useful, as it covers all the expectations one has about modeling tasks. Its iterative structure is simple and software development-friendly[3], and it clearly separates concerns in three distinct steps.

---

[3] As my work is both statistical and computer-science related, I wanted an exhaustive methodology that could unite both fields as much as possible, with as much emphasis on theoretical concerns as on practical concerns.

The building step corresponds to the actual modeling, when specifications are set so as to capture *the aspects of data that matter to the practitioner.* The freedom the framework allows enables to design parsimonious solutions that give control over the expected results, thus eliminating irrelevant information. This freedom, however, is a double-edged blade that can lead to a validation bias regarding prior, naive assumptions about the dataset at hand. This situation is particularly easily reached in topic modeling due to the very nature of the object of study, and words can be arranged in a way that seem meaningful without them being reflective of the actual corpus. To circumvent the issue, a Bayesian tool comes in handy: *posterior predictive checking (PPC) through realized discrepancies* [14] . *This procedure enables evaluating the whole probabilistic setting of an algorithm while focusing on the important aspects of the model, whatever the underlying probabilistic distributions, thanks to the generative properties of the models (RQ6).* Despite this ability to evaluate a Bayesian setting as a whole, PPCs per-se fail on evaluating the interpretability aspect of topic models. To include semantics -and therefore, interpretability- in the evaluation, *I suggest to regularize the training objective of a VAE with proper regularizers for semantics and coherence* [12] *, so as to twist the inference accordingly with all the objectives, then to perform a PPC (RQ6).* Following this procedure, the PPC should evaluate all the aspects, but the trick is still not enough to integrate all the elements I need to my models.

The hybrid nature of the VAEs, i.e., the fact that these models are neural network-powered probabilistic settings, *allows benefitting from the best properties and advances of both worlds, including embeddings (topics, words, graphs, etc.), transfer learning (RQ1), and recurrent neural networks (temporal aspects, hierarchies, etc.) (RQ1, RQ2),* for instance. Concerning noise, I believe that it follows Harris' distributional hypothesis, which is the exact one that underlies Mikolov et al.'s works on word embeddings [18]. According to this hypothesis, semantically similar words tend to occur in the same contexts. I think that the statement is also valid for what I call noise, e.g., that code will most likely appear with code if data parsing has failed. *I'm much more inclined to the solution of isolation instead of filtering as it is an additional tool for refining data cleaning processes (RQ3).* As for treating multilingual corpora, embeddings are useful in two ways: the first is that they allow to get distributed representations of words in a classical way[4], as for monolingual contexts; secondly, they allow for the emergence of a pivot language. *A combination between embeddings and adding a categorical distribution in the generative process could help examining crosslingual similarities (RQ2).*

Last but not least, and to circumvent any annotator bias, *I need to restrict knowledge injection to the objectively observable elements: text itself, document structure, time stamps, and the language a document is written in*[5] *(RQ5).*

---

[4] Except that, words in a given language are much more likely to appear within contexts in the same language.

[5] Language detection is out of the scope of this project, so I either rely on datasets' existing annotations or use off-the-shelf tools.

4   Related Works

Topic extraction is a well established task in the landscape of text mining [6], but the neural flavour of topic models is much more recent, as it originates from the works on black-box variational inference [16,22]. Gaussian VAEs [16] are one of the most famous realization of this line of research. To my knowledge, the *Neural Variational Document Model* (NVDM) [17] is the first VAE-based topic model. Due to its Gaussian prior, however, the NVDM is prone to posterior collapses. To circumvent the issue, Srivastava & Sutton developed the *ProdLDA* [23]. The ProdLDA tries to approximate a Dirichlet prior with a logistic normal distribution to cope with the original reparameterization trick used in the VAE. The ProdLDA makes topic modeling with variational autoencoders stabler, thus highlighting the importance of Dirichlet-like priors in the field. Other Gaussian-based developments include the *TopicRNN* [9], and the Embedded Topic Model (ETM) [10] and its time-dynamic extension [11] (by chronological order). These are particular, in the sense that both train word embeddings directly and conjointly with topic extraction. Prior to these algorithms, an extensive research aimed at linking words in (non-neural) topic models, sometimes while trying to address specific issues such as data mining on short texts [26]. Most of these works treat semantic units as auxiliary information. Other methods switch priors to include some linking between words, tweak word assignment to the topics [8,15,25], or use pre-trained embeddings [1]. Last but not least, and much later, Ning et al. [20] have proposed unsupervised settings that build on stick-breaking VAEs [19], thus automatically finding the number of topics from data.

5   Research Progress and Future Works

My first step for this project has been to adapt the VAE framework to non-parametric settings regarding the number of topics. The DPs have already been applied to topic modeling in non-neural settings. Nalisnick et al. [19] have successfully adapted an approximation of the stick-breaking take on the DPs thanks to a Kumaraswamy variational distribution instead of a Beta for the stick-breaks so as to enforce compliance of the setting with the *original reparameterization trick* [16]. This replacement, however, makes the VAE proner to posterior collapses. To solve the issue, I used Figurnov et al.'s *implicit reparameterization trick* [13] to use a Beta variational distribution, thus making my setting an exact, fully-fledged DP, as shown in [21]. I also included two kinds of embeddings, one for topics, and one for words, to capture similarities in the same embedding space. Finally, I added a Gamma prior on the concentration parameter of my single level DP [4] to learn it from data as it controls the number of topics the model extracts. The whole setting not only efficiently captures similarities and outperforms other state of the art approaches; it also tends to confirm my hypothesis on both crosslingual and noise aspects, as it isolates words by language and noise in separate topics. My approaches were tested on the 20 Newsgroups and on the Humanitarian Assistance and Disaster Relief datasets. On industrial datasets,

I also tried pre-initializing the word embedding matrix with Word2Vec to see if I could get an improvement, without much success regarding getting better results: my settings per-se are at least equivalent to Word2Vec in terms of word embeddings. In the future, I would like to try other kind of prior initializations, with a transformer's embeddings, for instance. Other kinds of future work will mainly consist at improving models' precision, by further improvement of the priors. I'll then move on to adding document linking, with a particular emphasis on nested hierarchies in documents. My last step on the project will consist in adding a categorical distribution in the generative process to capture crosslingual aspects.

I am currently working on extending my models to capture time dynamics, following the approaches of Dieng et al.'s D-ETM [11]. To achieve this, I rely on my models' capacity to use Gamma and Beta distributions to include elements of survival analysis. When its first parameter is set to 1, the Gamma distribution is equivalent to the exponential distribution, which is a common hazard function in survival analysis. Additionally, the Gamma distribution is usable as a conjugate prior for both Gamma distributions with a fixed parameter, and for a DP [4]. As a consequence, the generative process includes a chain of interdependent exponentials (one per time-slice), that will in turn act as conjugate priors for the DPs. I also included the two embedding matrices, but generalized word embeddings to a tensor to get a word distribution per time slice, so as to capture semantic evolution. Besides, I have devised a regularization term to encourage the training procedure to compute neatly separated topics, as per what I learnt on how humans interpret topic models from Chang et al.'s work [7]. I intend to use PPC to assess the whole setting.

6 Specific Research Issues for Discussion at the Doctoral Consortium

The specific research issues I would like to discuss are the following:

- **Issue 1:** The way probabilistic topic models treat corpora end up with a single topic-wise word distribution for all documents in the corpus. However, documents put varying emphasis on words. I would like to "personalize" word distributions with respect to the documents I treat by adding document-wise information, without losing the ability to generalize and predict about future documents.
- **Issue 2:** Some corpora's structures are complex, and modeling through means of a graph or of a hierarchy can help, but some sources can be both at the same time. It is particularly true for Twitter, where users can link to another tweet through retweets and start discussion threads.

References

1. Batmanghelich, K., et al.: Nonparametric spherical topic modeling with word embeddings. In: Proceedings of the 54th Annual Meeting of the Association for Computational Linguistics (Volume 2: Short Papers), pp. 537–542. Association for Computational Linguistics, Berlin, August 2016. https://doi.org/10.18653/v1/P16-2087
2. Blei, D., et al.: Latent Dirichlet allocation. J. Mach. Learn. Res. **3**, 993–1022 (2003)
3. Blei, D.M.: Build, compute, critique, repeat: data analysis with latent variable models. Ann. Rev. Stat. Appl. **1**(1), 203–232 (2014)
4. Blei, D.M., Jordan, M.I.: Variational inference for Dirichlet process mixtures. Bayesian Analysis **1**(1), March 2006. https://doi.org/10.1214/06-BA104
5. Blei, D.M., et al.: Variational inference: a review for statisticians. J. Am. Stat. Assoc. **112**(518), 859–877 (2017)
6. Boyd-Graber, J., et al.: Applications of topic models. Found. Trends Inf. Retrieval **11**, 143–296 (2017)
7. Chang, J., et al.: Reading tea leaves: how humans interpret topic models. In: Bengio, Y., Schuurmans, D., Lafferty, J., Williams, C., Culotta, A. (eds.) Advances in Neural Information Processing Systems, vol. 22. Curran Associates, Inc. (2009)
8. Das, R., et al.: Gaussian LDA for topic models with word embeddings. In: Proceedings of the 53rd Annual Meeting of the Association for Computational Linguistics and the 7th International Joint Conference on Natural Language Processing (Volume 1: Long Papers), pp. 795–804. Association for Computational Linguistics, Beijing, July 2015
9. Dieng, A.B., et al.: TopicRNN: a recurrent neural network with long-range semantic dependency. arXiv:1611.01702 [cs, stat], February 2017
10. Dieng, A.B., et al.: Topic modeling in embedding spaces. Trans. Assoc. Comput. Linguistics **8**, 439–453 (2020)
11. Dieng, A.B., Ruiz, F.J.R., Blei, D.M.: The dynamic embedded topic model. CoRR abs/1907.05545 (2019)
12. Ding, R., Nallapati, R., Xiang, B.: Coherence-aware neural topic modeling. In: Proceedings of the 2018 Conference on Empirical Methods in Natural Language Processing. pp. 830–836. Association for Computational Linguistics, Brussels, Belgium, October–November 2018
13. Figurnov, M., others: Implicit reparameterization gradients. In: Proceedings of the 32nd International Conference on Neural Information Processing Systems, NIPS 2018, pp. 439–450. Curran Associates Inc., Red Hook (2018)
14. Gelman, A., Meng, X.L., Stern, H.: Posterior predictive assessment of model fitness via realized discrepancies, p. 76
15. Hu, Y., et al.: Interactive topic modeling. Mach. Learn. **95**(3), 423–469 (2014)
16. Kingma, D.P., Welling, M.: Auto-encoding variational bayes. CoRR (2014)
17. Miao, Y., et al.: Neural variational inference for text processing. In: Proceedings of the 33rd International Conference on International Conference on Machine Learning - Volume 48, ICML 2016, pp. 1727–1736. JMLR.org (2016)
18. Mikolov, T., et al.: Efficient estimation of word representations in vector space. In: Proceedings of Workshop at ICLR 2013, January 2013
19. Nalisnick, E.T., Smyth, P.: Stick-breaking variational autoencoders. In: 5th International Conference on Learning Representations, ICLR 2017, Toulon, France, 24–26 April, 2017, Conference Track Proceedings. OpenReview.net (2017)

20. Ning, X., et al.: Nonparametric topic modeling with neural inference. Neurocomputing **399**, 296–306 (2020)
21. Palencia-Olivar, M., Bonnevay, S., Aussem, A., Canitia, B.: Neural embedded Dirichlet processes for topic modeling. In: Torra, V., Narukawa, Y. (eds.) MDAI 2021. LNCS (LNAI), vol. 12898, pp. 299–310. Springer, Cham (2021). https://doi.org/10.1007/978-3-030-85529-1_24
22. Rezende, D.J., Mohamed, S., Wierstra, D.: Stochastic backpropagation and approximate inference in deep generative models. In: Xing, E.P., Jebara, T. (eds.) Proceedings of the 31st International Conference on Machine Learning. Proceedings of Machine Learning Research, vol. 32, pp. 1278–1286. PMLR, Bejing, 22–24 June 2014
23. Srivastava, A., Sutton, C.: Autoencoding Variational Inference For Topic Models, p. 12 (2017)
24. Teh, Y.W., et al.: Hierarchical Dirichlet processes. J. Am. Stat. Assoc. **101**(476), 1566–1581 (2006)
25. Xun, G., et al.: A correlated topic model using word embeddings. In: Proceedings of the Twenty-Sixth International Joint Conference on Artificial Intelligence, IJCAI 2017, pp. 4207–4213 (2017)
26. Yan, X., et al.: A biterm topic model for short texts. In: WWW 2013 - Proceedings of the 22nd International Conference on World Wide Web, pp. 1445–1456 (2013)

Towards Explainable Search in Legal Text

Sayantan Polley[✉]

Otto von Guericke University, Magdeburg, Germany
sayantan.polley@ovgu.de

Abstract. Assume a non-AI expert user like a lawyer using an AI driven text retrieval (IR) system. A user is not always sure why a certain document is at the bottom of the ranking list although it seems quite relevant and is expected at the top. Is it due to the proportion of matching terms, semantically related topics, or unknown reasons? This can be confusing and leading to lack of trust and transparency in AI systems. Explainable AI (XAI) is currently a vibrant research topic which is being investigated from various perspectives in the IR and ML community. While a major focus of the ML community is to explain a classification decision, a key focus in IR is to explain the notion of similarity that is used to estimate relevance rankings. Relevance in IR is a complex entity based on various notions of similarity (e.g. semantic, syntactic, contextual) in text. This is often subjective and ranking is an estimation of the relevance. In this work, we attempt to explore the notion of similarity in text with regard to aspects such as semantics, law cross references and arrive at interpretable facets of evidence which can be used to explain rankings. The idea is to explain non-AI experts that why a certain document is relevant to a query, for legal domain. We present our preliminary findings, outline future work and discuss challenges.

Keywords: Explainable AI · Explainable search · XIR · XAI

1 Motivation

Consider a lawyer who is searching for documents related to European Union laws on export of commodities. The search system returns a ranked list of laws for ad-hoc keywords. Besides investigating the top two-to-three documents, the lawyer notices that one document X at a much lower rank, although it appears to be quite relevant to him/her, since it is cited by other laws. The lawyer is a bit confused by the behavior of the AI driven text search system. How is the document X at a much lower rank although it seems to be relevant for the search terms? Is it because of proportion of matching keywords? Or was it due to the topics manifested by the keywords? Was the search terms formulated correctly to capture the information need? Not understanding how the system retrieves results can often be a cause of frustration leading to lack of trust on AI systems. In this doctoral work, we attempt to investigate explainable AI or XAI [11], in rankings to support non-AI experts such as lawyers.

M. Hagen et al. (Eds.): ECIR 2022, LNCS 13186, pp. 528–536, 2022.
https://doi.org/10.1007/978-3-030-99739-7_65

AI and data driven platforms today have important consequences like swaying democratic elections, fake-news and processing of automated loan applications. This is due to the growth and usage of AI systems like smartphones, search engines, decision support systems along with availability of data and computing prowess. It is important to explain non-expert users how AI systems arrive at a result. Explainable AI (XAI) is currently a vibrant and multidisciplinary research topic which attempts to address this problem. In a classification setting, the idea is often related to identification of key features or add on methods on classification models to explain a decision. With respect to IR systems, we focus on the challenge of explaining how we arrive at the rankings or relevance list. The idea of relevance in IR is a complex entity in itself. It depends on multiple factors like context, application scenario and is often subjective based on the information need of the user. Ranking is often derived using a similarity measure to estimate relevance. In this work we focus on the notion of similarity in text in an ad-hoc query setting. We attempt to extract the interpretable facets that govern similarity in legal text. The idea is to expose such facets to non-AI experts to understand the notion of similarity used by the IR model and comprehend the rankings (Part D and E in Fig. 1, [23]).

Fig. 1. The ExDocS [23] Search Interface. Local Textual explanation, marked (D), explains the rank of a document with a simplified mathematical score (E) used for re-ranking. A query-term bar, marked (C), for each document signifies the contribution of each query term. A running column in the left marked (B) shows a gradual fading of color shade with decreasing rank. Global explanation via document comparison can done by comparing documents, marked here as (A). Showing search results for a sample query - 'wine market' on EUR-Lex [19] dataset.

2   Related Work

Early attempts for supporting users to understand search results can be found in [12,13], making use of visual elements to highlight textual features with distributions. Search result summaries [20] in text retrieval is a simple but effective attempt to explain results. Knowledge graphs on user logs have been used for generation of explanations in the content of product search [1] and recommendation systems. There is a sudden growth of XAI research in the ML community driven by development of model agnostic (e.g. LIME [27], SHAP [16]) and model specific methods (e.g. LRP [2] for CNNs) to explain a classification decision. Research in IR adapted LIME [30] and SHAP [10] values to highlight key features that contribute to ranking of documents by neural rankers like DRMM. Although these lack in explaining the rationale how the relevance decreases as a user moves from top to bottom of the ranking list, which we aim to focus in the current work. Works such as [7] provide visualization of term statistics and highlighting important passages within the documents retrieved. [28] offers a plug-in tool built upon Apache Lucene to explain the inner workings of the Vector Space Model, BM25, however these are aimed at assisting advanced users.

XAI in rankings is related to the notion of fairness, bias and ethics when an application attempts to rank sensitive subjects like people in job portals, ride sharing apps. Due care needs to be taken in data pre-processing or model processing [5], to ensure fair representation of attention. In a broader context of IR, [6] provides a categorization of strategies with respect to the idea of fairness and transparency in rankings - using data pre-processing [25], model in-processing [5] and ranking list post-processing [23,30] methods. Recent work [31] has used post-hoc interpretability of rankings to answer the question - why was the document ranked higher or lower. There is a plethora of research [3,33] on aspects that govern the notion of relevance in legal text and information extraction from case laws and prior cases [14]. Legal IR scenarios tend to have longer query text and is often driven by prior cases called precedents [29]. Recent research [32] have encoded summaries from law text to make retrieval effective. Most XAI methods in ML and IR suffer from the problem of quantitative evaluation due to a lack of ground truth explanations. Annotations by humans [9,34] have been used to generate explanations which can be used for evaluation. Explanations are often evaluated based on subjective factors such as trustworthiness, reliability [21], typically employing user-studies or counterfactual arguments. A recent work [17] employed ground truth explanation annotated by lawyers, and explored masking methods on BiGRU (Gated Recurrent Units) to detect sentences that explain the classification of case judgements. Benchmarks [9] by human annotations having 'reasoning' behind NLP models have been created to aid interpretability.

3   Approach and Preliminary Results

Approach. There are three broad approaches in the XAI community to make AI systems explainable. The first approach is to pre-process and extract an interpretable feature space and use inherently explainable models (decision trees). Secondly, use a complicated model (which often provides better performance) and post process or attribute the output (like LRP [2] for ANNs). Thirdly, infuse the fairness or XAI criteria directly in the model ranking or objective function [5]. We started with the first approach, with a focus of supporting users to understand the notion of similarity in text. The IR goal was to develop a ranker and explain the individual search results, such as "Why is X relevant for query Y".

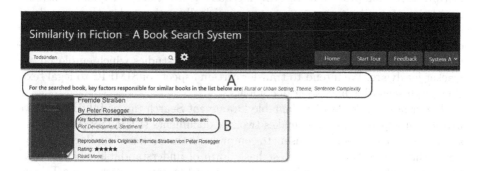

Fig. 2. SIMFIC 2.0 User Interface. Part A shows the Global Explanation. Part B shows a Local Explanation.

Our initial research [22, 25] involved creating a query-by-example book search system (available online [24]) which also explains how a certain book is similar to a query book. We named our first work as SIMFIC (*Sim*ilarity in *Fic*tion), related to retrieval of fiction books based on semantic similarity in fiction. We adapted XAI definitions from the community [21] like global and local explanations for a non-AI expert user. Global explanations (Part A in Fig. 2) attempt to use those features that 'globally contribute" to discriminate the top-K items from all other data points. Local explanations (Part B in Fig. 2) attempt to explain how each item in the retrieved result set is relevant to the query. We used domain knowledge from digital humanities [15] and extracted handcrafted features using NLP techniques from 19th Century English and German fiction books (Gutenberg corpus). For example, "writing style" is quantified using a combination of paragraph count, average sentence length, pronoun count and others [25]. We divide a book into shorter portions called chunks, feature vectors are extracted at chunks and rolled up to a book. We devise a novel ranking method that "rewards" while accumulating similarity values, only when it exceeds a threshold and "penalizes" multiple matches (long books). Consider $Ch_{Q,i}$ as i-th chunk (feature vector) of query Book Q with K chunks and $Ch_{X,j}$ as j-th chunk of a corpus Book X with M chunks. Then the similarity between Q and X is given in equation number 1.

$$sim(Book_Q, Book_X) = \frac{\sum_{i=1}^{K} \sum_{j=1}^{M} \frac{1}{1+L2(Ch_{Q,i}, Ch_{X,j})}}{K + M} \quad (1)$$

Converting bag-of-words into a compact and semantic feature space allowed us to explore research on creation of explanations. For example, we fitted linear regression [25] and classifiers [22] on the top-K results to find out the features that contribute globally. Local explanations were created in a variety of methods - visual [8] and textual [25]. For textual explanations we employed various distance metrics to compute local similarity. The choice of hyper-parameters in the feature extraction process using NLP methods, generations of explanations, and the choice of similarity measures were empirically estimated with a sample of around fifty relatively "known popular books", where we have prior knowledge on similarity based on commonly known consensus. The evaluation of explanation quality was evaluated in laboratory based user studies which employed eye tracker hardware. We intend to transfer certain aspects of SIMFIC to legal text, since both scenarios have long documents.

In our next work [23], **Ex**plainable **Doc**ument **S**earch (named 'ExDocS'), we focused on explaining - what makes two law documents (EUR-Lex [19] corpus), similar? How can we support users to comprehend the notion of similarity in legal text? Matching tokens is the most basic form of understandable evidence why two documents are similar, which is characterised by term statistics like TF and IDF. However, explaining TF-IDF scores is not very intuitive for a non-AI expert. We identified certain domain aspects that can make sense for lawyers. Often, law documents are referenced, and an important document is cited often by others. We explored usage of various known IR techniques and models that can exploit this aspect. This led to making use of simple inlink-outlink ratios, page rank and HITs score. The other aspect was detecting co-occurring words that appear in context of query words. Such words were identified by word-embeddings. We identify three major facets of evidence, term-statistics, citation-based-analysis and contextual words. As a next step, we devise a re-ranking algorithm which uses primary retrieved results of any ranker (BM25, VSM or neural ranker) and post-process them by these explainable facets to generate explanations (see Fig. 1). Note that neither TF-IDF nor a page rank score number is intuitive to an end user. Hence we encapsulate them with an interpretable facets called "factors" in the search interface. We evaluate our work in two ways. The re-ranking performance is evaluated on TREC-6 benchmarks. The explanations are evaluated in a user study, by comparing an explainable IR baseline [30] that adopted LIME used in conjunction with a neural ranker.

4   Proposed Research Questions and Methodology

Based on the preliminary results we propose the following research questions that will be investigated in this doctoral thesis:

1. **How explanations can help a lawyer in searching case priors?**
   From the general IR perspective, two major search strategies are fact-finding and exploration [18]. However in legal IR, common search scenarios are retrieval of case precedents or priors. We wish to explore - which search scenarios deserve an explanation to increase search efficiency? When explanations are not needed? To address this, we are conducting a literature survey on search scenarios that lie in the intersection of general text retrieval and legal text. We wish to explore search scenarios where an explanation makes sense and where it does not. From the law text perspective, there can be various users like lawyers, judges, notaries, law students and legal aid workers. We will restrict our focus on lawyers. The goal is to understand the search process of a lawyer and validate if explanations enhance search efficiency.

2. **How can we incorporate prior knowledge to learn feature representations that are interpretable and useful for legal IR?**
   Query terms for legal text is often longer, involving cases. This is helpful since more data often indicates an opportunity to capture the context and information need. For RQ2, we are in the process of formulating an objective function that will trade off aspects to learn feature representations [4], that make sense in legal text. One idea is to use hierarchy [4] of legal topics. Models like BERT trained over sentence to paragraph level is poised to capture the semantics [29]. The idea is to learn vectors in an unsupervised manner, for query phrases or sentences that will also be explainable, so that this evidence can be presented later to the user in the search interface.

3. **How can we explain search results individually and for a cluster of law text documents?**
   We plan to devise an additive evidence accumulation ranking scheme that will combine the various facets of evidence based on the interpretable representations. There can be two options here - model specific and model agnostic ranking approaches. For neural rankers, it will be more pragmatic to post-process the output. From the experimentation purpose, we plan to experiment on TREC-6 datasets along with legal corpus such as EUR-Lex [19] and COL-IEE [26].

4. **How can we evaluate the explanations?**
   For RQ4, we plan to create a gold standard of explanations over the next one year, starting with general news text (by students) and extending it to law by domain users. Since it is always challenging to arrange specialised users like lawyers, one idea is to leverage the recent explanations [17] annotated by lawyers on case judgements, in a classification setting. However, the challenge is the very small number of documents annotated with explanations. Other ideas involve formulating an quantitative measure based on counterfactual explanations or adapt recent benchmarks on various NLP tasks [9].

5   Research Issues for Discussion

1. Is it reasonable to explore and extend the classic relevance feedback with a focus on tuning personalized explanations. This is aligned with explanations often found in recommendation systems (e.g. users who wanted this, also wanted that)?
2. Can we develop a gold standard explanations like gold standard relevance, by extrapolating a small set of human labelled case judgements with semi-supervised learning? or using approaches such as [9,17] in case human users are difficult to arrange.
3. Can we devise a metric that will quantify the credibility of explanations? This will help to quantitatively evaluate the quality of explanations.

References

1. Ai, Q., Zhang, Y., Bi, K., Croft, W.B.: Explainable product search with a dynamic relation embedding model. ACM Trans. Inf. Syst. (TOIS) **38**(1), 1–29 (2019)
2. Bach, S., Binder, A., Montavon, G., Klauschen, F., Müller, K.R., Samek, W.: On pixel-wise explanations for non-linear classifier decisions by layer-wise relevance propagation. PLoS ONE **10**(7), e0130140 (2015)
3. Bench-Capon, T., et al.: A history of ai and law in 50 papers: 25 years of the international conference on ai and law. Artificial Intelligence and Law (2012)
4. Bengio, Y., Courville, A., Vincent, P.: Representation learning: a review and new perspectives. IEEE Trans. Pattern Anal. Mach. Intell. **35**(8), 1798–1828 (2013)
5. Biega, A.J., Gummadi, K.P., Weikum, G.: Equity of attention: Amortizing individual fairness in rankings. In: The 41st International ACM sigir Conference on Research & Development in Information Retrieval, pp. 405–414 (2018)
6. Castillo, C.: Fairness and transparency in ranking. In: ACM SIGIR Forum, vol. 52. ACM New York (2019)
7. Chios, V.: Helping results assessment by adding explainable elements to the deep relevance matching model. In: Proceedings of the 43rd International ACM SIGIR Conference on Research and Development in Information Retrieval (2020)
8. Dey, A., et al.: Evaluating reliability in explainable search. In: 2021 IEEE 2nd International Conference on Human-Machine Systems (ICHMS) (2021)
9. DeYoung, J., et al.: Eraser: a benchmark to evaluate rationalized nlp models. arXiv preprint arXiv:1911.03429 (2019)
10. Fernando, Z.T., Singh, J., Anand, A.: A study on the interpretability of neural retrieval models using DeepSHAP. In: Proceedings of the 42nd International ACM SIGIR Conference on Research and Development in Information Retrieval (2019)
11. Gunning, D., Aha, D.: Darpa's explainable artificial intelligence (xai) program. AI Mag. **40**(2), 44–58 (2019)
12. Hearst, M.A.: Visualization of term distribution information in full text information access. In: CHI '95 Conference Proceedings, pp. 59–66 (1995)
13. Hoeber, O., Brooks, M., Schroeder, D., Yang, X.D.: Thehotmap. com: enabling flexible interaction in next-generation web search interfaces. In: 2008 IEEE/WIC/ACM International Conference on Web Intelligence and Intelligent Agent Technology, vol. 1, pp. 730–734. IEEE (2008)
14. Jackson, P., Al-Kofahi, K., Tyrrell, A., Vachher, A.: Information extraction from case law and retrieval of prior cases. Artificial Intelligence (2003)

15. Jockers, M.L.: Macroanalysis: Digital Methods and Literary History, 1st edn. University of Illinois Press, USA (2013)
16. Lundberg, S.M., Lee, S.I.: A unified approach to interpreting model predictions. In: Proceedings of the 31st International Conference on Neural Information Processing Systems. NIPS (2017)
17. Malik, V., et al.: ILDC for CJPE: Indian legal documents corpus for court judgment prediction and explanation. In: Proceedings of the 59th Annual Meeting of the ACL (2021)
18. Marchionini, G.: Exploratory search: from finding to understanding. Communications of the ACM (2006)
19. Mencia, E.L., Fürnkranz, J.: Efficient multilabel classification algorithms for large-scale problems in the legal domain. In: Semantic Processing of Legal Texts, pp. 192–215. Springer (2010)
20. Mi, S., Jiang, J.: Understanding the interpretability of search result summaries. In: Proceedings of the 42nd International ACM SIGIR Conference on Research and Development in Information Retrieval, pp. 989–992 (2019)
21. Mohseni, S., Zarei, N., Ragan, E.D.: A multidisciplinary survey and framework for design and evaluation of explainable ai systems. ACM Trans. Interactive Intell. Syst. (TiiS) **11**(3–4), 1–45 (2021)
22. Polley, S., Ghosh, S., Thiel, M., Kotzyba, M., Nürnberger, A.: Simfic: an explainable book search companion. In: Proceedings of the IEEE International Conference on Human-Machine Systems (ICHMS) (2020)
23. Polley, S., Janki, A., Thiel, M., Hoebel-Mueller, J., Nuernberger, A.: Exdocs: evidence based explainable document search. In: Workshop on Causality in Search and Recommendations co-located with 44th International ACM SIGIR Conference on Research and Development in Information Retrieval (2021)
24. Polley, S., Koparde, R.R., Gowri, A.B., Perera, M., Nuernberger, A.: Simfic book search system. https://simfic-falcon.herokuapp.com/. Accessed 11 Nov 2021
25. Polley, S., Koparde, R.R., Gowri, A.B., Perera, M., Nuernberger, A.: Towards trustworthiness in the context of explainable search. In: Proceedings of the 44th International ACM SIGIR Conference on Research and Development in Information Retrieval, pp. 2580–2584 (2021)
26. Rabelo, J., Kim, M.-Y., Goebel, R., Yoshioka, M., Kano, Y., Satoh, K.: A summary of the COLIEE 2019 competition. In: Sakamoto, M., Okazaki, N., Mineshima, K., Satoh, K. (eds.) JSAI-isAI 2019. LNCS (LNAI), vol. 12331, pp. 34–49. Springer, Cham (2020). https://doi.org/10.1007/978-3-030-58790-1_3
27. Ribeiro, M.T., Singh, S., Guestrin, C.: Model-agnostic interpretability of machine learning. arXiv preprint arXiv:1606.05386 (2016)
28. Roy, D., Saha, S., Mitra, M., Sen, B., Ganguly, D.: I-REX: a lucene plugin for explainable IR. In: Proceedings of the 28th ACM International Conference on Information and Knowledge Management (2019)
29. Shao, Y.: Towards legal case retrieval. In: Proceedings of the 43rd International ACM SIGIR Conference on Research and Development in Information Retrieval (2020)
30. Singh, J., Anand, A.: Exs: explainable search using local model agnostic interpretability. In: Proceedings of the Twelfth ACM International Conference on Web Search and Data Mining, pp. 770–773 (2019)
31. Singh, J., Anand, A.: Model agnostic interpretability of rankers via intent modelling. In: Proceedings of the 2020 Conference on Fairness, Accountability, and Transparency, pp. 618–628 (2020)

32. Tran, V., Le Nguyen, M., Tojo, S., Satoh, K.: Encoded summarization: summarizing documents into continuous vector space for legal case retrieval. Artif. Intell. Law **28**(4), 441–467 (2020). https://doi.org/10.1007/s10506-020-09262-4
33. Van Opijnen, M., Santos, C.: On the concept of relevance in legal information retrieval. Kluwer Academic Publishers (2017)
34. Zhang, Z., Rudra, K., Anand, A.: Explain and predict, and then predict again. In: Proceedings of the 14th ACM International Conference on Web Search and Data Mining, pp. 418–426 (2021)

End to End Neural Retrieval for Patent Prior Art Search

Vasileios Stamatis$^{(\boxtimes)}$ (iD)

International Hellenic University, Nea Moudania, Greece
vstamatis@the.ihu.gr

Abstract. This research will examine neural retrieval methods for patent prior art search. One research direction is the federated search approach, where we proposed two new methods that solve the results merging problem in federated patent search using machine learning models. The methods are based on a centralized index containing samples of documents from all potential resources, and they implement machine learning models to predict comparable scores for the documents retrieved by different resources. The other research direction is the adaptation of end-to-end neural retrieval approaches to the patent characteristics such that the retrieval effectiveness will be increased. Off-the-self neural methods like BERT have lower effectiveness for patent prior art search. So, we adapt the BERT model to patent characteristics in order to increase retrieval performance. We propose a new gate-based document retrieval method and examine it in patent prior art search. The method combines a first-stage retrieval method using BM25 and a re-ranking approach where the BERT model is used as a gating function that operates on the BM25 score and modifies it according to the BERT relevance score. These experiments are based on two-stage retrieval approaches as neural models like BERT requires lots of computing power to be used. Eventually, the final part of the research will examine first-stage neural retrieval methods such as dense retrieval methods adapted to patent characteristics for prior art search.

Keywords: Patent search · Federated search · Machine learning · BERT · Dense retrieval

1 Introduction

This research is in the field of Information Retrieval (IR), specifically on the subfield of the professional search in the patent domain. Nowadays, the number of patents related to artificial intelligence, big data, and the internet of things has tremendously grown [2]. The increase of patent applications filed every year makes the need for better patent search systems inevitable. Patent and other innovation-related documents can be found in patent offices, online datasets, and resources that typically must be searched using various patent search systems and other online services such as espacenet, Google patents, bibliographic search, and many more [3]. From an information task perspective, patent retrieval tasks are typically recall-oriented [4]; therefore, retrieving all the patent documents related to a

---

The original version of this chapter was revised: this chapter was previously published non-open access. The correction to this chapter is available at
https://doi.org/10.1007/978-3-030-99739-7_76

patent application is crucially important otherwise, there might be a significant economic impact due to the lawsuits for patent infringement [5]. Thus, in professional search, it is vital to search effectively in all the potentially distributed resources containing patents or other patent-related data.

To that end, the Federated Search (FS) approach aims to solve the problem of effectively searching at all resources containing patent information. FS systems implement a Distributed Information Retrieval (DIR) scenario that permits the simultaneous search of multiple searchable, remote, and potentially physically distributed resources.

There are different patent search tasks with different purposes, such as prior-art search, infringement search, freedom to operate search etc. In this work, the focus will be on prior art search. Prior art search is a task where the novelty of an idea is examined [6]. Typically users use the boolean queries model to express their information need [7]. I plan to investigate methods and architectures in patent retrieval and use Artificial Intelligence (AI) end-to-end processes to improve patent search and retrieval effectiveness and propose future search engines.

2 Patent Search Characteristics

Patent search can be considered a specific example of Information Retrieval, i.e., finding relevant information of unstructured nature in huge collections [8] and has been considered a complex area. Patent text differs from regular text. Sentences used in patent documents are usually longer than general-use sentences [9]. More specifically, Iwayama in [10] found that the length of patent documents is 24 times the respective length of news documents. The syntactic structure of patent language is also a big challenge as founded by Verberne in [9]. The same study also found that patent authors tend to use multi-words to introduce novel terms. Another challenge in patent search is the vocabulary mismatch problem, i.e., the non-existence of common words between two relevant documents. Magdy et al. [11] showed that 12% of relevant documents for topics from the CLEF-IP 2009 have no words in common with the respective topics. All these make patent search a complicated process.

Researchers have categorized methodologies for patent search and retrieval. Lupu & Hanbury [12] summarized methods for patent retrieval, divided into text-based methodologies (Bag of Words, Latent Semantic Analysis, Natural Language Processing), Query Creation/Modification methodologies, Metadata-based methodologies, and Drawing-based methodologies. Khode & Jambhorkar [13] split the procedures for patent retrieval into IPC based and those based on patent features and query formulation. More recently, Shalaby et al. in [14] broke patent retrieval into the following categories. Keyword-based methods, Pseudo Relevance Feedback Methods, Semantic-based methods, Metadata based methods, Interactive methods.

In the last years, there has been a shift in research to neural approaches for IR. Neural approaches for IR are a new and developing field [15]. Transformer models like BERT [16] have achieved impressive results on various NLP tasks. The use of a BERT model for patent retrieval has not been investigated enough though. While BERT has drawn lots of attention in research in the patent industry, it is either used for classification [2, 17] or didn't work as expected for patent retrieval [18]. Dense retrieval [19] is a new neural method for search and given the particular characteristics of the patent industry, it is expected to solve problems like vocabulary mismatch and improve retrieval effectiveness. Generally, the use of AI techniques in the patent industry has drawn lots of attention and is currently an active area of research [7, 20, 21].

3 Research Questions

The research questions I will address in my PhD are the following.

1. Federated Search

   - What is machine learning algorithms? effectiveness on result merging when searching for patents in federated environments?

2. To what extent does an end-to-end neural retrieval approach that is adapted to patent characteristics improve retrieval effectiveness?

   - Why do the off-the-shelf methods have lower effectiveness for patent search?
   - -How can BERT be adapted to improve retrieval effectiveness in patent search?
   - How can first-stage retrieval or dense retrieval be adapted to improve the retrieval effectiveness in patent search?

4 Summary of My Research so Far

1a) What is machine learning algorithms? effect on result merging when searching for patents in federated environments?
The result merging problem was studied as a general DIR problem and not in the specific context of the patent domain. The result merging problem appeared in research many years ago. One of the first works that conducted experiments in results merging is [22]. After that many algorithms were presented in the relevant literature.

A very widely used and very robust estimation method is the collection inference retrieval network CORI [23]. CORI uses a linear combination of the score of the document returned by the collection and the source selection score and applies a simple heuristic formula. It finally normalizes the collection-specific scores and produces global comparable scores.

One more effective estimation algorithm is the semi-supervised learning algorithm (SSL) [24] which is based on linear regression. The SSL algorithm proposed by Si and Callan applies linear regression to assign the local collection scores to the global comparable scores. To achieve that, the algorithm functions on the common documents returned every time, between a collection and a centralized index created by samples from all the collections.

SAFE (sample-agglomerate fitting estimate) is a more recent algorithm designed to function on uncooperative environments [25]. SAFE is based on the principle that the results of the sampled documents for each query are a sub-ranking of the original collection, so this sub-ranking can be used to conduct curve fitting in order to predict the original scores.

In download methods, the results are downloaded locally to calculate their relevance. Hung in [32] proposed a technique in which the best documents are downloaded to re-rank and create the final merged list. He used machine learning and genetic programming to re-rank the final merged results. Whilst download methods seem to perform better than estimation approaches in the context tested by in [33], they have essential disadvantages such as increased computation, download time, and bandwidth overhead during the retrieval process.

Hybrid methods are combinations of estimation and download methods. Paltoglou et al. [34] proposed a hybrid method that combines download and estimation methods. More specifically it downloads a limited number of documents, and based on them, it trains a linear regression model for calculating the relevance of the rest documents. The results showed that this method achieved a good balance between the time and performance for the download and estimation approaches respectively.

Taylor et al. [35] published a patent about a machine learning process for conducting results merging. Another patent was published by [36] which uses the scores assigned to the lists and the documents to complete the final merging.

I started my research journey working on the first research question by implementing an idea that solves the results merging process in federated search scenarios. The initial idea of my work was published at the PCI 2020 conference [37]. This work proposes two new methods that solve the results merging problem in federated patent search using machine learning models. The methods are based on a centralized index containing samples of documents from all potential resources, and they implement machine learning models to predict comparable scores for the documents retrieved by different resources. The effectiveness of the new results merging methods was measured against very robust models and was found to be superior to them in many cases.

2a) Why do the off-the-shelf methods have lower effectiveness for patent search?
As some initial results show, the BERT model does not perform well in patent search as an off-the-shelf method and this is also consistent with the literature [18].

Patent documents have specific characteristics and differences compared with regular text, where BERT model approaches have achieved impressive results [16] [38]. Thus, patent search is different than other types of searches such as web search. For example, in a typical patent prior art search, the starting point is a patent application as a topic [39] that needs to be transformed to search queries [39, 40]. BERT can only take an input of up to 512 tokens, so the whole extensive patent documents cannot be used for direct feed to the model. Also, the diversity of the language, as well as the usual use of vague terms, makes the need for huge amounts of data for training inevitably important in order to effectively train BERT.

Another notable characteristic of patent documents is their structural information. A patent document is a summary of different fields describing the invention. These are title, abstract, description, claims, metadata, and figures. There are also language differences between them. For example, in abstract and description, it is usually used technical language while in the claims section legal jargon is used [40]. We need to choose which parts will be used to train a BERT model and for what task. As already mentioned, each part has its characteristics, and we need to look deep into them and decide how to adapt BERT model to them.

Another big challenge is the lack of data for training the BERT model for patent retrieval. Deep learning models, in general, require lots of data. BERT as well requires big datasets to take advantage of its power [16]. For example, CLEF-IP is a popular dataset used in patent retrieval research which is an extract of the more extensive MAREC, but its structure does not offer use for training models like BERT.

2b) How can BERT be adapted to improve retrieval effectiveness in patent search?
The re-ranker architecture where the final results list in document ranking comes from an initial classical retrieval model followed by a neural re-ranker is the state-of-the-art search process [41]. Transformer models like BERT [16] have achieved impressive results on various NLP tasks. While BERT has drawn lots of attention in research in the patent industry, it is either used for classification [2, 17] or didn't work as expected for patent retrieval [18]. This recommends that it needs more research when working with patents as patent language has many specific features, as already mentioned.

Lee & Hsiang [42] implemented a re-ranking approach for patent prior art search using a BM25 model for the first retrieval and then a re-ranker using the cosine similarity and BERT embeddings. As they only used the BERT embeddings, they train BERT using the plain text file architecture which has one sentence per line, and all the examples are positive. Their re-ranking effectiveness was satisfactory, but they found that calculating semantic similarities between longer texts is still challenging.

Althammer et al. [18] trained a BERT model using patent documents, and they used the BERT paragraph-level interaction architecture [43] and compared the retrieval performance with BM25. They found BM25 to perform better than BERT.

Dai & Callan [44] found that BERT-based re-rankers performed better on longer queries than short keyword queries. Therefore, as patent retrieval involves long queries, it makes sense to train a BERT re-ranker for the patent domain. Padaki et al. [45] worked on query expansion for BERT re-ranking. They found that queries need to have a rich set of concepts and grammar structures to take advantage of BERT-base re-rankers. The traditional word-based query expansion that results in short queries is not sufficient, and they found that BERT achieved higher accuracy when using longer queries.

Beltagy et al. [46] presented longformer, a BERT-like model designed to work with long documents. It combines a local attention mechanism in combination with a global one allowing the processing of longer documents. Longformer can take as input documents up to 4096 tokens long, eight times more than the BERT's maximum input.

Kang et al. [2] worked on prior art search performance by solving the binary classification problem of classifying patent documents as noisy and not relevant in order to be removed from the search and find valid patents using BERT model.

Lee & Hsiang [17] worked on patent classification using the BERT model. They fine-tuned a BERT model and used it for CPC classification. They also showed that using only the claims is sufficient for patent classification.

We use a BERT re-ranker along with a BM25 model for the first-stage retrieval. BERT model is used a gate-based function that modifies BM25 score according to BERT's relevance score. The main challenge is the lack of appropriate data for training such a model. Also, BERT can take a maximum input of 512 tokens. We use only the abstract, as the abstract is mandatory for every patent document and is a good description of the invention. The first step is to create a dataset of the relevant abstracts. We used the MAREC dataset [47], and from each patent document, we will found its citations and used them to create a dataset of relevant abstracts. This result in 80 million pair of abstracts 50% of which are positive and 50% are negative. We then trained the BERT model using this data and compare with BM25 and found the method to be superior to BM25.

2c) How can first-stage retrieval or dense retrieval be adapted to improve the retrieval effectiveness in patent search?

The two-stage retrieval process is an important step in order to solve patent retrieval-related problems analyzed in the previous section. However, transformer models like BERT are expensive as they require enough computing power to function. After solving the previous research question, we plan to move to the first stage retrieval and examine ways and architectures to include them on patent prior art search. Dense retrieval [19] is a new approach to search and has not been investigated enough for patent search. We plan to train dense retrieval models, adapt them for patent documents and investigate their performance comparing with state-of-the-art models and architectures. Also, term importance prediction is another approach for first stage retrieval, which could also be combined with dense retrieval. DeepCT model [44] for example use BERT model's term representations and predicts weights for terms in a sentences, which will used for bag of words retrieval afterward. We plan to train a DeepCt model, adapt it to patent documents and combine it with dense retrieval. The combination of two independent architectures for first-stage retrieval has not been investigated before. We also plan to examine further how the structure of patent documents can affect the performance of these models and explore ways to include this information in the retrieval process.

Acknowledgments. This work has received funding from the European Union's Horizon 2020 research and innovation programme under the Marie Sk'odowska-Curie grant agreement No: 860721 (DoSSIER Project, https://dossier-project.eu/).

References

1. Kang, D.M., Lee, C.C., Lee, S., Lee, W.: Patent prior art search using deep learning language model. In: 24th International Database Application & Enginnerring Symosium (IDEAS 2020), ACM, New York, NY, USA, pp. 5 (2020)
2. Salampasis, M.: Federated Patent Search. In: Lupu, M., Mayer, K., Kando, N., Trippe, A.J. (eds.) Current Challenges in Patent Information Retrieval, pp. 213–240. Springer, Berlin, Heidelberg (2017). https://doi.org/10.1007/978-3-662-53817-3_8
3. Mahdabi, P., Gerani, S., Huang, J.X., Crestani, F.: Leveraging conceptual lexicon: query disambiguation using proximity information for patent retrieval. In SIGIR (2013)
4. Khode, A., Jambhorkar, S.: Effect of technical domains and patent structure on patent information retrieval. Int. J. Eng. Adv. Technol. **9**(1), 6067–6074 (2019)
5. Clarke, N.: The basics of patent searching. World Patent Inf. **54**, S4–S10 (2018)
6. Setchi, R., Spasic, I., Morgan, J., Harrison, C., Corken, R.: Artificial intelligence for patent prior art searching. World Patent Inf. **64**, 102021 (2021)
7. Manning, C.D., Raghavan, P., Schutze, H.: Introduction to Information Retrieval. Cambridge University Press, Cambridge (2008)

8. Verberne, S., D'hondt, E., Oostdijk, N., Koster, C.H.: Quantifying the challenges in parsing patent claims. In: 1st International Workshop on Advances in Patent Information Retrieval (2010)
9. Iwayama, M., Fujii, A., Kando, N., Marukawa, Y.: An empirical study on retrieval models for different document genres: patents and newspaper articles. In: SIGIR '03: Proceedings of the 26th annual international ACM SIGIR conference on Research and development in Informaion Retrieval (2003)
10. Magdy, W., Leveling, J., Jones, G.J.F.: Exploring structured documents and query formulation techniques for patent retrieval. In: Peters, C., et al. (eds.) Multilingual Information Access Evaluation I. Text Retrieval Experiments. CLEF 2009. Lecture Notes in Computer Science, vol. 6241. Springer, Berlin, Heidelberg (2009).https://doi.org/10.1007/978-3-642-15754-7
11. Lupu, M., Hanbury, A.: Patent retrieval. Found. Trends Inf. Retreival 7(1), 1–97 (2013)
12. Khode, A., Jambhorkar, S.: A literature review on patent information retrieval techniques. Indian J. Sci. Technol. 10(37), 1–13 (2017)
13. Shalaby, W., Zadrozny, W.: Patent retrieval: a literature review. Knowl. Inf. Syst. 61(2), 631–660 (2019). https://doi.org/10.1007/s10115-018-1322-7
14. Mitra, B., Craswell, N.: An introduction to neural information retrieval. Found. Trends Inf. Retr. 13(1), 1–126 (2018)
15. Devlin, J., Chang, M.-W., Lee, K., Toutanova, K.: BERT: Pre-training of Deep Bidirectional Transformers for Language Understanding, in arXiv:1810.04805v2 (2019)
16. Lee, J.-S., Jieh, H.: Patent classification by fine-tuning BERT language model. World Patent Inf. 61, 101965 (2020)
17. Althammer, S., Hofstatter, S., Hanbury, A.: Cross-domain retrieval in the legal and patent domains: a reproducability study. In: ECIR (2021)
18. Karpukhin, V., et al.: Dense passage retrieval for open-domain question answering. In: Empirical Methods in Natural Language Processing (EMNLP) (2020)
19. Alderucci, D., Sicker, D.: Applying artificial intelligence to the patent system. Technol. Innov. 20, 415–425 (2019)
20. Aristodemou, L., Tietze, F.: The state-of-the-art on Intellectual Property Analytics (IPA): a literature review on artificial intelligence, machine learning and deep learning methods for analysing intellectual property (IP) data. World Patent Inf. 55, 37–51 (2018)
21. Voorhees, E.M., Gupta, N.K., Laird, B.J.: The collection fusion problem. In: Harman, D.K. (ed.) The third text retrieval conference(TREC-3), National Institute of Standards and Technology (1994).https://doi.org/10.1057/9780230379411_4
22. Callan, J., Lu, Z., Croft, B.: Searching distributed collections with inference networks. In: Proceedings of the 18th Annual International ACM SIGIR Conference on Research and Development in Information Retrieval – SIGIR 95 (1995)
23. Si, L., Callan, J.: A semisupervised learning method to merge search engine results. ACM Trans. Inf. Syst. 21(4), 457–491 (2003)
24. Shokouhi, M., Zobel, J.: Robust result merging using sample-based score estimates. ACM Trans. Inform. Syst. 27(3), 1–29 (2009)
25. Hung, V.T.: New re-ranking approach in merging search results. Informatica 43, 2 (2019)
26. Craswell, N., Hawking, D., Thistlewaite, P.: Merging results from isolated search engines. In: Australasian Database Conference (1999)
27. Paltoglou, G., Salampasis, M., Satratzemi, M.: A Results merging algorithm for distributed information retrieval environments that combines regression methodologies with a selective download phase. Inf. Process. Manage. 44, 1580–1599 (2008)
28. Taylor, M., Radlinski, F., Shokouhi, M.: Merging search results. US Patent US 9,495.460 B2, 15 November (2016)
29. Mao, J., Mukherjee, R., Raghavan, P., Tsaparas, P.: Method and apparatus for merging. US Patent US 6,728,704 B2, 27 April (2004)

30. Stamatis, V., Salampasis, M.: Results merging in the patent domain. In: PCI, Athens (2020)
31. Nogueira, R., Cho, K.: Passage re-ranking with BERT. In arXiv:1901.04085v5 (2020)
32. Mahdabi, P., Keikha, M., Gerani, S., Landoni, M., Crestani, F.: Building queries for prior-art search. In: Hanbury, A., Rauber, A., de Vries, A.P. (eds.) IRFC 2011. LNCS, vol. 6653, pp. 3–15. Springer, Heidelberg (2011). https://doi.org/10.1007/978-3-642-21353-3_2
33. Xue, X., Croft, W.B.: Automatic query generation for patent search. In: CIKM (2009)
34. Sekulic, I., Soleimani, A., Aliannejadi M., Crestani, F.: Longformer for MS MARCO Document Re-ranking Task. In: arXiv:2009.09392 (2020)
35. Lee, J.-S., Hsiang, J.: Prior art search and reranking for generated patent text. In: PatentSemTech Workshop, SIGIR21 (2021)
36. Shao, Y., et al: BERT-PLI: modeling paragraph-level interactions for legal case retrieval. In: Twenty-Ninth International Joint Conference on Artificial Intelligence (2020)
37. Dai, Z., Callan, J.: Context-Aware Sentence/Passage Term Importance Estimation For First Stage Retrieval. In arXiv:1910.10687v2 (2019)
38. Padaki, R., Dai, Z., Callan, J.: Rethinking query expansion for BERT Reranking. In: Jose, J.M., et al. (eds.) ECIR 2020. LNCS, vol. 12036, pp. 297–304. Springer, Cham (2020). https://doi.org/10.1007/978-3-030-45442-5_37
39. Beltagy, I., Peters, M.E., Cohan, A.: Longformer: The Long-Document Transformer. In: arXiv:2004.05150v2 (2020)
40. MAREC data set? [Online]. http://www.ifs.tuwien.ac.at/imp/marec.shtml

Workshops

Third International Workshop on Algorithmic Bias in Search and Recommendation (BIAS@ECIR2022)

Ludovico Boratto[1], Stefano Faralli[2], Mirko Marras[1]([✉]),
and Giovanni Stilo[3]

[1] University of Cagliari, Cagliari, Italy
{ludovico.boratto,mirko.marras}@acm.org
[2] Sapienza University of Rome, Rome, Italy
stefano.faralli@uniroma1.it
[3] University of L'Aquila, L'Aquila, Italy
giovanni.stilo@univaq.it

Abstract. Creating search and recommendation algorithms that are efficient and effective has been the main goal for the industry and the academia for years. However, recent research has shown that these algorithms lead to models, trained on historical data, that might exacerbate existing biases and generate potentially negative outcomes. Defining, assessing and mitigating these biases throughout experimental pipelines is hence a core step for devising search and recommendation algorithms that can be responsibly deployed in real-world applications. The Bias 2022 workshop aims to collect novel contributions in this field and offer a common ground for interested researchers and practitioners. The workshop website is available at https://biasinrecsys.github.io/ecir2022/.

Keywords: Bias · Algorithms · Search · Recommendation · Fairness

1 Introduction

Ranking algorithms facilitate our interaction with Web content on a daily basis, e.g., via *search* and *recommendation* systems. They can achieve high effectiveness thanks to the patterns extracted from our historical data, which allow these systems to learn who we are in terms of our preferences. These learnt patterns can however easily embed *biases*. If the models we train capture the biases that exist in the learned patterns, they are likely to exacerbate them. This can lead to possible *disparities* in the outputs generated by these systems. When a system generates a form of disparity that is associated to a sensitive attribute of the users (such as their race, gender, or religion), the consequences go beyond the output and touch on *legal and societal perspectives*, by causing discrimination and unfairness [6,12]. Other forms of disparity can generate biases that affect the success of certain categories of items, such those that are unpopular [4].

M. Hagen et al. (Eds.): ECIR 2022, LNCS 13186, pp. 547–551, 2022.
https://doi.org/10.1007/978-3-030-99739-7_67

The different types of bias, that can be captured by the models and then conveyed in the results, imply that different challenges emerge when trying to avoid them. This includes avoiding that popular items are over-recommended [11,13], enabling fairness for both consumers and item providers [3,8–10], and generating explanations of the conditions triggering a biased output [5,7]. For this reason, countermeasures can go in several directions, from *metrics* to assess a given phenomenon, via a *characterization* of a given form of bias, to the *mitigation* of the existing disparities, possibly without affecting the model effectiveness.

The growing adoption of search and recommendation technologies in the real world and the rapidly-changing techniques driving search and recommendation are constantly requiring novel and evolving definitions, techniques, and applications that timely address contexts, challenges, and issues that are emerging. Having another *dedicated event* allows the European IR community to stay updated and continue fostering a core contribution to this field. The Bias 2022 workshop is therefore aimed at collecting advances in this emerging field and providing a common ground for interested researchers and practitioners. This workshop represents a follow up to our ECIR 2020 and ECIR 2021 editions. The two past workshop editions saw more than 35 submitted papers each, an acceptance rate of around 40%, well-renowned keynote speakers, and a participation of around 70 attendants. Both workshops resulted in their proceedings published as volumes into the Springer's CCIS series in 2020 [1] and 2021 [2]. The workshop and the related initiatives are being supported by the ACM Conference on Fairness, Accountability, and Transparency (ACM FAccT) Network.

2 Workshop Vision and Topics

Our vision is that responsibly deploying search and recommendation algorithms in the real world requires methods and applications that put people first, inspect social and ethical impacts, and uplift the public's trust on the resulting technologies. The goal is hence to favor a community-wide dialogue on new research perspectives in this field through a workshop having the following objectives:

1. Providing means for the IR community to keep the pace with the recent advances in algorithmic bias in search and recommendation systems.
2. Assess the social and human dimensions that can be impacted by modern IR systems, with a focus on search and recommendation.
3. Provide a forum where novel advances in algorithmic bias in IR focusing on search and recommendation can be presented and discussed;
4. Identify the current open issues, considering the recent advances at the state of the art in biases for search and recommendation systems;
5. Allow the IR community to become familiar with modern practices of dealing with biases in search and recommendation systems;
6. Bridge academic research and the real-world, to provide insights and possibly disclose gaps in this emerging field.

To promote this vision and objectives, this workshop collects new contributions on emerging aspects in this research area. Submitted papers, written in English, fall into one of the following categories: full papers, reproducibility papers, short paper, and position papers. These papers are accompanied by presentation talks, whose slides and video recordings are also disseminated. Between sessions and in the final session of the workshop, the workshop includes a vivid discussion between the participants of the workshop, the presenters of the papers, and the keynote speakers, from which depicting the current state of the art and future research lines. The community of researchers working on algorithmic bias in IR that can consequently foster ideas and sparks for the current challenges, in view of possible collaboration on future projects and initiatives.

Current perspectives on this vision are accepted in the form of workshop contributions, that cover an extended list of topics related to algorithmic bias and fairness in search and recommendation, focused (but not limited) to:

- *Data Set Collection and Preparation*:
  - Studying the interplay between bias and imbalanced data or rare classes;
  - Designing methods for dealing with imbalances and inequalities in data.
  - Creating collection pipelines that lead to fair and less unbiased data sets;
  - Collecting data sets useful for the analysis of biased and unfair situations;
  - Designing collection protocols for data sets tailored to research on bias.
- *Countermeasure Design and Development*:
  - Formalizing and operationalizing bias and fairness concepts;
  - Conducting exploratory analysis that uncover novel types of bias;
  - Designing treatments that mitigate biases in pre-/in-/post-processing;
  - Devising methods for explaining bias in search and recommendation;
  - Studying causal and counterfactual reasoning for bias and fairness.
- *Evaluation Protocol and Metric Formulation*:
  - Performing auditing studies with respect to bias and fairness;
  - Conducting quantitative experimental studies on bias and unfairness;
  - Defining objective metrics that consider fairness and/or bias;
  - Formulating bias-aware protocols to evaluate existing algorithms;
  - Evaluating existing mitigation strategies in unexplored domains;
  - Comparative studies of existing evaluation protocols and strategies;
  - Analysing efficiency and scalability issues of debiasing methods.
- *Case Study Exploration*:
  - E-commerce platforms;
  - Educational environments;
  - Entertainment websites;
  - Healthcare systems;
  - Social media;
  - News platforms;
  - Digital libraries;
  - Job portals.
  - Dating platforms.

3   Workshop Organizers

Ludovico Boratto is an Assistant Professor at the Department of Mathematics and Computer Science of the University of Cagliari (Italy). His research interests focus on recommender systems and their impact on stakeholders, both considering accuracy and beyond-accuracy evaluation metrics. He has a wide experience in workshop organizations, with 10+ events organized at ECIR, IEEE ICDM, ECML-PKDD, and ACM EICS[1] and has given tutorials on bias in recommender systems at UMAP and ICDM 2020, and WSDM, ICDE, and ECIR 2021.

Stefano Faralli is an Assistant Professor at Sapienza University of Rome (Italy). His research interests include Ontology Learning, Distributional Semantics, Word Sense Disambiguation/Induction, Recommender Systems, Linked Open Data. He co-organized the International Workshop: Taxonomy Extraction Evaluation (TexEval) Task 17 of Semantic Evaluation (SemEval-2015), the International Workshop on Social Interaction-based Recommendation (SIR 2018), and the ECIR 2020 and 2021 BIAS workshops.

Mirko Marras is an Assistant Professor at the Department of Mathematics and Computer Science of the University of Cagliari (Italy). His research interests focus on responsible machine learning, with a special applicative focus on education and biometrics. He has taken a leading role when chairing the first editions of the ECIR BIAS workshop (2020 and 2021) and has also experience in organizing workshops held in conjunction with other top-tier venues, such as WSDM and ICCV[2]. He is currently giving tutorials on bias in recommender systems at UMAP and ICDM 2020, and WSDM, ICDE, and ECIR 2021.

Giovanni Stilo is an Associate Professor at the Department of Information Engineering, Computer Science and Mathematics of the University of L'Aquila (Italy). His research interests focus on machine learning and data mining, and specifically temporal mining, social network analysis, network medicine, semantics-aware recommender systems, and anomaly detection. He has organized several international workshops in conjunction with top-tier conferences (ICDM, CIKM, and ECIR), with the ECIR 2020 and 2021 BIAS workshops being two of them.

References

1. Boratto, L., Faralli, S., Marras, M., Stilo, G. (eds.): BIAS 2020. CCIS, vol. 1245. Springer, Cham (2020). https://doi.org/10.1007/978-3-030-52485-2
2. Boratto, L., Faralli, S., Marras, M., Stilo, G. (eds.): BIAS 2021. CCIS, vol. 1418. Springer, Cham (2021). https://doi.org/10.1007/978-3-030-78818-6
3. Boratto, L., Fenu, G., Marras, M.: Interplay between upsampling and regularization for provider fairness in recommender systems. User Model. User-Adapt. Interact **31**(3), 421–455 (2021). https://doi.org/10.1007/s11257-021-09294-8

---

[1] https://www.ludovicoboratto.com/activities/.
[2] https://www.mirkomarras.com/.

4. Chen, J., Dong, H., Wang, X., Feng, F., Wang, M., He, X.: Bias and debias in recommender system: A survey and future directions. CoRR abs/2010.03240 (2020). https://arxiv.org/abs/2010.03240

5. Deldjoo, Y., Bellogín, A., Noia, T.D.: Explaining recommender systems fairness and accuracy through the lens of data characteristics. Inf. Process. Manag. **58**(5), 102662 (2021)

6. Ekstrand, M.D., Das, A., Burke, R., Diaz, F.: Fairness and discrimination in information access systems. CoRR abs/2105.05779 (2021). https://arxiv.org/abs/2105.05779

7. Fu, Z., et al.: Fairness-aware explainable recommendation over knowledge graphs. In: SIGIR 2020: Proceedings of the 43rd International ACM SIGIR Conference on Research and Development in Information Retrieval, Virtual Event, China, 25–30 July 2020, pp. 69–78. ACM (2020)

8. Ge, Y., et al.: Towards long-term fairness in recommendation. In: WSDM 2021, The Fourteenth ACM International Conference on Web Search and Data Mining, Virtual Event, Israel, 8–12 March 2021, pp. 445–453. ACM (2021)

9. Gómez, E., Zhang, C.S., Boratto, L., Salamó, M., Marras, M.: The winner takes it all: geographic imbalance and provider (un)fairness in educational recommender systems. In: SIGIR 2021: The 44th International ACM SIGIR Conference on Research and Development in Information Retrieval, Virtual Event, Canada, 11–15 July 2021, pp. 1808–1812. ACM (2021)

10. Marras, M., Boratto, L., Ramos, G., Fenu, G.: Equality of learning opportunity via individual fairness in personalized recommendations. Int. J. Artif. Intell. Educ. 1–49 (2021)

11. Wei, T., Feng, F., Chen, J., Wu, Z., Yi, J., He, X.: Model-agnostic counterfactual reasoning for eliminating popularity bias in recommender system. In: KDD 2021: The 27th ACM SIGKDD Conference on Knowledge Discovery and Data Mining, Virtual Event, Singapore, 14–18 August 2021, pp. 1791–1800. ACM (2021)

12. Zehlike, M., Yang, K., Stoyanovich, J.: Fairness in ranking: A survey. CoRR abs/2103.14000 (2021). https://arxiv.org/abs/2103.14000

13. Zhu, Z., He, Y., Zhao, X., Zhang, Y., Wang, J., Caverlee, J.: Popularity-opportunity bias in collaborative filtering. In: WSDM 2021, The Fourteenth ACM International Conference on Web Search and Data Mining, Virtual Event, Israel, 8–12 March 2021, pp. 85–93. ACM (2021)

The 5th International Workshop on Narrative Extraction from Texts: Text2Story 2022

Ricardo Campos[1,2](✉) ⓘ, Alípio Jorge[1,3] ⓘ, Adam Jatowt[4] ⓘ, Sumit Bhatia[5],
and Marina Litvak[6] ⓘ

[1] LIAAD – INESCTEC, Porto, Portugal
[2] Ci2 - Smart Cities Research Center - Polytechnic Institute of Tomar, Tomar, Portugal
ricardo.campos@ipt.pt
[3] FCUP, University of Porto, Porto, Portugal
amjorge@fc.up.pt
[4] University of Innsbruck, Innsbruck, Austria
adam.jatowt@uibk.ac.at
[5] MDSR Lab, Adobe Systems, Noida, India
Sumit.Bhatia@adobe.com
[6] Shamoon College of Engineering, Ashdod, Israel
marinal@sce.ac.il

Abstract. Narrative extraction, understanding, verification, and visualization are currently popular topics for users interested in achieving a deeper understanding of text, researchers who want to develop accurate methods for text mining, and commercial companies that strive to provide efficient tools for that. Information Retrieval (IR), Natural Language Processing (NLP), Machine Learning (ML) and Computational Linguistics (CL) already offer many instruments that aid the exploration of narrative elements in text and within unstructured data. Despite evident advances in the last couple of years, the problem of automatically representing narratives in a structured form and interpreting them, beyond the conventional identification of common events, entities and their relationships, is yet to be solved. This workshop held virtually on April 10th, 2022 in conjunction with the 44th European Conference on Information Retrieval (ECIR'22) aims at presenting and discussing current and future directions for IR, NLP, ML and other computational linguistics-related fields capable of improving the automatic understanding of narratives. It includes sessions devoted to research, demo, position papers, work-in-progress, project description, nectar, and negative results papers, keynote talks and space for an informal discussion of the methods, of the challenges and of the future of this research area.

1 Motivation

Narratives have long been studied in the computational field as a sequence or chain of events (happening) communicated by words (oral and written) and/or visually (through images, videos or other forms of representations). Over the years several methods borrowed from different computational areas, including Information Retrieval (IR), Natural Language Processing (NLP), Computational Linguistics (CL), and Machine Learning

M. Hagen et al. (Eds.): ECIR 2022, LNCS 13186, pp. 552–556, 2022.
https://doi.org/10.1007/978-3-030-99739-7_68

(ML) have been applied as a means to better understand the constituents of a narrative, their actors, events, entities, facts and their relationships on time and space. Industries such as finance [2, 8, 9, 25], business [11], news outlets [17], and health care [19] have been the main beneficiaries of the investment in this kind of technology. The ultimate goal is to offer users tools allowing to quickly understand the information conveyed in economic and financial reports, patient records, verify the extracted information, and to offer them more appealing and alternative formats of exploring common narratives through interactive visualizations [10]. Timelines [21] and infographics for instance, can be employed to represent in a more compact way automatically identified narrative chains in a cloud of news articles [16] or keywords [4], assisting human readers in grasping complex stories with different key turning points and networks of characters. Also, the automatic generation of text [24] shows impressive results towards computational creativity. However, it still needs to develop means for controlling the narrative intent of the output and a profound understanding of the applied methods by humans (explainable AI). There are various open problems and challenges in this area, such as hallucination in generated text [18], bias in text [12], transparency and explainability of the generation techniques [1], reliability of the extracted facts [22, 23], and efficient and objective evaluation of generated narratives [3, 7].

The Text2Story workshop, now in its fifth edition, aims to provide a common forum to consolidate the multi-disciplinary efforts and foster discussions to identify the wide-ranging issues related to the narrative extraction task. In the first four editions [5, 6, 13, 14], we had an approximate number of 220 participants in total, 80 of which took part in the last edition of the workshop. This adds to the fact that a Special Issue on IPM Journal [15] devoted to this matter has also been proposed in the past demonstrating the growing activity of this research area. In this year's edition, we welcomed contributions from interested researchers on all aspects related to narrative understanding, including the extraction and formal representation of events, their temporal aspects and intrinsic relationships, verification of extracted facts, evaluation of the generated texts, and more. In addition to this, we seek contributions related to alternative means of presenting the information and on the formal aspects of evaluation, including the proposal of new datasets. Special attention will be given to multilingual approaches and resources. Finally, we challenge the interested researchers to consider submitting a paper that makes use of the tls-covid19 dataset [20], which consists of a number of curated topics related to the Covid-19 outbreak, with associated news articles from Portuguese and English news outlets and their respective reference timelines as gold-standard. While it was designed to support timeline summarization research tasks, it can also be used for other tasks including the study of news coverage about the COVID-19 pandemic. A list of all the topics can be found on the Text2Story 2022 webpage [https://text2story22.inesctec.pt/].

2 Scientific Objectives

The workshop has the following main objectives: (1) raise awareness within the IR community to the problem of narrative extraction and understanding; (2) shorten the gap between academic research, practitioners and industrial application; (3) obtain insight on new methods, recent advances and challenges, as well on future directions; (4) share experiences of research projects, case studies and scientific outcomes, (5) identify dimensions potentially affected by the automatization of the narrative process, (6) highlight tested hypotheses that did not result in the expected outcomes. Our topics are organized around the basic research questions related to narrative generation, which are as follows: How to efficiently extract/generate reliable and accurate narrative from a large multi-genre and multi-lingual data? How to annotate data and evaluate new approaches? Is a new approach transparent, explainable and easily reproducible? Is it adjustable to new tasks, genres and languages without much effort required?

3 Organizing Team

Ricardo Campos is an assistant professor at the Polytechnic Institute of Tomar. He is an integrated researcher of LIAAD-INESC TEC, the Artificial Intelligence and Decision Support Lab of U. Porto, and a collaborator of Ci2.ipt, the Smart Cities Research Center of the Polytechnic of Tomar. He is PhD in Computer Science by the University of Porto (U. Porto). He has over ten years of research experience in IR and NLP. He is an editorial board member of the IPM Journal (Elsevier), co-chaired international conferences and workshops, being also a PC member of several international conferences. More in http://www.ccc.ipt.pt/~ricardo.

Alípio M. Jorge works in the areas of data mining, ML, recommender systems and NLP. He is a PhD in Comp. Science (CS) by the University of Porto (UP). He is an Associate Professor of the dep. of CS of the UP since 2009 and is the head of that dep. since 2017. He is the coordinator of the research lab LIAAD-INESC TEC. He has projects in NLP, web automation, recommender systems, IR, text mining and decision support for the management of public transport. He represents Portugal in the Working Group on Artificial Intelligence at the European Commission and was the coordinator for the Portuguese Strategy on Artificial Intelligence "AI Portugal 2030".

Adam Jatowt is Full Professor at the University of Innsbruck. He has received his Ph.D. in Information Science and Technology from the University of Tokyo, Japan in 2005. His research interests lie in an area of IR, knowledge extraction from text and in digital history. Adam has been serving as a PC co-chair of IPRES2011, SocInfo2013, ICADL2014, JCDL2017 and ICADL2019 conferences and a general chair of ICADL2020, TPDL2019 and a tutorial co-chair of SIGIR2017. He was also a co-organizer of 3 NTCIR evaluation tasks and co-organizer of over 20 international workshops at WWW, CIKM, ACL, ECIR, IJCAI, IUI, SOCINFO, TPDL and DH conferences.

Sumit Bhatia is a Senior Machine Learning Scientist at Media and Data Science Research Lab, Adobe Systems, India. He received his Ph.D. from the Pennsylvania State University in 2013. His doctoral research focused on enabling easier information access in online discussion forums followed by a post-doc at Xerox Research Labs on event detection and customer feedback monitoring in social media. With primary research interests in the fields of Knowledge Management, IR and Text Analytics, Sumit is a co-inventor of more than a dozen patents. He has served on program committees of multiple conferences and journals including WWW, CIKM, ACL, EMNLP, NAACL, TKDE, TOIS, WebDB, JASIST, IJCAI, and AAAI.

Marina Litvak is a Senior Lecturer at the department of Software Engineering, Shamoon College of Engineering (SCE), Israel. Marina received her PhD degree from Information Sciences dept. at Ben Gurion University at the Negev, Israel in 2010. Marina's research focuses mainly on Multilingual Text Analysis, Social Networks, Knowledge Extraction from Text, and Summarization. Marina published over 70 academic papers, including journal and top-level conference publications. She constantly serves on program committees and editorial boards in multiple journals and conferences. Marina is a co-organizer of the MultiLing (2011, 2013, 2015, 2017, and 2019) and the FNP (2020, 2021, and 2022) workshop series.

Acknowledgements. Ricardo Campos and Alípio Jorge were financed by the ERDF – European Regional Development Fund through the North Portugal Regional Operational Programme (NORTE 2020), under the PORTUGAL 2020 and by National Funds through the Portuguese funding agency, FCT - Fundação para a Ciência e a Tecnologia within project PTDC/CCI-COM/31857/2017 (NORTE-01-0145-FEDER-03185). This funding fits under the research line of the Text2Story project.

References

1. Alonso, J. M., et al: Interactive natural language technology for explainable artificial intelligence. In: TAILOR, pp. 63–70 (2020)
2. Athanasakou, V., et al.: Proceedings of the 1st Joint Workshop on Financial Narrative Processing and MultiLing Financial Summarisation (FNP-FNS'20) co-located to Coling '20, Barcelona, Spain (Online), pp. 1–245 (2020)
3. Ayed, A.B., Biskri, I., Meunier, J.G.: An efficient explainable artificial intelligence model of automatically generated summaries evaluation: a use case of bridging cognitive psychology and computational linguistics. In: Explainable AI Within the Digital Transformation and Cyber Physical Systems, pp. 69–90. Springer, Cham (2021). https://doi.org/10.1007/978-3-030-76409-8_5
4. Campos, R., Mangaravite, V., Pasquali, A., Jorge, A.M., Nunes, C., Jatowt, A.: A text feature based automatic keyword extraction method for single documents. In: Pasi, G., Piwowarski, B., Azzopardi, L., Hanbury, A. (eds.) ECIR 2018. LNCS, vol. 10772, pp. 684–691. Springer, Cham (2018). https://doi.org/10.1007/978-3-319-76941-7_63
5. Campos, R., Jorge, A., Jatowt, A., Bhatia, S., Finlayson, M.: The 4th International Workshop on Narrative Extraction from Texts: Text2Story 2021. In: European Conference on Information Retrieval, pp. 701–704. Springer, Cham (2021). https://doi.org/10.1007/978-3-030-72240-1_84
6. Campos, R., Jorge, A., Jatowt, A., Sumit, B.: Third International Workshop on Narrative Extraction from Texts (Text2Story'20). In: Jose, J., et al., (eds.) ECIR 2020, LNCS, vol. 12036, pp. 648–653 (2020)
7. Celikyilmaz, A., Clark, E., Gao, J. Evaluation of text generation: a survey. arXiv preprint arXiv:2006.14799 (2020)
8. El-Haj, M., Litvak, M., Pittaras, N., Giannakopoulos, G.: The Financial Narrative Summarisation Shared Task (FNS 2020). In: Proceedings of the 1st Joint Workshop on Financial Narrative Processing and MultiLing Financial Summarisation, pp. 1–12 (2020)
9. El-Haj, M., Rayson, P., Zmandar, N.: Proceedings of the 3rd Financial Narrative Processing Workshop (2021)

10. Figueiras, A.: How to tell stories using visualization: strategies towards Narrative Visualization. Ph.D. Dissertation. Universidade Nova de Lisboa, Lisboa, Portugal (2016)
11. Grobelny, J., Smierzchalska, J., Krzysztof, K.: Narrative Gamification as a method of increasing sales performance: a field experimental study. Int. J. Acad. Res. Bus. Soc. Sci. **8**(3), 430–447 (2018)
12. Guo, W., Caliskan, A.: Detecting emergent intersectional biases: Contextualized word embeddings contain a distribution of human-like biases. In: Proceedings of the 2021 AAAI/ACM Conference on AI, Ethics, and Society, pp. 122–133 (2021)
13. Jorge, A., Campos, R., Jatowt, A., Bhatia, S.: Second International Workshop on Narrative Extraction from Texts (Text2Story'19). In: Azzopardi, L., Stein, B., Fuhr, N., Mayr, P., Hau, C., Hiemstra, D., (eds.) ECIR 2019, LNCS, vol. 11438, pp. 389–393 (2019)
14. Jorge, A., Campos, R., Jatowt, A., Nunes, S.: First International Workshop on Narrative Extraction from Texts (Text2Story'18). In: Pasi, G., Piwowarski, B., Azzopardi, L., Hanbury, A., (eds.) ECIR 2018, LNCS, vol. 10772, pp. 833–834 (2018)
15. Jorge, A., Campos, R., Jatowt, A., Nunes, S.: Special issue on narrative extraction from texts (Text2Story): preface. IPM J. **56**(5), 1771–1774 (2019)
16. Liu, S., et al.: TIARA: interactive, topic-based visual text summarization and analysis. ACM Trans. Intell. Syst. Technol. **3**(2), 28 (2012)
17. Martinez-Alvarez, M., et al.: First International Workshop on Recent Trends in News Information Retrieval (NewsIR'16). In: Nicola, F., et al. (eds.) ECIR 2016, LNCS, vol. 9626, pp. 878–882 (2016)
18. Maynez, J., Narayan, S., Bohnet, B., McDonald, R.: On faithfulness and factuality in abstractive summarization. In: Proceedings of the 58th Annual Meeting of the Association for Computational Linguistics, pp. 1906–1919 (2020)
19. Özlem, U., Amber, S., Weiyi, S.: Chronology of your health events: approaches to extracting temporal relations from medical narratives. Biomed. Inf. **46**, 1–4 (2013)
20. Pasquali, A., Campos, R., Ribeiro, A., Santana, B., Jorge, A., Jatowt, A.: TLS-Covid19: a new annotated corpus for timeline summarization. In: Hiemstra, D., Moens, M.-F., Mothe, J., Perego, R., Potthast, M., Sebastiani, F. (eds.) ECIR 2021. LNCS, vol. 12656, pp. 497–512. Springer, Cham (2021). https://doi.org/10.1007/978-3-030-72113-8_33
21. Pasquali, A., Mangaravite, V., Campos, R., Jorge, A.M., Jatowt, A.: Interactive system for automatically generating temporal narratives. In: Azzopardi, L., Stein, B., Fuhr, N., Mayr, P., Hauff, C., Hiemstra, D. (eds.) ECIR 2019. LNCS, vol. 11438, pp. 251–255. Springer, Cham (2019). https://doi.org/10.1007/978-3-030-15719-7_34
22. Saakyan, A., Chakrabarty, T., Muresan, S.: COVID-Fact: Fact Extraction and Verification of Real-World Claims on COVID-19 Pandemic. arXiv preprint arXiv:2106.03794 (2021)
23. Vo, N., Lee, K.: Learning from fact-checkers: analysis and generation of fact-checking language. In: Proceedings of the 42nd International ACM SIGIR Conference on Research and Development in Information Retrieval, pp. 335–344 (2019)
24. Wu, Y.: Is automated journalistic writing less biased? An experimental test of auto-written and human-written news stories. J. Pract. **14**(7), 1–21 (2019)
25. Zmandar, N., El-Haj, M., Rayson, P., Litvak, M., Giannakopoulos, G., Pittaras, N. The financial narrative summarisation shared task FNS 2021. In: Proceedings of the 3rd Financial Narrative Processing Workshop, pp. 120–125 (2021)

Augmented Intelligence in Technology-Assisted Review Systems (ALTARS 2022): Evaluation Metrics and Protocols for eDiscovery and Systematic Review Systems

Giorgio Maria Di Nunzio[1](✉) ⓘ, Evangelos Kanoulas[2] ⓘ,
and Prasenjit Majumder[3]

[1] Department of Information Engineering, University of Padova, Padova, Italy
giorgiomaria.dinunzio@unipd.it
[2] Faculty of Science, Informatics Institute, University of Amsterdam,
Amsterdam, The Netherlands
E.Kanoulas@uva.nl
[3] DAIICT, Gandhinagar, India
prasenjit_t@isical.ac.in

Abstract. In this workshop, we aim to fathom the effectiveness of Technology-Assisted Review Systems from different viewpoints. In fact, despite the number of evaluation measures at our disposal to assess the effectiveness of a "traditional" retrieval approach, there are additional dimensions of evaluation for these systems. For example, it is true that an effective high-recall system should be able to find the majority of relevant documents using the least number of assessments. However, this kind of evaluation usually discards the resources used to achieve this goal, such as the total time spent on those assessments, or the amount of money spent for the experts judging the documents.

Keywords: Technology-assisted review systems · Augmented intelligence · Evaluation · Systematic reviews · eDiscovery

1 Motivations

Augmented Intelligence is "a subsection of AI machine learning developed to enhance human intelligence rather than operate independently of or outright replace it. It is designed to do so by improving human decision-making and, by extension, actions taken in response to improved decisions."[1] In this sense, users are supported, not replaced, in the decision-making process by the filtering capabilities of the Augmented Intelligence solutions, but the final decision will always be taken by the users who are still accountable for their actions.

---

[1] https://digitalreality.ieee.org/publications/what-is-augmented-intelligence.

M. Hagen et al. (Eds.): ECIR 2022, LNCS 13186, pp. 557–560, 2022.
https://doi.org/10.1007/978-3-030-99739-7_69

Given these premises, we focus on High-recall Information Retrieval (IR) systems which tackle challenging tasks that require the finding of (nearly) all the relevant documents in a collection. Electronic discovery (eDiscovery) and systematic review systems are probably the most important examples of such systems where the search for relevant information with limited resources, such as time and money, is necessary.

In this field, Technology-assisted review (TAR) systems use a kind of human-in-the-loop approach where classification and/or ranking algorithms are continuously trained according to the relevance feedback from expert reviewers, until a substantial number of the relevant documents are identified. This approach, named Continuous Active Learning (CAL), has been shown to be more effective and more efficient than traditional e-discovery and systematic review practices, which typically consists of a mix of keyword search and manual review of the search results.

In order to achieve high recall values, machine-learning methods need large numbers of human relevance assessments which represent the primary cost of such methods. It is therefore necessary to evaluate these systems not only in terms of "batch"/off-line performances, but also in terms of the time spent per assessment, the hourly pay rate for assessors, and the quality of the assessor. For example, by reducing the amount of work by using sentence-level assessments in place of document-level assessments to reduce the time to read the document and the number of judgments needed. In addition, it would be also necessary to include in the validation of the system the feedback of the users by asking direct questions about the information carried in the missing documents instead of just asking about their relevance [2,5–7].

In the context of High Recall Information Retrieval Systems, we believe that it is necessary to compare 1) the vetting approach that use evaluation collections to optimize systems and carry out pre-hoc evaluation, 2) the validation of the system to measure the actual outcome of the system in real situations.

2 Topics of Interest

In this workshop, we aim to fathom the effectiveness of these systems which is a research challenge itself. In fact, despite the number of evaluation measures at our disposal to assess the effectiveness of a "traditional" retrieval approach, there are additional dimensions of evaluation for TAR systems. For example, it is true that an effective high-recall system should be able to find the majority of relevant documents using the least number of assessments. However, this type of evaluation discards the resources used to achieve this goal, such as the total time spent on those assessments, or the amount of money spent for the experts judging the documents.

The topics of the workshop are:

- Novel evaluation approaches and measures for e-Discovery;
- Novel evaluation approaches and measures for Systematic reviews;
- Reproducibility of experiments with test collections;

- Design and evaluation of interactive high-recall retrieval systems;
- Study of evaluation measures;
- User studies in high-recall retrieval systems;
- Novel evaluation protocols for Continuous Active Learning;
- Evaluation of sampling bias.

3 Organizing Team

Giorgio Maria Di Nunzio is Associate Professor at the Department of Information Engineering of the Universiryt of Padova. He has been the co-organizer of the ongoing Covid-19 Multilingual Information Access Evaluation forum,[2] in particular for the evaluation of high-recall systems and high-precision systems tasks. He will bring to this workshop the perspective of alternative (to the standard) evaluation measures and multilingual challenges.

Evangelos Kanoulas is Full Professor at the Faculty of Science of the Informatics Institute at the University of Amsterdam. He has been the co-organizer CLEF eHealth Lab and of the Technologically Assisted Reviews in Empirical Medicine task.[3] He will bring to the workshop the perspective of the evaluation of the costs in eHealth TAR systems, in particular of the early stopping strategies.

Prasenjit Majumder is Associate Professor at the Dhirubhai Ambani Institute of Information and Communication Technology (DA-IICT), Gandhinagar and TCG CREST, Kolkata, India. He has been the co-organizer of the Forum for Information Retrieval Evaluation and, in particular, the Artificial Intelligence for Legal Assistance (AILA) task.[4] He will bring to the workshop the perspective of the evaluation of the costs of eDiscovery, in particular of the issues related to legal precedence findings.

All the three organizing committee members have been active participants in the past editions of the TREC, CLEF and FIRE evaluation forum for the Total Recall and Precision Medicine TREC Tasks, TAR in eHealth tasks, and AI for Legal Assistance.[5,6,7] The committee members have strong research record with a total of more than 400 papers in international journals and conferences. They and have been doing research in technology assisted review systems and problems related to document distillation both in the eHealth and eDiscovery domain and made significant contributions in this specific research area [1, 3, 4].

---

[2] http://eval.covid19-mlia.eu.
[3] https://clefehealth.imag.fr.
[4] https://sites.google.com/view/aila-2021.
[5] https://scholar.google.it/citations?user=Awl_HDoAAAAJ.
[6] https://scholar.google.com/citations?user=0HybxV4AAAAJ.
[7] https://scholar.google.co.in/citations?user=3xIpiKEAAAAJ.

References

1. Di Nunzio, G.M.: A study on a stopping strategy for systematic reviews based on a distributed effort approach. In: Arampatzis, A., et al. (eds.) CLEF 2020. LNCS, vol. 12260, pp. 112–123. Springer, Cham (2020). https://doi.org/10.1007/978-3-030-58219-7_10

2. Lewis, D.D., Yang, E., Frieder, O.: Certifying one-phase technology-assisted reviews. In: CIKM 2021: The 30th ACM International Conference on Information and Knowledge Management, Virtual Event, Queensland, Australia, 1–5 November 2021, pp. 893–902 (2021). https://doi.org/10.1145/3459637.3482415

3. Li, D., Kanoulas, E.: When to stop reviewing in technology-assisted reviews: Sampling from an adaptive distribution to estimate residual relevant documents. ACM Trans. Inf. Syst. **38**(4), 41:1–41:36 (2020). https://doi.org/10.1145/3411755

4. Mehta, P., Mandl, T., Majumder, P., Gangopadhyay, S.: Report on the FIRE 2020 evaluation initiative. SIGIR Forum **55**(1), 3:1–3:11 (2021). https://doi.org/10.1145/3476415.3476418

5. Pickens, J., III, T.C.G.: On the effectiveness of portable models versus human expertise under continuous active learning. In: Joint Proceedings of the Workshops on Automated Semantic Analysis of Information in Legal Text (ASAIL 2021) & AI and Intelligent Assistance for Legal Professionals in the Digital Workplace (LegalAIIA 2021) held online in conjunction with 18th International Conference on Artificial Intelligence and Law (ICAIL 2021), São Paolo, Brazil (held online), 21 & 25 June 2021,pp. 69–76 (2021). http://ceur-ws.org/Vol-2888/paper10.pdf

6. Sneyd, A., Stevenson, M.: Stopping criteria for technology assisted reviews based on counting processes. In: SIGIR 2021: The 44th International ACM SIGIR Conference on Research and Development in Information Retrieval, Virtual Event, Canada, 11–15 July 2021, pp. 2293–2297 (2021). https://doi.org/10.1145/3404835.3463013

7. Zhang, H., Cormack, G.V., Grossman, M.R., Smucker, M.D.: Evaluating sentence-level relevance feedback for high-recall information retrieval. Inf. Retr. J. **23**(1), 1–26 (2020)

Bibliometric-enhanced Information Retrieval: 12th International BIR Workshop (BIR 2022)

Ingo Frommholz[1]([✉]), Philipp Mayr[2], Guillaume Cabanac[3], and Suzan Verberne[4]

[1] School of Mathematics and Computer Science, University of Wolverhampton, Wolverhampton, UK
ifrommholz@acm.org
[2] GESIS – Leibniz-Institute for the Social Sciences, Cologne, Germany
philipp.mayr@gesis.org
[3] Computer Science Department, IRIT UMR, University of Toulouse, 5505 Toulouse, France
guillaume.cabanac@univ-tlse3.fr
[4] Leiden University, LIACS, Leiden, The Netherlands
s.verberne@liacs.leidenuniv.nl

Abstract. The 12th iteration of the Bibliometric-enhanced Information Retrieval (BIR) workshop series is a full-day ECIR 2022 workshop. BIR tackles issues related to, for instance, academic search and recommendation, at the intersection of Information Retrieval, Natural Language Processing, and Bibliometrics. As an interdisciplinary scientific event, BIR brings together researchers and practitioners from the Scientometrics/Bibliometrics community on the one hand and the Information Retrieval community on the other hand. BIR is an ever-growing topic investigated by both academia and the industry.

Keywords: Academic search · Information retrieval · Digital libraries · Bibliometrics · Scientometrics

1 Motivation and Relevance to ECIR

The aim of the BIR workshop series is to bring together researchers and practitioners from Scientometrics/Bibliometrics as well as Information Retrieval (IR). Scientometrics is a sub-field of Bibliometrics which, like IR, is, in turn, a sub-field of Information Science. Bibliometrics and Scientometrics are concerned with all quantitative aspects of information and academic literature [6], which naturally make them interesting for IR research, in particular when it comes to academic search and recommendation. In the 1960s, Salton was already striving to enhance IR by including clues inferred from bibliographic citations [7]. In the course of decades, both disciplines (Bibliometrics and IR) evolved apart from each other over time, leading to the two loosely connected fields we know of today [8].

© The Author(s), under exclusive license to Springer Nature Switzerland AG 2022
M. Hagen et al. (Eds.): ECIR 2022, LNCS 13186, pp. 561–565, 2022.
https://doi.org/10.1007/978-3-030-99739-7_70

However, the exploding number of scholarly publications and the need to satisfy scholars' specific information needs led Bibliometric-enhanced IR to receive growing recognition in the IR as well as the Scientometrics communities. Challenges in academic search and recommendation became particularly apparent during the COVID-19 crisis and the need for effective and efficient solutions for scholarly search for and discovery of high-quality publications on that topic. Bibliometric-enhanced IR tries to provide these solutions to the peculiar needs of scholars to keep on top of the research in their respective fields, utilising the wide range of suitable relevance signals that come with academic scientific publications, such as keywords provided by authors, topics extracted from the full-texts, co-authorship networks, citation networks, bibliometric figures, and various classification schemes of science. Bibliometric-enhanced IR systems must deal with the multifaceted nature of scientific information by searching for or recommending academic papers, patents, venues (i.e., conference proceedings, journals, books, manuals, grey literature), authors, experts (e.g., peer reviewers), references (to be cited to support an argument), and datasets.

To this end, the BIR workshop series was founded in 2014 [5] to tackle these challenges by tightening up the link between IR and Bibliometrics. We strive to bring the 'retrievalists' and 'citationists' [8] active in both academia and industry together. The success of past BIR events, as shown in Table 1, evidences that BIR@ECIR is a much needed interdisciplinary scientific event that attracts researchers and practitioners from IR and Bibliometrics alike.

Table 1. Overview of the BIR workshop series and CEUR proceedings

Year	Conference	Venue	Country	Papers	Proceedings
2014	ECIR	Amsterdam	NL	6	Vol-1143
2015	ECIR	Vienna	AT	6	Vol-1344
2016	ECIR	Padua	IT	8	Vol-1567
2016	JCDL	Newark	US	$10 + 10^a$	Vol-1610
2017	ECIR	Aberdeen	UK	12	Vol-1823
2017	SIGIR	Tokyo	JP	11	Vol-1888
2018	ECIR	Grenoble	FR	9	Vol-2080
2019	ECIR	Cologne	DE	14	Vol-2345
2019	SIGIR	Paris	FR	$16 + 10^b$	Vol-2414
2020	ECIR	Lisbon (Online)	PT	9	Vol-2591
2021	ECIR	Lucca (Online)	IT	9	Vol-2847

[a] with CL-SciSumm 2016 Shared Task; [b] with CL-SciSumm 2019 Shared Task

2   Workshop Goals/Objectives

Our vision is to bring together researchers and practitioners from Scientometrics/Bibliometrics on the one hand and IR on the other hand to create better methods and systems for instance for academic search and recommendation. Our view is to expose people from one community to the work of the respective other community and to foster fruitful interaction across communities. Therefore, in the call for papers for the 2022 BIR workshop at ECIR, we will address, but are not limited to, current research issues regarding 3 aspects of the academic search/recommendation process:

1. User needs and behaviour regarding scientific information, such as:
   - Finding relevant papers/authors for a literature review.
   - Filtering high-quality research papers, e.g. in preprint servers.
   - Measuring the degree of plagiarism in a paper.
   - Identifying expert reviewers for a given submission.
   - Flagging predatory conferences and journals.
   - Understanding information-seeking behaviour and HCI in academic search.
2. Mining the scientific literature, such as:
   - Information extraction, text mining and parsing of scholarly literature.
   - Natural language processing (e.g., citation contexts).
   - Discourse modelling and argument mining.
3. Academic search/recommendation systems, such as:
   - Modelling the multifaceted nature of scientific information.
   - Building test collections for reproducible BIR.
   - System support for literature search and recommendation.

3   Target Audience and Dissemination

The target audience of the BIR workshops is researchers and practitioners, junior and senior, from Scientometrics as well as IR and Natural Language Processing (NLP). These could be IR/NLP researchers interested in potential new application areas for their work as well as researchers and practitioners working with bibliometric data and interested in how IR/NLP methods can make use of such data. BIR 2022 will be open for anyone interested in the topic.

The 10th-anniversary edition in 2020 ran online with an audience peaking at 97 online participants [1]. BIR 2021, the 11th edition [3,4], attracted around 57 participants at peak times but a larger number throughout due to participants dropping in and out.

In December 2020, we published our third special issue emerging from the past BIR workshops [2]. More special issues based on BIR workshops are planned.

As a follow-up of the workshop and following the tradition of previous years, the co-chairs will write a report summing up the main themes and discussions to *SIGIR Forum* [4, for instance] and BCS Informer, as a way to advertise our research topics as widely as possible among the IR community. As in the past, we plan to publish our accepted papers open-access in a CEUR Workshop Proceedings volume.

4   Peer Review Process and Workshop Format

Our peer review process will be supported by Easychair. Each submission is assigned to 2 to 3 reviewers, preferably at least one expert in IR and one expert in Bibliometrics or NLP. The programme committee for 2022 will consist of peer reviewers from all participating communities. Accepted papers are either long papers (15-min talks) or short papers (5-min talks). Two interactive sessions close the morning and afternoon sessions with posters and demos, allowing attendees to discuss the latest developments in the field and opportunities (e.g., shared tasks such as CL-SciSumm). We also invite attendees to demonstrate prototypes during flash presentations (5 min).

These interactive sessions serve as ice-breakers, sparking interesting discussions that, in non-pandemic times, usually continue during lunch and the evening social event. The sessions are also an opportunity for our speakers to further discuss their work. BIR has a friendly and open atmosphere where there is an opportunity for participants (including students) to share their ideas and current work and to receive feedback from the community.

5   Organisers and Programme Commitee

- Ingo Frommholz is Reader in Data Science at the University of Wolverhampton, UK.
- Philipp Mayr is a team leader at the GESIS – Leibniz-Institute for the Social Sciences department Knowledge Technologies for the Social Sciences, Germany.
- Guillaume Cabanac is an Associate Professor at the University of Toulouse, France.
- Suzan Verberne is an Associate professor at the Leiden Institute of Advanced Computer Science (LIACS) Assocoate Professor at Leiden University and group leader of Text Mining and Retrieval Leiden.

The list of PC members will be available on the BIR 2022 page.

References

1. Cabanac, G., Frommholz, I., Mayr, P.: Report on the 10th anniversary workshop on bibliometric-enhanced information retrieval (BIR 2020). SIGIR Forum **54**(1) (2020). https://doi.org/10.1145/3451964.3451974
2. Cabanac, G., Frommholz, I., Mayr, P.: Scholarly literature mining with information retrieval and natural language processing: preface. Scientometrics **125**(3), 2835–2840 (2020). https://doi.org/10.1007/s11192-020-03763-4
3. Cabanac, G., Frommholz, I., Mayr, P., Verberne, S. (eds.): Proceedings of the 11th International Workshop on Bibliometric-enhanced Information Retrieval (BIR). No. 2847 in CEUR Workshop Proceedings, Aachen (2021). http://ceur-ws.org/Vol-2847/

4. Frommholz, I., Cabanac, G., Mayr, P., Verberne, S.: Report on the 11th Bibliometric-enhanced Information Retrieval Workshop (BIR 2021). SIGIR Forum **55**(1) (2021). https://doi.org/10.1145/3476415.3476426
5. Mayr, P., Scharnhorst, A., Larsen, B., Schaer, P., Mutschke, P.: Bibliometric-enhanced information retrieval. In: de Rijke, M., et al. (eds.) ECIR 2014. LNCS, vol. 8416, pp. 798–801. Springer, Cham (2014). https://doi.org/10.1007/978-3-319-06028-6_99
6. Pritchard, A.: Statistical bibliography or bibliometrics? [Documentation notes]. J. Doc. **25**(4), 348–349 (1969). https://doi.org/10.1108/eb026482
7. Salton, G.: Associative document retrieval techniques using bibliographic information. J. ACM **10**(4), 440–457 (1963). https://doi.org/10.1145/321186.321188
8. White, H.D., McCain, K.W.: Visualizing a discipline: an author co-citation analysis of information science, 1972–1995. J. Am. Soc. Inf. Sci. **49**(4), 327–355 (1998)

ROMCIR 2022: Overview of the 2nd Workshop on Reducing Online Misinformation Through Credible Information Retrieval

Marinella Petrocchi[1] and Marco Viviani[2(✉)]

[1] National Research Council (CNR) Institute of Informatics and Telematics (IIT),
Pisa, Italy
marinella.petrocchi@iit.cnr.it
[2] Department of Informatics, Systems, and Communication (DISCo),
University of Milano-Bicocca, Milan, Italy
marco.viviani@unimib.it

Abstract. The ROMCIR 2022 workshop is focused on discussing and addressing issues related to information disorder, a new term that holistically encompasses all forms of communication pollution. In particular, the aim of ROMCIR is reducing such clutter, from false content to incorrect correlations, from misinformation to disinformation, through Information Retrieval solutions, by providing users with access to genuine information. This topic is very broad, as it concerns different contents (e.g., Web pages, news, reviews, medical information, online accounts, etc.), different Web and social media platforms (e.g., microblogging platforms, social networking services, social question-answering systems, etc.), and different purposes (e.g., identifying false information, accessing and retrieving information based on its genuineness, providing explainable solutions to users, etc.). Therefore, interdisciplinary input to ROMCIR is more than welcome.

Keywords: Credible information retrieval · Information disorder · Communication pollution · Disinformation · Misinformation

1 Introduction

"All war is based on deception". These are the words of the general and philosopher Sun Tzu, who lived between 544 and 496 BC, author of the work *The Art of War* [7]. With the advent of the Social Web, we are constantly and more than ever deceived by *information disorder* propagating online [14]. By this expression we mean all forms of communication pollution, from *misinformation* made out of ignorance, to intentional sharing of *disinformation* [15]. Deception is more successful as more refined are the techniques of manipulation of those who create and disseminate false information, and may lead to severe issues for society

M. Hagen et al. (Eds.): ECIR 2022, LNCS 13186, pp. 566–571, 2022.
https://doi.org/10.1007/978-3-030-99739-7_71

[2,11]. False news can, for example, guide public opinion in political and financial choices; false reviews can promote or, on the contrary, discredit economic activities; unverified medical information, especially in the *consumer health search* scenario [6], can lead people to follow behaviors that can be harmful both to their own health and to that of society as a whole (let us think, for example, of the set of unverified news stories that have been disseminated in recent years with respect to Covid-19) [3,5,13].

In this context, it becomes essential to ensure that users have access to truthful information that does not distort their perception of reality. Hence, advances in Information Retrieval become crucial to investigate and tackle this issue, by providing users with automatic but understandable tools to help them come into contact with genuine information. To that end, all approaches that can serve to combat online information disorder find a place at ROMCIR. The workshop is also the ideal plaza where people from different fields can promote their research ideas and discuss them. The distinct forms of communication pollution are so nuanced that not only will we need technical tools to mitigate information disorder, but also the support of cognitive and social scientists, lawyers, and sociologists, to name a few.

2 Topics of Interest

The topics of interest of ROMCIR 2022 include, but are not limited to:

- Access to genuine information;
- Bias detection;
- Bot/spam/troll detection;
- Computational fact-checking;
- Crowdsourcing for credibility;
- Deep fakes;
- Disinformation/misinformation detection;
- Evaluation strategies for disinformation/misinformation detection;
- Fake news detection;
- Fake reviews detection;
- Filter bubble and echo chamber detection and analysis;
- Hate speech/harassment/bullying detection;
- Information polarization in online communities;
- Propaganda identification and analysis;
- Retrieval of genuine information;
- Security, privacy and information genuineness;
- Sentiment/emotional analysis;
- Stance detection;
- Trust and reputation systems;
- Trustworthy AI for information disorder;
- Understanding and guiding the societal reaction under information disorder.

Data-driven approaches, supported by publicly available datasets, are more than welcome.

3   Previous Edition

The first edition of the Workshop, namely ROMCIR 2021, received 15 submissions, of which 6 were accepted, so with an acceptance rate of 40%. The accepted articles, collected in CEUR Proceedings [12], considered different problems. There were issues tangentially related to Credible Information Retrieval, such as those of *authorship verification* [16] and *bias detection* in science evaluation [1]. Furthermore, the problems of opinion mining and misinformation identification were tackled, such as those of *hate speech detection* [8] and *claim verification* [10]. Finally, the problem of the access to genuine information was considered, by proposing the definition of systems to support users in *retrieving genuine news* [9], and the study of new IR methods able to consider the *credibility* of the data collected in the retrieval process [4].

4   Organizers

The following people contributed in different capacities to the organization of the Workshop and to the verification of the quality of the submitted work.

4.1   Workshop Chairs

 **Marinella Petrocchi** is a Senior Researcher at the National Research Council (CNR), Institute of Informatics and Telematics (IIT), Pisa, Italy, under the Trust, Security and Privacy Research Unit. She also collaborates with the Sysma Unit at the IMT School for Advanced Studies, Lucca, Italy. Her field of research lies between Cybersecurity and Data Science. Specifically, she studies novel techniques for online fake news/fake accounts detection. She is in the core team of the TOFFEe project (TOols for Fighting FakEs), funded by IMT, and WP leader in H2020 Medina, where she studies automatic translation from NL to machine-readable languages for cloud certification schemes. *Website*: https://www.iit.cnr.it/en/marinella.petrocchi/.

**Marco Viviani** is an Associate Professor at the University of Milano-Bicocca, Department of Informatics, Systems, and Communication (DISCo). He is currently working in the Information and Knowledge Representation, Retrieval and Reasoning (IKR3) Lab. He is involved in the organization of several research initiatives at the international level. He was General Co-chair of MDAI 2019, and organized several Workshops and Special Tracks at International Conferences. He is Associate Editor of "Social Network Analysis and Mining", Editorial Board Member of "Online Social Networks and Media", and Guest Editor of several Special Issues in International Journals related to information disorder detection. His main research activities include Social Computing, Information Retrieval, Text Mining, Natural Language Processing, Trust and Reputation Management, User Modeling. On these topics, he has published more than 80 research works in International Journals, at International Conferences, as Monographs, and Book Chapters. *Website*: https://ikr3.disco.unimib.it/people/marco-viviani/.

4.2 Proceedings Chair

Rishabh Upadhyay is a Research Fellow at the University of Milano-Bicocca, Department of Informatics, Systems, and Communication (DISCo). His research interests are related to Machine and Deep Learning, Information Retrieval, and Social Computing. He is currently working within the EU Horizon 2020 ITN/ETN DoSSIER project on Domain-Specific Systems for Information Extraction and Retrieval, in particular on the project: "Assessing Credibility, Value, and Relevance". He was one of the co-organizers of Task 2: Consumer Health Search, at CLEF 2021 eHealth Lab Series. He has recently published papers at International Conferences on the topic of health misinformation detection.

4.3 Program Committee

- Rino Falcone, ISTC–CNR, Rome, Italy;
- Carlos A. Iglesias, Universidad Politécnica de Madrid, Spain;
- Petr Knoth, The Open University, London, UK;
- Udo Kruschwitz, University of Regensburg, Germany;
- Yelena Mejova, ISI Foundation, Turin, Italy;
- Preslav Nakov, Qatar Computing Research Institute, HBKU, Doha, Qatar;
- Symeon Papadopoulos, ITI, Thessaloniki, Greece;
- Gabriella Pasi, University of Milano-Bicocca, Milan, Italy;
- Marinella Petrocchi, IIT– CNR, Pisa, Italy;
- Francesco Pierri, Politecnico di Milano, Milan, Italy
- Adrian Popescu, CEA LIST, Gif-sur-Yvette, France;
- Paolo Rosso, Universitat Politècnica de València, Spain;
- Fabio Saracco, IMT School for Advanced Studies, Lucca, Italy;

– Marco Viviani, University of Milano-Bicocca, Milan, Italy;
– Xinyi Zhou, Syracuse University, NY, USA;
– Arkaitz Zubiaga, Queen Mary University of London, UK.

Acknowledgements. The ROMCIR 2022 Workshop is supported by the National Research Council (CNR) – Institute of Informatics and Telematics (IIT), Pisa, Italy, by the University of Milano-Bicocca – Department of Informatics, Systems, and Communication (DISCo), Milan, Italy, and by the EU Horizon 2020 ITN/ETN on Domain Specific Systems for Information Extraction and Retrieval (H2020-EU.1.3.1., ID: 860721).

References

1. Bethencourt, A.M., Luo, J., Feliciani, T.: Bias and truth in science evaluation: a simulation model of grant review panel discussions. In: ROMCIR 2021 CEUR Workshop Proceedings, vol. 2838, pp. 16–24 (2021). http://ceur-ws.org/Vol-2838/paper2.pdf
2. Caldarelli, G., De Nicola, R., Petrocchi, M., Del Vigna, F., Saracco, F.: The role of bot squads in the political propaganda on Twitter. Commun. Phys. **3**(81) (2020). https://doi.org/10.1038/s42005-020-0340-4
3. Caldarelli, G., De Nicola, R., Petrocchi, M., Pratelli, M., Saracco, F.: Flow of online misinformation during the peak of the COVID-19 pandemic in Italy. EPJ Data Sci. **10**(1), 34 (2021). https://doi.org/10.1140/epjds/s13688-021-00289-4
4. Denaux, R., Gomez-Perez, J.M.: Sharing retrieved information using linked credibility reviews. In: ROMCIR 2021 CEUR Workshop Proceedings, vol. 2838, pp. 59–65 (2021). http://ceur-ws.org/Vol-2838/paper6.pdf
5. Di Sotto, S., Viviani, M.: Assessing health misinformation in online content. In: Proceedings of the 37th Annual ACM Symposium on Applied Computing (2022). to appear
6. Goeuriot, L., et al.: CLEF eHealth evaluation lab 2021. In: Hiemstra, D., Moens, M.-F., Mothe, J., Perego, R., Potthast, M., Sebastiani, F. (eds.) ECIR 2021. LNCS, vol. 12657, pp. 593–600. Springer, Cham (2021). https://doi.org/10.1007/978-3-030-72240-1_69
7. Griffith, S.B.: Sun Tzu: The Art of War, vol. 39. Oxford University Press London, Oxford (1963)
8. Gupta, S., Nagar, S., Nanavati, A.A., Dey, K., Barbhuiya, F.A., Mukherjea, S.: Consumption of hate speech on Twitter: a topical approach to capture networks of hateful users. In: ROMCIR 2021 CEUR Workshop Proceedings, vol. 2838, pp. 25–34 (2021). http://ceur-ws.org/Vol-2838/paper3.pdf
9. Gupta, V., Beckh, K., Giesselbach, S., Wegener, D., Wi, T.: Supporting verification of news articles with automated search for semantically similar articles. In: ROMCIR 2021 CEUR Workshop Proceedings, vol. 2838, pp. 47–58 (2021). http://ceur-ws.org/Vol-2838/paper5.pdf
10. Hatua, A., Mukherjee, A., Verma, R.M.: On the feasibility of using GANs for claim verification- experiments and analysis. In: ROMCIR 2021 CEUR Workshop Proceedings, vol. 2838, pp. 35–46 (2021). http://ceur-ws.org/Vol-2838/paper4.pdf
11. Pasi, G., Viviani, M.: Information credibility in the social web: Contexts, approaches, and open issues. arXiv preprint arXiv:2001.09473 (2020)

12. Saracco, F., Viviani, M.: Overview of ROMCIR 2021: workshop on reducing online misinformation through credible information retrieval. In: ROMCIR 2021 CEUR Workshop Proceedings, vol. 2838, pp. i–vii (2021). http://ceur-ws.org/Vol-2838/xpreface.pdf
13. Upadhyay, R., Pasi, G., Viviani, M.: Health misinformation detection in web content: a structural-, content-based, and context-aware approach based on web2vec. In: Proceedings of the Conference on Information Technology for Social Good, pp. 19–24 (2021). https://doi.org/10.1145/3462203.3475898
14. Wardle, C., Derakhshan, H.: Information disorder: toward an interdisciplinary framework for research and policy making. Council of Europe, no. 27 (2017). https://tinyurl.com/edoc-informationdisorder
15. Wardle, C., Derakhshan, H., et al.: Thinking about 'information disorder': formats of misinformation, disinformation, and mal-information. Ireton, Cherilyn; Posetti, Julie. Journalism, 'fake news' & disinformation, pp. 43–54. Unesco, Paris (2018). https://en.unesco.org/sites/default/files/f._jfnd_handbook_module_2.pdf
16. Zhang, Y., Boumber, D., Hosseinia, M., Yang, Mukherjee, A.: Improving authorship verification using linguistic divergence. In: ROMCIR 2021 CEUR Workshop Proceedings, vol. 2838, pp. 1–15 (2021). http://ceur-ws.org/Vol-2838/paper1.pdf

Tutorials

Online Advertising Incrementality Testing: Practical Lessons, Paid Search and Emerging Challenges

Joel Barajas[4(✉)], Narayan Bhamidipati[1], and James G. Shanahan[2,3]

[1] Yahoo Research, Sunnyvale, CA 94089, USA
narayanb@yahooinc.com
[2] Church and Duncan Group Inc, San Francisco, CA, USA
[3] UC Berkeley, Berkeley, CA, USA
[4] Amazon, Sunnyvale, CA 94086, USA
joelbz@amazon.com

Abstract. Online advertising has historically been approached as an ad-to-user matching problem within sophisticated optimization algorithms. As the research and ad tech industries have progressed, advertisers have increasingly emphasized the causal effect estimation of their ads (incrementality) using controlled experiments (A/B testing). With low lift effects and sparse conversion, the development of incrementality testing platforms at scale suggests tremendous engineering challenges in measurement precision. Similarly, the correct interpretation of results addressing a business goal requires significant data science and experimentation research expertise.

We propose a practical tutorial in the incrementality testing landscape, including:
- The business need
- Literature solutions and industry practices
- Designs in the development of testing platforms
- The testing cycle, case studies, and recommendations
- Paid search effectiveness in the marketplace
- Emerging privacy challenges for incrementality testing and research solutions

We provide first-hand lessons based on the development of such a platform in a major combined DSP and ad network, and after running several tests for up to two months each over recent years. With increasing privacy constraints, we survey literature and current practices. These practices include private set union and differential privacy for conversion modeling, and geo-testing combined with synthetic control techniques.

1 Learning Objectives and Scope

Even though there are currently solutions to evaluate the advertising effectiveness with randomized experiments, many details and recommendations rarely appear in papers. This tutorial provides a 360-degree view of the topic, from engineering designs to experiment planning and business use cases.

J. Barajas—Work done while the author was employed at Yahoo.

M. Hagen et al. (Eds.): ECIR 2022, LNCS 13186, pp. 575–581, 2022.
https://doi.org/10.1007/978-3-030-99739-7_72

The key benefits to participants include:

- Specific recommendations to the correct execution of A/B testing
- Online advertising testing engineering designs and econometric evaluation approaches
- Marketing use cases for online advertising incrementality testing
- Review of paid search effectiveness evaluation literature and challenges to operationalizing the estimations
- Review of emerging challenges fo incrementality testing within privacy constraints

Participants will:

1. Identify and formulate key approaches to measuring the effectiveness of online advertising.
2. Execute relevant statistics for hypothesis testing, power analysis in experiment planning and simulate experiment scenarios.
3. Be able to define key ingredients of an operational incrementality testing platform and their trade-offs.
4. Understand the business need for incrementality testing.
5. Identify the necessary conditions to increase the likelihood of successful test given minimum detectable lift, conversion type, test duration among others.
6. Differentiate between demand generation advertising (display, social ads) and demand capture advertising (paid search) incrementality measurement.

2   Tutorial Outline

Part 1 The basics: context and challenges [8, 12, 15, 19]
- The problem
  - Online Advertising spend trends between performance and brand
  - Big picture problem: quarterly/yearly budget allocation
  - Budget allocation practices based on financial models
  - The need for testing combined with industry attribution practices
- How channel-level testing fits within other forms of testing
  - Real-time decision making in targeting engines
  - Tactic testing: A/B testing with last-touch attribution
  - Multi-cell testing A/B testing + Incrementality testing
  - CMO decision making at the end of the quarter/semester/year
- Business Use cases
  - Advertiser joining new partners
  - Testing to calibrate and rebase financial models
  - Media Mix Models calibration
  - Last-touch attribution multipliers
  - The Marketing component: Growth marketing vs CRM marketing

Part 2 Incrementality Testing: concepts, solutions and literature [2, 3, 14, 22]
- Literature and Industry practices
  - Placebo based testing: practice and issues

- Intention to treat testing
- Ghost ads testing proposal
  - Estimation Frameworks
    - Econometric causality
    - Potential Outcomes Causal Framework
    - Pitfalls

Part 3 From concept to production: platform building, challenges, case studies [3, 16]
- Building the experiment platform journey
- The identity graph and treatment groups
  - Cookie-based experiments
  - Device-based experiments
  - Logged-in users based experiments
  - Household-level experiments
- User holdout design within modern Ad tech serving systems
  - The hashing functions
  - The challenge with targeting and scoring algorithms
  - How to avoid targeting bias
  - The role of look-back windows in last-touch attribution engines
- Data Logging and Analysis

Part 4 Deployment at Scale: test cycle and case studies
- Experiment execution cycle
  - Experiment Design and Planning
  - Intervention Execution
  - Experiment Tracking and Metrics
  - End of experiment readout
- Case Studies
  - Insurance quotes and comparison with post-click conversions
  - Online food ordering revenue: CRM versus New audiences
  - Online acquisition signup

Part 5 Paid Search Incrementality Testing [7, 9, 21]
- Evaluating Demand Capture Channels
  - The challenge with demand capture ads in paid search
  - Organic versus paid search results
  - The effects on the search marketplace
- Tehniques with Aggregate Data
  - Differences-in-Differences
  - Synthetic Control

Part 6 Emerging trends: identity challenges, industry trends and solutions [1, 6, 7]
- Advertisers Testing without Ad Network holdouts
  - Spend as experiment intervention
  - Methodologies: Time series based testing
- Geo-testing
  - Geo units specification
  - Geo unit treatment assignment

- • The power of A/A tests in the experiment design
- – Emerging challenges with user ids
  - • Private set Intersection
  - • Differential Privacy
  - • Identity fragmentation challenges

3   Authors Biography

Joel Barajas, Sr Research Scientist, has over 11 years of experience in the online advertising industry with research contributions at the intersection of Ad tech, Marketing Science, and Experimentation. He has experience with Ad load personalization and experimentation in a publisher marketplace. Within Marketing Data Science, he has supported regular budget allocation and Media Mix Models in multi-channel advertising. With a PhD dissertation focussed on ad incrementality testing, his published work has appeared in top outlets including INFORMS Marketing Science Journal, ACM CIKM, ACM WWW, SIAM SDM. He led the science development and marketing analytics of the incrementality testing platform in a multidisciplinary team. He currently oversees most incrementality tests in Verizon Media ad network (previously yahoo!) and DSP (previously AOL advertising.com). Joel also leads the science development in CTV and linear TV measurement modeling. He holds a B.S. (with honors) in Electrical and Electronics Engineering from the Tecnológico de Monterrey, and a PhD in Electrical Engineering (with emphasis on statistics) from UC Santa Cruz.

Narayan Bhamidipati, Sr Director of Research, has over 14 years of experience in Computational Advertising and Machine Learning. He currently leads a team of researchers focused on providing state-of-the-art ad targeting solutions to help ads be more effective and relevant. This includes creating various contextual targeting products to reduce the company's reliance on user profiles and help improve monetization in a more privacy aware world. Alongside that, Narayan ensures that the user profile based ad targeting products continue to improve despite the decline of tracking data. In addition, Narayan is keen on developing the most accurate ad effectiveness measurement platform which would help the company attract more revenue by proving the true value of the ad spend on our platforms. He holds B.Stat(Hons), M.Stat and PhD(CS) degrees, all from the Indian Statistical Institute, Kolkata.

Dr. James G. Shanahan has spent the past 30 years developing and researching cutting-edge artificial intelligence systems, splitting his time between industry and academia. For the academic year 2019–2020, Jimi held the position of Rowe Professor of Data Science at Bryant University, Rhode Island. He has (co) founded several companies that leverage AI/machine learning/deep learning/computer vision in verticals such as digital advertising, web search, local search, and smart cameras. Previously he has held appointments at AT&T (Executive Director of Research), NativeX (SVP of data science), Xerox Research (staff research scientist), and Mitsubishi. He is on the board of Anvia, and he also advises several high-tech startups including Aylien, ChartBoost, Digital-Bank, LucidWorks, and others. Dr. Shanahan received his PhD in engineering mathematics and computer vision from the University of Bristol, U. K. Jimi has been involved with KDD since 2004 as an author, as a tutorial presenter, and as a workshop co-chair; he has actively been involved as a PC/SPC member over the years also.

4   List of References by Topic

4.1   The need for Incrementality Testing Solutions

- *A comparison of approaches to advertising measurement: Evidence from big field experiments at Facebook* by Gordon et al. (2019) [12]
- *Do display ads influence search? Attribution and dynamics in online advertising* by Kireyev et al. (2016) [15].
- *Attributing conversions in a multichannel online marketing environment: An empirical model and a field experiment* by Li and Kannan (2014) [19].
- *Evaluating online ad campaigns in a pipeline: causal models at scale* by Chan et al. (2010) [8].

4.2   Incrementality Testing Solutions

- *Incrementality Testing in Programmatic Advertising: Enhanced Precision with Double-Blind Designs* by Barajas and Bhamidipati (2021) [3]
- *Ghost ads: Improving the economics of measuring online ad effectiveness* by Johnson et al. (2017) [14].
- *Experimental designs and estimation for online display advertising attribution in marketplaces* by Barajas et al. (2016) [2].
- *Here, there, and everywhere: correlated online behaviors can lead to overestimates of the effects of advertising* by Lewis et al. (2011) [18].

4.3   Causal Inference

- *Causal inference using potential outcomes: Design, modeling, decisions* by Rubin (2005) [22]
- *Principal stratification in causal inference* by Frangakis and Rubin (2002) [11].
- *Bayesian inference for causal effects in randomized experiments with noncompliance* by Imbens and Rubin (1997) [13].

4.4   Operationalization and Practical Recommendations

- *Incrementality Testing in Programmatic Advertising: Enhanced Precision with Double-Blind Designs* by Barajas and Bhamidipati (2021) [3]
- *Trustworthy online controlled experiments: A practical guide to a/b testing* by Kohavi et al. (2020) [16].
- *The unfavorable economics of measuring the returns to advertising* by Lewis et al. (2015) [17].

4.5 Paid Search Incrementality Testing

- *Consumer heterogeneity and paid search effectiveness: A large-scale field experiment* by Blake *et al.* (2015) [7].
- *Sponsored Search in Equilibrium: Evidence from Two Experiments* by Moshary (2021) [21]
- *Effectiveness of Paid Search Advertising: Experimental Evidence* by Dai and Luca (2016) [9].

4.6 Geo-testing and Synthetic Control and Identity Challenges

- *Advertising Incrementality Measurement using Controlled Geo-Experiments: The Universal App Campaign Case Study* by Barajas *et al.* (2020) [6]
- *Consumer heterogeneity and paid search effectiveness: A large-scale field experiment* by Blake *et al.* (2015) [7].
- *Synthetic control methods for comparative case studies: Estimating the effect of California's tobacco control program* by Abadie *et al.* (2010) [1].
- *The identity fragmentation bias* by Lin and Misra (2020) [20].

References

1. Abadie, A., Diamond, A., Hainmueller, J.: Synthetic control methods for comparative case studies: estimating the effect of California's tobacco control program. J. Am. Stat. Assoc. **105**(490), 493–505 (2010)
2. Barajas, J., Akella, R., Holtan, M., Flores, A.: Experimental designs and estimation for online display advertising attribution in marketplaces. Mark. Sci. **35**(3), 465–483 (2016)
3. Barajas, J., Bhamidipati, N.: Incrementality testing in programmatic advertising: enhanced precision with double-blind designs. In: Proceedings of the Web Conference 2021, pp. 2818–2827. WWW 2021, Association for Computing Machinery, New York, NY, USA (2021). https://doi.org/10.1145/3442381.3450106
4. Barajas, J., Bhamidipati, N., Shanahan, J.G.: Online advertising incrementality testing and experimentation: industry practical lessons. In: Proceedings of the 27th ACM SIGKDD Conference on Knowledge Discovery & Data Mining, pp. 4027–4028. KDD 2021, Association for Computing Machinery, New York, NY, USA (2021). https://doi.org/10.1145/3447548.3470819
5. Barajas, J., Bhamidipati, N., Shanahan, J.G.: Online advertising incrementality testing: practical lessons and emerging challenges. In: Proceedings of the 30th ACM International Conference on Information & Knowledge Management, pp. 4838–4841. CIKM 2021, Association for Computing Machinery, New York, NY, USA (2021). https://doi.org/10.1145/3459637.3482031
6. Barajas, J., Zidar, T., Bay, M.: Advertising incrementality measurement using controlled geo-experiments: the universal app campaign case study (2020)
7. Blake, T., Nosko, C., Tadelis, S.: Consumer heterogeneity and paid search effectiveness: a large-scale field experiment. Econometrica **83**(1), 155–174 (2015)
8. Chan, D., Ge, R., Gershony, O., Hesterberg, T., Lambert, D.: Evaluating online ad campaigns in a pipeline: causal models at scale. In: Proceedings of the 16th ACM SIGKDD International Conference on Knowledge Discovery and Data Mining, pp. 7–16. KDD 2010, ACM, New York, NY, USA (2010). https://doi.org/10.1145/1835804.1835809,http://doi.acm.org/10.1145/1835804.1835809

9. Dai, D., Luca, M.: Effectiveness of paid search advertising: Experimental evidence. Technical report, Harvard Business School (October 2016). workin Paper No. 17–025

10. Farahat, A., Shanahan, J.: Econometric analysis and digital marketing: how to measure the effectiveness of an ad. In: Proceedings of the sixth ACM International Conference on Web Search and Data Mining, pp. 785–785 (2013)

11. Frangakis, C., Rubin, D.: Principal stratification in causal inference. Biometrics **58**(1), 21–29 (2002). https://doi.org/10.1111/j.0006-341X.2002.00021.x

12. Gordon, B.R., Zettelmeyer, F., Bhargava, N., Chapsky, D.: A comparison of approaches to advertising measurement: evidence from big field experiments at Facebook. Mark. Sci. **38**(2), 193–225 (2019)

13. Imbens, G.W., Rubin, D.B.: Bayesian inference for causal effects in randomized experiments with noncompliance. Ann. Stat. **25**(1), 305–327 (1997). http://www.jstor.org/stable/2242722

14. Johnson, G.A., Lewis, R.A., Nubbemeyer, E.I.: Ghost ads: improving the economics of measuring online ad effectiveness. J. Mark. Res. **54**(6), 867–884 (2017). https://doi.org/10.1509/jmr.15.0297

15. Kireyev, P., Pauwels, K., Gupta, S.: Do display ads influence search? Attribution and dynamics in online advertising. Int. J. Res. Mark. **33**(3), 475–490 (2016)

16. Kohavi, R., Tang, D., Xu, Y.: Trustworthy Online Controlled Experiments: A Practical Guide to A/B Testing. Cambridge University Press, Cambridge (2020)

17. Lewis, R.A., Rao, J.M.: The unfavorable economics of measuring the returns to advertising *. Q. J. Econ. **130**(4), 1941–1973 (2015)

18. Lewis, R.A., Rao, J.M., Reiley, D.H.: Here, there, and everywhere: correlated online behaviors can lead to overestimates of the effects of advertising. In: Proceedings of the 20th International Conference on World Wide Web, pp. 157–166. WWW 2011, ACM, New York, NY, USA (2011). https://doi.org/10.1145/1963405.1963431,http://doi.acm.org/10.1145/1963405.1963431

19. Li, H.A., Kannan, P.: Attributing conversions in a multichannel online marketing environment: an empirical model and a field experiment. J. Mark. Res. **51**(1), 40–56 (2014)

20. Lin, T., Misra, S.: The identity fragmentation bias (2020)

21. Moshary, S.: Sponsored search in equilibrium: evidence from two experiments. Available at SSRN 3903602 (2021)

22. Rubin, D.B.: Causal inference using potential outcomes. J. Am. Stat. Assoc. **100**(469), 322–331 (2005). https://doi.org/10.1198/016214504000001880

From Fundamentals to Recent Advances: A Tutorial on Keyphrasification

Rui Meng[1(✉)], Debanjan Mahata[2], and Florian Boudin[3]

[1] Salesforce Research, Palo Alto, USA
ruimeng@salesforce.com
[2] Moody's Analytics, New York, USA
Debanjan.Mahata@moodys.com
[3] LS2N, Nantes Université, Nantes, France
florian.boudin@univ-nantes.fr

Abstract. Keyphrases represent the most important information of text which often serve as a surrogate for efficiently summarizing text documents. With the advancement of deep neural networks, recent years have witnessed rapid development in automatic identification of keyphrases. The performance of keyphrase extraction methods has been greatly improved by the progresses made in natural language understanding, enable models to predict relevant phrases not mentioned in the text. We name the task of summarizing texts with phrases *keyphrasification*.

In this half-day tutorial, we provide a comprehensive overview of keyphrasification as well as hands-on practice with popular models and tools. This tutorial covers important topics ranging from basics of the task to the advanced topics and applications. By the end of the tutorial, participants will have a better understanding of 1) classical and state-of-the-art keyphrasification methods, 2) current evaluation practices and their issues, and 3) current trends and future directions in keyphrasification research. Tutorial-related resources are available at https://keyphrasification.github.io/.

Keywords: Tutorial · Keyphrasification · Keyphrase extraction · Keyphrase generation · Automatic identification of keyphrases

1 Presenters

Three researchers will tutor this tutorial. They will contribute to the tutorial equally and will be presenting at the tutorial (upon acceptance of this proposal).

- **Rui Meng** is a research scientist at Salesforce, USA. He did his doctoral study in Information Science at University of Pittsburgh and his research focuses on keyphrase generation and text representation learning. His seminal work on deep keyphrase generation led to signifigant future developments and progress in the area of applying deep learning models to the problem of identifying keyphrases from text documents.

M. Hagen et al. (Eds.): ECIR 2022, LNCS 13186, pp. 582–588, 2022.
https://doi.org/10.1007/978-3-030-99739-7_73

- **Debanjan Mahata** is a director of AI at Moody's Analytics and an adjunct faculty at IIIT-Delhi. He obtained his PhD from University of Arkansas at Little Rock and was previously a research scientist at Bloomberg AI. He is an experienced industry researcher with a major focus on document understanding and information extraction tasks that includes keyphrase extraction and generation. He has published several research articles on keyphrase extraction and applied them in an industry setting.
- **Florian Boudin** is an associate professor of Computer Science at the University of Nantes, France. His research lies in the intersection of natural language processing and information retrieval, and focuses on weakly supervised and un-supervised learning with applications including keyphrase extraction and generation, summarization and document retrieval in scholarly collections. He has authored several articles related to keyphrase extraction including the popular method of TopicRank [6]. He is also the author of PKE[1]: an open source python-based keyphrase extraction toolkit, which is one of the most popular open source library for trying out state-of-the-art keyphrase extraction methods.

2   Why is This Tutorial Important?

Keyphrases play an important role in various real-world applications and often serves as the fulcrum between users and the vast amount of unstructured data on the Internet. The automatic methods for identifying keyphrases – keyphrasification – has received growing attention from the NLP and IR communities in the recent past. Various models and tools have been proposed that have reported regular increase in performance of keyphrasification and facilitated their practical use in a broad range of applications, e.g. information retrieval, question answering, recommendation, summarization and many other NLP tasks. However, compared to other popular NLP tasks, resources and knowledge of keyphrasification are less available to the community. This tutorial aims at filling the gap between research and practice, by providing a walk-through of state-of-the-art research and a step-by-step guidance to users and researchers who are interested in using tools or conducting research in keyphrasification.

In this tutorial, we overview recent developments of this task and will cover three major themes: (1) basics of keyphrasification, including its definition, evaluation, and classic automatic methods; (2) recent progress in keyphrasification based on neural networks, including extraction/generation models as well as methods using pretrained models; and (3) advanced topics on keyphrasification, in which we will cover recent studies concerning several substantial issues in this task and discuss promising research directions.

3   Target Audience and Prerequisite Knowledge

This tutorial is targeted towards researchers and practitioners in the fields of natural language processing, information retrieval, and machine learning.

---

[1] https://github.com/boudinfl/pke.

Participants are expected to have basic knowledge on these topics, as well as some experience in Python programming. Our tutorial does not require any prerequisite knowledge on keyphrasification. Fundamental concepts and related algorithms such as noun phrase detection, graph-based ranking will be introduced throughout the tutorial, interleaved with interactive practice of relevant softwares and tools. In the later half of the tutorial, particular focus will be placed on contemporary neural methods for keyphrasification, with techniques such as sequence-to-sequence learning, sequence labeling and text pretraining.

4   Format of the Tutorial

The tutorial will be a half day session (3 h), divided into three parts: (1) introduction and basics of keyphrasification; (2) neural network based methods; and (3) advanced topics and applications. Besides a standard presentation of each part of the tutorial, we interleave the principles and method introductions with a hands-on practice session to familiarize participants with the common softwares and tools for keyphrasification. The demos will show how newly developed methods work with various real-world datasets. The detailed outline of the topics that will be covered in the tutorial is presented below.

- **Introduction**
  - **Motivation of the tutorial**
  - **An overview of history and applications:** Keyphrases were initially introduced as a means for cataloguing and indexing documents in digital libraries [10,17]. Because they distill the important information from documents, keyphrases are useful for many applications such as summarization [27,33], document classification [14,16], opinion mining [4] and recommendation [9,11].
  - **Taxonomy of Methods for Keyphrasification**
    - ⋆ **Extraction:** Unsupervised term weighting-based methods (e.g. TF × IDF, Yake [8]), graph-based ranking methods (e.g. TextRank [21], PositionRank [12], TopicRank [6]), supervised classification methods (e.g. Kea [28]), neural network-based models (e.g. DivGraphPointer [26]).
    - ⋆ **Generation:** Thesaurus-based methods (e.g. Kea++ [19]), weakly-supervised methods (e.g. TopicCoRank [7], neural network based generative methods (e.g. CopyRNN [20]).
    - ⋆ **Tagging:** Multi-label classification methods [2,15,25].
  - **Evaluation Metrics and Hands-on Session with PKE**
- **Neural Networks based Methods**
  - **Neural Keyphrase Generation:** *One2One* [20], *One2Seq* [30,32] and *One2Set* [31].
  - **Neural Keyphrase Extraction:** Sequence Labeling-based Methods [1], Embedding-based Methods [3,18], Ranking-based Methods [29] and Pointer-based Methods [26].

- Keyphrasification with Pre-trained Models [22–24]
- Practice of Neural Models with Colab.
- **Advanced Topics**
  Keyphrase generation for document retrieval [5]; Keyphrasification beyond the scientific domain (e.g. news articles [13], webpages [29] and QA communities [32]); Language models for learning better keyphrase representations from text.
- **Summary and Future Directions**
  - Challenges and Future Research Directions
  - Interaction with the Audience
    ⋆ How to identify keyphrases on your own data and applications?

5   Participation Encouragement

We will actively promote our tutorial in the natural language processing, information retrieval, and machine learning communities. Specifically, we will (1) prepare the slides and build the tutorial website earlier; (2) collect and implement standard tools/APIs to facilitate the practical use of keyphrasificaion; (3) post the preliminary slides and briefs on social media (e.g., Twitter and LinkedIn); and (4) advertise our tutorial in open-source communities (e.g., GitHub). Besides, we plan to write a survey paper to review the research of keyphrasification alongside this tutorial.

Demos & Software: We will show demos right after each part of our tutorial. These demos are built based on the techniques we covered in the previous sections. Moreover, we will share the GitHub links to these tools so the audience can try by themselves. We believe this will strongly improve the interactions during the tutorial and lead to helpful takeaways.

Acknowledgments. Florian Boudin is partially supported by the French National Research Agency through the DELICES project (ANR-19-CE38-0005-01). Rui Meng was partially supported by the Amazon Research Awards for the project "Transferable, Controllable, Applicable Keyphrase Generation" and by the University of Pittsburgh Center for Research Computing through the resources provided.

References

1. Alzaidy, R., Caragea, C., Giles, C.L.: BI-LSTM-CRF sequence labeling for keyphrase extraction from scholarly documents. In: The World Wide Web Conference, pp. 2551–2557 (2019)
2. Belém, F., Almeida, J., Gonçalves, M.: Tagging and tag recommendation (September 2019). https://doi.org/10.5772/intechopen.82242
3. Bennani-Smires, K., Musat, C., Hossmann, A., Baeriswyl, M., Jaggi, M.: Simple unsupervised keyphrase extraction using sentence embeddings. In: Proceedings of the 22nd Conference on Computational Natural Language Learning, pp. 221–229 (2018)

4. Berend, G.: Opinion expression mining by exploiting keyphrase extraction. In: Proceedings of 5th International Joint Conference on Natural Language Processing, pp. 1162–1170. Asian Federation of Natural Language Processing, Chiang Mai, Thailand (November 2011). https://aclanthology.org/I11-1130

5. Boudin, F., Gallina, Y., Aizawa, A.: Keyphrase generation for scientific document retrieval. In: Proceedings of the 58th Annual Meeting of the Association for Computational Linguistics, pp. 1118–1126. Association for Computational Linguistics, Online (July 2020). https://doi.org/10.18653/v1/2020.acl-main.105, https://aclanthology.org/2020.acl-main.105

6. Bougouin, A., Boudin, F., Daille, B.: TopicRank: graph-based topic ranking for keyphrase extraction. In: Proceedings of the Sixth International Joint Conference on Natural Language Processing, pp. 543–551. Asian Federation of Natural Language Processing, Nagoya, Japan (October 2013). https://aclanthology.org/I13-1062

7. Bougouin, A., Boudin, F., Daille, B.: Keyphrase annotation with graph co-ranking. In: Proceedings of COLING 2016, the 26th International Conference on Computational Linguistics: Technical Papers, pp. 2945–2955. The COLING 2016 Organizing Committee, Osaka, Japan (December 2016). https://aclanthology.org/C16-1277

8. Campos, R., Mangaravite, V., Pasquali, A., Jorge, A., Nunes, C., Jatowt, A.: Yake! keyword extraction from single documents using multiple local features. Inf. Sci. **509**, 257–289 (2020)

9. Collins, A., Beel, J.: Document embeddings vs. keyphrases vs. terms for recommender systems: a large-scale online evaluation. In: Proceedings of the 18th Joint Conference on Digital Libraries, pp. 130–133. JCDL 2019, IEEE Press (2019). https://doi.org/10.1109/JCDL.2019.00027

10. Fagan, J.: Automatic phrase indexing for document retrieval. In: Proceedings of the 10th Annual International ACM SIGIR Conference on Research and Development in Information Retrieval, pp. 91–101. SIGIR 1987, Association for Computing Machinery, New York, NY, USA (1987). https://doi.org/10.1145/42005.42016

11. Ferrara, F., Pudota, N., Tasso, C.: A keyphrase-based paper recommender system. In: Agosti, M., Esposito, F., Meghini, C., Orio, N. (eds.) IRCDL 2011. CCIS, vol. 249, pp. 14–25. Springer, Heidelberg (2011). https://doi.org/10.1007/978-3-642-27302-5_2

12. Florescu, C., Caragea, C.: Positionrank: an unsupervised approach to keyphrase extraction from scholarly documents. In: Proceedings of the 55th Annual Meeting of the Association for Computational Linguistics (Volume 1: Long Papers), pp. 1105–1115 (2017)

13. Gallina, Y., Boudin, F., Daille, B.: KPTimes: a large-scale dataset for keyphrase generation on news documents. In: Proceedings of the 12th International Conference on Natural Language Generation, pp. 130–135. Association for Computational Linguistics, Tokyo, Japan, Oct-Nov 2019. https://doi.org/10.18653/v1/W19-8617, https://aclanthology.org/W19-8617

14. Han, J., Kim, T., Choi, J.: Web document clustering by using automatic keyphrase extraction. In: Proceedings of the 2007 IEEE/WIC/ACM International Conferences on Web Intelligence and Intelligent Agent Technology - Workshops, pp. 56–59. WI-IATW 2007, IEEE Computer Society, USA (2007)

15. Heymann, P., Ramage, D., Garcia-Molina, H.: Social tag prediction. In: Proceedings of the 31st Annual International ACM SIGIR Conference on Research and Development in Information Retrieval, pp. 531–538 (2008)

16. Hulth, A., Megyesi, B.B.: A study on automatically extracted keywords in text categorization. In: Proceedings of the 21st International Conference on Computational Linguistics and 44th Annual Meeting of the Association for Computational Linguistics, pp. 537–544. Association for Computational Linguistics, Sydney, Australia (July 2006). https://doi.org/10.3115/1220175.1220243, https://aclanthology.org/P06-1068

17. Jones, S., Staveley, M.S.: Phrasier: a system for interactive document retrieval using keyphrases. In: Proceedings of the 22nd Annual International ACM SIGIR Conference on Research and Development in Information Retrieval, pp. 160–167. SIGIR 1999, Association for Computing Machinery, New York, NY, USA (1999). https://doi.org/10.1145/312624.312671

18. Mahata, D., Kuriakose, J., Shah, R., Zimmermann, R.: Key2vec: automatic ranked keyphrase extraction from scientific articles using phrase embeddings. In: Proceedings of the 2018 Conference of the North American Chapter of the Association for Computational Linguistics: Human Language Technologies, vol. 2 (Short Papers), pp. 634–639 (2018)

19. Medelyan, O., Witten, I.H.: Thesaurus based automatic keyphrase indexing. In: Proceedings of the 6th ACM/IEEE-CS Joint Conference on Digital Libraries, pp. 296–297. JCDL 2006, Association for Computing Machinery, New York, NY, USA (2006). https://doi.org/10.1145/1141753.1141819

20. Meng, R., Zhao, S., Han, S., He, D., Brusilovsky, P., Chi, Y.: Deep keyphrase generation. In: Proceedings of the 55th Annual Meeting of the Association for Computational Linguistics (Volume 1: Long Papers), pp. 582–592. Association for Computational Linguistics, Vancouver, Canada (July 2017). https://doi.org/10.18653/v1/P17-1054, https://aclanthology.org/P17-1054

21. Mihalcea, R., Tarau, P.: TextRank: bringing order into text. In: Proceedings of the 2004 Conference on Empirical Methods in Natural Language Processing, pp. 404–411. Association for Computational Linguistics, Barcelona, Spain (July 2004). https://aclanthology.org/W04-3252

22. Mu, F., et al.: Keyphrase extraction with span-based feature representations. arXiv preprint arXiv:2002.05407 (2020)

23. Park, S., Caragea, C.: Scientific keyphrase identification and classification by pretrained language models intermediate task transfer learning. In: Proceedings of the 28th International Conference on Computational Linguistics, pp. 5409–5419 (2020)

24. Sahrawat, D.: Keyphrase extraction as sequence labeling using contextualized embeddings. Adv. Inf. Retr. **12036**, 328 (2020)

25. Song, Y., Zhang, L., Giles, C.L.: Automatic tag recommendation algorithms for social recommender systems. ACM Trans. Web (TWEB) **5**(1), 1–31 (2011)

26. Sun, Z., Tang, J., Du, P., Deng, Z.H., Nie, J.Y.: Divgraphpointer: a graph pointer network for extracting diverse keyphrases. In: Proceedings of the 42nd International ACM SIGIR Conference on Research and Development in Information Retrieval, pp. 755–764. SIGIR 2019, Association for Computing Machinery, New York, NY, USA (2019). https://doi.org/10.1145/3331184.3331219

27. Wan, X., Yang, J., Xiao, J.: Towards an iterative reinforcement approach for simultaneous document summarization and keyword extraction. In: Proceedings of the 45th Annual Meeting of the Association of Computational Linguistics, pp. 552–559. Association for Computational Linguistics, Prague, Czech Republic (June 2007). https://aclanthology.org/P07-1070

28. Witten, I.H., Paynter, G.W., Frank, E., Gutwin, C., Nevill-Manning, C.G.: Kea: practical automatic keyphrase extraction. In: Proceedings of the Fourth ACM Conference on Digital Libraries, p. 254–255. DL 1999, Association for Computing Machinery, New York, NY, USA (1999). https://doi.org/10.1145/313238.313437

29. Xiong, L., Hu, C., Xiong, C., Campos, D., Overwijk, A.: Open domain web keyphrase extraction beyond language modeling. In: Proceedings of the 2019 Conference on Empirical Methods in Natural Language Processing and the 9th International Joint Conference on Natural Language Processing (EMNLP-IJCNLP), pp. 5175–5184 (2019)

30. Ye, H., Wang, L.: Semi-supervised learning for neural keyphrase generation. In: Proceedings of the 2018 Conference on Empirical Methods in Natural Language Processing, pp. 4142–4153. Association for Computational Linguistics, Brussels, Belgium, Oct-Nov 2018. https://doi.org/10.18653/v1/D18-1447, https://aclanthology.org/D18-1447

31. Ye, J., Gui, T., Luo, Y., Xu, Y., Zhang, Q.: One2Set: generating diverse keyphrases as a set. In: Proceedings of the 59th Annual Meeting of the Association for Computational Linguistics and the 11th International Joint Conference on Natural Language Processing (Volume 1: Long Papers), pp. 4598–4608. Association for Computational Linguistics, Online (August 2021). https://doi.org/10.18653/v1/2021.acl-long.354, https://aclanthology.org/2021.acl-long.354

32. Yuan, X., et al.: One size does not fit all: generating and evaluating variable number of keyphrases. In: Proceedings of the 58th Annual Meeting of the Association for Computational Linguistics, pp. 7961–7975. Association for Computational Linguistics, Online (July 2020). https://doi.org/10.18653/v1/2020.acl-main.710, https://aclanthology.org/2020.acl-main.710

33. Zha, H.: Generic summarization and keyphrase extraction using mutual reinforcement principle and sentence clustering. In: Proceedings of the 25th Annual International ACM SIGIR Conference on Research and Development in Information Retrieval, pp. 113–120. SIGIR 2002, Association for Computing Machinery, New York, NY, USA (2002). https://doi.org/10.1145/564376.564398

Information Extraction from Social Media: A Hands-On Tutorial on Tasks, Data, and Open Source Tools

Shubhanshu Mishra[1]([✉]) [ID], Rezvaneh Rezapour[2] [ID], and Jana Diesner[3] [ID]

[1] Twitter, Inc., Chicago, USA
mishra@shubhanshu.com
[2] Drexel's College of Computing and Informatics, Philadelphia, USA
shadi.rezapour@drexel.edu
[3] University of Illinois at Urbana-Champaign, Champaign, USA
jdiesner@illinois.edu
https://shubhanshu.com/

Abstract. Information extraction (IE) is a common sub-area of natural language processing that focuses on identifying structured data from unstructured data. The community of Information Retrieval (IR) relies on accurate and high-performance IE to be able to retrieve high quality results from massive datasets. One example of IE is to identify named entities in a text, e.g., "Barack Obama served as the president of the USA". Here, Barack Obama and USA are named entities of types of PERSON and LOCATION, respectively. Another example is to identify sentiment expressed in a text, e.g., "This movie was awesome". Here, the sentiment expressed is positive. Finally, identifying various linguistic aspects of a text, e.g., part of speech tags, noun phrases, dependency parses, etc., which can serve as features for additional IE tasks. This tutorial introduces participants to a) the usage of Python based, open-source tools that support IE from social media data (mainly Twitter), and b) best practices for ensuring the reproducibility of research. Participants will learn and practice various semantic and syntactic IE techniques that are commonly used for analyzing tweets. Additionally, participants will be familiarized with the landscape of publicly available tweet data, and methods for collecting and preparing them for analysis. Finally, participants will be trained to use a suite of open source tools (SAIL for active learning, TwitterNER for named entity recognition3, and SocialMediaIE for multi task learning), which utilize advanced machine learning techniques (e.g., deep learning, active learning with human-in-the-loop, multi-lingual, and multi-task learning) to perform IE on their own or existing datasets. Participants will also learn how social context can be integrated in Information Extraction systems to make them better. The tools introduced in the tutorial will focus on the three main stages of IE, namely, collection of data (including annotation), data processing and analytics, and visualization of the extracted information. More details can be found at: https://socialmediaie.github. io/tutorials/.

M. Hagen et al. (Eds.): ECIR 2022, LNCS 13186, pp. 589–596, 2022.
https://doi.org/10.1007/978-3-030-99739-7_74

Keywords: Information extraction · Multi-task learning · Natural language processing · Social media data · Twitter · Machine learning bias

1 Introduction

1.1 Aims and Learning Objectives

In this hands-on tutorial (details and material at: https://socialmediaie.github. io/tutorials/), we introduce the participants to working with social media data, which are an example of Digital Social Trace Data (DSTD). The DSTD abstraction allows us to model social media data with rich information associated with social media text, such as authors, topics, and time stamps. We introduce the participants to several Python-based, open-source tools for performing Information Extraction (IE) on social media data. Furthermore, the participants will be familiarized with a catalogue of more than 30 publicly available social media corpora for various IE tasks such as named entity recognition (NER), part of speech (POS) tagging, chunking, super sense tagging, entity linking, sentiment classification, and hate speech identification. We will also show how these approaches can be expanded to word in a multi-lingual setting. Finally, the participants will be introduced to the following applications of extracted information: (i) combining network analysis and text-based signals to rank accounts, and (ii) correlation between sentiment and user-level attributes in existing corpora. The tutorial aims to serve the following use cases for social media researchers: (iii) high accuracy IE on social media text via multi-task and semi-supervised learning, including the recent transformer-based tools which work across languages, (iv) rapid annotation of new data for text classification via active human-in-the-loop learning, (v) temporal visualization of the communication structure in social media corpora via social communication temporal graph visualization technique, and (vi) detecting and prioritizing needs during crisis events (e.g., COVID19). (vii) Furthermore, the participants will be familiarized with a catalogue of more than 30 publicly available social media corpora for various IE tasks, e.g., named entity recognition (NER), part of speech (POS) tagging, chunking, super sense tagging, entity linking, sentiment classification, and hate speech identification. We propose a full day tutorial session using Python based open-source tools. This tutorial builds upon our previous tutorials on this topic at ACM Hypertext 2019, IC2S2 2020, WWW 2021.

1.2 Scope and Benefit to the ECIR Community

Information extraction (IE) is a common sub-area of natural language processing that focuses on identifying structured data from unstructured data. While many open source tools are available for performing IE on newswire and academic publication corpora, there is a lack of such tool when dealing with social media corpora, which tends to exhibit very different linguistic patterns compared to the other corpora. It has also been found that publicly available tools for IE, which are trained on news and academic corpora do not tend to perform very well on social media corpora. Topics of interest include: (i) Machine learning for social media IE (ii) Generating annotated text classification data using active human-in-the-loop learning (iii) Public corpora for social media IE (iv) Open source tools for social media IE (v) Visualizing social media corpora (vi) Bias in social media IE systems.

Scholars in Information Retrieval community who work with social media text can benefit from the recent machine learning advances in information extraction and retrieval in this domain, especially the knowledge of its difference from regular newswire text. This tutorial will help them learn state-of-the-art methods for processing social media text which can help them improve their information retrieval systems on social media text. They will also learn how social media text has a social context, which can be included as part of the analysis.

1.3   Presenter Bios

Shubhanshu Mishra, Twitter, Inc. Shubhanshu Mishra is a Machine Learning Researcher at Twitter. He earned his Ph.D. in Information Sciences from the University of Illinois at Urbana-Champaign in 2020 His thesis was titled "Information extraction from digital social trace data: applications in social media and scholarly data analysis". His current work is at the intersection of machine learning, information extraction, social network analysis, and visualizations. His research has led to the development of open source tools of open source information extraction solutions from large scale social media and scholarly data. He has finished his Integrated Bachelor's and Master's degree in Mathematics and Computing from the Indian Institute of Technology, Kharagpur in 2012.

Rezvaneh (Shadi) Rezapour, Department of Information Science at Drexel's College of Computing and Informatics, USA Shadi is an Assistant Professor in the Department of Information Science at Drexel's College of Computing and Informatics. Her research interests lie at the intersection of Computational Social Science and Natural Language Processing (NLP). More specifically, she is interested in bringing computational models and social science theories together, to analyze texts and better understand and explain real-world behaviors, attitudes, and cultures. Her research goal is to develop "socially-aware" NLP models that bring social and cultural contexts in analyzing (human) language to better capture attributes, such as social identities, stances, morals, and power from language, and understand real-world communication. Shadi completed her Ph.D. in Information Sciences at University of Illinois at Urbana-Champaign (UIUC) where she was advised by Dr. Jana Diesner.

Jana Diesner, The iSchool at University of Illinois Urbana-Champaign, USA Jana is an Associate Professor at the School of Information Sciences (the iSchool) at the University of Illinois at Urbana-Champaign, where she leads the Social Computing Lab. Her research in social computing and human-centered data science combines methods from natural language processing, social network analysis and machine learning with theories from the social sciences to advance knowledge and discovery about interaction-based and information-based systems. Jana got her PhD (2012) in Societal Computing from the School of Computer Science at Carnegie Mellon University.

2   Tutorial Details

- **Duration of the tutorial:** 1 day (full day)
- **Interaction Style:** Hands-on-tutorial with live coding session.
- **Target audience:** We expect the participants to have familiarity with python programming and social media platforms like Twitter and Facebook.

Setup and Introduction (1 h) (i) Introducing the differences between social media data versus newswire and academic data, (ii) Digital Social Trace Data abstraction for social media data, (iii) Introduction to information extraction tasks for social media data, e.g., sequence tagging (named entity, part of speech tagging, chunking, and super-sense tagging), and text classification (sentiment prediction, sarcasm detection, and abusive content detection).

Applications of information extraction (1 h) (i) Indexing social media corpora in database, (ii) Network construction from text corpora, (iii) Visualizing temporal trends in social media corpora using social communication temporal graphs, (iv) Aggregating text-based signals at the user-level, (v) Improving text classification using user-level attributes, (vi) Analyzing social debate using sentiment and political identity signals otherwise, (vii) Detecting and Prioritizing Needs during Crisis Events (e.g., COVID19), (viii) Mining and Analyzing Public Opinion Related to COVID-19, and (ix) Detecting COVID-19 Misinformation in Videos on YouTube.

Collecting and distributing social media data (30 min) (i) Overview on available annotated tweet datasets, (ii) Respecting API terms and user privacy considerations for collecting & sharing social media data, (iii) Demo on collecting data from a few social media APIs, such as Twitter and Reddit.

Break 30 min

Improving IE on social media data via Machine Learning (2 h 30 min) (i) Semi-supervised learning for Twitter NER, (ii) Multi-task learning for social media IE, (iii) Active learning for annotating social media data for text classification via SAIL (another version pySAIL to be released soon), (iv) Finetuning transformer models for monolingual and multi-lingual social media NLP tasks. (v) Biases in social media NER. (vi) Utilizing Social Context for improving NLP Models.

Conclusion and future directions (10 min) (i) Open questions in social media IE, (ii) Tutorial feedback and additional questions.

References

1. Addawood, A., Rezapour, R., Mishra, S., Schneider, J., Diesner, J.: Developing an information source lexicon. In: Prioritising Online Content Workshop Co-located at NIPS (2017)
2. Collier, D., Mishra, S., Houston, D., Hensley, B., Mitchell, S., Hartlep, N.: Who is most likely to oppose federal tuition-free college policies? Investigating variable interactions of sentiments to America's college promise. SSRN Electron. J. (2019). https://doi.org/10.2139/ssrn.3423054
3. Collier, D.A., Mishra, S., Houston, D.A., Hensley, B.O., Hartlep, N.D.: Americans 'support' the idea of tuition-free college: an exploration of sentiment and political identity signals otherwise. J. Furth. High. Educ. **43**(3), 347–362 (2019). https://doi.org/10.1080/0309877X.2017.1361516
4. Diesner, J., Carley, K.M.: Relation extraction from texts (in German: Extraktion relationaler Daten aus Texten). In: Stegbauer, C., Häußling, R. (eds.) Handbook network research (Handbuch Netzwerkforschung), pp. 507–521. VS Verlag (2010)
5. Diesner, J., Kumaraguru, P., Carley, K.M.: Mental models of data privacy and security extracted from interviews with Indians. In: Proceedings of 55th Annual Conference of International Communication Association (ICA). New York, NY (2005)
6. Diesner, J., Chin, C.L.: Usable ethics: practical considerations for responsibly conducting research with social trace data. In: Proceedings of Beyond IRBs: Ethical Review Processes for Big Data Research (2015)
7. Diesner, J., Chin, C.L.: Seeing the forest for the trees: considering applicable types of regulation for the responsible collection and analysis of human centered data. In: Human-Centered Data Science (HCDS) Workshop at 19th ACM Conference on Computer-Supported Cooperative Work and Social Computing (2016)
8. Eisenstein, J.: What to do about bad language on the internet. In: Proceedings of the 2013 Conference of the North American Chapter of the Association for Computational Linguistics: Human Language Technologies, pp. 359–369. Association for Computational Linguistics, Atlanta, Georgia (June 2013)
9. Han, K., Yang, P., Mishra, S., Diesner, J.: WikiCSSH: extracting computer science subject headings from Wikipedia. In: Workshop on Scientific Knowledge Graphs (SKG 2020) (2020)
10. Hutto, C.J., Gilbert, E.: Vader: A parsimonious rule-based model for sentiment analysis of social media text. In: International AAAI Conference on Web and Social Media. Ann Arbor, Michigan, USA (2014)
11. Kaplan, A.M., Haenlein, M.: Users of the world, unite! The challenges and opportunities of social media. Bus. Horiz. **53**(1), 59–68 (2010). https://doi.org/10.1016/j.bushor.2009.09.003
12. Kosinski, M., Matz, S.C., Gosling, S.D., Popov, V., Stillwell, D.: Facebook as a research tool for the social sciences: opportunities, challenges, ethical considerations, and practical guidelines. Am. Psychol. **70**(6), 543–556 (2015). https://doi.org/10.1037/a0039210
13. Kulkarni, V., Mishra, S., Haghighi, A.: LMSOC: an approach for socially sensitive pretraining. In: Findings of the Association for Computational Linguistics: EMNLP 2021, pp. 2967–2975. Association for Computational Linguistics, Stroudsburg, PA, USA (November 2021). https://doi.org/10.18653/v1/2021.findings-emnlp.254

14. Kwak, H., Lee, C., Park, H., Moon, S.: What is Twitter, a social network or a news media? In: Proceedings of the 19th international conference on World wide web–WWW 2010, p. 591. ACM Press, New York, New York, USA (April 2010). https://doi.org/10.1145/1772690.1772751

15. Mishra, S.: SCTG: social communications temporal graph - a novel approach to visualize temporal communication graphs from social data. In: UIUC Data Science Day (October 2017)

16. Mishra, S.: Multi-dataset-multi-task neural sequence tagging for information extraction from tweets. In: Proceedings of the 30th ACM Conference on Hypertext and Social Media - HT 2019, pp. 283–284. ACM Press, New York, New York, USA (2019). https://doi.org/10.1145/3342220.3344929

17. Mishra, S.: Information extraction from digital social trace data with applications to social media and scholarly communication data. ACM SIGIR Forum **54**(1), 1–2 (2020). https://doi.org/10.1145/3451964.3451981

18. Mishra, S.: Information Extraction from Digital Social Trace Data with Applications to Social Media and Scholarly Communication Data. Ph.D. thesis, University of Illinois at Urbana-Champaign (2020)

19. Mishra, S.: Non-neural structured prediction for event detection from news in Indian languages. In: Mehta, P., Mandl, T., Majumder, P., Mitra, M. (eds.) Working Notes of FIRE 2020–Forum for Information Retrieval Evaluation. CEUR Workshop Proceedings, CEUR-WS.org, Hyderabad, India (2020)

20. Mishra, S., Agarwal, S., Guo, J., Phelps, K., Picco, J., Diesner, J.: Enthusiasm and support: alternative sentiment classification for social movements on social media. In: Proceedings of the 2014 ACM conference on Web science - WebSci 2014, pp. 261–262. ACM Press, Bloomington, Indiana, USA (June 2014). https://doi.org/10.1145/2615569.2615667

21. Mishra, S., Collier, D.: A framework for generating annotated social media corpora with demographics, stance, civility, and topicality. SSRN Electron. J. (2020). https://doi.org/10.2139/ssrn.3757554

22. Mishra, S., Diesner, J.: Semi-supervised named entity recognition in noisy-text. In: Proceedings of the 2nd Workshop on Noisy User-generated Text (WNUT), pp. 203–212. The COLING 2016 Organizing Committee, Osaka, Japan (2016)

23. Mishra, S., Diesner, J.: Detecting the correlation between sentiment and user-level as well as text-level meta-data from benchmark corpora. In: Proceedings of the 29th on Hypertext and Social Media - HT 2018, pp. 2–10. ACM Press, New York, New York, USA (2018). https://doi.org/10.1145/3209542.3209562

24. Mishra, S., Diesner, J.: Capturing signals of enthusiasm and support towards social issues from Twitter. In: Proceedings of the 5th International Workshop on Social Media World Sensors - SIdEWayS 2019, pp. 19–24. ACM Press, New York, New York, USA (2019). https://doi.org/10.1145/3345645.3351104

25. Mishra, S., Diesner, J., Byrne, J., Surbeck, E.: Sentiment analysis with incremental human-in-the-loop learning and lexical resource customization. In: Proceedings of the 26th ACM Conference on Hypertext & Social Media - HT 2015, pp. 323–325. ACM Press, New York, New York, USA (2015). https://doi.org/10.1145/2700171.2791022

26. Mishra, S., Haghighi, A.: Improved multilingual language model pretraining for social media text via translation pair prediction. In: Proceedings of the Seventh Workshop on Noisy User-generated Text (W-NUT 2021), pp. 381–388. Association for Computational Linguistics, Stroudsburg, PA, USA (November 2021). https://doi.org/10.18653/v1/2021.wnut-1.42

27. Mishra, S., He, S., Belli, L.: Assessing demographic bias in named entity recognition. In: Bias in Automatic Knowledge Graph Construction–A Workshop at AKBC 2020 (August 2020)
28. Mishra, S., Mishra, S.: 3Idiots at HASOC 2019: fine-tuning transformer neural networks for hate speech identification in Indo-European languages. In: Proceedings of the 11th Annual Meeting of the Forum for Information Retrieval Evaluation, pp. 208–213. Kolkata, India (2019)
29. Mishra, S., Mishra, S.: Scubed at 3C task a–a simple baseline for citation context purpose classification. In: Proceedings of the 8th International Workshop on Mining Scientific Publications, pp. 59–64. Association for Computational Linguistics, Wuhan, China (2020)
30. Mishra, S., Mishra, S.: Scubed at 3C task b–a simple baseline for citation context influence classification. In: Proceedings of the 8th International Workshop on Mining Scientific Publications, pp. 65–70. Association for Computational Linguistics, Wuhan, China (2020)
31. Mishra, S., Prasad, S., Mishra, S.: Multilingual joint fine-tuning of transformer models for identifying trolling, aggression and cyberbullying at TRAC 2020. In: Proceedings of the Second Workshop on Trolling, Aggression and Cyberbullying, pp. 120–125. European Language Resources Association (ELRA), Marseille, France (2020)
32. Mishra, S., Prasad, S., Mishra, S.: Exploring multi-task multi-lingual learning of transformer models for hate speech and offensive speech identification in social media. SN Comput. Sci. 2(2), 1–19 (2021). https://doi.org/10.1007/s42979-021-00455-5
33. Mohammad, S.M., Kiritchenko, S., Zhu, X.: NRC-Canada: building the state-of-the-art in sentiment analysis of tweets. In: Second Joint Conference on Lexical and Computational Semantics (*SEM), Volume 2: Proceedings of the Seventh International Workshop on Semantic Evaluation (SemEval 2013), vol. 2, pp. 321–327. Association for Computational Linguistics, Atlanta, Georgia, USA (2013)
34. Pang, B., Lee, L.: Opinion mining and sentiment analysis. Found. Trends® Inf. Retr. 2(1–2), 1–135 (2008). https://doi.org/10.1561/1500000011
35. Rezapour, R., Dinh, L., Diesner, J.: Incorporating the measurement of moral foundations theory into analyzing stances on controversial topics. In: Proceedings of the 32st ACM Conference on Hypertext and Social Media, pp. 177–188. ACM, New York, NY, USA (August 2021). https://doi.org/10.1145/3465336.3475112
36. Rezapour, R., Shah, S.H., Diesner, J.: Enhancing the measurement of social effects by capturing morality. In: Proceedings of the Tenth Workshop on Computational Approaches to Subjectivity, Sentiment and Social Media Analysis, pp. 35–45. Association for Computational Linguistics, Stroudsburg, PA, USA (2019). https://doi.org/10.18653/v1/W19-1305
37. Rezapour, R., Wang, L., Abdar, O., Diesner, J.: Identifying the overlap between election result and candidates' ranking based on hashtag-enhanced, lexicon-based sentiment analysis. In: 2017 IEEE 11th International Conference on Semantic Computing (ICSC), pp. 93–96. IEEE (2017). https://doi.org/10.1109/ICSC.2017.92
38. Sarawagi, S.: Information extraction. Found. Trends® Databases 1(3), 261–377 (2007). https://doi.org/10.1561/1900000003
39. Sarol, M.J., Dinh, L., Rezapour, R., Chin, C.L., Yang, P., Diesner, J.: An empirical methodology for detecting and prioritizing needs during crisis events. In: Findings of the Association for Computational Linguistics: EMNLP 2020, pp. 4102–4107. Association for Computational Linguistics, Stroudsburg, PA, USA (2020). https://doi.org/10.18653/v1/2020.findings-emnlp.366

40. Schwartz, H.A., et al.: Personality, gender, and age in the language of social media: the open-vocabulary approach. PLoS ONE **8**(9), e73791 (2013). https://doi.org/10.1371/journal.pone.0073791
41. Yee, K., Tantipongpipat, U., Mishra, S.: Image cropping on twitter: fairness metrics, their limitations, and the importance of representation, design, and agency. Proc. ACM Hum. Comput. Interact. **5**(CSCW2), 1–24 (2021). https://doi.org/10.1145/3479594

ECIR 2022 Tutorial: Technology-Assisted Review for High Recall Retrieval

Eugene Yang[1]([✉]), Jeremy Pickens[2], and David D. Lewis[3]

[1] Human Language Technology Center of Excellence, Johns Hopkins University,
Baltimore, USA
eugene.yang@jhu.edu
[2] OpenText, San Mateo, USA
jpickens@opentext.com
[3] Redgrave Data, San Francisco, USA
ecir2022paper@davelewis.com

1 Introduction

Basics:

> *Length*: Half day.
> *Target audience*: Intermediate.
> *Expected prerequisite knowledge*: Some exposure to basics of information retrieval and machine learning.

Scope of the Tutorial: Human-in-the-loop (HITL) IR workflows are being applied to an increasing range of tasks in the law, medicine, social media, and other areas. These tasks differ from ad hoc retrieval in their focus on high recall, and differ from text categorization in their need for extensive human judgment. These tasks also differ from both in their industrial scale and, often, their use of teams of multiple reviewers. In the research literature, these tasks have been variously referred to as *review, moderation, annotation*, or *high recall retrieval* (*HRR*) tasks. Technologies applied to these tasks have also been referred to by many names, but *technology-assisted review* (*TAR*) has emerged as a consensus term, so these tasks are also referred to as *TAR tasks*.

The growth in the deployment of TAR systems, combined with the many open research problems in this area, suggest this is an appropriate time for a TAR tutorial at a major IR conference. Such a tutorial would also serve as background for attendees of the TAR workshop that has been approved for ECIR 2022.

Aims and Learning Objectives: This tutorial will introduce students to the key application areas, technologies, and evaluation methods in technology-assisted review. After taking the tutorial, attendees will be able to

- Recognize real-world applications appropriate for TAR technology

M. Hagen et al. (Eds.): ECIR 2022, LNCS 13186, pp. 597–600, 2022.
https://doi.org/10.1007/978-3-030-99739-7_75

- Apply well-known information retrieval and machine learning approaches to TAR problems
- Design basic TAR workflows
- Identify levers for cost minimization in real-world TAR tasks
- Apply standard TAR evaluation measures
- Find publications on TAR technology, evaluation methods, HCI issues, ethical implications, and open problems in a range of literatures

Relevance to IR Community: Identifying of relevant documents is of course a central concern of IR. HITL approaches to IR have grown in prominence, both as IR is applied to increasingly complex tasks, and as ethics concerns have arisen around full automation of tasks using AI methods. TAR methods have seen intense development over the past 15 years in the law and medicine, but these developments are not widely known outside of these practice communities. Conversely, numerous unsolved algorithmic and statistical problems have arisen in these areas which pose interesting challenges for IR researchers.

Tutorial History: One of the presenters presented tutorials on TAR applications in the law at SIGIR 2010 and SIGIR 2012. This tutorial is broader in application scope, and incorporates numerous developments in technology, evaluation, and workflow design since that time.

2   Full Description

- Introduction to TAR
  - What is a TAR task
  - Comparison to other IR tasks
  - Application areas
    * Law: litigation, antitrust, investigations
    * Systematic reviews in medicine
    * Content moderation
    * Data set annotation
    * Sunshine laws, declassification, and archival tasks
    * Patent search and other high recall review tasks
  - History
- Dimensions of TAR Tasks
  - Volume and temporal characteristics of data
  - Time constraints
  - Reviewer characteristics
  - Cost structure and constraints
  - Nature of classification task (single, multiple, and cascaded classifications)
- TAR Workflows
  - Importance of workflow design in HITL system
  - One-phase vs. two-phase workflows

- Quality vs. quantity of training
- Pipeline workflows
- Collection segmentation and multi-technique workflows
- When to stop?
- Technology: Basics
  - Review software and traditional review workflows
  - Duplicate detection, aggregation, and propagation
  - Search and querying
  - Unsupervised learning and visual analytics
- Technology: Supervised Learning
  - Basics of text classification
  - Data modeling and task definition
  - Prioritization vs. classification
  - Reviewing and labeling in TAR workflows
  - Relevance feedback and other active learning approaches
  - Implications of transductive context
  - Classifier reuse and transfer learning
  - Research questions
- Evaluation
  - Effectiveness measures
  - Sample-based estimation of effectiveness
  - Impact of category prevalence
  - Cost measures
  - Evaluating progress within a TAR project
  - Collection segmentation and evaluation
  - Choosing and tuning methods across multiple projects
  - Research questions
- Stopping rules
  - Stopping rules, cutoffs, and workflow design
  - Cost targets and effectiveness targets
  - The cost landscape
  - Distinctions among stopping rules
    * Interventional, standoff, and hybrid rules
    * Gold standard vs. self-evaluation rules
    * Certification vs. heuristic rules
  - Example stopping rules
    * Knee, Target, and Budget Methods [1]
    * Quant and QuantCI Rule [3]
    * QBCB Rule [2]
  - Research questions
- Societal context
  - TAR and the ethical obligations of attorneys
  - Bias and ethics issues in TAR for monitoring and surveillance
  - Implications of TAR for evidence-based medicine
  - Controversies in automated content moderation
  - Research questions
- Summary and Future
  - TAR research and industry practice
  - Challenges in access to data
  - The potential for interdisciplinary TAR research

3 Presenters and Their Credentials

Eugene Yang is a Research Associate at the Human Language Technology Center of Excellence at Johns Hopkins University. He has been developing state-of-the-art approaches for technology-assisted review. His Ph.D. dissertation focuses on cost reduction and cost analysis for TAR, including cost modeling and stopping rules for one- and two-phase workflows. He is currently working on cross-lingual human-in-the-loop retrieval approaches.

Jeremy Pickens is a pioneer in the field of collaborative exploratory search, a form of information seeking in which a group of people who share a common information need actively collaborate to achieve it. As Principal Data Scientist at OpenText, he has spearheaded the development of Insight Predict. His ongoing research and development focuses on methods for continuous learning, and the variety of real world technology-assisted review workflows that are only possible with this approach. Dr. Pickens earned his doctoral degree at the University of Massachusetts, Amherst, Center for Intelligent Information Retrieval. Before joining Catalyst Repository Systems and later OpenText, he spent five years as a research scientist at FX Palo Alto Lab, Inc.

David D. Lewis is Chief Scientific Officer for Redgrave Data, a legal technology services company. He has researched, designed, and consulted on human-in-the-loop document classification and review systems since the early 1990's. His 1994 paper with Gale introduced uncertainty sampling, a core technique used in commercial TAR systems. This paper won an ACM SIGIR Test of Time Award in 2017. In 2005, Dave co-founded the TREC Legal Track, the first open evaluation of TAR technology. He was elected a Fellow of the American Association for the Advancement of Science in 2006 for foundational work on algorithms, data sets, and evaluation in text analytics.

References

1. Cormack, G.V., Grossman, M.R.: Engineering quality and reliability in technology-assisted review. In: SIGIR, pp. 75–84. ACM Press, Pisa, Italy (2016). https://doi.org/10.1145/2911451.2911510, http://dl.acm.org/citation.cfm?doid=2911451.2911510, 00024
2. Lewis, D.D., Yang, E., Frieder, O.: Certifying one-phase technology-assisted reviews. In: Proceedings of 30th ACM International Conference on Information and Knowledge Management (2021)
3. Yang, E., Lewis, D.D., Frieder, O.: Heuristic stopping rules for technology-assisted review. In: Proceedings of the 21st ACM Symposium on Document Engineering (2021)

Correction to: End to End Neural Retrieval for Patent Prior Art Search

Vasileios Stamatis (iD)

Correction to:
Chapter "End to End Neural Retrieval for Patent Prior Art Search" in: M. Hagen et al. (Eds.): *Advances in Information Retrieval*, **LNCS 13186,**
https://doi.org/10.1007/978-3-030-99739-7_66

Chapter End to End Neural Retrieval for Patent Prior Art Search was previously published non-open access. It has now been changed to open access under a CC BY 4.0 license and the copyright holder updated to 'The Author(s)'. The book has also been updated with this change.

The updated original version of this chapter can be found at
https://doi.org/10.1007/978-3-030-99739-7_66

Author Index